SOLUTIONS TO RED EXERCISES

CHEMISTRY

THE CENTRAL SCIENCE

14TH EDITION

SOLUTIONS TO RED EXERCISES

Roxy Wilson
University of Illinois, Urbana-Champaign

CHEMISTRY

THE CENTRAL SCIENCE

14TH EDITION

BROWN | LeMAY
BURSTEN | MURPHY
WOODWARD | STOLTZFUS

Courseware Portfolio Manager: Terry Haugen
Managing Producer, Science: Kristen Flatham
Product Marketing Manager: Elizabeth Bell
Content Producer, Science: Beth Sweeten
Full Service Vendor: Cenveo® Publisher Services
Main Text Cover Designer: Jeff Puda
Supplement Cover Designer: 17th Street Studios
Buyer: Maura Zaldivar-Garcia

Credits and acknowledgments borrowed from other sources and reproduced, with permission, in this textbook appear on the appropriate page within the text.

---

1 17

ISBN-10: 0-13-455223-7
ISBN-13: 978-0-13-455223-1

Contents

Introduction

Chemistry: The Central Science, 14th edition, contains over 2700 end-of-chapter exercises. Considerable attention has been given to these exercises because one of the best ways for students to master chemistry is by solving problems. Grouping the exercises according to subject matter is intended to aid the student in selecting and recognizing particular types of problems. Within each subject matter group, similar problems are arranged in pairs. This provides the student with an opportunity to reinforce a particular kind of problem. There are also a substantial number of general exercises in each chapter to supplement those grouped by topic. Integrative exercises, which require students to integrate concepts from several chapters, are a continuing feature of the 14th edition. Answers to the odd-numbered topical exercises plus selected general and integrative exercises, about 1150 in all, are provided in the textbook. These appendix answers help to make the textbook a useful self-contained vehicle for learning.

This manual, **Solutions to Red Exercises in Chemistry: The Central Science, 14th edition**, was written to enhance the end-of-chapter exercises by providing documented solutions. The manual assists the instructor by saving time spent generating solutions for assigned problem sets and aids the student by offering a convenient independent source to check their understanding of the material. Most solutions have been worked in the same detail as the in-chapter sample exercises to help guide students in their studies.

To reinforce the *Analyze, Plan, Solve, Check* problem-solving method used extensively in the text, this strategy has also been incorporated into the solutions manual. Solutions to most red topical exercises and selected Additional and Integrative exercises feature this four-step approach. We strongly encourage students to master this powerful and totally general method.

When using this manual, keep in mind that the numerical result of any calculation is influenced by the precision of the numbers used in the calculation. In this manual, for example, atomic masses and physical constants are typically expressed to four significant figures, or at least as precisely as the data given in the problem. If students use slightly different values to solve problems, their answers will differ slightly from those listed in the appendix of the text or this manual. This is a normal and a common occurrence when comparing results from different calculations or experiments.

Rounding methods are another source of differences between calculated values. In this manual, when a solution is given in steps, intermediate results will be rounded to the correct number of significant figures; however, unrounded numbers will be used in subsequent calculations. By following this scheme, calculators need not be cleared to re-enter rounded intermediate results in the middle of a calculation sequence. The final answer will appear with the correct number of significant figures. This may result in a small discrepancy in the last significant digit between student-calculated answers and those given in this manual. Variations due to rounding can occur in any analysis of numerical data.

The first step in checking your solution and resolving differences between your answer and the listed value is to look for similarities and differences in problem-solving methods. Ultimately, resolving the small numerical differences described above is less important than understanding the general method for solving a problem. The goal of this manual is to provide a reference for sound and consistent problem-solving methods in addition to accurate answers to text exercises.

Extraordinary efforts have been made to keep this manual as error-free as possible. All exercises were worked and proofread by at least three chemists to ensure clarity in methods and accuracy in mathematics. The ongoing work and advice of Dr. Richard Helmich, Ms. Rene Musto, and Dr. Christopher Musto continue to be invaluable to this project. In any written work as technically challenging as this manual, typos and errors inevitably creep in, despite our combined efforts. Please help us find and eliminate them. We hope that both instructors and students will find this manual accurate, helpful, and instructive.

Roxy B. Wilson, Ph.D.
1829 Maynard Dr.
Champaign, IL 61822
rbwilson@illinois.edu

1 Introduction: Matter, Energy, and Measurement

Visualizing Concepts

1.1 *Pure elements* contain only one kind of atom. Atoms can be present singly or as tightly bound groups called molecules. *Compounds* contain two or more kinds of atoms bound tightly into molecules. *Mixtures* contain more than one kind of atom and/or molecule, not bound into discrete particles.

(a) pure element: i

(b) mixture of elements: v, vi

(c) pure compound: iv

(d) mixture of an element and a compound: ii, iii

1.3 (a) Brass is composed of two different kinds of atoms, so it is a mixture. The mixture appears homogeneous under an optical microscope, so it is a homogeneous mixture.

(b) Because brass is a homogeneous mixture, it is a solution. We usually think of solutions as liquids, but they can be solids, liquids, or gases.

1.5 Filtration. When brewing a cup of coffee, hot water contacts the coffee grounds and dissolves components of the coffee bean that are water-soluble. This creates a heterogeneous mixture of undissolved coffee bean solids and liquid coffee solution; this mixture is separated by filtration. Undissolved grounds remain on the filter paper and liquid coffee drips into the container below.

1.7 Density is the ratio of mass to volume. For a sphere, size is like volume; both are determined by the radius of the sphere.

(a) For spheres of the same size or volume, the denominator of the density relationship is the same. The denser the sphere, the heavier it is. A list from lightest to heaviest is in order of increasing density and mass. The aluminum sphere (density = 2.70 g/cm^3) is lightest, then nickel (density = 8.90 g/cm^3), then silver (density = 10.49 g/cm^3).

(b) For cubes of equal mass, the numerator of the density relationship is the same. The denser the sphere, the smaller its volume or size. A list from smallest to largest is in order of decreasing density. The platinum sphere (density = 21.45 g/cm^3) is smallest, then gold (density = 19.30 g/cm^3), then lead (density = 11.35 g/cm^3).

1.9 (a) 7.5 cm. There are two significant figures in this measurement; the number of cm can be read precisely, but there is some estimating (uncertainty) required to read tenths of a centimeter. Listing two significant figures is consistent with the convention that measured quantities are reported so that there is uncertainty in only the last digit.

(b) The speed is 72 mi/hr (inner scale, two significant figures) or 115 km/hr (outer scale, three significant figures). Both scales are read with certainty in the "hundreds" and "tens" place, and some uncertainty in the "ones" place. The km/hr speed has one more significant figure because its magnitude is in the hundreds.

1.11 Given: masses of six jelly beans, mass of full jar, mass of empty jar. Find: number of jelly beans in the jar. The total mass of jelly beans is the mass of the jar full minus the mass of the jar empty. The mass of an "average" jelly bean is the average of the six masses. Then, the number of jelly beans is the total mass of beans divided by the average mass of a single bean.

Total mass of beans = 2082 g – 653 g = 1429 g

Average mass of a bean = (3.15 + 3.12 + 2.98 + 3.14 + 3.02 + 3.09) / 6 = 3.08 g

Number of beans = 1429 g total / 3.08 g per bean = 463.96 = 464 beans

By applying the significant figure rules for addition and subtraction, the total mass of beans has 0 decimal places and thus 4 significant figures. The mass of an average bean has 2 decimal places and 3 sig figs. The number of beans then has 3 sig figs, by the rules for multiplication and division. This makes sense, because we expect an integer number of beans in the jar. (Note that using an unrounded average bean mass of 3.0833 g predicts the number of beans to be 463.46, which rounds to 463 beans. The difference in these two values shows uncertainty in the last significant figure of the number of beans, as we expect in an experimental result.)

Classification and Properties of Matter (Sections 1.2 and 1.3)

1.13 (a) heterogeneous mixture

(b) homogeneous mixture (If there are undissolved particles, such as sand or decaying plants, the mixture is heterogeneous.)

(c) pure substance

(d) pure substance

1.15 (a) S (b) Au (c) K (d) Cl (e) Cu (f) uranium (g) nickel (h) sodium (i) aluminum (j) silicon

1.17 A(s) → B(s) + C(g)

Substances A and C are definitely compounds; B is probably a compound. When solid carbon is burned in excess oxygen gas, the two elements combine to form a gaseous compound, carbon dioxide. Clearly substance C is this compound. Because C is produced when A is heated in the absence of oxygen (from air), both the carbon and the oxygen in C must have been present in A originally. A is, therefore, a compound composed of two or more elements chemically combined. Without more information on the chemical or physical properties of B, we cannot determine absolutely whether it is an element or a compound. However, few if any elements exist as white solids, so B is probably also a compound.

1 Matter, Energy, and Measurement

Solutions to Red Exercises

1.19 *Physical properties*: silvery white (color); lustrous; melting point = 649 °C; boiling point = 1105 °C; density at 20 °C = 1.738 g/cm^3; pounded into sheets (malleable); drawn into wires (ductile); good conductor. *Chemical properties*: burns in air to give intense white light; reacts with Cl_2 to produce brittle white solid.

1.21 (a) chemical (b) physical (c) physical (d) chemical (e) chemical

1.23 Distillation. A solution of sugar and water is a homogeneous mixture that cannot be separated by filtration. Distillation takes advantage of the much lower boiling point of water.

The Nature of Energy (Section 1.4)

1.25 (a) *Plan.* $E_k = 1/2\ mv^2$; m = 1200 kg; v = 18 m/s; $1\ kg\text{-}m^2/s^2 = 1\ J$
Solve. $E_k = 1/2 \times 1200\ kg \times (18)^2\ m^2/s^2 = 1.944 \times 10^5 = 1.9 \times 10^5\ J$

(b) $1\ cal = 4.184\ J;\ 1.944 \times 10^5\ J \times \dfrac{1\ cal}{4.184\ J} = 4.646 \times 10^4\ cal = 4.6 \times 10^4\ cal$

(c) As the automobile brakes to a stop, its speed (and hence its kinetic energy) drops to 0. This "lost" kinetic energy is mostly converted to heat. (The heat shows up in the brake parts, tires, and road.) It is not converted to some form of potential energy.

1.27 (a) Kinetic energy; the particles move apart.

(b) Potential energy decreases. The greater the separation prior to release, the smaller the electrostatic repulsion and potential energy.

1.29 *Analyze/Plan.* Use results from Solution 1.4 and energy relationships discussed in Section 1.4 to solve for kinetic energy and velocity.

Solve. From Solution 1.4 we have the work required to lift the spheres. This is equal to the potential energy of the spheres at 2.2 m. As a sphere hits the floor, all potential energy is changed to kinetic energy. The kinetic energy of the Al sphere is then 11 J.
$E_k = 11.41 = 11\ J$; mass(Al) = 0.5292 = 0.529 kg; $E_k = 1/2\ mv^2$; $v = (2\ E_k/m)^{1/2}$;

$$v(Al) = \left(2 \times \frac{11.41\ kg\text{-}m^2}{s^2} \times \frac{1}{0.5292\ kg}\right)^{1/2} = 6.567 = 6.6\ m/s$$

Units and Measurement (Section 1.5)

1.31 (a) 1×10^{-1} (b) 1×10^{-2} (c) 1×10^{-15} (d) 1×10^{-6} (e) 1×10^{6}
(f) 1×10^{3} (g) 1×10^{-9} (h) 1×10^{-3} (i) 1×10^{-12}

1.33 (a) °C = 5/9 (°F – 32 °); 5/9 (72 – 32) = 22 °C
(b) °F = 9/5 (°C) + 32 °; 9/5 (216.7) + 32 = 422.1 °F
(c) K = °C + 273.15; 233 °C + 273.15 = 506 K
(d) °C = 315 K – 273.15 = 41.85 = 42 °C; °F = 9/5 (41.85 °C) + 32 = 107 °F
(e) °C = 5/9 (°F – 32); 5/9 (2500 – 32) = 1371 = 1400 °C
K = 1371 °C + 273.15 = 1644 = 1600 K

(f) $^\circ C = 0\ K - 273.15 = -273.15\ ^\circ C$; $^\circ F = 9/5\,(-273.15\ ^\circ C) + 32 = -459.67\ ^\circ F$ (assuming 0 K has infinite sig figs)

1.35 (a) $\text{density} = \frac{\text{mass}}{\text{volume}} = \frac{40.55\ g}{25.0\ mL} = 1.62\ g/mL \text{ or } 1.62\ g/cm^3$

(The units cm^3 and mL will be used interchangeably in this manual.)

Tetrachloroethylene, 1.62 g/mL, is more dense than water, 1.00 g/mL; tetrachloroethylene will sink rather than float on water.

(b) $25.0\ cm^3 \times 0.469 \frac{g}{cm^3} = 11.7\ g$

1.37 (a) $\text{calculated density} = \frac{38.5\ g}{45\ mL} = 0.86\ g/mL$

The substance is probably toluene, density = 0.866 g/mL.

(b) $45.0\ g \times \frac{1\ mL}{1.114\ g} = 40.4\ mL$ ethylene glycol

(c) A 100 mL graduated cylinder such as the one in Figure 1.21 usually has 1-mL markings. One can read the volume with certainty to the nearest mL, and estimate tenths of a mL. The volume calculated in (b) has uncertainty in the tenths place, so a graduated cylinder like this will provide the appropriate accuracy of measurement.

(d) $(5.00)^3\ cm^3 \times \frac{8.90\ g}{1\ cm^3} = 1.11 \times 10^3\ g\ (1.11\ kg)$ nickel

1.39 $36 \text{ billion metric tons} \times \frac{1 \times 10^9 \text{ metric tons}}{1 \text{ billion metric tons}} \times \frac{1000\ kg}{1 \text{ metric ton}} \times \frac{1000\ g}{1\ kg} = 3.6 \times 10^{16}\ g$

The metric prefix for 1×10^{15} is peta, abbreviated P.

$3.6 \times 10^{16}\ g \times \frac{1\ Pg}{1 \times 10^{15}\ g} = 36\ Pg$

1.41 *Analyze.* Given: heat capacity of water = 1 Btu/lb-°F Find: J/Btu

Plan. heat capacity of water = $\frac{1\ cal}{g\text{-}^\circ C}$; $\frac{cal}{g\text{-}^\circ C} \rightarrow \frac{J}{g\text{-}^\circ C} \rightarrow \frac{J}{lb\text{-}^\circ F} \rightarrow \frac{J}{Btu}$

This strategy requires changing °F to °C. Because this involves the magnitude of a degree on each scale, rather than a specific temperature, the 32 in the temperature relationship is not needed. 100 °C = 180 °F; 5 °C = 9 °F

Solve. $\frac{1\ cal}{g\text{-}^\circ C} \times \frac{4.184\ J}{cal} \times \frac{453.6\ g}{lb} \times \frac{5\ ^\circ C}{9\ ^\circ F} \times \frac{1\ lb\text{-}^\circ F}{1\ Btu} = 1054\ J/Btu$

Uncertainty in Measurement (Section 1.6)

1.43 Exact: (b), (d), and (f) (All others depend on measurements and standards that have margins of error, e.g., the length of a week as defined by the Earth's rotation.)

1.45 (a) 3 (b) 2 (c) 5 (d) 3 (e) 5 (f) 1 [See Sample Exercise 1.7 (c)]

1.47 (a) 1.025×10^{2} (b) 6.570×10^{2} (c) 8.543×10^{-3}
(d) 2.579×10^{-4} (e) -3.572×10^{-2}

1.49 (a) 14.3505 + 2.65 = 17.0005 = 17.00 (For addition and subtraction, the minimum number of decimal places, here two, determines decimal places in the result.)

(b) 952.7 – 140.7389 = 812.0

(c) $(3.29 \times 10^{4})(0.2501) = 8.23 \times 10^{3}$ (For multiplication and division, the minimum number of significant figures, here three, determines sig figs in the result.)

(d) $0.0588/0.677 = 8.69 \times 10^{-2}$

1.51 The mass 21.427 g has 5 significant figures.

Dimensional Analysis (Section 1.7)

1.53 In each conversion factor, the old unit appears in the denominator, so it cancels, and the new unit appears in the numerator.

(a) $\text{mm} \rightarrow \text{nm}: \dfrac{1\times 10^{-3}\ \text{m}}{1\ \text{mm}} \times \dfrac{1\ \text{nm}}{1\times 10^{-9}\ \text{m}} = 1\times 10^{6}\ \text{nm/mm}$

(b) $\text{mg} \rightarrow \text{kg}: \dfrac{1\times 10^{-3}\ \text{g}}{1\ \text{mg}} \times \dfrac{1\ \text{kg}}{1000\ \text{g}} = 1\times 10^{-6}\ \text{kg/mg}$

(c) $\text{km} \rightarrow \text{ft}: \dfrac{1000\ \text{m}}{1\ \text{km}} \times \dfrac{1\ \text{cm}}{1\times 10^{-2}\ \text{m}} \times \dfrac{1\ \text{in}}{2.54\ \text{cm}} \times \dfrac{1\ \text{ft}}{12\ \text{in}} = 3.28 \times 10^{3}\ \text{km/ft}$

(d) $\text{in}^3 \rightarrow \text{cm}^3: \dfrac{(2.54)^3\ \text{cm}^3}{1^3\ \text{in}^3} = 16.4\ \text{cm}^3/\text{in}^3$

1.55 (a) $\dfrac{15.2\ \text{m}}{\text{s}} \times \dfrac{1\ \text{km}}{1000\ \text{m}} \times \dfrac{60\ \text{s}}{1\ \text{min}} \times \dfrac{60\ \text{min}}{1\ \text{hr}} = 54.7\ \text{km/hr}$

(b) $5.0\times 10^{3}\ \text{L} \times \dfrac{1\ \text{gal}}{3.7854\ \text{L}} = 1.3\times 10^{3}\ \text{gal}$

(c) $151\ \text{ft} \times \dfrac{1\ \text{yd}}{3\ \text{ft}} \times \dfrac{1\ \text{m}}{1.0936\ \text{yd}} = 46.025 = 46.0\ \text{m}$

(d) $\dfrac{60.0\ \text{cm}}{\text{d}} \times \dfrac{1\ \text{in}}{2.54\ \text{cm}} \times \dfrac{1\ \text{d}}{24\ \text{hr}} = 0.984\ \text{in/hr}$

1.57 (a) $5.00 \text{ days} \times \frac{24 \text{ hr}}{1 \text{ day}} \times \frac{60 \text{ min}}{1 \text{ hr}} \times \frac{60 \text{ s}}{1 \text{ min}} = 4.32 \times 10^5 \text{ s}$

(b) $0.0550 \text{ mi} \times \frac{1.6093 \text{ km}}{\text{mi}} \times \frac{1000 \text{ m}}{1 \text{ km}} = 88.5 \text{ m}$

(c) $\frac{\$1.89}{\text{gal}} \times \frac{1 \text{ gal}}{3.7854 \text{ L}} = \frac{\$0.499}{\text{L}}$

(d) $\frac{0.510 \text{ in}}{\text{ms}} \times \frac{2.54 \text{ cm}}{1 \text{ in}} \times \frac{1 \times 10^{-2} \text{ m}}{1 \text{ cm}} \times \frac{1 \text{ km}}{1000 \text{ m}} \times \frac{1 \text{ ms}}{1 \times 10^{-3} \text{ s}} \times \frac{60 \text{ s}}{1 \text{ min}} \times \frac{60 \text{ min}}{1 \text{ hr}} = \frac{46.6 \text{ km}}{\text{hr}}$

Estimate: $0.5 \times 2.5 = 1.25$; $1.25 \times 0.01 \approx 0.01$; $0.01 \times 60 \times 60 \approx 36$ km/hr

(e) $\frac{22.50 \text{ gal}}{\text{min}} \times \frac{3.7854 \text{ L}}{\text{gal}} \times \frac{1 \text{ min}}{60 \text{ s}} = 1.41953 = 1.420 \text{ L/s}$

Estimate: $20 \times 4 = 80$; $80/60 \approx 1.3$ L/s

(f) $0.02500 \text{ ft}^3 \times \frac{12^3 \text{ in}^3}{1 \text{ ft}^3} \times \frac{2.54^3 \text{ cm}^3}{1 \text{ in}^3} = 707.9 \text{ cm}^3$

Estimate: $10^3 = 1000$; $3^3 = 27$; $1000 \times 27 = 27{,}000$; $27{,}000/0.04 \approx 700 \text{ cm}^3$

1.59 (a) $31 \text{ gal} \times \frac{4 \text{ qt}}{1 \text{ gal}} \times \frac{1 \text{ L}}{1.057 \text{ qt}} = 1.2 \times 10^2 \text{ L}$

Estimate: $(30 \times 4)/1 \approx 120$ L

(b) $\frac{6 \text{ mg}}{\text{kg (body)}} \times \frac{1 \text{ kg}}{2.205 \text{ lb}} \times 185 \text{ lb} = 5 \times 10^2 \text{ mg}$

Estimate: $6/2 = 3$; $3 \times 180 = 540$ mg

(c) $\frac{400 \text{ km}}{47.3 \text{ L}} \times \frac{1 \text{ mi}}{1.6093 \text{ km}} \times \frac{1 \text{ L}}{1.057 \text{ qt}} \times \frac{4 \text{ qt}}{1 \text{ gal}} = \frac{19.9 \text{ mi}}{\text{gal}}$

(2×10^1 mi/gal for 1 sig fig)

Estimate: $400/50 = 8$; $8/1.6 = 5$; $5/1 = 5$; $5 \times 4 \approx 20$ mi/gal

(d) $200 \text{ cups} \times \frac{1 \text{ lb}}{50 \text{ cups}} \times \frac{1 \text{ kg}}{2.205 \text{ lb}} = 1.81 \text{ kg}$

(2 kg for 1 sig fig)

Estimate: 1 lb = 50 cups, 4 lb = 200 cups; 4 lb ≈ 2 kg

1.61 $14.5 \text{ ft} \times 16.5 \text{ ft} \times 8.0 \text{ ft} = 1914 = 1.9 \times 10^3 \text{ ft}^3$ (2 sig figs)

$$1914 \text{ ft}^3 \times \frac{(1 \text{ yd})^3}{(3 \text{ ft})^3} \times \frac{(1 \text{ m})^3}{(1.0936)^3 \text{ yd}^3} \times \frac{10^3 \text{ dm}^3}{1 \text{ m}^3} \times \frac{1 \text{ L}}{1 \text{ dm}^3} \times \frac{1.19 \text{ g}}{\text{L}} \times \frac{1 \text{ kg}}{1000 \text{ g}} = 64.4985 = 64 \text{ kg air}$$

Estimate: $1900/27 \approx 60$; $(60 \times 1)/1 \approx 60$ kg

1.63 Strategy: 1) Calculate volume of gold (Au) in cm^3 in the sheet

2) Mass = density × volume

3) Change g → troy oz and $

$$100 \text{ ft} \times 82 \text{ ft} \times \frac{(12)^2 \text{ in}^2}{1 \text{ ft}^2} \times 5 \times 10^{-6} \text{ in} \times \frac{(2.54)^3 \text{ cm}^3}{1 \text{ in}^3} = 96.75 = 1 \times 10^2 \text{ cm}^3 \text{ Au}$$

$$96.75 \text{ cm}^3 \text{ Au} \times \frac{19.32 \text{ g}}{1 \text{ cm}^3} \times \frac{1 \text{ troy oz}}{31.1034768 \text{ g}} \times \frac{\$1654}{\text{troy oz}} = \$99,399 = \$1 \times 10^5$$

(Strictly speaking, the datum 100 ft has 1 sig fig, so the result has 1 sig fig.)

Additional Exercises

1.67 According to the law of constant composition, any sample of vitamin C has the same relative amount of carbon and oxygen; the ratio of oxygen to carbon in the isolated sample is the same as the ratio in synthesized vitamin C.

$$\frac{2.00 \text{ g O}}{1.50 \text{ g C}} = \frac{\text{x g O}}{6.35 \text{ g C}}; \quad \text{x} = \frac{(2.00 \text{ g O})(6.35 \text{ g C})}{1.50 \text{ g C}} = 8.47 \text{ g O}$$

1.69 (a) I. $(22.52 + 22.48 + 22.54)/3 = 22.51$

II. $(22.64 + 22.58 + 22.62)/3 = 22.61$

Based on the average, set I is more accurate. That is, it is closer to the true value of 22.52%.

(b) Average deviation = Σ | value-average | /3

I. $|22.52 - 22.51| + |22.48 - 22.51| + |22.54 - 22.51|/3 = 0.02$

II. $|22.64 - 22.61| + |22.58 - 22.61| + |22.62 - 22.61|/3 = 0.02$

Based on average deviations, the two sets display the same precision, even though set I is more accurate. [According to Section 1.5, standard deviation is a measure that is often used to determine precision. Using the formula for calculating standard deviation given in Appendix A.5, the values for the two sets are 0.03 and 0.03, respectively. The standard deviations of the two sets are also the same, confirming that the two sets are equally precise.]

1.71 (a) volume (b) area (c) volume (d) density

(e) time (f) length (g) temperature

1.74 (a) Baking powder is not a pure substance. It is a mixture of basic and acidic compounds that, in the presence of water, react to form CO_2 which causes baked goods to rise.

(b) Lemon juice is not a pure substance. It is a mixture of water, citric acid, and other natural flavors. Its exact composition depends on the characteristics of the lemon that produces the juice.

(c) Propane is a nearly pure substance. Propane itself is odorless, but tank propane contains trace amounts of an odor-producing substance so that leaks can easily be detected.

(d) Aluminum foil is a nearly pure substance.

(e) Ibuprofen itself is a pure substance. Ibuprofen tablets probably contain the active ingredient and other binders, drying agents and coatings.

(f) Bourbon whiskey is not a pure substance. It is a mixture of water, alcohol, and flavoring compounds.

(g) Helium gas is a pure substance.

(h) Clear water pumped from a deep aquifer is a nearly pure substance. All natural water (not distilled or deionized) contains trace minerals.

1.75 (a) $575\text{ ft}\times\frac{12\text{ in}}{1\text{ ft}}\times\frac{2.54\text{ cm}}{1\text{ in}}\times\frac{10\text{ mm}}{1\text{ cm}}\times\frac{1\text{ quarter}}{1.55\text{ mm}}=1.1307\times10^5=1.13\times10^5\text{ quarters}$

(b) $1.1307\times10^5\text{ quarters}\times\frac{5.67\text{ g}}{1\text{ quarter}}=6.41\times10^5\text{ g (641 kg)}$

(c) $1.1307\times10^5\text{ quarters}\times\frac{1\text{ dollar}}{4\text{ quarters}}=\$28{,}268=\$2.83\times10^4$

(d) $\$16{,}213{,}166{,}914{,}811.11\times\frac{1\text{ stack}}{\$28{,}268}=5.7355\times10^8=5.74\times10^8\text{ stacks}$

1.79 The most dense liquid, Hg, will sink; the least dense, cyclohexane, will float; H_2O will be in the middle.

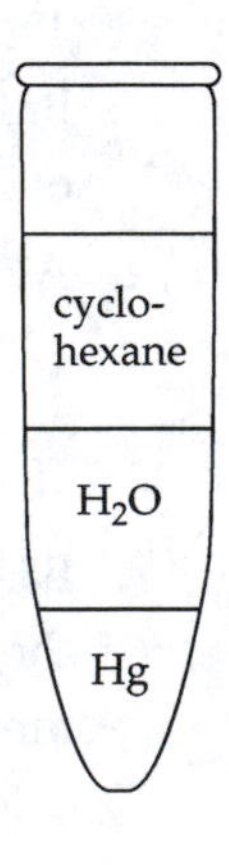

1.82 mass of toluene = 58.58 g – 32.65 g = 25.93 g

volume of toluene $=25.93\text{ g}\times\frac{1\text{ mL}}{0.864\text{ g}}=30.0116=30.0\text{ mL}$

volume of solid = 50.00 mL – 30.0116 mL = 19.9884 = 20.0 mL

density of solid $=\frac{32.65\text{ g}}{19.9884\text{ mL}}=1.63\text{ g/mL}$

1.85 (a) $\frac{40\text{ lb peat}}{14\times20\times30\text{ in}^3}\times\frac{1\text{ in}^3}{(2.54)^3\text{ cm}^3}\times\frac{453.6\text{ g}}{1\text{ lb}}=0.13\text{ g/cm}^3\text{ peat}$

$\frac{40\text{ lb soil}}{1.9\text{ gal}}\times\frac{1\text{ gal}}{4\text{ qt}}\times\frac{1.057\text{ qt}}{1\text{ L}}\times\frac{1\times10^{-3}\text{ L}}{1\text{ mL}}\times\frac{1\text{ mL}}{1\text{ cm}^3}\times\frac{453.6\text{ g}}{1\text{ lb}}=2.5\text{ g/cm}^3\text{ soil}$

No. Volume must be specified to compare mass. The densities tell us that a certain volume of peat moss is "lighter" (weighs less) than the same volume of top soil.

(b) $1 \text{ bag peat} = 14 \times 20 \times 30 = 8.4 \times 10^3 \text{ in}^3$

$$15.0 \text{ ft} \times 20.0 \text{ ft} \times 3.0 \text{ in} \times \frac{12^2 \text{ in}^2}{\text{ft}^2} = 129{,}600 = 1.3 \times 10^5 \text{ in}^3 \text{ peat needed}$$

$$129{,}600 \text{ in}^3 \times \frac{1 \text{ bag}}{8.4 \times 10^3 \text{ in}^3} = 15.4 = 15 \text{ bags (Buy 16 bags of peat.)}$$

1.89 $$45.23 \text{ g ethanol} \times \frac{1 \text{ cm}^3}{0.789 \text{ g ethanol}} = 57.3257 = 57.33 \text{ cm}^3\text{, volume of cylinder}$$

$$V = \pi r^2 h; r = (V/\pi h)^{1/2} = \left[\frac{57.3257 \text{ cm}^3}{\pi \times 25.0 \text{ cm}}\right]^{1/2} = 0.854338 = 0.854 \text{ cm}$$

$d = 2\,r = 1.71 \text{ cm}$

1.91 The separation with distinctly separated red and blue spots is more successful. The procedure that produced the purple blur did not separate the two dyes. To quantify the characteristics of the separation, calculate a reference value for each spot that is

$$\frac{\text{distance traveled by spot}}{\text{distance traveled by solvent}}$$

If the values for the two spots are fairly different, the separation is successful. (One could measure the distance between the spots, but this would depend on the length of paper used and be different for each experiment. The values suggested above are independent of the length of paper.)

1.94 (a) $$10.0 \text{ mg} \times \frac{1 \times 10^{-3} \text{ g}}{1 \text{ mg}} \times \frac{1 \text{ cm}^3}{0.20 \text{ g}} = 0.050 \text{ cm}^3 = 0.050 \text{ mL volume}$$

(b) $$10.0 \text{ mg} \times \frac{1 \times 10^{-3} \text{ g}}{1 \text{ mg}} \times \frac{1242 \text{ m}^2}{1 \text{ g}} = 12.42 = 12.4 \text{ m}^2 \text{ surface area}$$

(c) 7.748 mg Hg initial – 0.001 mg Hg remain = 7.747 mg Hg removed

$$\frac{7.747 \text{ mg Hg removed}}{7.748 \text{ mg Hg initial}} \times 100 = 99.99\% \text{ Hg removed}$$

(d) 10.0 mg "spongy" initial + 7.747 mg Hg removed = 17.747 = 17.7 mg after exposure

2 Atoms, Molecules, and Ions

Visualizing Concepts

2.1 (a) Like charges repel and opposite charges attract, so the sign of the electrical charge on the particle is negative.

(b) The greater the magnitude of the charges, the greater the electrostatic repulsion or attraction. As the charge on the plates is increased, the bending will increase.

(c) As the mass of the particle increases and speed stays the same, linear momentum (mv) of the particle increases and bending decreases. (See **A Closer Look**: The Mass Spectrometer.)

2.4 Because the number of electrons (negatively charged particles) does not equal the number of protons (positively charged particles), the particle is an ion. The charge on the ion is 2–.

Atomic number = number of protons = 16. The element is S, sulfur.

Mass number = protons + neutrons = 32

${}^{32}_{16}S^{2-}$

2.6 Formula: IF_5 Name: iodine pentafluoride

Because the compound is composed of elements that are all nonmetals, it is molecular.

2.8 Cations (red spheres) have positive charges; anions (blue spheres) have negative charges. There are twice as many anions as cations, so the formula has the general form CA_2. Only $Ca(NO_3)_2$, calcium nitrate, is consistent with the diagram.

2.10 (a) In the absence of an electric field, there is no electrostatic interaction between the oil drops and the apparatus, so the rate of fall of the oil drops is determined solely by the force of gravity. In the presence of an electric field, there is electrostatic attraction between the negatively charged oil drops and the positively charged plate, as well as electrostatic repulsion between the negatively charged oil drops and the negative plate. These electrostatic forces oppose the force of gravity and change the rate of fall of the drops.

(b) Each individual drop has a different number of electrons associated with it. The greater the accumulated negative charge on the drop, the greater the electrostatic forces between the oil drop and the plates. If the combined electrostatic forces are greater than the force of gravity, the drop moves up.

The Atomic Theory of Matter and the Discovery of Atomic Structure (Sections 2.1 and 2.2)

2.11 (a) ratio of masses $= \dfrac{0.727 \text{ g O}}{0.273 \text{ g C}} = 2.663 = 2.66$

(b) ratio of masses $= \dfrac{0.571 \text{ g O}}{0.429 \text{ g C}} = 1.331 = 1.33$

(c) The two mass ratios are related by a factor of 2. In the first compound, CO_2, twice as much O is bound to one gram of C as in the second compound. The empirical formula of the second compound is then CO.

2.13 (a) $\dfrac{17.60 \text{ g oxygen}}{30.82 \text{ g nitrogen}} = \dfrac{0.5711 \text{ g O}}{1 \text{ g N}}$; $0.5711/0.5711 = 1.0$

$\dfrac{35.20 \text{ g oxygen}}{30.82 \text{ g nitrogen}} = \dfrac{1.142 \text{ g O}}{1 \text{ g N}}$; $1.142/0.5711 = 2.0$

$\dfrac{70.40 \text{ g oxygen}}{30.82 \text{ g nitrogen}} = \dfrac{2.284 \text{ g O}}{1 \text{ g N}}$; $2.284/0.5711 = 4.0$

$\dfrac{88.00 \text{ g oxygen}}{30.82 \text{ g nitrogen}} = \dfrac{2.855 \text{ g O}}{1 \text{ g N}}$; $2.855/0.5711 = 5.0$

(b) These masses of oxygen per one gram nitrogen are in the ratio of 1:2:4:5 and thus obey the *law of multiple proportions*. Multiple proportions arise because atoms are the indivisible entities combining, as stated in Dalton's theory. Because atoms are indivisible, they must combine in ratios of small whole numbers.

2.15 Electrons, in the form of cathode rays, were discovered first. Neutrons were discovered last.

2.17 *Analyze.* We are given the diameters of a gold atom and its nucleus, and a gold foil that is two atoms thick. What fraction of alpha particles in Rutherford's experiment are deflected at large angles?

Plan. In order to be deflected at a large angle, an alpha particle must directly strike a gold nucleus. Assume that the gold atoms in a single row touch. Consider the cross-sectional area of the gold foil exposed to the beam of alpha particles. Calculate the percentage of this area occupied by the nucleus. But, there are two rows of gold particles, offset relative to one another (Figure 2.9). Assume each alpha particle has two chances to hit a gold nucleus, so the fraction deflected at large angles is twice the ratio of areas. [This approach ignores empty space in the arrangement of gold atoms, which is about 9% of the total cross-sectional area.]

Solve.

fraction of alpha particles deflected at large angles $= \dfrac{\text{area of Au nucleus}}{\text{area of Au atom}} \times 2$

The cross-sectional area of a spherical atom is a circle. Area = πr^2

fraction deflected at large angles $= \dfrac{\pi[r(\text{nucleus})]^2}{\pi[r(\text{atom})]^2} \times 2$

fraction deflected at large angles $= \dfrac{(1.0 \times 10^{-4}\ \text{Å})^2}{(2.7\ \text{Å})^2} \times 2 = 2.7 \times 10^{-9}$

That is, 1 out of approximately 365 million alpha particles is deflected at a large angle.

The Modern View of Atomic Structure; Atomic Weights (Sections 2.3 and 2.4)

2.19 (a) $1.35\ \text{Å} \times \dfrac{1 \times 10^{-10}\ \text{m}}{1\ \text{Å}} \times \dfrac{1\ \text{nm}}{1 \times 10^{-9}\ \text{m}} = 0.135\ \text{nm}$

$1.35\ \text{Å} \times \dfrac{1 \times 10^{-10}\ \text{m}}{1\ \text{Å}} \times \dfrac{1\ \text{pm}}{1 \times 10^{-12}\ \text{m}} = 1.35 \times 10^2$ or 135 pm (1 Å = 100 pm)

(b) Aligned Au atoms have **diameters** touching. d = 2r = 2(1.35 Å) = 2.70 Å

$1.0\ \text{mm} \times \dfrac{1\ \text{m}}{1000\ \text{mm}} \times \dfrac{1\ \text{Å}}{1 \times 10^{-10}\ \text{m}} \times \dfrac{1\ \text{Au atom}}{2.70\ \text{Å}} = 3.70 \times 10^6$ Au atoms

(c) $V = 4/3\,\pi r^3.\ \ r = 1.35\ \text{Å} \times \dfrac{1 \times 10^{-10}\ \text{m}}{1\ \text{Å}} \times \dfrac{100\ \text{cm}}{\text{m}} = 1.35 \times 10^{-8}\ \text{cm}$

$V = (4/3)(\pi)(1.35 \times 10^{-8})^3\ \text{cm}^3 = 1.03 \times 10^{-23}\ \text{cm}^3$

2.21 (a) proton, neutron, electron

(b) proton = +1, neutron = 0, electron = –1

(c) The neutron is most massive. (The neutron and proton have very similar masses.)

(d) The electron is least massive.

2.23 (a) 5 protons, 5 neutrons, 5 electrons. Every neutral ^{10}B atom has 5 protons and 5 neutrons. The mass number of this B atom is 10, so it has (10 – 5) = 5 neutrons.

(b) $^{11}_{6}$C. Adding a proton increases the atomic number of the atom to 6 and the mass number to 11. The element with atomic number 6 is carbon.

(c) $^{11}_{5}$B. Adding a neutron increases the mass number by 1, but does not change the identity of the atom.

(d) The atom in part (c) is an isotope of ^{10}B. Isotopes are atoms of the same element with different masses.

2.25 (a) *Atomic number* is the number of protons in the nucleus of an atom. *Mass number* is the total number of nuclear particles, protons plus neutrons, in an atom.

(b) The mass number can vary without changing the identity of the atom, but the atomic number of every atom of a given element is the same.

2.27 p = protons, n = neutrons, e = electrons

(a) ^{40}Ar has 18 p, 22 n, 18 e (b) ^{65}Zn has 30 p, 35 n, 30 e

(c) ^{70}Ga has 31 p, 39 n, 31 e (d) ^{80}Br has 35 p, 45 n, 35 e

(e) ^{184}W has 74 p, 110 n, 74 e (f) ^{243}Am has 95 p, 148 n, 95 e

2.29

Symbol	^{79}Br	^{55}Mn	^{112}Cd	^{222}Rn	^{207}Pb
Protons	35	25	48	86	82
Neutrons	44	30	64	136	125
Electrons	35	25	48	86	82
Mass no.	79	55	112	222	207

2.31 (a) $^{196}_{78}Pt$ (b) $^{84}_{36}Kr$ (c) $^{75}_{33}As$ (d) $^{24}_{12}Mg$

2.33 (a) $^{12}_{6}C$

(b) Atomic weights are really average atomic masses, the sum of the mass of each naturally occurring isotope of an element times its fractional abundance. Each B atom will have the mass of one of the naturally occurring isotopes, whereas the "atomic weight" is an average value. The naturally occurring isotopes of B, their atomic masses, and relative abundances are:
^{10}B, 10.012937, 19.9%; ^{11}B, 11.009305, 80.1%.

2.35 Atomic weight (average atomic mass) = Σ fractional abundance × mass of isotope
Atomic weight = 0.6917(62.9296) + 0.3083(64.9278) = 63.5456 = 63.55 amu

2.37 (a) In Thomson's cathode ray tube, the charged particles are electrons. In a mass spectrometer, the charged particles are positively charged ions (cations).

(b) The x-axis label (independent variable) is atomic mass (or particle mass) and the y-axis label (dependent variable) is signal intensity.

(c) The Cl^{2+} ion will be deflected more. The greater the charge on the positive ion, the larger its interaction with the electric and magnetic fields. (For this reason, the x-axis label of a mass spectrum is usually mass-to-charge ratio of the particles.)

2.39 (a) Average atomic mass = 0.7899(23.98504) + 0.1000(24.98584) + 0.1101(25.98259)
= 24.31 amu

(b)

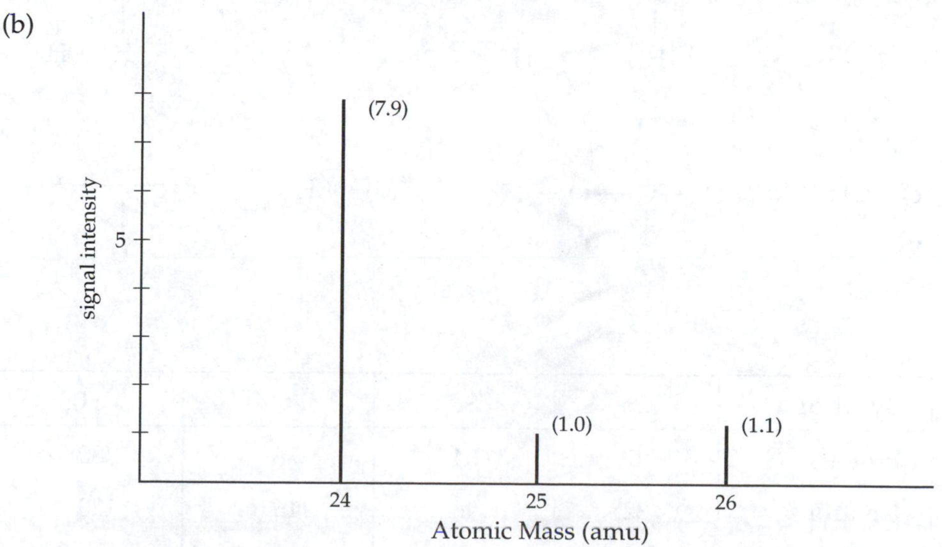

The relative intensities of the peaks in the mass spectrum are the same as the relative abundances of the isotopes. The abundances and peak heights are in the ratio ^{24}Mg: ^{25}Mg: ^{26}Mg as 7.8 : 1.0 : 1.1.

The Periodic Table, Molecules and Molecular Compounds, and Ions and Ionic Compounds (Sections 2.5, 2.6, and 2.7)

2.41 (a) Cr, 24 (metal) (b) He, 2 (nonmetal) (c) P, 15 (nonmetal)
(d) Zn, 30 (metal) (e) Mg, 12 (metal) (f) Br, 35 (nonmetal)
(g) As, 33 (metalloid)

2.43 (a) K, alkali metals (metal) (b) I, halogens (nonmetal)
(c) Mg, alkaline earth metals (metal) (d) Ar, noble gases (nonmetal)
(e) S, chalcogens (nonmetal)

2.45 (a) C_4H_{10} is the molecular formula for both compounds. For the molecular formula, count the total number of each kind of atom in the structural formula.

(b) C_2H_5. Starting with the molecular formula, divide subscripts by any common factors to determine the simplest ratio of atom types in the molecule. In this example the common factor for both molecules is 2.

(c) Structural. In this example, the molecules are *structural isomers* and only the structural formulas allow us to determine that the molecules are different.

2.47 From left to right, the molecular and empirical formulas are: N_2H_4, NH_2; N_2H_2, NH; NH_3, NH_3

2.49 (a) $AlBr_3$ (b) C_4H_5 (c) C_2H_4O
(d) P_2O_5 (e) C_3H_2Cl (f) BNH_2

2.51 (a) 6 (b) 10 (c) 12

2.53 (a) C_2H_6O

```
    H       H
    |       |
H — C — O — C — H
    |       |
    H       H
```

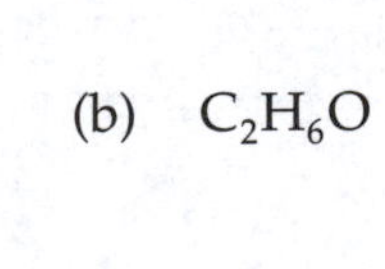

(b) C_2H_6O

```
    H   H
    |   |
H — C — C — O — H
    |   |
    H   H
```

(c) CH_4O

```
    H
    |
H — C — O — H
    |
    H
```

(d) PF_3

```
F — P — F
    |
    F
```

2.55

Symbol	$^{59}Co^{3+}$	$^{80}Se^{2-}$	$^{192}Os^{2+}$	$^{200}Hg^{2+}$
Protons	27	34	76	80
Neutrons	32	46	116	120
Electrons	24	36	74	78
Net Charge	3+	2–	2+	2+

2.57 (a) Mg^{2+} (b) Al^{3+} (c) K^{+} (d) S^{2-} (e) F^{-}

2.59 (a) GaF_3, gallium(III) fluoride (b) LiH, lithium hydride
(c) AlI_3, aluminum iodide (d) K_2S, potassium sulfide

2.61 (a) $CaBr_2$ (b) K_2CO_3 (c) $Al(CH_3COO)_3$ (d) $(NH_4)_2SO_4$ (e) $Mg_3(PO_4)_2$

2.63

Ion	K^{+}	NH_4^{+}	Mg^{2+}	Fe^{3+}
Cl^{-}	KCl	NH_4Cl	$MgCl_2$	$FeCl_3$
OH^{-}	KOH	NH_4OH*	$Mg(OH)_2$	$Fe(OH)_3$
CO_3^{2-}	K_2CO_3	$(NH_4)_2CO_3$	$MgCO_3$	$Fe_2(CO_3)_3$
PO_4^{3-}	K_3PO_4	$(NH_4)_3PO_4$	$Mg_3(PO_4)_2$	$FePO_4$

*Equivalent to $NH_3(aq)$.

2.65 Molecular (all elements are nonmetals):
(a) B_2H_6 (b) CH_3OH (f) NOCl (g) NF_3
Ionic (formed by a cation and an anion, usually contains a metal cation):
(c) $LiNO_3$ (d) Sc_2O_3 (e) CsBr (h) Ag_2SO_4

Naming Inorganic Compounds; Some Simple Organic Compounds (Sections 2.8 and 2.9)

2.67 (a) ClO_2^{-} (b) Cl^{-} (c) ClO_3^{-} (d) ClO_4^{-} (e) ClO^{-}

2.69 (a) calcium, 2+; oxide, 2– (b) sodium, 1+; sulfate, 2–
(c) potassium, 1+; perchlorate, 1– (d) iron, 2+; nitrate, 1–
(e) chromium, 3+; hydroxide, 1–

2.71 (a) lithium oxide (b) iron(III) chloride (ferric chloride)
(c) sodium hypochlorite (d) calcium sulfite
(e) copper(II) hydroxide (cupric hydroxide) (f) iron(II) nitrate (ferrous nitrate)
(g) calcium acetate (h) chromium(III) carbonate (chromic carbonate)
(i) potassium chromate (j) ammonium sulfate

2.73 (a) $Al(OH)_3$ (b) K_2SO_4 (c) Cu_2O (d) $Zn(NO_3)_2$
(e) $HgBr_2$ (f) $Fe_2(CO_3)_3$ (g) NaBrO

2.75 (a) bromic acid (b) hydrobromic acid (c) phosphoric acid
(d) HClO (e) HIO_3 (f) H_2SO_3

2.77 (a) sulfur hexafluoride (b) iodine pentafluoride (c) xenon trioxide

(d) N_2O_4 (e) HCN (f) P_4S_6

2.79 (a) $ZnCO_3$, ZnO, CO_2 (b) HF, SiO_2, SiF_4, H_2O (c) SO_2, H_2O, H_2SO_3

(d) PH_3 (e) $HClO_4$, Cd, $Cd(ClO_4)_2$ (f) VBr_3

2.81 (a) A hydrocarbon is a compound composed of the elements hydrogen and carbon only.

(b)

```
    H   H   H   H   H
    |   |   |   |   |
H — C — C — C — C — C — H
    |   |   |   |   |
    H   H   H   H   H
```

molecular and empirical formulas: C_5H_{12}

2.83 (a) A *functional group* is a group of specific atoms that are constant (arranged the same way) from one molecule to the next.

(b) The characteristic alcohol functional group is an –OH. Another way to say this is that whenever a molecule is called an alcohol, it contains the –OH group.

(c)

```
    H   H   H   H   H
    |   |   |   |   |
H — C — C — C — C — C — OH
    |   |   |   |   |
    H   H   H   H   H
```

2.85 (a)

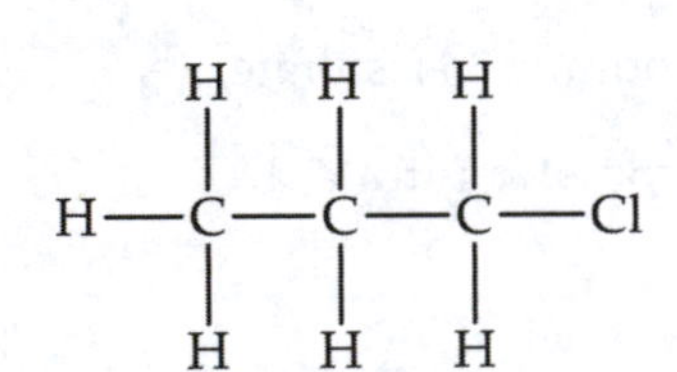

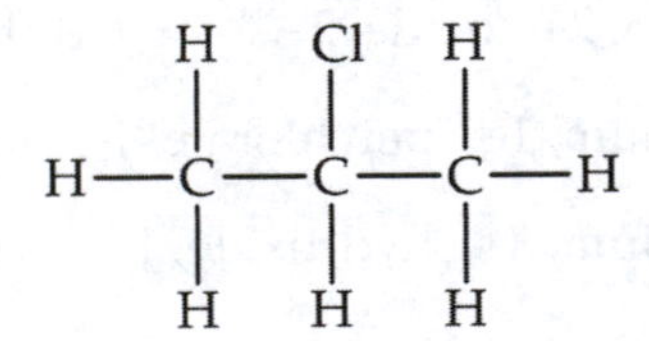

(b) 1-chloropropane 2-chloropropane

Additional Exercises

2.88 (a) ^3He has 2 protons, 1 neutron, and 2 electrons.

(b) ^3H has 1 proton, 2 neutrons, and 1 electron.

^3He: $2(1.672621673 \times 10^{-24}\ g) + 1.674927211 \times 10^{-24}\ g + 2(9.10938215 \times 10^{-28}\ g)$

$= 5.021992 \times 10^{-24}\ g$

^3H: $1.672621673 \times 10^{-24}\ g + 2(1.674927211 \times 10^{-24}\ g) + 9.10938215 \times 10^{-28}\ g$

$= 5.023387 \times 10^{-24}\ g$

Tritium, ^3H, is more massive.

(c) The masses of the two particles differ by 0.0014×10^{-24} g. Each particle loses 1 electron to form the +1 ion, so the difference in the masses of the ions is still 1.4×10^{-27}. A mass spectrometer would need precision to 1×10^{-27} g to differentiate $^3He^+$ and ^3H.

2.90 (a) In arrangement A, the number of atoms in 1 cm^2 is just the square of the number that fit linearly in 1 cm.

$$1.0\ \text{cm} \times \frac{1\ \text{atom}}{4.95\ \text{Å}} \times \frac{1 \times 10^{10}\ \text{Å}}{1\ \text{m}} \times \frac{1\ \text{m}}{100\ \text{cm}} = 2.02 \times 10^7 = 2.0 \times 10^7\ \text{atoms/cm}$$

$$1.0\ \text{cm}^2 = (2.02 \times 10^7)^2 = 4.081 \times 10^{14} = 4.1 \times 10^{14}\ \text{atoms/cm}^2$$

(b) In arrangement B, the atoms in the horizontal rows are touching along their diameters, as in arrangement A. The number of Rb atoms in a 1.0 cm row is then 2.0×10^7 Rb atoms. Relative to arrangement A, the vertical rows are offset by 1/2 of an atom. Atoms in a "column" are no longer touching along their vertical diameter. We must calculate the vertical distance occupied by a row of atoms, which is now less than the diameter of one Rb atom.

Consider the triangle shown below. This is an isosceles triangle (equal side lengths, equal interior angles) with a side-length of 2d and an angle of 60°. Drop a bisector to the uppermost angle so that it bisects the opposite side.

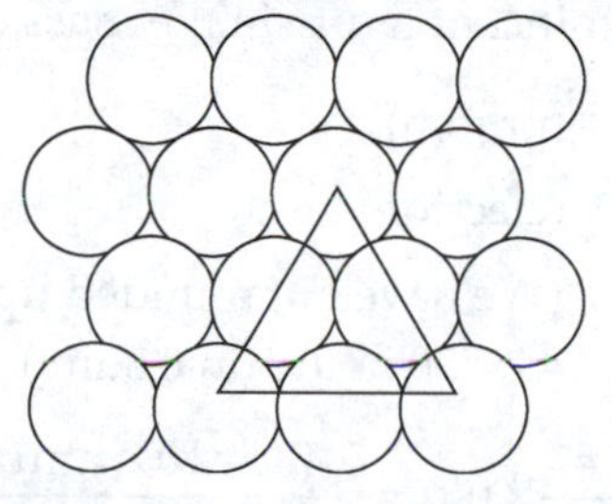

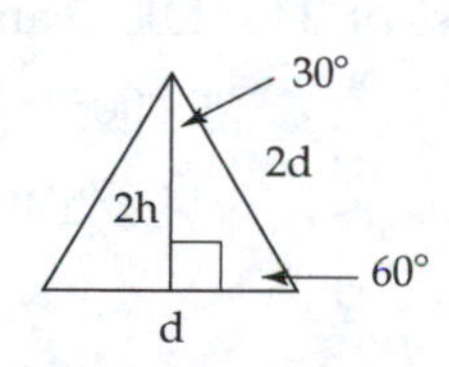

The result is a right triangle with two known side lengths. The length of the unknown side (the angle bisector) is 2h, two times the vertical distance occupied by a row of atoms. Solve for h, the "height" of one row of atoms.

$(2h)^2 + d^2 = (2d)^2$; $4h^2 = 4d^2 - d^2 = 3d^2$; $h^2 = 3d^2/4$

$h = (3d^2/4)^{1/2} = (3(4.95\ \text{Å})^2/4)^{1/2} = 4.2868 = 4.29\ \text{Å}$

The number of rows of atoms in 1 cm is then

$$1.0\ \text{cm} \times \frac{1\ \text{row}}{4.2868\ \text{Å}} \times \frac{1 \times 10^{10}\ \text{Å}}{1\ \text{m}} \times \frac{1\ \text{m}}{100\ \text{cm}} = 2.333 \times 10^7 = 2.3 \times 10^7$$

The number of atoms in a 1.0 cm^2 square area is then

$$\frac{2.020 \times 10^7\ \text{atoms}}{1\ \text{row}} \times 2.333 \times 10^7\ \text{rows} = 4.713 \times 10^{14} = 4.7 \times 10^{14}\ \text{atoms}$$

Note that we have ignored the loss of "1/2" atom at the end of each horizontal row. Out of 2.0×10^7 atoms per row, one atom is not significant.

(c) The ratio of atoms in arrangement B to arrangement A is then 4.713×10^{14} atoms/4.081×10^{14} = 1.155 = 1.2:1. Clearly, arrangement B results in less empty space per unit area or volume. If extended to three dimensions, arrangement B would lead to a greater density for Rb metal.

2.94 (a) $^{16}_{8}O$, $^{17}_{8}O$, $^{18}_{8}O$,

(b) All isotopes are atoms of the same element, oxygen, with the same atomic number (Z = 8), 8 protons in the nucleus and 8 electrons. Elements with similar electron arrangements have similar chemical properties (Section 2.5). Because the 3 isotopes all have 8 electrons, we expect their electron arrangements to be the same and their chemical properties to be very similar, perhaps identical. Each has a different number of neutrons (8, 9, or 10), a different mass number (A = 16, 17, or 18) and thus a different atomic mass.

2.96 (a) The 68.926 amu isotope has a mass number of 69, with 31 protons, 38 neutrons and the symbol $^{69}_{31}Ga$. The 70.925 amu isotope has a mass number of 71, 31 protons, 40 neutrons, and symbol $^{71}_{31}Ga$. (All Ga atoms have 31 protons.)

(b) The average mass of a Ga atom is 69.72 amu. Let x = abundance of the lighter isotope, 1–x = abundance of the heavier isotope. Then x(68.926) + (1–x)(70.925) = 69.72; x = 0.6028 = 0.603, ^{69}Ga = 60.3%, ^{71}Ga = 39.7%.

2.99 (a) Five significant figures. $^{1}H^{+}$ is a bare proton with mass 1.0073 amu. ^{1}H is a hydrogen atom, with 1 proton and 1 electron. The mass of the electron is 5.486×10^{-4} or 0.0005486 amu. Thus the mass of the electron is significant in the fourth decimal place or fifth significant figure in the mass of ^{1}H.

(b) Mass of ^{1}H = 1.0073 amu (proton)

0.0005486 amu (electron)

1.0078 amu (We have not rounded up to 1.0079 because 49 < 50 in the final sum.)

$$\text{Mass \% of electron} = \frac{\text{mass of e}^-}{\text{mass of }^1\text{H}} \times 100 = \frac{5.486 \times 10^{-4}\ \text{amu}}{1.0078\ \text{amu}} \times 100 = 0.05444\%$$

2.106 (a) nickel(II) oxide, 2+ (b) manganese(IV) oxide, 4+

(c) chromium(III) oxide, 3+ (d) molybdenium(VI) oxide, 6+

2.109 (a) perbromate ion (b) selenite ion

(c) AsO_4^{3-} (d) $HTeO_4^{-}$

2.112 (a) potassium nitrate (b) sodium carbonate (c) calcium oxide

(d) hydrochloric acid (e) magnesium sulfate (f) magnesium hydroxide

3 Chemical Reactions and Reaction Stoichiometry

Visualizing Concepts

3.1 Reactant A = blue, reactant B = red

Overall, 4 blue A_2 molecules + 4 red B atoms $\rightarrow$ 4 A_2B molecules

Because 4 is a common factor, this equation reduces to equation (a).

3.3 (a) There are twice as many O atoms as N atoms, so the empirical formula of the original compound is NO_2.

(b) No, because we have no way of knowing whether the empirical and molecular formulas are the same. NO_2 represents the simplest ratio of atoms in a molecule but not the only possible molecular formula.

3.5 (a) *Analyze.* Given the molecular model, write the molecular formula.

Plan. Use the colors of the atoms (spheres) in the model to determine the number of atoms of each element.

Solve. Observe 2 gray C atoms, 5 white H atoms, 1 blue N atom, 2 red O atoms. $C_2H_5NO_2$

(b) *Plan.* Follow the method in Sample Exercise 3.9. Calculate formula weight in amu and molar mass in grams.

2 C atoms = 2(12.0 amu) = 24.0 amu

5 H atoms = 5(1.0 amu) = 5.0 amu

1 N atoms = 1(14.0 amu) = 14.0 amu

2 O atoms = 2(16.0 amu) = <u>32.0 amu</u>

75.0 amu

Formula weight = 75.0 amu, molar mass = 75.0 g/mol

(c) *Plan.* The molar mass of a substance provides the factor for converting grams to moles (or moles to grams).

Solve.

Because the mass of glycine has 4 significant figures, use a molar mass of glycine that has at least 4 significant figures. Using molar masses of the component elements from the periodic chart, the molar mass of glycine is

[2(12.0107) + 5(1.00794) + 14.0067 + 2(15.9994)] = 75.0666 = 75.067 g/mol

$$100.0 \text{ g glycine} \times \frac{1 \text{ mol glycine}}{75.067 \text{ g glycine}} = 1.3321 = 1.332 \text{ mol glycine}$$

(d) *Plan.* Use the definition of mass % and the results from parts (a) and (b) above to find mass % N in glycine.

Solve. $\text{mass \% N} = \dfrac{\text{g N}}{\text{g } C_2H_5NO_2} \times 100$

Assume 1 mol $C_2H_5NO_2$. From the molecular formula of glycine [part (a)], there is 1 mol N/mol glycine.

$$\text{mass \% N} = \frac{1 \times (\text{molar mass N})}{\text{molar mass glycine}} \times 100 = \frac{14.0 \text{ g}}{75.0 \text{ g}} \times 100 = 18.7\%$$

3.7 *Analyze.* Given a box diagram and formulas of reactants, answer questions about the reaction mixture in the box.

Plan. Write and balance the chemical equation. Determine combining ratios of elements and decide on limiting reactant. Decide the maximum number of NH_3 molecules that can be produced and the number of leftover reactant molecules.

Solve.

(a) $N_2 + 3\,H_2 \rightarrow 2\,NH_3$

(b) H_2 is the limiting reactant. There are 4 N_2 molecules and 9 H_2 molecules in the box. According to the chemical reaction, each N_2 molecule requires 3 H_2 molecules for complete reaction. If all N_2 molecules were to react, 12 H_2 molecules would be required. There are only 9 H_2 molecules, so H_2 is the limiting reactant.

(c) 6 NH_3 molecules. Because H_2 is the limiting reactant, the amount of H_2 available determines the amount of NH_3 produced. Three H_2 molecules produce 2 NH_3 molecules, so 9 H_2 molecules produce 6 NH_3 molecules.

(d) One N_2 molecule is left over. The 9 H_2 molecules react with 3 N_2 molecules, leaving one N_2 molecule unreacted. All H_2 molecules are used up.

Chemical Equations and Simple Patterns of Chemical Reactivity (Sections 3.1 and 3.2)

3.9 (a) False. We balance chemical equations as we do because mass must be conserved.

(b) True. Mass is conserved.

(c) False. Subscripts in chemical formulas cannot be changed when balancing an equation. Changing a subscript changes the identity of a compound, which changes the overall reaction.

3.11 (a) $2\,CO(g) + O_2(g) \rightarrow 2\,CO_2(g)$

(b) $N_2O_5(g) + H_2O(l) \rightarrow 2\,HNO_3(aq)$

(c) $CH_4(g) + 4\,Cl_2(g) \rightarrow CCl_4(l) + 4\,HCl(g)$

(d) $Zn(OH)_2(s) + 2\,HNO_3(aq) \rightarrow Zn(NO_3)_2(aq) + 2\,H_2O(l)$

3.13 (a) $Al_4C_3(s) + 12\ H_2O(l) \rightarrow 4\ Al(OH)_3(s) + 3\ CH_4(g)$

(b) $2\ C_5H_{10}O_2(l) + 13\ O_2(g) \rightarrow 10\ CO_2(g) + 10\ H_2O(g)$

(c) $2\ Fe(OH)_3(s) + 3\ H_2SO_4(aq) \rightarrow Fe_2(SO_4)_3(aq) + 6\ H_2O(l)$

(d) $Mg_3N_2(s) + 4\ H_2SO_4(aq) \rightarrow 3\ MgSO_4(aq) + (NH_4)_2SO_4(aq)$

3.15 (a) $CaC_2(s) + 2\ H_2O(l) \rightarrow Ca(OH)_2(aq) + C_2H_2(g)$

(b) $2\ KClO_3(s) \xrightarrow{\Delta} 2\ KCl(s) + 3\ O_2(g)$

(c) $Zn(s) + H_2SO_4(aq) \rightarrow H_2(g) + ZnSO_4(aq)$

(d) $PCl_3(l) + 3\ H_2O(l) \rightarrow H_3PO_3(aq) + 3\ HCl(aq)$

(e) $3\ H_2S(g) + 2\ Fe(OH)_3(s) \rightarrow Fe_2S_3(s) + 6\ H_2O(g)$

Patterns of Chemical Reactivity (Section 3.2)

3.17 (a) NaBr. When a metal reacts with a nonmetal, an ionic compound forms. The combining ratio of the atoms is such that the total positive charge on the metal cation(s) is equal to the total negative charge on the nonmetal anion(s). Determine the formula by balancing the positive and negative charges in the ionic product.

(b) The product is a solid at room temperature. All ionic compounds are solids.

(c) In the balanced chemical equation, the coefficient in front of the product is two.

$2\ Na(s) + Br_2(l) \rightarrow 2\ NaBr(s)$

3.19 (a) $Mg(s) + Cl_2(g) \rightarrow MgCl_2(s)$

(b) $BaCO_3(s) \xrightarrow{\Delta} BaO(s) + CO_2(g)$

(c) $C_8H_8(l) + 10\ O_2(g) \rightarrow 8\ CO_2(g) + 4\ H_2O(l)$

(d) CH_3OCH_3 is C_2H_6O. $C_2H_6O(g) + 3\ O_2(g) \rightarrow 2\ CO_2(g) + 3\ H_2O(l)$

3.21 (a) $2\ C_3H_6(g) + 9\ O_2(g) \rightarrow 6\ CO_2(g) + 6\ H_2O(g)$ combustion

(b) $NH_4NO_3(s) \rightarrow N_2O(g) + 2\ H_2O(g)$ decomposition

(c) $C_5H_6O(l) + 6\ O_2(g) \rightarrow 5\ CO_2(g) + 3\ H_2O(g)$ combustion

(d) $N_2(g) + 3\ H_2(g) \rightarrow 2\ NH_3(g)$ combination

(e) $K_2O(s) + H_2O(l) \rightarrow 2\ KOH(aq)$ combination

Formula Weights (Section 3.3)

3.23 *Analyze.* Given molecular formula or name, calculate formula weight.

Plan. If a name is given, write the correct molecular formula. Then, follow the method in Sample Exercise 3.5. *Solve.*

(a) HNO_3: 1(1.0) + 1(14.0) + 3(16.0) = 63.0 amu

(b) $KMnO_4$: 1(39.1) + 1(54.9) + 4(16.0) = 158.0 amu

(c) $Ca_3(PO_4)_2$: 3(40.1) + 2(31.0) + 8(16.0) = 310.3 amu

(d) SiO_2: 1(28.1) + 2(16.0) = 60.1 amu

(e) Ga_2S_3: 2(69.7) + 3(32.1) = 235.7 amu

(f) $Cr_2(SO_4)_3$: 2(52.0) + 3(32.1) + 12(16.0) = 392.3 amu

(g) PCl_3: 1(31.0) + 3(35.5) = 137.5 amu

3.25 *Plan*. Calculate the formula weight (FW), then the mass % oxygen in the compound. *Solve*.

(a) $C_{17}H_{19}NO_3$: FW = 17(12.0) + 19(1.0) + 1(14.0) + 3(16.0) = 285.0 amu

$$\% \text{ O} = \frac{3(16.0) \text{ amu}}{285.0 \text{ amu}} \times 100 = 16.842 = 16.8\%$$

(b) $C_{18}H_{21}NO_3$: FW = 18(12.0) + 21(1.0) + 1(14.0) + 3(16.0) = 299.0 amu

$$\% \text{ O} = \frac{3(16.0) \text{ amu}}{299.0 \text{ amu}} \times 100 = 16.054 = 16.1\%$$

(c) $C_{17}H_{21}NO_4$: FW = 17(12.0) + 21(1.0) + 1(14.0) + 4(16.0) = 303.0 amu

$$\% \text{ O} = \frac{4(16.0) \text{ amu}}{303.0 \text{ amu}} \times 100 = 21.122 = 21.1\%$$

(d) $C_{22}H_{24}N_2O_8$: FW = 22(12.0) + 24(1.0) + 2(14.0) + 8(16.0) = 444.0 amu

$$\% \text{ O} = \frac{8(16.0) \text{ amu}}{444.0 \text{ amu}} \times 100 = 28.829 = 28.8\%$$

(e) $C_{41}H_{64}O_{13}$: FW = 41(12.0) + 64(1.0) + 13(16.0) = 764.0 amu

$$\% \text{ O} = \frac{13(16.0) \text{ amu}}{764 \text{ amu}} \times 100 = 27.225 = 27.2\%$$

(f) $C_{66}H_{75}Cl_2N_9O_{24}$: FW = 66(12.0) + 75(1.0) + 2(35.5) + 9(14.0) + 24(16.0) = 1448.0 amu

$$\% \text{ O} = \frac{24(16.0) \text{ amu}}{1448.0 \text{ amu}} \times 100 = 26.519 = 26.5\%$$

3.27 *Plan*. Follow the logic for calculating mass % C given in Sample Exercise 3.6. *Solve*.

(a) C_7H_6O: FW = 7(12.0) + 6(1.0) + 1(16.0) = 106.0 amu

$$\% \text{ C} = \frac{7(12.0) \text{ amu}}{106.0 \text{ amu}} \times 100 = 79.2\%$$

(b) $C_8H_8O_3$: FW = 8(12.0) + 8(1.0) + 3(16.0) = 152.0 amu

$$\% \text{ C} = \frac{8(12.0) \text{ amu}}{152.0 \text{ amu}} \times 100 = 63.2\%$$

(c) $C_7H_{14}O_2$: FW = 7(12.0) + 14(1.0) + 2(16.0) = 130.0 amu

$$\% \text{ C} = \frac{7(12.0) \text{ amu}}{130.0 \text{ amu}} \times 100 = 64.6\%$$

Avogadro's Number and the Mole (Section 3.4)

3.29 (a) False. A mole of horses contains four moles of horse legs.

(b) True.

(c) False. The mass of one mole of water is 18.0 g.

(d) True. Electrically neutral NaCl is composed of Na^+ cations and Cl^- anions.

3.31 *Plan.* Because the mole is a counting unit, use it as a basis of comparison; determine the total moles of atoms in each given quantity. *Solve.*

23 g Na contains 1 mol of atoms

0.5 mol H_2O contains (3 atoms × 0.5 mol) = 1.5 mol atoms

6.0×10^{23} N_2 molecules contains (2 atoms × 1 mol) = 2 mol atoms

3.33 *Analyze.* Given: 160 lb/person; Avogadro's number of people, 6.022×10^{23} people. Find: mass in kg of Avogadro's number of people; compare with mass of Earth.

Plan. people → mass in lb → mass in kg; mass of people/mass of Earth

Solve. $6.022 \times 10^{23} \text{ people} \times \frac{160 \text{ lb}}{\text{person}} \times \frac{1 \text{ kg}}{2.2046 \text{ lb}} = 4.370 \times 10^{25} = 4.37 \times 10^{25} \text{ or } 4.4 \times 10^{25} \text{ kg}$

$$\frac{4.370 \times 10^{25} \text{ kg of people}}{5.98 \times 10^{24} \text{ kg Earth}} = 7.31 \text{ or } 7.3$$

One mole of people weighs 7.3 times as much as Earth.

Check. This mass of people is reasonable because Avogadro's number is large.

Estimate: 160 lb ≈ 70 kg; $6 \times 10^{23} \times 70 = 420 \times 10^{23} = 4.2 \times 10^{25}$ kg

3.35 (a) *Analyze.* Given: 0.105 mol sucrose, $C_{12}H_{22}O_{11}$. Find: mass in g.

Plan. Use molar mass (g/mol) of $C_{12}H_{22}O_{11}$ to find g $C_{12}H_{22}O_{11}$.

Solve. molar mass = 12(12.0107) + 22(1.00794) + 11(15.9994) = 342.296 = 342.30

$$0.105 \text{ mol sucrose} \times \frac{342.30 \text{ g}}{1 \text{ mol}} = 35.942 = 35.9 \text{ g } C_{12}H_{22}O_{11}$$

Check. 0.1(342) = 34.2 g. The calculated result is reasonable.

(b) *Analyze.* Given: mass. Find: moles. *Plan.* Use molar mass of $Zn(NO_3)_2$.

Solve. molar mass = 1(65.39) + 2(14.0067) + 6(15.9994) = 189.3998 = 189.40

$$143.50 \text{ g } Zn(NO_3)_2 \times \frac{1 \text{ mol}}{189.40 \text{ g } Zn(NO_3)_2} = 0.75766 \text{ mol } Zn(NO_3)_2$$

Check. 140/180 ≈ 7/9 = 0.78 mol

(c) *Analyze.* Given: moles. Find: molecules. *Plan.* Use Avogadro's number.

Solve. $1.0 \times 10^{-6} \text{ mol } CH_3CH_2OH \times \frac{6.022 \times 10^{23} \text{ molecules}}{1 \text{ mol}} = 6.022 \times 10^{17}$

6.0×10^{17} CH_3CH_2OH molecules

Check. $(1.0 \times 10^{-6})(6 \times 10^{23}) = 6 \times 10^{17}$

(d) *Analyze.* Given: mol NH_3. Find: N atoms.

Plan. mol NH_3 → mol N atoms → N atoms

Solve. $0.410 \text{ mol } NH_3 \times \frac{1 \text{ mol N atoms}}{1 \text{ mol } NH_3} \times \frac{6.022 \times 10^{23} \text{ atoms}}{1 \text{ mol}}$

$= 2.47 \times 10^{23}$ N atoms

Check. $(0.4)(6 \times 10^{23}) = 2.4 \times 10^{23}$.

3.37 *Analyze/Plan.* See 3.35 for stepwise problem-solving approaches. *Solve.*

(a) $(NH_4)_3PO_4$ molar mass = 3(14.007) + 12(1.008) + 1(30.974) + 4(16.00) = 149.091

= 149.1 g/mol

$$2.50 \times 10^{-3} \text{ mol } (NH_4)_3PO_4 \times \frac{149.1 \text{ g } (NH_4)_3PO_4}{1 \text{ mol}} = 0.373 \text{ g } (NH_4)_3PO_4$$

(b) $AlCl_3$ molar mass = 26.982 + 3(35.453) = 133.341 = 133.34 g/mol

$$0.2550 \text{ g } AlCl_3 \times \frac{1 \text{ mol}}{133.34 \text{ g } AlCl_3} \times \frac{3 \text{ mol } Cl^-}{1 \text{ mol } AlCl_3} = 5.737 \times 10^{-3} \text{ mol } Cl^-$$

(c) $C_8H_{10}N_4O_2$ molar mass = 8(12.01) + 10(1.008) + 4(14.01) + 2(16.00) = 194.20

= 194.2 g/mol

$$7.70 \times 10^{20} \text{ molecules} \times \frac{1 \text{ mol}}{6.022 \times 10^{23} \text{ molecules}} \times \frac{194.2 \text{ g } C_8H_{10}N_4O_2}{1 \text{ mol caffeine}}$$

$= 0.248 \text{ g } C_8H_{10}N_4O_2$

(d) $$\frac{0.406 \text{ g cholesterol}}{0.00105 \text{ mol}} = 387 \text{ g cholesterol/mol}$$

3.39 (a) $C_6H_{10}OS_2$ molar mass = 6(12.01) + 10(1.008) + 1(16.00) + 2(32.07) = 162.28

= 162.3 g/mol

(b) *Plan.* mg → g → mol *Solve.*

$$5.00 \text{ mg allicin} \times \frac{1 \times 10^{-3} \text{ g}}{1 \text{ mg}} \times \frac{1 \text{ mol}}{162.3 \text{ g}} = 3.081 \times 10^{-5} = 3.08 \times 10^{-5} \text{ mol allicin}$$

Check. 5.00 mg is a small mass, so the small answer is reasonable.

$(5 \times 10^{-3})/200 = 2.5 \times 10^{-5}$

(c) *Plan.* Use mol from part (b) and Avogadro's number to calculate molecules.

Solve. $3.081 \times 10^{-5} \text{ mol allicin} \times \frac{6.022 \times 10^{23} \text{ molecules}}{\text{mol}} = 1.855 \times 10^{19}$

$= 1.86 \times 10^{19}$ allicin molecules

Check. $(3 \times 10^{-5})(6 \times 10^{23}) = 18 \times 10^{18} = 1.8 \times 10^{19}$

(d) *Plan.* Use molecules from part (c) and molecular formula to calculate S atoms.

Solve. 1.855×10^{19} allicin molecules $\times \dfrac{2 \text{ S atoms}}{1 \text{ allicin molecule}} = 3.71 \times 10^{19}$ S atoms

Check. Obvious.

3.41 (a) *Analyze.* Given: $C_6H_{12}O_6$, 1.250×10^{21} C atoms. Find: H atoms.

Plan. Use molecular formula to determine number of H atoms that are present with 1.250×10^{21} C atoms. *Solve.*

$$\frac{12 \text{ H atoms}}{6 \text{ C atoms}} = \frac{2 \text{ H}}{1 \text{ C}} \times 1.250 \times 10^{21} \text{ C atoms} = 2.500 \times 10^{21} \text{ H atoms}$$

Check. $(2 \times 1 \times 10^{21}) = 2 \times 10^{21}$

(b) *Plan.* Use molecular formula to find the number of glucose molecules that contain 1.250×10^{21} C atoms. *Solve.*

$$\frac{1 \text{ } C_6H_{12}O_6 \text{ molecule}}{6 \text{ C atoms}} \times 1.250 \times 10^{21} \text{ C atoms} = 2.0833 \times 10^{20}$$

$$= 2.083 \times 10^{20} \text{ } C_6H_{12}O_6 \text{ molecules}$$

Check. $(12 \times 10^{20}/6) = 2 \times 10^{20}$

(c) *Plan.* Use Avogadro's number to change molecules $\rightarrow$ mol. *Solve.*

$$2.0833 \times 10^{20} \text{ } C_6H_{12}O_6 \text{ molecules} \times \frac{1 \text{ mol}}{6.022 \times 10^{23} \text{ molecules}}$$

$$= 3.4595 \times 10^{-4} = 3.460 \times 10^{-4} \text{ mol } C_6H_{12}O_6$$

Check. $(2 \times 10^{20})/(6 \times 10^{23}) = 0.33 \times 10^{-3} = 3.3 \times 10^{-4}$

(d) *Plan.* Use molar mass to change mol $\rightarrow$ g. *Solve.*

1 mole of $C_6H_{12}O_6$ weighs 180.0 g (Sample Exercise 3.9)

$$3.4595 \times 10^{-4} \text{ mol } C_6H_{12}O_6 \times \frac{180.0 \text{ g } C_6H_{12}O_6}{1 \text{ mol}} = 0.06227 \text{ g } C_6H_{12}O_6$$

Check. $3.5 \times 180 = 630$; $630 \times 10^{-4} = 0.063$

3.43 *Analyze.* Given: g C_2H_3Cl/L. Find: mol/L, molecules/L.

Plan. The /L is constant throughout the problem, so we can ignore it. Use molar mass for g $\rightarrow$ mol, Avogadro's number for mol $\rightarrow$ molecules. *Solve.*

$$\frac{2.0 \times 10^{-6} \text{ g } C_2H_3Cl}{1 \text{ L}} \times \frac{1 \text{ mol } C_2H_3Cl}{62.50 \text{ g } C_2H_3Cl} = 3.20 \times 10^{-8} = 3.2 \times 10^{-8} \text{ mol } C_2H_3Cl/L$$

$$\frac{3.20 \times 10^{-8} \text{ mol } C_2H_3Cl}{1 \text{ L}} \times \frac{6.022 \times 10^{23} \text{ molecules}}{1 \text{ mol}} = 1.9 \times 10^{16} \text{ molecules/L}$$

Check. $(200 \times 10^{-8})/60 = 2.5 \times 10^{-8}$ mol

$(2.5 \times 10^{-8}) \times (6 \times 10^{23}) = 15 \times 10^{15} = 1.5 \times 10^{16}$

Empirical Formulas from Analyses (Section 3.5)

3.45 (a) *Analyze.* Given: moles. Find: empirical formula.

Plan. Find the **simplest ratio of moles** by dividing by the smallest number of moles present.

Solve. 0.0130 mol C / 0.0065 = 2

0.0390 mol H / 0.0065 = 6

0.0065 mol O / 0.0065 = 1

The empirical formula is C_2H_6O.

Check. The subscripts are simple integers.

(b) *Analyze.* Given: grams. Find: empirical formula.

Plan. Calculate the moles of each element present, then the simplest ratio of moles.

Solve. $11.66 \text{ g Fe} \times \frac{1 \text{ mol Fe}}{55.85 \text{ g Fe}} = 0.2088 \text{ mol Fe}; \ 0.2088/0.2088 = 1$

$5.01 \text{ g O} \times \frac{1 \text{ mol O}}{16.00 \text{ g O}} = 0.3131 \text{ mol O}; \ 0.3131/0.2088 \approx 1.5$

Multiplying by two, the integer ratio is 2 Fe : 3 O; the empirical formula is Fe_2O_3.

Check. The subscripts are simple integers.

(c) *Analyze.* Given: mass %. Find: empirical formulas.

Plan. Assume 100 g sample, calculate moles of each element, find the simplest ratio of moles.

Solve. $40.0 \text{ g C} \times \frac{1 \text{ mol C}}{12.01 \text{ g C}} = 3.33 \text{ mol C}; \ 3.33/3.33 = 1$

$6.7 \text{ g H} \times \frac{1 \text{ mol H}}{1.008 \text{ mol H}} = 6.65 \text{ mol H}; \ 6.65/3.33 \approx 2$

$53.3 \text{ g O} \times \frac{1 \text{ mol O}}{16.00 \text{ mol O}} = 3.33 \text{ mol O}; \ 3.33/3.33 = 1$

The empirical formula is CH_2O.

Check. The subscripts are simple integers.

3.47 *Analyze/Plan.* The procedure in all these cases is to assume 100 g of sample, calculate the number of moles of each element present in that 100 g, then obtain the ratio of moles as smallest whole numbers. *Solve.*

(a) $10.4 \text{ g C} \times \frac{1 \text{ mol C}}{12.01 \text{ g C}} = 0.866 \text{ mol C}; \ 0.866/0.866 = 1$

$27.8 \text{ g S} \times \frac{1 \text{ mol S}}{32.07 \text{ g S}} = 0.867 \text{ mol S}; \ 0.867/0.866 \approx 1$

$61.7 \text{ g Cl} \times \frac{1 \text{ mol Cl}}{35.45 \text{ g Cl}} = 1.74 \text{ mol Cl}; \ 1.74/0.866 \approx 2$

The empirical formula is $CSCl_2$.

(b) $$21.7\text{ g C} \times \frac{1\text{ mol C}}{12.01\text{ g C}} = 1.81\text{ mol C}; \;\; 1.81/0.600 \approx 3$$

$$9.6\text{ g O} \times \frac{1\text{ mol O}}{16.00\text{ g O}} = 0.600\text{ mol O}; \;\; 0.600/0.600 = 1$$

$$68.7\text{ g F} \times \frac{1\text{ mol F}}{19.00\text{ g F}} = 3.62\text{ mol F}; \;\; 3.62/0.600 \approx 6$$

The empirical formula is C_3OF_6.

(c) The mass of F is [100 g total – (32.79 g Na + 13.02 Al)] = 54.19 g F

$$32.79\text{ g Na} \times \frac{1\text{ mol Na}}{22.99\text{ g Na}} = 1.426\text{ mol Na}; \;\; 1.426/0.4826 \approx 3$$

$$13.02\text{ g Al} \times \frac{1\text{ mol Al}}{26.98\text{ g Al}} = 0.4826\text{ mol Al}; \;\; 0.4826/0.4826 = 1$$

$$54.19\text{ g F} \times \frac{1\text{ mol F}}{19.00\text{ g F}} = 2.852\text{ mol F}; \;\; 2.852/0.4826 \approx 6$$

The empirical formula is Na_3AlF_6.

3.49 *Analyze.* Given: mass% F; empirical formula XF_3 implies 3:1 ratio of mol F to mol X. Find: atomic mass (AM) of X.

Plan. Calculate mol F. This is 3 times mol X. mol X = 35 g X/AM X.
mol F/3 = 35 g X/AM X. Solve for AM X.

Solve. Mol F = 65/19.0 = 3.421 = 3.4; mol X = 3.421/3 = 1.14035 = 1.1

1.14035 mol X = 35 g X/AM X; AM X = 35 g X/1.14035 mol X = 30.69 = 31 g/mol
The element is likely phosphorus and the compound is then PF_3.

3.51 *Analyze.* Given: empirical formula, molar mass. Find: molecular formula.

Plan. Calculate the empirical formula weight (FW); divide FW by molar mass (MM) to calculate the integer that relates the empirical and molecular formulas. Check. If FW/MM is an integer, the result is reasonable. *Solve.*

(a) $\text{FW } CH_2 = 12.0 + 2(1.01) = 14.0 \quad \frac{MM}{FW} = \frac{84.0}{14.0} = 6.00 = 6$

The subscripts in the empirical formula are multiplied by 6. The molecular formula is C_6H_{12}.

(b) $\text{FW } NH_2Cl = 14.01 + 2(1.008) + 35.45 = 51.48. \quad \frac{MM}{FW} = \frac{51.5}{51.5} = 1$

The empirical and molecular formulas are NH_2Cl.

3.53 *Analyze.* Given: mass %, molar mass. Find: molecular formula.

Plan. Use the plan detailed in Solution 3.47 to find an empirical formula from mass % data. Then use the plan detailed in Solution 3.51 to find the molecular formula. Note that some indication of molar mass must be given, or the molecular formula cannot be determined. *Check.* If there is an integer ratio of moles and MM/ FW is an integer, the result is reasonable. *Solve.*

(a) $92.3 \text{ g C} \times \frac{1 \text{ mol C}}{12.01 \text{ g C}} = 7.685 \text{ mol C}; \ 7.685/7.639 = 1.006 \approx 1$

$7.7 \text{ g H} \times \frac{1 \text{ mol H}}{1.008 \text{ g H}} = 7.639 \text{ mol H}; \ 7.639/7.639 = 1$

The empirical formula is CH, FW = 13.

$\frac{MM}{FW} = \frac{104}{13} = 8$; the molecular formula is C_8H_8.

(b) $49.5 \text{ g C} \times \frac{1 \text{ mol C}}{12.01 \text{ g C}} = 4.12 \text{ mol C}; \ 4.12 / 1.03 \approx 4$

$5.15 \text{ g H} \times \frac{1 \text{ mol H}}{1.008 \text{ g H}} = 5.11 \text{ mol H}; \ 5.11 / 1.03 \approx 5$

$28.9 \text{ g N} \times \frac{1 \text{ mol N}}{14.01 \text{ g N}} = 2.06 \text{ mol N}; \ 2.06 / 1.03 \approx 2$

$16.5 \text{ g O} \times \frac{1 \text{ mol O}}{16.00 \text{ g O}} = 1.03 \text{ mol O}; \ 1.03 / 1.03 = 1$

Thus, $C_4H_5N_2O$, FW = 97. If the molar mass is about 195, a factor of 2 gives the molecular formula $C_8H_{10}N_4O_2$.

(c) $35.51 \text{ g C} \times \frac{1 \text{ mol C}}{12.01 \text{ g C}} = 2.96 \text{ mol C}; \ 2.96 / 0.592 = 5$

$4.77 \text{ g H} \times \frac{1 \text{ mol H}}{1.008 \text{ g H}} = 4.73 \text{ mol H}; 4.73 / 0.592 = 7.99 \approx 8$

$37.85 \text{ g O} \times \frac{1 \text{ mol O}}{16.00 \text{ g O}} = 2.37 \text{ mol O}; 2.37 / 0.592 = 4$

$8.29 \text{ g N} \times \frac{1 \text{ mol N}}{14.01 \text{ g N}} = 0.592 \text{ mol N}; \ 0.592 / 0.592 = 1$

$13.60 \text{ g Na} \times \frac{1 \text{ mol Na}}{22.99 \text{ g Na}} = 0.592 \text{ mol Na}; \ 0.592 / 0.592 = 1$

The empirical formula is $C_5H_8O_4NNa$, FW = 169 g. Because the empirical formula weight and molar mass are approximately equal, the empirical and molecular formulas are both $NaC_5H_8O_4N$.

3.55 (a) *Analyze.* Given: mg CO_2, mg H_2O Find: empirical formula of hydrocarbon, C_xH_y

Plan. Upon combustion, all $C \rightarrow CO_2$, all $H \rightarrow H_2O$.

mg $CO_2 \rightarrow$ g $CO_2 \rightarrow$ mol C; mg $H_2O \rightarrow$ g H_2O, mol H

Find simplest ratio of moles and empirical formula. *Solve.*

$5.86 \times 10^{-3} \text{ g } CO_2 \times \frac{1 \text{ mol } CO_2}{44.01 \text{ g } CO_2} \times \frac{1 \text{ mol C}}{1 \text{ mol } CO_2} = 1.33 \times 10^{-4} \text{ mol C}$

$1.37 \times 10^{-3} \text{ g } H_2O \times \frac{1 \text{ mol } H_2O}{18.02 \text{ g } H_2O} \times \frac{2 \text{ mol H}}{1 \text{ mol } H_2O} = 1.52 \times 10^{-4} \text{ mol H}$

Dividing both values by 1.33×10^{-4} gives C:H of 1:1.14. This is not "close enough" to be considered 1:1. No obvious multipliers (2, 3, 4) produce an integer ratio. Testing other multipliers (trial and error!), the correct factor seems to be 7. The empirical formula is C_7H_8.

Check. See discussion of C:H ratio above.

(b) *Analyze.* Given: g of menthol, g CO_2, g H_2O, molar mass. Find: molecular formula.

Plan/Solve. Calculate mol C and mol H in the sample.

$$0.2829\text{ g } CO_2 \times \frac{1\text{ mol } CO_2}{44.01\text{ g } CO_2} \times \frac{1\text{ mol C}}{1\text{ mol } CO_2} = 0.0064281 = 0.006428\text{ mol C}$$

$$0.1159\text{ g } H_2O \times \frac{1\text{ mol } H_2O}{18.02\text{ g } H_2O} \times \frac{2\text{ mol H}}{1\text{ mol } H_2O} = 0.012863 = 0.01286\text{ mol H}$$

Calculate g C, g H and get g O by subtraction.

$$0.0064281\text{ mol C} \times \frac{12.01\text{ g C}}{1\text{ mol C}} = 0.07720\text{ g C}$$

$$0.012863\text{ mol H} \times \frac{1.008\text{ g H}}{1\text{ mol H}} = 0.01297\text{ g H}$$

mass O = 0.1005 g sample – (0.07720 g C + 0.01297 g H) = 0.01033 g O

Calculate mol O and find integer ratio of mol C: mol H: mol O.

$$0.01033\text{ g O} \times \frac{1\text{ mol O}}{16.00\text{ g O}} = 6.456 \times 10^{-4}\text{ mol O}$$

Divide moles by 6.456×10^{-4}.

$$\text{C: } \frac{0.006428}{6.456 \times 10^{-4}} \approx 10; \quad \text{H: } \frac{0.01286}{6.456 \times 10^{-4}} \approx 20; \quad \text{O: } \frac{6.456 \times 10^{-4}}{6.456 \times 10^{-4}} = 1$$

The empirical formula is $C_{10}H_{20}O$.

$$\text{FW} = 10(12) + 20(1) + 16 = 156; \quad \frac{\text{M}}{\text{FW}} = \frac{156}{156} = 1$$

The molecular formula is the same as the empirical formula, $C_{10}H_{20}O$.

Check. The mass of O wasn't negative or greater than the sample mass; empirical and molecular formulas are reasonable.

3.57 *Analyze.* Given: mass of H_2O and mass of CO_2 from combustion of 0.165 g of valproic acid; molar mass of valproic acid. Find: empirical and molecular formulas of valproic acid.

Plan. Calculate mol C and mol H, then g C and g H; get g O by subtraction.

Solve.

$$0.403 \text{ g } CO_2 \times \frac{1 \text{ mol } CO_2}{44.01 \text{ g } CO_2} \times \frac{1 \text{ mol C}}{1 \text{ mol } CO_2} = 9.157 \times 10^{-3} = 9.16 \times 10^{-3} \text{ mol C}$$

$$0.166 \text{ g } H_2O \times \frac{1 \text{mol } H_2O}{18.02 \text{ g } H_2O} \times \frac{2 \text{ mol H}}{1 \text{ mol } H_2O} = 0.018424 = 0.0184 \text{ mol H}$$

$$9.157 \times 10^{-3} \text{ mol C} \times \frac{12.01 \text{ g C}}{1 \text{ mol C}} = 0.10998 \text{ g C} = 0.110 \text{ g C}$$

$$0.018424 \text{ mol H} \times \frac{1.008 \text{ g H}}{1 \text{ mol H}} = 0.01857 \text{ g H} = 0.0186 \text{ g H}$$

mass of O = 0.165 g sample – (0.10998 g C + 0.01857 g H) = 0.03645 = 0.036 g O

$$0.03645 \text{ g O} \times \frac{1 \text{ mol O}}{16.00 \text{ g O}} = 2.278 \times 10^{-3} = 2.3 \times 10^{-3} \text{mol O. Divide moles by } 2.278 \times 10^{-3}.$$

$$\text{C: } \frac{9.157 \times 10^{-3}}{2.278 \times 10^{-3}} \approx 4; \quad \text{H: } \frac{0.018424}{2.278 \times 10^{-3}} \approx 8; \quad \text{O: } \frac{2.278 \times 10^{-3}}{2.278 \times 10^{-3}} = 1$$

The empirical formula is C_4H_8O, FW = 72. A molar mass of 144 g/mol indicates a factor of two and a molecular formula of $C_8H_{16}O_2$.

3.59 *Analyze.* Given 2.558 g $Na_2CO_3 \bullet xH_2O$, 0.948 g Na_2CO_3. Find: x.

Plan. The reaction involved is $Na_2CO_3 \bullet xH_2O(s) \rightarrow Na_2CO_3(s) + xH_2O(g)$.

Calculate the mass of H_2O lost and then the mole ratio of Na_2CO_3 and H_2O.

Solve. g H_2O lost = 2.558 g sample – 0.948 g Na_2CO_3 = 1.610 g H_2O

$$0.948 \text{ g } Na_2CO_3 \times \frac{1 \text{ mol } Na_2CO_3}{106.0 \text{ g } Na_2CO_3} = 0.00894 \text{ mol } Na_2CO_3$$

$$1.610 \text{ g } H_2O \times \frac{1 \text{ mol } H_2O}{18.02 \text{ g } H_2O} = 0.08935 \text{ mol } H_2O$$

The formula is $Na_2CO_3 \bullet$ **<u>10</u>** H_2O.

Check. x is an integer.

Quantitative Information from Balanced Equations (Section 3.6)

3.61 $Na_2SiO_3(s) + 8\,HF(aq) \rightarrow H_2SiF_6(aq) + 2\,NaF(aq) + 3\,H_2O(l)$

(a) *Analyze.* Given: mol Na_2SiO_3. Find: mol HF. *Plan.* Use the mole ratio 8 HF: 1 Na_2SiO_3 from the balanced equation to relate moles of the two reactants.

Solve. $0.300 \text{ mol } Na_2SiO_3 \times \frac{8 \text{ mol HF}}{1 \text{ mol } Na_2SiO_3} = 2.40 \text{ mol HF}$

Check. Mol HF should be greater than mol Na_2SiO_3.

(b) *Analyze.* Given: mol HF. Find: g NaF. *Plan.* Use the mole ratio 2 NaF : 8 HF to change mol HF to mol NaF, then molar mass to get NaF. *Solve.*

$$0.500 \text{ mol HF} \times \frac{2 \text{ mol NaF}}{8 \text{ mol HF}} \times \frac{41.99 \text{ g NaF}}{1 \text{ mol NaF}} = 5.25 \text{ g NaF}$$

Check. $(0.5/4) = 0.125;\ 0.13 \times 42 > 4$ g NaF

(c) *Analyze.* Given: g HF Find: g Na_2SiO_3.

Plan. $g\ HF \rightarrow mol\ HF \begin{pmatrix} mol \\ ratio \end{pmatrix} \rightarrow mol\ Na_2SiO_3 \rightarrow g\ Na_2SiO_3$

The mole ratio is at the heart of every stoichiometry problem. Molar mass is used to change to and from grams. *Solve.*

$$0.800\ g\ HF \times \frac{1\ mol\ HF}{20.01 g\ HF} \times \frac{1\ mol\ Na_2SiO_3}{8\ mol\ HF} \times \frac{122.1\ g\ Na_2SiO_3}{1\ mol\ Na_2SiO_3} = 0.610\ g\ Na_2SiO_3$$

Check. 0.8 (120/160) < 0.75 mol

3.63 (a) $Al(OH)_3(s) + 3\ HCl(aq) \rightarrow AlCl_3(aq) + 3\ H_2O(l)$

(b) *Analyze.* Given mass of one reactant, find stoichiometric mass of other reactant and products.

Plan. Follow the logic in Sample Exercise 3.16. Calculate mol $Al(OH)_3$ in 0.500 g $Al(OH_3)_3$ separately, because it will be used several times.

Solve. $$0.500\ g\ Al(OH)_3 \times \frac{1\ mol\ Al(OH)_3}{78.00\ g\ Al(OH)_3} = 6.410 \times 10^{-3} = 6.41 \times 10^{-3}\ mol\ Al(OH)_3$$

$$6.410 \times 10^{-3}\ mol\ Al(OH)_3 \times \frac{3\ mol\ HCl}{1\ mol\ Al(OH)_3} \times \frac{36.46\ g\ HCl}{1\ mol\ HCl} = 0.7012 = 0.701\ g\ HCl$$

$$6.410 \times 10^{-3}\ molAl(OH)_3 \times \frac{1\ mol\ HCl}{1\ mol\ Al(OH)_3} \times \frac{133.34\ g\ AlCl_3}{1\ mol\ AlCl_3} = 0.8547$$

$$= 0.855\ g\ AlCl_3$$

(c) $$6.410 \times 10^{-3}\ mol\ Al(OH)_3 \times \frac{3\ mol\ H_2O}{1\ mol\ Al(OH)_3} \times \frac{18.02\ g\ H_2O}{1\ mol\ H_2O} = 0.3465 = 0.347\ g\ H_2O$$

(d) Conservation of mass: mass of products = mass of reactants

reactants: $Al(OH)_3 + HCl$, 0.500 g + 0.701 g = 1.201 g

products: $AlCl_3 + H_2O$, 0.855 g + 0.347 g = 1.202 g

The 0.001 g difference is due to rounding (0.8547 + 0.3465 = 1.2012). This is an excellent *check* of results.

3.65 (a) $Al_2S_3(s) + 6\ H_2O(l) \rightarrow 2\ Al(OH)_3(s) + 3\ H_2S(g)$

(b) *Plan.* $g\ A \rightarrow mol\ A \rightarrow mol\ B \rightarrow g\ B$. See Solution 3.61 (c). *Solve.*

$$14.2\ g\ Al_2S_3 \times \frac{1\ mol\ Al_2S_3}{150.2\ g\ Al_2S_3} \times \frac{2\ mol\ Al(OH)_3}{1\ mol\ Al_2S_3} \times \frac{78.00\ g\ Al(OH)_3}{1\ mol\ Al(OH)_3} = 14.7\ g\ Al(OH)_3$$

Check. $14\left(\frac{2 \times 78}{150}\right) \approx 14(1) \approx 14\ g\ Al(OH)_3$

3.67 (a) *Analyze.* Given: mol NaN_3. Find: mol N_2.

Plan. Use mole ratio from balanced equation. *Solve.*

$$1.50\ mol\ NaN_3 \times \frac{3\ mol\ N_2}{2\ mol\ NaN_3} = 2.25\ mol\ N_2$$

Check. The resulting mol N_2 should be greater than mol NaN_3, (the N_2:NaN_3 ratio is > 1), and it is.

(b) *Analyze.* Given: g N_2 Find: g NaN_3.

Plan. Use molar masses to get from and to grams, mol ratio to relate moles of the two substances. *Solve.*

$$10.0\text{ g } N_2 \times \frac{1\text{ mol } N_2}{28.01\text{ g } N_2} \times \frac{2\text{ mol } NaN_3}{3\text{ mol } N_2} \times \frac{65.01\text{ g } NaN_3}{1\text{ mol } NaN_3} = 15.5\text{ g } NaN_3$$

Check. Mass relations are less intuitive than mole relations. Estimating the ratio of molar masses is sometimes useful. In this case, 65 g NaN_3/28 g N_2 ≈ 2.25. Then, $(10 \times 2/3 \times 2.25) \approx 15$ g NaN_3. The calculated result looks reasonable.

(c) *Analyze.* Given: vol N_2 in ft^3, density N_2 in g/L. Find: g NaN_3.

Plan. First determine how many g N_2 are in 10.0 ft^3, using the density of N_2. Then proceed as in part (b).
Solve.

$$\frac{1.25\text{ g}}{1\text{ L}} \times \frac{1\text{ L}}{1000\text{ cm}^3} \times \frac{(2.54)^3\text{ cm}^3}{1\text{ in}^3} \times \frac{(12)^3\text{ in}^3}{1\text{ ft}^3} \times 10.0\text{ ft}^3 = 354.0 = 354\text{ g } N_2$$

$$354.0\text{ g } N_2 \times \frac{1\text{ mol } N_2}{28.01\text{ g } N_2} \times \frac{2\text{ mol } NaN_3}{3\text{ mol } N_2} \times \frac{65.01\text{ g } NaN_3}{1\text{ mol } NaN_3} = 548\text{ g } NaN_3$$

Check. 1 ft^3 ~ 28 L; 10 ft^3 ~ 280 L; 280 L × 1.25 ~ 350 g N_2
Using the ratio of molar masses from part (b), $(350 \times 2/3 \times 2.25) \approx 525$ g NaN_3

3.69 (a) *Analyze.* Given: dimensions of Al foil. Find: mol Al.

Plan. Dimensions → vol $\xrightarrow{\text{density}}$ mass $\xrightarrow[\text{mass}]{\text{molar}}$ mol Al

Solve. $1.00\text{ cm} \times 1.00\text{ cm} \times 0.550\text{ mm} \times \frac{1\text{ cm}}{10\text{ mm}} = 0.0550\text{ cm}^3$ Al

$$0.0550\text{ cm}^3\text{ Al} \times \frac{2.699\text{ g Al}}{1\text{ cm}^3} \times \frac{1\text{ mol Al}}{26.98\text{ g Al}} = 5.502 \times 10^{-3} = 5.50 \times 10^{-3}\text{ mol Al}$$

Check. $2.699/26.98 \approx 0.1$; $(0.055\text{ cm}^3 \times 0.1) = 5.5 \times 10^{-3}$ mol Al

(b) *Plan.* Write the balanced equation to get a mole ratio; change mol Al → mol $AlBr_3$ → g $AlBr_3$.

Solve. $2\text{ Al(s)} + 3\text{ Br}_2\text{(l)} \rightarrow 2\text{ AlBr}_3\text{(s)}$

$$5.502 \times 10^{-3}\text{ mol Al} \times \frac{2\text{ mol AlBr}_3}{2\text{ mol Al}} \times \frac{266.69\text{ g AlBr}_3}{1\text{ mol AlBr}_3} = 1.467 = 1.47\text{ g AlBr}_3$$

Check. $(0.006 \times 1 \times 270) \approx 1.6$ g $AlBr_3$

3.71 *Analyze/Plan.* We are given heat produced by combustion of one mole of ethanol and a mass of ethanol to burn. Change mass ethanol to moles ethanol and multiply by kJ/mol.

Solve. The molar mass of CH_3CH_2OH is [2(12.0107) + 6(1.00794 + 15.9994] = 46.06844 = 46.07 g/mol

$$235.0\text{ g } CH_3CH_2OH \times \frac{1\text{ mol } CH_3CH_2OH}{46.068\text{ g } CH_3CH_2OH} \times \frac{1367\text{ kJ}}{1\text{ mol } CH_3CH_2OH} = 6973.3 = 6.973 \times 10^3\text{ kJ}$$

Limiting Reactants (Section 3.7)

3.73 (a) The *limiting reactant* determines the maximum number of product moles resulting from a chemical reaction; any other reactant is an *excess reactant*.

(b) The limiting reactant regulates the amount of products, because it is completely used up during the reaction; no more product can be made when one of the reactants is unavailable.

(c) Combining ratios are molecule and mole ratios. Since different molecules have different masses, equal masses of different reactants will not have equal numbers of molecules. By comparing initial moles, we compare numbers of available reactant molecules, the fundamental combining units in a chemical reaction.

3.75 (a) $2\,C_2H_5OH + 6\,O_2 \rightarrow 4\,CO_2 + 6\,H_2O$

The equation above corresponds to the contents of the diagram shown in Exercise 3.75, but does not have the simplest ratio of coefficients. Divide all integer coefficients by 2 to obtain $C_2H_5OH + 3\,O_2 \rightarrow 2\,CO_2 + 3\,H_2O$

(b) C_2H_5OH is the limiting reactant. According to the balanced equation, two molecules of C_2H_5OH require six molecules of O_2 for reaction. In the box there are two molecules of C_2H_5OH and seven molecules of O_2. O_2 is present in excess and C_2H_5OH limits.

(c) If the reaction goes to completion, there will be four molecules of CO_2, six molecules of H_2O, zero molecules of C_2H_5OH (the limiting reactant is completely consumed), and one molecule of O_2 (excess reactant).

3.77 *Analyze.* Given: 1.85 mol NaOH, 1.00 mol CO_2. Find: mol Na_2CO_3.

Plan. Amounts of more than one reactant are given, so we must determine which reactant regulates (limits) product. Then apply the appropriate mole ratio from the balanced equation.

Solve. The mole ratio is 2 NaOH: 1 CO_2, so 1.00 mol CO_2 requires 2.00 mol NaOH for complete reaction. Less than 2.00 mol NaOH are present, so NaOH is the limiting reactant.

$$1.85 \text{ mol NaOH} \times \frac{1 \text{ mol } Na_2CO_3}{2 \text{ mol NaOH}} = 0.925 \text{ mol } Na_2CO_3 \text{ can be produced}$$

The Na_2CO_3:CO_2 ratio is 1:1, so 0.925 mol Na_2CO_3 produced requires 0.925 mol CO_2 consumed. (Alternately, 1.85 mol NaOH × 1 mol CO_2/2 mol NaOH = 0.925 mol CO_2 reacted). 1.00 mol CO_2 initial – 0.925 mol CO_2 reacted = 0.075 mol CO_2 remain.

Check.	2 NaOH(s)	+	CO_2(g)	→	Na_2CO_3(s)	+	H_2O(l)
initial	1.85 mol		1.00 mol		0 mol		
change (reaction)	–1.85 mol		–0.925 mol		+0.925 mol		
final	0 mol		0.075 mol		0.925 mol		

Note that the "change" line (but not necessarily the "final" line) reflects the mole ratios from the balanced equation.

3.79 $3\ NaHCO_3(aq) + H_3C_6H_5O_7(aq) \rightarrow 3\ CO_2(g) + 3\ H_2O(l) + Na_3C_6H_5O_7(aq)$

(a) *Analyze/Plan*. Abbreviate citric acid as H_3Cit. Follow the approach in Sample Exercise 3.19. *Solve*.

$$1.00\ g\ NaHCO_3 \times \frac{1\ mol\ NaHCO_3}{84.01\ g\ NaHCO_3} = 1.190 \times 10^{-2} = 1.19 \times 10^{-2}\ mol\ NaHCO_3$$

$$1.00\ g\ H_3C_6H_5O_7 \times \frac{1\ mol\ H_3Cit}{192.1\ g\ H_3Cit} = 5.206 \times 10^{-3} = 5.21 \times 10^{-3}\ mol\ H_3Cit$$

But $NaHCO_3$ and H_3Cit react in a 3:1 ratio, so 5.21×10^{-3} mol H_3Cit require $3(5.21 \times 10^{-3}) = 1.56 \times 10^{-2}$ mol $NaHCO_3$. We have only 1.19×10^{-2} mol $NaHCO_3$, so $NaHCO_3$ is the limiting reactant.

(b) $$1.190 \times 10^{-2}\ mol\ NaHCO_3 \times \frac{3\ mol\ CO_2}{3\ mol\ NaHCO_3} \times \frac{44.01\ g\ CO_2}{1\ mol\ CO_2} = 0.524\ g\ CO_2$$

(c) $$1.190 \times 10^{-2}\ mol\ NaHCO_3 \times \frac{1\ mol\ H_3Cit}{3\ mol\ NaHCO_3} = 3.968 \times 10^{-3} = 3.97 \times 10^{-3}\ mol\ H_3Cit\ react$$

$$5.206 \times 10^{-3}\ mol\ H_3Cit - 3.968 \times 10^{-3}\ mol\ react = 1.238 \times 10^{-3} = 1.24 \times 10^{-3}\ mol\ H_3Cit\ remain$$

$$1.238 \times 10^{-3}\ mol\ H_3Cit \times \frac{192.1\ g\ H_3Cit}{mol\ H_3Cit} = 0.238\ g\ H_3Cit\ remain$$

3.81 *Analyze*. Given: initial g Na_2CO_3, g $AgNO_3$. Find: final g Na_2CO_3, $AgNO_3$, Ag_2CO_3, $NaNO_3$

Plan. Write balanced equation; determine limiting reactant; calculate amounts of excess reactant remaining and products, based on limiting reactant.

Solve. $2\ AgNO_3(aq) + Na_2CO_3(aq) \rightarrow Ag_2CO_3(s) + 2\ NaNO_3(aq)$

$$3.50\ g\ Na_2CO_3 \times \frac{1\ mol\ Na_2CO_3}{106.0\ g\ Na_2CO_3} = 0.03302 = 0.0330\ mol\ Na_2CO_3$$

$$5.00\ g\ AgNO_3 \times \frac{1\ mol\ AgNO_3}{169.9\ g\ AgNO_3} = 0.02943 = 0.0294\ mol\ AgNO_3$$

$$0.02943\ mol\ AgNO_3 \times \frac{1\ mol\ Na_2CO_3}{2\ mol\ AgNO_3} = 0.01471 = 0.0147\ mol\ Na_2CO_3\ required$$

$AgNO_3$ is the limiting reactant and Na_2CO_3 is present in excess.

	$2\ AgNO_3(aq)$	+	$Na_2CO_3(aq)$	$\rightarrow$	$Ag_2CO_3(s)$	+	$2\ NaNO_3(aq)$
initial	0.0294 mol		0.0330 mol		0 mol		0 mol
reaction	–0.0294 mol		–0.0147 mol		+0.0147 mol		+0.0294 mol
final	0 mol		0.0183 mol		0.0147 mol		0.0294 mol

$0.01830\ mol\ Na_2CO_3 \times 106.0\ g/mol = 1.940 = 1.94\ g\ Na_2CO_3$

$0.01471\ mol\ Ag_2CO_3 \times 275.8\ g/mol = 4.057 = 4.06\ g\ Ag_2CO_3$

$0.02943\ mol\ NaNO_3 \times 85.00\ g/mol = 2.502 = 2.50\ g\ NaNO_3$

Check. The initial mass of reactants was 8.50 g, and the final mass of excess reactant and products is 8.50 g; mass is conserved.

3.83 *Analyze*. Given: amounts of two reactants. Find: theoretical yield.

Plan. Determine the limiting reactant and the maximum amount of product it could produce. Then calculate % yield. *Solve*.

(a) $$30.0 \text{ g } C_6H_6 \times \frac{1 \text{ mol } C_6H_6}{78.11 \text{ g } C_6H_6} = 0.3841 = 0.384 \text{ mol } C_6H_6$$

$$65.0 \text{ g } Br_2 \times \frac{1 \text{ mol } Br_2}{159.8 \text{ g } Br_2} = 0.4068 = 0.407 \text{ mol } Br_2$$

Because C_6H_6 and Br_2 react in a 1:1 mole ratio, C_6H_6 is the limiting reactant and determines the theoretical yield.

$$0.3841 \text{ mol } C_6H_6 \times \frac{1 \text{ mol } C_6H_5Br}{1 \text{ mol } C_6H_6} \times \frac{157.0 \text{ g } C_6H_5Br}{1 \text{ mol } C_6H_5Br} = 60.30 = 60.3 \text{ g } C_6H_5Br$$

Check. 30/78 ~ 3/8 mol C_6H_6. 65/160 ~ 3/8 mol Br_2. Because moles of the two reactants are similar, a precise calculation is needed to determine the limiting reactant. $3/8 \times 160 \approx 60$ g product

(b) $$\% \text{ yield} = \frac{42.3 \text{ g } C_6H_5Br \text{ actual}}{60.3 \text{ g } C_6H_5Br \text{ theoretical}} \times 100 = 70.149 = 70.10\%$$

3.85 *Analyze*. Given: g of two reactants, % yield. Find: g S_8.

Plan. Determine limiting reactant and theoretical yield. Use definition of % yield to calculate actual yield. *Solve*.

$$30.0 \text{ g } H_2S \times \frac{1 \text{ mol } H_2S}{34.08 \text{ g } H_2S} = 0.8803 = 0.880 \text{ mol } H_2S$$

$$50.0 \text{ g } O_2 \times \frac{1 \text{ mol } O_2}{32.00 \text{ g } H_2S} = 1.5625 = 1.56 \text{ mol } O_2$$

$$0.8803 \text{ mol } H_2S \times \frac{4 \text{ mol } O_2}{8 \text{ mol } H_2S} = 0.4401 = 0.440 \text{ mol } O_2 \text{ required}$$

Because there is more than enough O_2 to react exactly with 0.880 mol H_2S, O_2 is present in excess and H_2S is the limiting reactant.

$$0.8803 \text{ mol } H_2S \times \frac{1 \text{ mol } S_8}{8 \text{ mol } H_2S} \times \frac{256.56 \text{ g } S_8}{1 \text{ mol } S_8} = 28.231 = 28.2 \text{ g } S_8 \text{ theoretical yield}$$

Check. 30/34 ≈ 1 mol H_2S; 50/32 ≈ 1.5 mol O_2. Twice as many mol H_2S as mol O_2 are required, so H_2S limits. $1 \times (260/8) \approx 30$ g S_8 theoretical.

$$\% \text{ yield} = \frac{\text{actual}}{\text{theoretical}} \times 100; \quad \frac{\% \text{ yield} \times \text{theoretical}}{100} = \text{actual yield}$$

$$\frac{98\%}{100} \times 28.231 \text{ g } S_8 = 27.666 = 28 \text{ g } S_8 \text{ actual}$$

Additional Exercises

3.87 (a) $CH_3COOH = C_2H_4O_2$. At room temperature and pressure, pure acetic acid is a liquid. $C_2H_4O_2(l) + 2\,O_2(g) \rightarrow 2\,CO_2(g) + 2\,H_2O(l)$

(b) $Ca(OH)_2(s) \rightarrow CaO(s) + H_2O(g)$

(c) $Ni(s) + Cl_2(g) \rightarrow NiCl_2(s)$

3.91 (a) *Analyze.* Given: diameter of CdSe sphere (dot), density of CdSe. Find: mass of dot.

Plan. Calculate volume of sphere in cm^3, use density to calculate mass of the sphere (dot).

Solve. $V = 4/3\pi r^3$; $r = d/2$

$$\text{radius of dot} = \frac{2.5\text{ nm}}{2} \times \frac{1\times10^{-9}\text{ m}}{1\text{ nm}} \times \frac{1\text{ cm}}{1\times10^{-2}\text{ m}} = 1.25\times10^{-7}\text{ cm}$$

$$\text{volume of dot} = (4/3)\times\pi\times(1.25\times10^{-7})^3 = 8.1812\times10^{-21} = 8.2\times10^{-21}\text{ cm}^3$$

$$8.1812\times10^{-21}\text{ cm}^3 \times \frac{5.82\text{ g CdSe}}{\text{cm}^3} = 4.7615\times10^{-20} = 4.8\times10^{-20}\text{ g/dot}$$

(b) *Plan.* Change g CdSe to mol Cd using molar mass, then mol Cd to atoms Cd using Avogadro's number. *Solve.*

$$4.7615\times10^{-20}\text{ g CdSe}\times\frac{1\text{ mol CdSe}}{191.385\text{ g CdSe}}\times\frac{1\text{ mol Cd}}{1\text{ mol CdSe}}\times\frac{6.0221\times10^{23}\text{ Cd atoms}}{\text{mol Cd}}$$

$$= 149.82 = 150\text{ Cd atoms}$$

(c) $$\text{volume of dot} = (4/3)\times\pi\times(3.25\times10^{-7})^3 = 1.4379\times10^{-19} = 1.4\times10^{-19}\text{ cm}^3$$

$$1.4379\times10^{-19}\text{ cm}^3\times\frac{5.82\text{ g CdSe}}{\text{cm}^3} = 8.3688\times10^{-19} = 8.4\times10^{-19}\text{ g/dot}$$

(d) $$8.3688\times10^{-19}\text{ g CdSe}\times\frac{1\text{ mol CdSe}}{191.385\text{ g CdSe}}\times\frac{1\text{ mol Cd}}{1\text{ mol CdSe}}\times\frac{6.022\times10^{23}\text{ Cd atoms}}{\text{mol Cd}}$$

$$= 2633.3 = 2.6\times10^3\text{ Cd atoms}$$

(e) We can calculate the number of 2.5-nm dots required to make a 6.5-nm dot using either the ratio of the volumes of the dots, or the ratio of the number of CdSe units in the two dots.

$$\frac{1.4379\times10^{-19}\text{ cm}^3}{8.1812\times10^{-21}\text{ cm}^3} = 17.5758\text{ 2.5-nm dots};\quad \frac{2633.3\text{ CdSe units}}{149.82\text{ CdSe units}} = 17.5764\text{ 2.5-nm dots}$$

The number of 2.5-nm dots required to make a 6.5-nm dot and the number of CdSe units left over are integer numbers. We have included extra figures in the ratio calculations to show that the results are amazingly close. Both ratios indicate that 18 of the smaller dots are needed to produce one 6.5 nm dot. There will be a few CdSe units left over, slightly less than half of one dot. Taking the average of the "leftovers,"

0.424 small dot x 150 CdSe units/small dot = 63.6 = 64 CdSe units left over

3.95 Because all the C in the vanillin must be present in the CO_2 produced, get g C from g CO_2.

$$2.43 \text{ g } CO_2 \times \frac{1 \text{ mol } CO_2}{44.01 \text{ g } CO_2} \times \frac{12.01 \text{ g C}}{1 \text{ mol C}} = 0.6631 = 0.663 \text{ g C}$$

Because all the H in vanillin must be present in the H_2O produced, get g H from g H_2O.

$$0.50 \text{ g } H_2O \times \frac{1 \text{ mol } H_2O}{18.02 \text{ g } H_2O} \times \frac{2 \text{ mol H}}{1 \text{ mol } H_2O} \times \frac{1.008 \text{ g H}}{1 \text{ mol H}} = 0.0559 = 0.056 \text{ g H}$$

Get g O by subtraction. (Because the analysis was performed by combustion, an unspecified amount of O_2 was a reactant, and thus not all the O in the CO_2 and H_2O produced came from vanillin.) 1.05 g vanillin – 0.663 g C – 0.056 g H = 0.331 g O

$$0.6631 \text{ g C} \times \frac{1 \text{ mol C}}{12.01 \text{ g C}} = 0.0552 \text{ mol C}; \; 0.0552 / 0.0207 = 2.67$$

$$0.0559 \text{ g H} \times \frac{1 \text{ mol H}}{1.008 \text{ g H}} = 0.0555 \text{ mol C}; \; 0.0555/0.0207 = 2.68$$

$$0.331 \text{ g O} \times \frac{1 \text{ mol O}}{16.00 \text{ g O}} = 0.0207 \text{ mol O}; \; 0.0207/0.0207 = 1.00$$

Multiplying the numbers above by **3** to obtain an integer ratio of moles, the empirical formula of vanillin is $C_8H_8O_3$.

3.99 $O_3(g) + 2 \text{ NaI}(aq) + H_2O(l) \rightarrow O_2(g) + I_2(s) + 2 \text{ NaOH}(aq)$

(a) $5.95 \times 10^{-6} \text{ mol } O_3 \times \frac{2 \text{ mol NaI}}{1 \text{ mol } O_3} = 1.19 \times 10^{-5} \text{ mol NaI}$

(b) $1.3 \text{ mg } O_3 \times \frac{1 \times 10^{-3} \text{ g}}{1 \text{ mg}} \times \frac{1 \text{ mol } O_3}{48.00 \text{ g } O_3} \times \frac{2 \text{ mol NaI}}{1 \text{ mol } O_3} \times \frac{149.9 \text{ g NaI}}{1 \text{ mol NaI}}$
$= 8.120 \times 10^{-3} = 8.1 \times 10^{-3} \text{ g NaI} = 8.1 \text{ mg NaI}$

3.103 $N_2(g) + 3 H_2(g) \rightarrow 2 NH_3(g)$

Determine the moles of N_2 and H_2 required to form the 3.0 moles of NH_3 present after the reaction has stopped.

$$3.0 \text{ mol } NH_3 \times \frac{3 \text{ mol } H_2}{2 \text{ mol } NH_3} = 4.5 \text{ mol } H_2 \text{ reacted}$$

$$3.0 \text{ mol } NH_3 \times \frac{1 \text{ mol } N_2}{2 \text{ mol } NH_3} = 1.5 \text{ mol } N_2 \text{ reacted}$$

mol H_2 initial = 3.0 mol H_2 remain + 4.5 mol H_2 reacted = 7.5 mol H_2

mol N_2 initial = 3.0 mol N_2 remain + 1.5 mol N_2 reacted = 4.5 mol N_2

In tabular form:	$N_2(g)$	+	$3 H_2(g)$	→	$2 NH_3(g)$
initial	4.5 mol		7.5 mol		0 mol
reaction	–1.5 mol		–4.5 mol		+3.0 mol
final	3.0 mol		3.0 mol		3.0 mol

(Tables like this will be extremely useful for solving chemical equilibrium problems in Chapter 15.)

Integrative Exercises

3.108 *Plan.* Volume cube $\xrightarrow{\text{density}}$ mass $CaCO_3$ $\rightarrow$ moles $CaCO_3$ $\rightarrow$ moles O $\rightarrow$ O atoms

Solve. $$(2.005)^3\ \text{in}^3 \times \frac{(2.54)^3\ \text{cm}^3}{1\ \text{in}^3} \times \frac{2.71\ \text{g CaCO}_3}{1\ \text{cm}^3} \times \frac{1\ \text{mol CaCO}_3}{100.1\ \text{g CaCO}_3} \times \frac{3\ \text{mol O}}{1\ \text{mol CaCO}_3}$$

$$\times \frac{6.022\times10^{23}\ \text{O atoms}}{1\ \text{mol O}} = 6.46\times10^{24}\ \text{O atoms}$$

3.112 (a) $S(s) + O_2(g) \rightarrow SO_2(g);\ SO_2(g) + CaO(s) \rightarrow CaSO_3(s)$

(b) If the coal contains 2.5% S, then 1 g coal contains 0.025 g S.

$$\frac{2000.0\ \text{tons coal}}{\text{day}} \times \frac{2000\ \text{lb}}{1\ \text{ton}} \times \frac{1\ \text{kg}}{2.2046\ \text{lb}} \times \frac{1000\ \text{g}}{1\ \text{kg}} \times \frac{0.025\ \text{g S}}{1\ \text{g coal}} \times \frac{1\ \text{mol S}}{32.065\ \text{g S}}$$

$$\times \frac{1\ \text{mol SO}_2}{1\ \text{mol S}} \times \frac{1\ \text{mol CaO}}{1\ \text{mol SO}_2} \times \frac{56.077\ \text{g CaO}}{1\ \text{mol CaO}} \times \frac{1\ \text{kg CaO}}{1000\ \text{g CaO}} =$$

$$79{,}327.49 = 7.9 \times 10^4\ \text{kg CaO or } 7.9 \times 10^7\ \text{g CaO}$$

(c) $1\ \text{mol CaO} = 1\ \text{mol CaSO}_3$

$$7.9327 \times 10^7\ \text{g CaO} \times \frac{1\ \text{mol CaO}}{56.077\ \text{g CaO}} \times \frac{1\ \text{mol CaSO}_3}{1\ \text{mol CaO}} \times \frac{120.14\ \text{g CaSO}_3}{1\ \text{mol CaSO}_3}$$

$$= 1.6995 \times 10^8 = 1.7 \times 10^8\ \text{g CaSO}_3$$

This corresponds to about 190 tons of $CaSO_3$ per day as a waste product.

4 Reactions in Aqueous Solution

Visualizing Concepts

4.1 *Analyze.* Correlate the formula of the solute with the charged spheres in the diagrams.

Plan. Determine the electrolyte properties of the solute and the relative number of cations, anions, or neutral molecules produced when the solute dissolves.

Solve. Li_2SO_4 is a strong electrolyte, a soluble ionic solid that dissociates into separate Li^+ and SO_4^{2-} when it dissolves in water. There are twice as many Li^+ cations as SO_4^{2-} anions. Diagram (c) represents the aqueous solution of a 2:1 electrolyte.

4.5 *Analyze/Plan.* From the names and/or formulas of the three possible solids, determine which exhibits the described solubility properties. Use Table 4.1.

Solve. The three possible compounds are $BaCl_2$, $PbCl_2$, and $ZnCl_2$. $PbCl_2$ does not dissolve in water to give a clear solution, so it can be eliminated. Of the remaining possibilities, Ba^{2+} has a sulfate precipitate, but Zn^{2+} does not. The compound is indeed $BaCl_2$.

4.7 *Analyze/Plan.* Given three metal powders and three 1 *M* solutions, use Table 4.5, the activity series of metals, to find a scheme to distinguish the metals.

Solve. In the activity series, any metal on the list can be oxidized by the ions of elements below it. The nitric acid solution contains $H^+(aq)$. This solution will oxidize and thus dissolve Zn(s) and Pb(s), which appear above $H_2(g)$ on the list. Platinum, Pt(s), is distinguished by its lack of reaction with nitric acid.

To distinguish between Zn and Pb, use a metal ion that occurs between them on the list. We have such an ion, $Ni^{2+}(aq)$ in the nickel nitrate solution. $Ni^{2+}(aq)$ will oxidize and thus dissolve Zn(s), which is above it on the list. $Ni^{2+}(aq)$ will not oxidize or dissolve Pb(s), which is below it on the list.

To summarize, Pt(s) will neither be oxidized by nor dissolve in any of the three available solutions. Pb(s) is oxidized by and will dissolve in the nitric acid solution, but not the nickel nitrate solution. Zn(s) is oxidized by and will dissolve in both nitric acid and nickel nitrate solutions.

4.9 The answer is: (c) a redox reaction. The "water-splitting" reaction is

$$2\,H_2O(l) \rightarrow 2\,H_2(g) + O_2(g)$$

In the reaction, the oxidation number of hydrogen decreases from +1 to 0, and the oxidation number of oxygen increases from –2 to 0. Hydrogen is reduced and oxygen is oxidized.

General Properties of Aqueous Solutions (Section 4.1)

4.13 (a) False. Electrolyte solutions conduct electricity because *ions* are moving through the solution.

(b) True. The conductivity is unchanged as long as the concentration of electrolytes is unchanged. Because ions are mobile in solution, the added presence of uncharged molecules does not inhibit conductivity.

4.15 Statement (b) is most correct. Statements (a) and (c) are incorrect because water is not a strong acid and the hydrogen and oxygen bonds of water are not broken by ionic solids.

4.17 *Analyze/Plan.* Given the solute formula, determine the separate ions formed upon dissociation. *Solve.*

(a) $FeCl_2(aq) \rightarrow Fe^{2+}(aq) + 2\ Cl^-(aq)$

(b) $HNO_3(aq) \rightarrow H^+(aq) + NO_3^-(aq)$

(c) $(NH_4)_2SO_4(aq) \rightarrow 2\ NH_4^+(aq) + SO_4^{2-}(aq)$

(d) $Ca(OH)_2(aq) \rightarrow Ca^{2+}(aq) + 2\ OH^-(aq)$

4.19 *Analyze/Plan.* Apply the definition of a weak electrolyte to HCOOH.

Solve. When HCOOH dissolves in water, neutral HCOOH molecules, H^+ ions and $HCOO^-$ ions are all present in the solution. $HCOOH(aq) \rightleftharpoons H^+(aq) + HCOO^-(aq)$

Precipitation Reactions (Section 4.2)

4.21 *Analyze.* Given: formula of compound. Find: solubility.

Plan. Follow the guidelines in Table 4.1, in light of the anion present in the compound and notable exceptions to the "rules." *Solve.*

(a) $Mg\mathbf{Br_2}$: soluble

(b) $Pb\mathbf{I_2}$: insoluble, Pb^{2+} is an exception to soluble iodides

(c) $(NH_4)_2\mathbf{CO_3}$: soluble, NH_4^+ is an exception to insoluble carbonates

(d) $Sr\mathbf{(OH)_2}$: soluble, Sr^{2+} is an exception to insoluble hydroxides

(e) $Zn\mathbf{SO_4}$: soluble

4.23 *Analyze.* Given: formulas of reactants. Find: balanced equation including precipitates.

Plan. Follow the logic in Sample Exercise 4.3.

Solve. In each reaction, the precipitate is in bold type.

(a) $Na_2CO_3(aq) + 2\ AgNO_3(aq) \rightarrow \mathbf{Ag_2CO_3(s)} + 2\ NaNO_3(aq)$

(b) No precipitate (all nitrates and most sulfates are soluble).

(c) $FeSO_4(aq) + Pb(NO_3)_2(aq) \rightarrow \mathbf{PbSO_4(s)} + Fe(NO_3)_2(aq)$

4.25 *Analyze/Plan.* Follow the logic in Sample Exercise 4.4. From the complete ionic equation, identify the ions that don't change during the reaction; these are the spectator ions. *Solve.*

(a) $2\,K^+(aq) + CO_3^{2-}(aq) + Mg^{2+}(aq) + SO_4^{2-}(aq) \rightarrow MgCO_3(s) + 2\,K^+(aq) + SO_4^{2-}(aq)$
Spectators: K^+, SO_4^{2-}

(b) $Pb^{2+}(aq) + 2\,NO_3^-(aq) + 2\,Li^+(aq) + S^{2-}(aq) \rightarrow PbS(s) + 2\,Li^+(aq) + 2\,NO_3^-(aq)$
Spectators: Li^+, NO_3^-

(c) $6\,NH_4^+(aq) + 2\,PO_4^{3-}(aq) + 3\,Ca^{2+}(aq) + 6\,Cl^-(aq) \rightarrow Ca_3(PO_4)_2(s) + 6\,NH_4^+(aq) + 6\,Cl^-(aq)$ Spectators: NH_4^+, Cl^-

4.27 *Analyze.* Given: reactions of unknown salt with HBr, H_2SO_4, NaOH. Find: Does the unknown salt contain K^+ or Pb^{2+} or Ba^{2+}?

Plan. Analyze solubility guidelines for Br^-, SO_4^{2-}, and OH^- and select the cation that produces a precipitate with each of the anions.

Solve. K^+ forms no precipitates with any of the anions. $BaSO_4$ is insoluble, but $BaCl_2$ and $Ba(OH)_2$ are soluble. Because the unknown salt forms precipitates with all three anions, it must contain Pb^{2+}.

Check. $PbBr_2$, $PbSO_4$, and $Pb(OH)_2$ are all insoluble according to Table 4.1, so our process of elimination is confirmed by the insolubility of the Pb^{2+} compounds.

4.29 *Analyze/Plan.* Using Table 4.1, determine the precipitates that could form when each of the unknowns is reacted with $Ba(NO_3)_2$ and NaCl. *Solve.*

(a) True. If the unknown is $Al_2(SO_4)_3$, $BaSO_4$ could precipitate.

(b) True. If the unknown is $AgNO_3$, AgCl could precipitate.

(c) False (two ways). Ag_2SO_4 is a soluble ionic compound, and no combination of the possible unknowns and the two reagents could produce Ag_2SO_4.

(d) True. This is the overall correct answer to the question.

(e) False, because (c) is false.

Acids, Bases, and Neutralization Reactions (Section 4.3)

4.31 *Analyze.* Given: solute and concentration of three solutions. Find: the solution that is most acidic.

Plan: See Sample Exercise 4.6 and Table 4.2. Determine whether solutes are strong or weak acids or bases or nonelectrolytes. For solutions of equal concentration, strong acids will have greatest concentration of solvated protons. Take varying concentration into consideration when evaluating the same class of solutions.

Solve. (a) LiOH is a strong base, (b) HI is a strong acid, (c) CH_3OH is a molecular compound and nonelectrolyte. Solution (b), 0.2 *M* HI, is the most acidic solution.

Check. The solution concentrations weren't needed to answer the question.

4.33 (a) False. Sulfuric acid, H_2SO_4, is a diprotic acid; it has two ionizable hydrogen atoms.

(b) False. According to Table 4.2, HCl is a strong acid.

(c) False. Methanol, CH_3OH, is a molecular nonelectrolyte.

4.35 *Analyze.* Given: chemical formulas. Find: classify as acid, base, salt; strong, weak, or nonelectrolyte.

Plan. See Tables 4.2 and 4.3. Ionic or molecular? Ionic, soluble: OH^-, strong base and strong electrolyte; otherwise, salt, strong electrolyte. Molecular: NH_3, weak base and weak electrolyte; H-first, acid; strong acid (Table 4.2), strong electrolyte; otherwise weak acid and weak electrolyte. *Solve.*

(a) HF: acid, mixture of ions and molecules (weak electrolyte)

(b) CH_3CN: none of the above, entirely molecules (nonelectrolyte)

(c) $NaClO_4$: salt, entirely ions (strong electrolyte)

(d) $Ba(OH)_2$: base, entirely ions (strong electrolyte)

4.37 *Analyze.* Given: chemical formulas. Find: electrolyte properties.

Plan. To classify as electrolytes, formulas must be identified as acids, bases, or salts as in Solution 4.35. *Solve.*

(a) H_2SO_3: H first, so acid; not in Table 4.2, so weak acid; therefore, weak electrolyte

(b) CH_3CH_2OH: not acid, not ionic (no metal cation), contains OH group, but not as anion so not a base; therefore, nonelectrolyte

(c) NH_3: common weak base; therefore, weak electrolyte

(d) $KClO_3$: ionic compound, so strong electrolyte

(e) $Cu(NO_3)_2$: ionic compound, so strong electrolyte

4.39 *Plan.* Follow Sample Exercise 4.7. *Solve.*

(a) $2\ HBr(aq) + Ca(OH)_2(aq) \rightarrow CaBr_2(aq) + 2\ H_2O(l)$

$H^+(aq) + OH^-(aq) \rightarrow H_2O(l)$

(b) $Cu(OH)_2(s) + 2\ HClO_4(aq) \rightarrow Cu(ClO_4)_2(aq) + 2\ H_2O(l)$

$Cu(OH)_2(s) + 2\ H^+(aq) \rightarrow 2\ H_2O(l) + Cu^{2+}(aq)$

(c) $Al(OH)_3(s) + 3\ HNO_3(aq) \rightarrow Al(NO_3)_3(aq) + 3\ H_2O(l)$

$Al(OH)_3(s) + 3\ H^+(aq) \rightarrow 3\ H_2O(l) + Al^{3+}(aq)$

4.41 *Analyze.* Given: names of reactants. Find: gaseous products.

Plan. Write correct chemical formulas for the reactants, complete and balance the metathesis reaction, and identify either H_2S or CO_2 products as gases. *Solve.*

(a) $CdS(s) + H_2SO_4(aq) \rightarrow CdSO_4(aq) + H_2S(g)$

$CdS(s) + 2\ H^+(aq) \rightarrow H_2S(g) + Cd^{2+}(aq)$

(b) $MgCO_3(s) + 2\ HClO_4(aq) \rightarrow Mg(ClO_4)_2(aq) + H_2O(l) + CO_2(g)$

$MgCO_3(s) + 2\ H^+(aq) \rightarrow H_2O(l) + CO_2(g) + Mg^{2+}(aq)$

4.43 *Analyze.* Given the formulas or names of reactants, write balanced molecular and net ionic equations for the reactions.

Plan. Write correct chemical formulas for all reactants. Predict products of the neutralization reactions by exchanging ion partners. Balance the complete molecular equation, identify spectator ions by recognizing strong electrolytes, write the corresponding net ionic equation (omitting spectators). *Solve.*

(a) $MgCO_3(s) + 2\,HCl(aq) \rightarrow MgCl_2(aq) + H_2O(l) + CO_2(g)$

$MgCO_3(s) + 2\,H^+(aq) \rightarrow Mg^{2+}(aq) + H_2O(l) + CO_2(g)$

$MgO(s) + 2\,HCl(aq) \rightarrow MgCl_2(aq) + H_2O(l)$

$MgO(s) + 2\,H^+(aq) \rightarrow Mg^{2+}(aq) + H_2O(l)$

$Mg(OH)_2(s) + 2\,HCl(aq) \rightarrow MgCl_2(aq) + 2\,H_2O(l)$

$Mg(OH)_2(s) + 2\,H^+(aq) \rightarrow Mg^{2+}(aq) + 2\,H_2O(l)$

(b) We can distinguish magnesium carbonate, $MgCO_3(s)$, because its reaction with acid produces $CO_2(g)$, which appears as bubbles. The other two compounds are indistinguishable because the products of the two reactions are exactly the same.

Oxidation–Reduction Reactions (Section 4.4)

4.45 (a) False. *Oxidation* is loss of electrons; *reduction* is gain of electrons. (LEO says GER.)

(b) True. When a substance is oxidized, its oxidation number increases. When a substance is reduced, its oxidation number decreases.

4.47 *Analyze.* Given the labeled periodic chart, determine which regions are most and least easily oxidized.

Plan. Review the definition of oxidation and apply it to the properties of elements in the indicated regions of the chart. *Solve.*

(a) Oxidation is loss of electrons. Elements easily oxidized form positive ions; these are metals. Elements in regions A and B are metals, and their ease of oxidation is shown in Table 4.5.

(b) Elements not readily oxidized tend to gain electrons and form negative ions; these are nonmetals. Elements in region D are nonmetals and are least easily oxidized.

4.49 *Analyze.* Given the chemical formula of a substance, determine the oxidation number of a particular element in the substance.

Plan. Follow the logic in Sample Exercise 4.8. *Solve.*

(a) +4 (b) +4 (c) +7 (d) +1 (e) +3 (f) –1 (O_2^{2-} is peroxide ion)

4.51 *Analyze.* Given: chemical reaction. Find: element oxidized or reduced. *Plan.* Assign oxidation numbers to all species. The element whose oxidation number increases (becomes more positive) is oxidized; the one whose oxidation number decreases (becomes more negative) is reduced. *Solve.*

(a) $N_2(g)$ [N, 0] $\rightarrow 2\,NH_3(g)$ [N, –3], N is reduced; $3\,H_2(g)$ [H, 0] $\rightarrow 2\,NH_3(g)$ [H, +1], H is oxidized.

(b) $Fe^{2+} \rightarrow Fe$, Fe is reduced; $Al \rightarrow Al^{3+}$, Al is oxidized

(c) $Cl_2 \rightarrow 2\,Cl^-$, Cl is reduced; $2\,I^- \rightarrow I_2$, I is oxidized

(d) $S^{2-} \rightarrow SO_4^{2-}$ (S, +6), S is oxidized; H_2O_2 (O, –1) $\rightarrow H_2O$ (O, –2); O is reduced

4.53 *Analyze.* Given: reactants. Find: balanced molecular and net ionic equations.

Plan. Metals oxidized by H^+ form cations. Predict products by exchanging cations and balance. The anions are the spectator ions and do not appear in the net ionic equations.

Solve.

(a) $Mn(s) + H_2SO_4(aq) \rightarrow MnSO_4(aq) + H_2(g)$;
$Mn(s) + 2\,H^+(aq) \rightarrow Mn^{2+}(aq) + H_2(g)$

Products with the metal in a higher oxidation state are possible, depending on reaction conditions and acid concentration.

(b) $2\,Cr(s) + 6\,HBr(aq) \rightarrow 2\,CrBr_3(aq) + 3\,H_2(g)$;
$2\,Cr(s) + 6\,H^+(aq) \rightarrow 2\,Cr^{3+}(aq) + 3\,H_2(g)$

(c) $Sn(s) + 2\,HCl(aq) \rightarrow SnCl_2(aq) + H_2(g)$; $Sn(s) + 2\,H^+(aq) \rightarrow Sn^{2+}(aq) + H_2(g)$

(d) $2\,Al(s) + 6\,HCOOH(aq) \rightarrow 2\,Al(HCOO)_3(aq) + 3\,H_2(g)$;
$2\,Al(s) + 6\,HCOOH(aq) \rightarrow 2\,Al^{3+}(aq) + 6\,HCOO^-(aq) + 3\,H_2(g)$

4.55 *Analyze.* Given: a metal and an aqueous solution. Find: balanced equation.

Plan. Use Table 4.5. If the metal is above the aqueous solution, reaction will occur; if the aqueous solution is higher, NR. If reaction occurs, predict products by exchanging cations (a metal ion or H^+), then balance the equation. *Solve.*

(a) $Fe(s) + Cu(NO_3)_2(aq) \rightarrow Fe(NO_3)_2(aq) + Cu(s)$

(b) $Zn(s) + MgSO_4(aq) \rightarrow$ NR

(c) $Sn(s) + 2\,HBr(aq) \rightarrow SnBr_2(aq) + H_2(g)$

(d) $H_2(g) + NiCl_2(aq) \rightarrow$ NR

(e) $2\,Al(s) + 3\,CoSO_4(aq) \rightarrow Al_2(SO_4)_3(aq) + 3\,Co(s)$

4.57 (a) i. $Zn(s) + Cd^{2+}(aq) \rightarrow Cd(s) + Zn^{2+}(aq)$

ii. $Cd(s) + Ni^{2+}(aq) \rightarrow Ni(s) + Cd^{2+}(aq)$

Observation (i) indicates that Cd is less active than Zn; observation (ii) indicates that Cd is more active than Ni. Cd is between Zn and Ni on the activity series.

(b) Chromium, iron, and cobalt, the three elements between Zn and Ni in Table 4.5, more closely define the position of Cd in the activity series.

(c) Place an iron strip in $CdCl_2(aq)$. If Cd(s) is deposited, Cd is less active than Fe; if there is no reaction, Cd is more active than Fe. Do the same test with Co if Cd is less active than Fe or with Cr if Cd is more active than Fe.

Concentrations of Solutions (Section 4.5)

4.59 (a) *Concentration* is an *intensive* property; it is the **ratio** of the amount of solute present in a certain quantity of solvent or solution. This ratio remains constant regardless of how much solution is present.

(b) The term *0.50 mol HCl* defines an amount (~18 g) of the pure substance HCl. The term 0.50 *M* HCl is a ratio; it indicates that there are 0.50 mol of HCl solute in 1.0 liter of solution. This same ratio of moles solute to solution volume is present regardless of the volume of solution under consideration.

4.61 *Analyze/Plan*. Follow the logic in Sample Exercises 4.11 and 4.12. *Solve.*

(a) $M = \frac{\text{mol solute}}{\text{L solution}}; \frac{0.175 \text{ mol ZnCl}_2}{150 \text{ mL}} \times \frac{1000 \text{ mL}}{1 \text{ L}} = 1.17\ M\ \text{ZnCl}_2$

Check. (0.175 / 0.150) > 1.0 *M*

(b) $\text{mol} = M \times \text{L}; \frac{4.50 \text{ mol HNO}_3}{1 \text{ L}} \times \frac{1 \text{ mol H}^+}{\text{mol HNO}_3} \times 0.0350 \text{ L} = 0.158 \text{ mol H}^+$

Check. $(4.5 \times .04) \approx 0.16$ mol

(c) $\text{L} = \frac{\text{mol}}{M}; \frac{0.350 \text{ mol NaOH}}{6.00 \text{ mol NaOH/L}} = 0.0583 \text{ L or } 58.3 \text{ mL of } 6.00\ M\ \text{NaOH}$

Check. (0.325/6.0) > 0.50 L.

4.63 *Analyze.* Given molarity, *M*, and volume, L, find mass of Na^+(aq) in the blood.

Plan. Calculate moles Na^+(aq) using the definition of molarity: $M = \frac{\text{mol}}{\text{L}}; \text{mol} = M \times \text{L}$.

Calculate mass Na^+(aq) using the definition moles: mol = g/MM; g = mol × MM. (MM is the symbol for molar mass in this manual.)

Solve. $\frac{0.135 \text{ mol}}{\text{L}} \times 5.0 \text{ L} \times \frac{23.0 \text{ g Na}^+}{\text{mol Na}^+} = 15.525 = 16 \text{ g Na}^+\text{(aq)}$

Check. Because there are more than 0.1 mol/L and we have 5.0 L, there should be more than half a mol (11.5 g) of Na^+. The calculation agrees with this estimate.

4.65 *Analyze.* Given: g alcohol/100 mL blood; molecular formula of alcohol. Find: molarity (mol/L) of alcohol. *Plan.* Use the molar mass (MM) of alcohol to change (g/100) mL to (mol/100 mL) then mL to L.

Solve. MM of alcohol = 2(12.01) + 6(1.1008) + 1(16.00) = 46.07 g alcohol/mol

$$\text{BAC} = \frac{0.08 \text{ g alcohol}}{100 \text{ mL blood}} \times \frac{1 \text{ mol alcohol}}{46.07 \text{ g alcohol}} \times \frac{1000 \text{ mL}}{1 \text{ L}} = 0.0174 = 0.02\ M \text{ alcohol}$$

4.67 *Plan.* Proceed as in Sample Exercises 4.13.

$M = \frac{\text{mol}}{\text{L}}; \ \text{mol} = \frac{\text{g}}{\text{MM}}$ (MM is the symbol for molar mass in this manual.)

Solve.

(a) $\frac{6.86 \text{ mol } CH_3CH_2OH}{1 \text{ L}} \times 1.00 \text{ L} \times \frac{46.07 \text{ g } CH_3CH_2OH}{1 \text{ mol } CH_3CH_2OH} = 316.04 = 316 \text{ g } CH_3CH_2OH$

Check. $(7 \times 50) \approx 350$ g ethanol (this is an upper limit)

(b) $316.04 \text{ g } CH_3CH_2OH \times \frac{1 \text{ mL}}{0.789 \text{ g } CH_3CH_2OH} = 400.56 = 401 \text{ mL} = 0.401 \text{ L } CH_3CH_2OH$

Check. $(320/0.8) \approx 400$ mL ethanol

4.69 *Analyze.* Given: formula and concentration of each solute. Find: concentration of K^+ in each solution. *Plan.* Note mol K^+/mol solute and compare concentrations or total moles.

Solve.

(a) $KCl \rightarrow K^+ + Cl^-$; $0.20\ M\ KCl = 0.20\ M\ K^+$

$K_2CrO_4 \rightarrow \mathbf{2}\ K^+ + CrO_4^{2-}$; $0.15\ M\ K_2CrO_4 = 0.30\ M\ K^+$

$K_3PO_4 \rightarrow \mathbf{3}\ K^+ + PO_4^{3-}$; $0.080\ M\ K_3PO_4 = 0.24\ M\ K^+$

$0.15\ M\ K_2CrO_4$ has the highest K^+ concentration.

(b) K_2CrO_4: $0.30\ M\ K^+ \times 0.0300 \text{ L} = 0.0090 \text{ mol } K^+$

K_3PO_4: $0.24\ M\ K^+ \times 0.0250 \text{ L} = 0.0060 \text{ mol } K^+$

30.0 mL of $0.15\ M\ K_2CrO_4$ has more K^+ ions.

4.71 *Analyze.* Given: molecular formula and solution molarity. Find: concentration (M) of each ion.

Plan. Follow the logic in Sample Exercise 4.12.

Solve.

(a) $NaNO_3 \rightarrow Na^+, NO_3^-$; $0.25\ M\ NaNO_3 = 0.25\ M\ Na^+$, $0.25\ M\ NO_3^-$

(b) $MgSO_4 \rightarrow Mg^{2+}, SO_4^{2-}$; $1.3 \times 10^{-2}\ M\ MgSO_4 = 1.3 \times 10^{-2}\ M\ Mg^{2+}$, $1.3 \times 10^{-2}\ M\ SO_4^{2-}$

(c) $C_6H_{12}O_6 \rightarrow C_6H_{12}O_6$ (molecular solute); $0.0150\ M\ C_6H_{12}O_6 = 0.0150\ M\ C_6H_{12}O_6$

(d) *Plan.* There is no reaction between NaCl and $(NH_4)_2CO_3$, so this is just a dilution problem, $M_1V_1 = M_2V_2$. Then account for ion stoichiometry.

Solve. 45.0 mL + 65.0 mL = 110.0 mL total volume

$\frac{0.272\ M\ NaCl \times 45.0 \text{ mL}}{110.0 \text{ mL}} = 0.111\ M\ NaCl$; $0.111\ M\ Na^+$, $0.111\ M\ Cl^-$

$\frac{0.0247\ M\ (NH_4)_2CO_3 \times 65.0 \text{ mL}}{110.0 \text{ mL}} = 0.0146\ M\ (NH_4)_2CO_3$

$2 \times (0.0146\ M) = 0.0292\ M\ NH_4^+$, $0.0146\ M\ CO_3^{2-}$

Check. By adding the two solutions (with no common ions or chemical reaction), we have approximately doubled the solution volume, and reduced the concentration of each ion by approximately a factor of two.

4.73 *Analyze/Plan.* Follow the logic of Sample Exercise 4.14.

Solve.

(a) $V_1 = M_2V_2/M_1; \quad \dfrac{0.250\ M\ NH_3 \times 1000.0\ mL}{14.8\ M\ NH_3} = 16.89 = 16.9\ mL\ 14.8\ M\ NH_3$

Check. $250/15 \approx 15\ M$

(b) $M_2 = M_1V_1/V_2; \quad \dfrac{14.8\ M\ NH_3 \times 10.0\ mL}{500\ mL} = 0.296\ M\ NH_3$

Check. $150/500 \approx 0.30\ M$

4.75 *Analyze/Plan.* Calculate the number of drug molecules in 1.00 mL of the stock solution, using $M \times L$ = moles and Avogadro's number. Then calculate the desired ratio.

Solve. 1.00 mL = 0.00100 L

$$\frac{1.5 \times 10^{-9}\ mol}{L} \times 0.0010\ L \times \frac{6.022 \times 10^{23}\ molecules}{mole} = 9.033 \times 10^{11} = 9.0 \times 10^{11}\ molecules$$

$$\frac{9.033 \times 10^{11}\ drug\ molecules}{2.0 \times 10^{5}\ cancer\ cells} = 4.517 \times 10^{6} = 4.5 \times 10^{6}$$

4.77 *Analyze.* Given: density of pure acetic acid, volume pure acetic acid, volume new solution. Find: molarity of new solution. *Plan.* Calculate the mass of acetic acid, CH_3COOH, present in 20.0 mL of the pure liquid. *Solve.*

$$20.00\ mL\ acetic\ acid \times \frac{1.049\ g\ acetic\ acid}{1\ mL\ acetic\ acid} = 20.98\ g\ acetic\ acid$$

$$20.98\ g\ CH_3COOH \times \frac{1\ mol\ CH_3COOH}{60.05\ g\ CH_3COOH} = 0.349375 = 0.3494\ mol\ CH_3COOH$$

$$M = mol/L = \frac{0.349375\ mol\ CH_3COOH}{0.2500\ L\ solution} = 1.39750 = 1.398\ M\ CH_3COOH$$

Check. $(20 \times 1) \approx 20$ g acid; $(20/60) \approx 0.33$ mol acid; $(0.33/0.25 = 0.33 \times 4) \approx 1.33\ M$

Solution Stoichiometry and Chemcial Analysis (Section 4.6)

4.79 *Analyze.* Given: volume and molarity $AgNO_3$, molarity HCl. Find: volume HCl or mass of KCl.

(a) *Plan.* $M \times L$ = mol $AgNO_3$ = mol Ag^+; balanced equation gives ratio mol HCl/mol $AgNO_3$; mol HCl → vol HCl. *Solve.*

$$\frac{0.200\ mol\ AgNO_3}{1\ L} \times 0.0150\ L = 3.00 \times 10^{-3}\ mol\ AgNO_3(aq)$$

$AgNO_3(aq) + HCl(aq) \rightarrow AgCl(s) + HNO_3(aq)$

mol HCl = mol $AgNO_3$ = 3.00×10^{-3} mol KCl

$$3.00 \times 10^{-3}\ mol\ HCl \times \frac{1\ L}{0.150\ mol\ HCl} = 0.0200\ L = 20.0\ mL\ 0.15\ M\ HCl$$

Check. $(0.2 \times 0.015) = 0.003$ mol; $(0.003/0.15) \approx 0.02$ L HCl

(b) *Plan.* $M \times L$ = mol $AgNO_3$ = mol Ag^+; balanced equation gives ratio mol KCl/mol $AgNO_3$; mol KCl $\rightarrow$ vol KCl. *Solve.*

$$\frac{0.200 \text{ mol AgNO}_3}{1 \text{ L}} \times 0.0150 \text{ L} = 3.00 \times 10^{-3} \text{ mol AgNO}_3\text{(aq)}$$

$$AgNO_3(aq) + KCl(aq) \rightarrow AgCl(s) + KNO_3(aq)$$

$$\text{mol KCl} = \text{mol AgNO}_3 = 3.00 \times 10^{-3} \text{ mol KCl}$$

$$3.00 \times 10^{-3} \text{ mol KCl} \times \frac{74.55 \text{ g KCl}}{1 \text{ mol KCl}} = 0.224 \text{ g KCl}$$

Check. $(0.2 \times 0.015) = 0.003$ mol; $(0.003 \times 75) \approx 0.225$ g KCl

(c) Clearly, the KCl reagent is virtually free relative to the HCl solution. The KCl analysis is more cost-effective.

4.81 (a) *Analyze.* Given: M and vol base, M acid. Find: vol acid

Plan/Solve. Write the balanced equation for the reaction in question:

$$HClO_4(aq) + NaOH(aq) \rightarrow NaClO_4(aq) + H_2O(l)$$

Calculate the moles of the known substance, in this case NaOH.

$$\text{moles NaOH} = M \times L = \frac{0.0875 \text{ mol NaOH}}{1 \text{ L}} \times 0.0500 \text{ L} = 0.004375 = 0.00438 \text{ mol NaOH}$$

Apply the mole ratio (mol unknown/mol known) from the chemical equation.

$$0.004375 \text{ mol NaOH} \times \frac{1 \text{ mol HClO}_4}{1 \text{ mol NaOH}} = 0.004375 \text{ mol HClO}_4$$

Calculate the desired quantity of unknown, in this case the volume of 0.115 M $HClO_4$ solution.

$$L = \text{mol}/M; \ L = 0.004375 \text{ mol HClO}_4 \times \frac{1 \text{ L}}{0.115 \text{ mol HClO}_4} = 0.0380 \text{ L} = 38.0 \text{ mL}$$

Check. $(0.09 \times 0.05) = 0.0045$ mol; $(0.0045/0.11) \approx 0.040$ L ≈ 40 mL

(b) Following the logic outlined in part (a):

$$2\,HCl(aq) + Mg(OH)_2(s) \rightarrow MgCl_2(aq) + 2\,H_2O(l)$$

$$2.87 \text{ g Mg(OH)}_2 \times \frac{1 \text{ mol Mg(OH)}_2}{58.32 \text{ g Mg(OH)}_2} = 0.049211 = 0.0492 \text{ mol Mg(OH)}_2$$

$$0.0492 \text{ mol Mg(OH)}_2 \times \frac{2 \text{ mol HCl}}{1 \text{ mol Mg(OH)}_2} = 0.0984 \text{ mol HCl}$$

$$L = \text{mol}/M = 0.09840 \text{ mol HCl} \times \frac{1 \text{ L HCl}}{0.128 \text{ mol HCl}} = 0.769 \text{ L} = 769 \text{ mL}$$

(c) $AgNO_3(aq) + KCl(aq) \rightarrow AgCl(s) + KNO_3(aq)$

$$785 \text{ mg KCl} \times \frac{1 \times 10^{-3} \text{ g}}{1 \text{ mg}} \times \frac{1 \text{ mol KCl}}{74.55 \text{ g KCl}} \times \frac{1 \text{ mol AgNO}_3}{1 \text{ mol KCl}} = 0.01053 = 0.0105 \text{ mol AgNO}_3$$

$$M = \text{mol/L} = \frac{0.01053 \text{ mol AgNO}_3}{0.0258 \text{ L}} = 0.408\ M \text{ AgNO}_3$$

(d) $HCl(aq) + KOH(aq) \rightarrow KCl(aq) + H_2O(l)$

$$\frac{0.108 \text{ mol HCl}}{1 \text{ L}} \times 0.0453 \text{ L} \times \frac{1 \text{ mol KOH}}{1 \text{ mol HCl}} \times \frac{56.11 \text{ g KOH}}{1 \text{ mol KOH}} = 0.275 \text{ g KOH}$$

4.83 *Analyze/Plan*. See Exercise 4.81(a) for a more detailed approach. *Solve.*

$$\frac{6.0 \text{ mol } H_2SO_4}{1 \text{ L}} \times 0.027 \text{ L} \times \frac{2 \text{ mol } NaHCO_3}{1 \text{ mol } H_2SO_4} \times \frac{84.01 \text{ g } NaHCO_3}{1 \text{ mol } NaHCO_3} = 27 \text{ g } NaHCO_3$$

4.85 *Analyze*. Given: *M* and vol HCl. Find: MM of base, an alkali metal hydroxide. *Plan*. Alkali metal ions have a 1+ charge, so the general formula of an alkali metal hydroxide is MOH. One mol of MOH requires one mol of HCl for neutralization.

(a) $M \times \text{L} = \text{mol HCl} = \text{mol MOH}$. $\text{MM} = \dfrac{\text{g MOH}}{1 \text{ mol MOH}}$. *Solve.*

$$\frac{2.50 \text{ mol HCl}}{1 \text{ L}} \times 0.0170 \text{ L} = 0.0425 \text{ mol HCl} = 0.0425 \text{ mol MOH}$$

$$\text{MM of MOH} = \frac{4.36 \text{ g MOH}}{0.0425 \text{ mol MOH}} = 102.59 = 103 \text{ g/mol}$$

(b) MM of alkali metal = MM of MOH – (17.01 g) *Solve.*

MM of alkali metal = (102.59 g/mol – 17.01 g/mol) = 85.58 = 86 g/mol

The experimental molar mass most closely fits that of Rb^+, 85.47 g/mol

Check. The experimental molar mass matches one of the alkali metals.

4.87 (a) $NiSO_4(aq) + 2\ KOH(aq) \rightarrow Ni(OH)_2(s) + K_2SO_4(aq)$

(b) The precipitate is $Ni(OH)_2$.

(c) *Plan*. Compare mol of each reactant; mol = $M \times \text{L}$

Solve. $0.200\ M$ KOH $\times$ 0.1000 L KOH = 0.0200 mol KOH

$0.150\ M\ NiSO_4 \times 0.2000 \text{ L } NiSO_4 = 0.0300 \text{ mol } NiSO_4$

1 mol $NiSO_4$ requires 2 mol KOH, so 0.0300 mol $NiSO_4$ requires 0.0600 mol KOH. Because only 0.0200 mol KOH is available, KOH is the limiting reactant.

(d) *Plan*. The amount of the limiting reactant (KOH) determines amount of product, in this case $Ni(OH)_2$.

Solve. $0.0200 \text{ mol KOH} \times \dfrac{1 \text{ mol } Ni(OH)_2}{2 \text{ mol KOH}} \times \dfrac{92.71 \text{ g } Ni(OH)_2}{1 \text{ mol } Ni(OH)_2} = 0.927 \text{ g } Ni(OH)_2$

(e) *Plan/Solve*. Limiting reactant: OH^-: no excess OH^- remains in solution.

Excess reactant: Ni^{2+}: M Ni^{2+} remaining = mol Ni^{2+} remaining/L solution

0.0300 mol Ni^{2+} initial – 0.0100 mol Ni^{2+} reacted = 0.0200 mol Ni^{2+} remaining

0.0200 mol Ni^{2+}/0.3000 L = 0.0667 M $Ni^{2+}(aq)$

Spectators: SO_4^{2-}, K^+. These ions do not react, so the only change in their concentration is dilution. The final volume of the solution is 0.3000 L.

$M_2 = M_1V_1/V_2$: 0.200 M $K^+ \times$ 0.1000 L/0.3000 L = 0.0667 M $K^+(aq)$

0.150 M $SO_4^{2-} \times$ 0.2000 L/0.3000 L = 0.100 M $SO_4^{2-}(aq)$

4.89 *Analyze*. Given: mass impure $Mg(OH)_2$; M and vol **excess** HCl; M and vol NaOH.

Find: mass % $Mg(OH)_2$ in sample. *Plan/Solve*. Write balanced equations.

$Mg(OH)_2(s) + 2\ HCl(aq) \rightarrow MgCl_2(aq) + 2\ H_2O(l)$

$HCl(aq) + NaOH(aq) \rightarrow NaCl(aq) + H_2O(l)$

Calculate total moles HCl = M HCl × L HCl

$$\frac{0.2050 \text{ mol HCl}}{1 \text{ L soln}} \times 0.1000 \text{ L} = 0.02050 \text{ mol HCl total}$$

mol excess HCl = mol NaOH used = M NaOH × L NaOH

$$\frac{0.1020 \text{ mol NaOH}}{1 \text{ L soln}} \times 0.01985 \text{ L} = 0.0020247 = 0.002025 \text{ mol NaOH}$$

mol HCl reacted with $Mg(OH)_2$ = total mol HCl – excess mol HCl

0.02050 mol total – 0.0020247 mol excess = 0.0184753 = 0.01848 mol HCl reacted

(The result has 5 decimal places and 4 sig. figs.)

Use mol ratio to get mol $Mg(OH)_2$ in sample, then molar mass of $Mg(OH)_2$ to get g pure $Mg(OH)_2$.

$$0.0184753 \text{ mol HCl} \times \frac{1 \text{ mol } Mg(OH)_2}{2 \text{ mol HCl}} \times \frac{58.32 \text{ g } Mg(OH)_2}{1 \text{ mol } Mg(OH)_2} = 0.53874 = 0.5387 \text{ g } Mg(OH)_2$$

$$\text{mass \% } Mg(OH)_2 = \frac{\text{g } Mg(OH)_2}{\text{g sample}} \times 100 = \frac{0.53874 \text{ g } Mg(OH)_2}{0.5895 \text{ g sample}} \times 100 = 91.39 \text{ \% } Mg(OH)_2$$

Additional Exercises

4.91 (a) $U(s) + 2\ ClF_3(g) \rightarrow UF_6(g) + Cl_2(g)$

(b) This is not a metathesis reaction. The compounds involved are molecular rather than ionic, so the reaction is more complex than ions changing "partners."

(c) It is a redox reaction. U is oxidized, Cl is reduced.

4.95 (a) $Al(OH)_3(s) + 3\ H^+(aq) \rightarrow Al^{3+}(aq) + 3\ H_2O(l)$

(b) $Mg(OH)_2(s) + 2\ H^+(aq) \rightarrow Mg^{2+}(aq) + 2\ H_2O(l)$

(c) $MgCO_3(s) + 2\ H^+(aq) \rightarrow Mg^{2+}(aq) + H_2O(l) + CO_2(g)$

(d) $NaAl(CO_3)(OH)_2(s) + 4\ H^+(aq) \rightarrow Na^+(aq) + Al^{3+}(aq) + 3\ H_2O(l) + CO_2(g)$

(e) $CaCO_3(s) + 2\ H^+(aq) \rightarrow Ca^{2+}(aq) + H_2O(l) + CO_2(g)$

[In (c), (d), and (e), one could also write the equation for formation of bicarbonate, e.g., $MgCO_3(s) + H^+(aq) \rightarrow Mg^{2+} + HCO_3^-(aq)$.]

4.99 *Plan.* Calculate moles KBr from the two quantities of solution (mol = $M \times$ L). Moles $AgNO_3$ required equals total moles KBr present. Change grams $AgNO_3$ to moles $AgNO_3$. *Solve.*

$$1.00\ M\ \text{KBr} \times 0.0350\ \text{L} = 0.0350\ \text{mol KBr};\quad 0.600\ M\ \text{KBr} \times 0.060\ \text{L} = 0.0360\ \text{mol KBr}$$

$$0.0350\ \text{mol KBr} + 0.0360\ \text{mol KBr} = 0.0710\ \text{mol KBr total requires } 0.0710\ \text{mol AgNO}_3$$

$$0.0710\ \text{mol AgNO}_3 \times \frac{169.87\ \text{g AgNO}_3}{\text{mol AgNO}_3} = 12.0607 = 12.1\ \text{g AgNO}_3$$

4.103 (a) $$12.50\ \text{g Sr(OH)}_2 \times \frac{1\ \text{mol Sr(OH)}_2}{121.64\ \text{g Sr(OH)}_2} \times \frac{1}{0.05000\ \text{L}} = 2.0552 = 2.055\ M\ \text{Sr(OH)}_2$$

(b) $2\ HNO_3(aq) + Sr(OH)_2(aq) \rightarrow Sr(NO_3)_2(aq) + 2\ H_2O(l)$

(c) *Plan.* mol $Sr(OH)_2 = M \times L \rightarrow$ mol ratio $\rightarrow$ mol $HNO_3 \rightarrow M\ HNO_3$. *Solve.*

$$\frac{2.0552\ \text{mol Sr(OH)}_2}{\text{L}} \times 0.0239\ \text{L Sr(OH)}_2 \times \frac{2\ \text{mol HNO}_3}{1\ \text{mol Sr(OH)}_2} \times \frac{1}{0.0375\ \text{L HNO}_3}$$
$$= 2.6197 = 2.62\ M\ \text{HNO}_3$$

Integrative Exercises

4.109 (a) $Mg(OH)_2(s) + 2\ HNO_3(aq) \rightarrow Mg(NO_3)_2(aq) + 2\ H_2O(l)$

(b) $$7.75\ \text{g Mg(OH)}_2 \times \frac{1\ \text{mol Mg(OH)}_2}{58.32\ \text{g Mg(OH)}_2} = 0.13289 = 0.133\ \text{mol Mg(OH)}_2$$

$$0.200\ M\ \text{HNO}_3 \times 0.0250\ \text{L} = 0.00500\ \text{mol HNO}_3$$

The 0.00500 mol HNO_3 would neutralize 0.00250 mol $Mg(OH)_2$ and much more $Mg(OH)_2$ is present, so HNO_3 is the limiting reactant.

(c) Because HNO_3 limits, 0 mol HNO_3 is present after reaction. 0.00250 mol $Mg(NO_3)_2$ is produced.

$$0.13289\ \text{mol Mg(OH)}_2\ \text{initial} - 0.00250\ \text{mol Mg(OH)}_2\ \text{react}$$
$$= 0.130\ \text{mol Mg(OH)}_2\ \text{remain}$$

4.113 (a) $\mathbf{As}O_4^{3-}$; +5

(b) $Ag_3\mathbf{P}O_4$ is silver phosphate; $Ag_3\mathbf{As}O_4$ is silver arsenate

(c) $$0.0250\ \text{L soln} \times \frac{0.102\ \text{mol Ag}^+}{1\ \text{L soln}} \times \frac{1\ \text{mol Ag}_3\text{AsO}_4}{3\ \text{mol Ag}^+} \times \frac{1\ \text{mol As}}{1\ \text{mol Ag}_3\text{AsO}_4} \times \frac{74.92\ \text{g As}}{1\ \text{mol As}}$$
$$= 0.06368 = 0.0637\ \text{g As}$$

$$\text{mass percent} = \frac{0.06368\ \text{g As}}{1.22\ \text{g sample}} \times 100 = 5.22\%\ \text{As}$$

5 Thermochemistry

Visualizing Concepts

5.1 (a) *Analyze/Plan.* The exercise gives the charges and separation of two particles. Use Equation 5.2 to calculate electrostatic potential energy, E_{el}. *Solve.*

$$E_{el} = \frac{\kappa Q_1 Q_2}{d};\ \kappa = 8.99 \times 10^9 \text{ J-m/C}^2;\ Q_1 = Q_2 = 2.0 \times 10^{-5} \text{ C};\ d = 1.0 \text{ cm}$$

$$E_{el} = \frac{8.99 \times 10^9 \text{ J-m}}{C^2} \times \frac{1}{1.0 \text{ cm}} \times \frac{100 \text{ cm}}{m} \times 2.0 \times 10^{-5} \text{ C} \times 2.0 \times 10^{-5} \text{ C} = 359.6 = 3.6 \times 10^2 \text{ J}$$

(b) The spheres are both positively charged, so they will move away from each other. (Like charged particles repel, oppositely charged particles attract.)

(c) As the like charged spheres move apart, the electrostatic potential energy of the system is converted to kinetic energy. As the distance between them approaches infinity, potential energy approaches zero and the kinetic energy of each particle is 1.8×10^2 J, one half of the initial potential energy calculated in part (a).

$E_k = 1/2\ mv^2$; $v = (2\ E_k\ /m)^{1/2}$; $E_k = ½(3.6 \times 10^2 \text{ J}) = 1.8 \times 10^2 \text{ J}$

$$v = \left(2 \times \frac{1.8 \times 10^2 \text{ kg-m}^2}{s^2} \times \frac{1}{1.0 \text{ kg}}\right)^{1/2} = 18.97 = 19 \text{ m/s}$$

5.4 (a) For an endothermic process, the sign of q is positive; the system gains heat. This is true only for system (iii).

(b) In order for ΔE to be less than 0, there is a net transfer of heat or work from the system to the surroundings. The magnitude of the quantity leaving the system is greater than the magnitude of the quantity entering the system. In system (i), the magnitude of the heat leaving the system is less than the magnitude of the work done on the system. In system (iii), the magnitude of the work done by the system is less than the magnitude of the heat entering the system. None of the systems has $\Delta E < 0$.

(c) In order for ΔE to be greater than 0, there is a net transfer of work or heat to the system from the surroundings. In system (i), the magnitude of the work done on the system is greater than the magnitude of the heat leaving the system. In system (ii), work is done on the system with no change in heat. In system (iii), the magnitude of the heat gained by the system is greater than the magnitude of the work done on the surroundings. $\Delta E > 0$ for all three systems.

5.7 (a) You, part of the surroundings, do work on the air, part of the system. Energy is transferred to the system via work and the sign of w is (+).

(b) The body of the pump (the system) is warmer than the surroundings. Heat is transferred from the warmer system to the cooler surroundings, and the sign of q is (–).

(c) The sign of w is positive, and the sign of q is negative, so we cannot absolutely determine the sign of ΔE. It is likely that the heat lost is much smaller than the work done on the system, so the sign of ΔE is probably positive.

5.10 (a) $N_2(g) + O_2(g) \rightarrow 2\,NO(g)$. Because $\Delta V = 0$, $w = 0$.

(b) The reaction of two elements to form one mole of a compound fits the definition of a formation reaction. Find the value for enthalpy of formation of NO(g) in Appendix C. $\Delta H = \Delta H_f = 90.37$ kJ for production of 1 mol of NO(g).

The Nature of Chemical Energy (Section 5.1)

5.13 *Analyze/Plan.* Use Equation 5.2 to calculate electrostatic potential energy, E_{el}. The distances between the particles are given in the exercise. The charge of an electron and a proton are given in Section 5.1. *Solve.*

(a) $E_{el} = \dfrac{\kappa Q_1 Q_2}{d}$; d = 53 pm; $Q_e = -1.60 \times 10^{-19}$ C; $Q_p = 1.60 \times 10^{-19}$ C

$$E_{el} = \frac{8.99\times10^9 \text{ J-m}}{C^2 \times 53 \text{ pm}} \times \frac{1 \text{ pm}}{1\times10^{-12} \text{ m}} \times -1.60\times10^{-19} \text{ C} \times 1.60\times10^{-19} \text{ C}$$

$$= -4.342\times10^{-18} = -4.3\times10^{-18} \text{ J}$$

(b) $$E_{el} = \frac{8.99\times10^9 \text{ J-m}}{C^2 \times 1.0 \text{ nm}} \times \frac{1 \text{ nm}}{1\times10^{-9} \text{ m}} \times -1.60\times10^{-19} \text{ C} \times 1.60\times10^{-19} \text{ C}$$

$$= -2.301\times10^{-19} = -2.3\times10^{-19} \text{ J}$$

The change in potential energy is $[-2.3 \times 10^{-19} \text{ J} - (-4.3 \times 10^{-18} \text{ J})] = 4.1 \times 10^{-18}$ J

(c) The electrostatic potential energy of the system increases (becomes less negative) as the separation between the oppositely charged particles increases.

5.15 (a) *Analyze/Plan.* Use the equation for electrostatic attractive force and the distance between particles given in the exercise. Find the charges of a proton and an electron in Section 5.1. *Solve.*

$F_{el} = \dfrac{\kappa Q_1 Q_2}{d^2}$; $d = 1.0 \times 10^2$ pm;

$Q_e = -1.60 \times 10^{-19}$ C; $Q_p = 1.60 \times 10^{-19}$ C; 1 J = 1 N-m

$$F_{el} = \frac{8.99\times10^9 \text{ J-m}}{C^2 \times (1.0\times10^2 \text{ pm})^2} \times \frac{(1 \text{ pm})^2}{(1\times10^{-12})^2 \text{ m}^2} \times -1.60\times10^{-19} \text{ C} \times 1.60\times10^{-19} \text{ C}$$

$$= -2.301\times10^{-8} \text{ J/m} = -2.3\times10^{-8} \text{ N}$$

(b) *Analyze/Plan.* Use the formula for gravitational force between two particles and the distance given in the exercise. Find the masses of a proton and an electron on the inside cover of the text. *Solve.*

$$F_g = \frac{Gm_1m_2}{d^2};\ G = \frac{6.674\times10^{-11}\ \text{N-m}^2}{\text{kg}^2};$$

$$d = 1.0\times10^2\ \text{pm};\ m_e = 9.109\times10^{-31}\ \text{kg};\ m_p = 1.673\times10^{-27}\ \text{kg}$$

$$F_g = \frac{6.674\times10^{-11}\ \text{N-m}^2}{\text{kg}^2\times(1.0\times10^2\ \text{pm})^2}\times\frac{(1\ \text{pm})^2}{(1\times10^{-12})^2\ \text{m}^2}\times9.109\times10^{-31}\ \text{kg}\times1.673\times10^{-27}\ \text{kg}$$

$$= 1.0171\times10^{-47} = 1.0\times10^{-47}\ \text{N}$$

(c) The magnitude of the electrostatic force of attraction is 2.3×10^{-8} N. (The negative sign of the electrostatic force indicates attraction.) The gravitational force is 1.0×10^{-47} N. The electrostatic attraction is 2.3×10^{39} time larger. (This is almost 40 orders of magnitude.)

5.17 *Analyze/Plan.* We must find the work required to completely separate two oppositely charged particles. Work is the energy required to move an object against a force (Section 1.4). At infinite separation, the electrostatic potential energy of the pair of ions is zero. The magnitude of the work required is equal to the electrostatic potential energy, E_{el}, of the pair of ions. Use Equation 5.2 to calculate E_{el} and work. The charges of the ions and the distance between them are given in the exercise. *Solve.*

$$E_{el} = \frac{\kappa Q_1Q_2}{d};\ d = 0.50\ \text{nm};\ Q_{Cl} = -1.6\times10^{-19}\ \text{C};\ Q_{Na} = 1.6\times10^{-19}\ \text{C}$$

$$E_{el} = \frac{8.99\times10^9\ \text{J-m}}{\text{C}^2\times0.50\ \text{nm}}\times\frac{1\ \text{nm}}{1\times10^{-9}\ \text{m}}\times-1.6\times10^{-19}\ \text{C}\times1.6\times10^{-19}\ \text{C} =$$

$$-4.603\times10^{-19} = -4.6\times10^{-19}\ \text{J}$$

The sign of E_{el} is negative, so the work required to separate the ions is 4.6×10^{-19} J.

5. 19 (a) Gravity; work is done because the force of gravity is opposed and the pencil is lifted a distance above the desk.

(b) Spring force; work is done because the force of the coiled spring is opposed as the spring is compressed over a distance.

The First Law of Thermodynamics (Section 5.2)

5.21 (a) Matter cannot leave a closed system. Energy in the form of heat or work can be transferred between a closed system and the surroundings.

(b) Neither matter nor energy can leave or enter an isolated system.

(c) Any part of the universe not part of the system is called the surroundings.

5.23 (a) According to the first law of thermodynamics, energy is conserved.

(b) The total *internal energy* (E) of a system is the sum of all the kinetic and potential energies of the system components.

(c) The internal energy of a closed system (where no matter exchange with surroundings occurs) increases when work is done on the system by the surroundings and/or when heat is transferred to the system from the surroundings (the system is heated).

5.25 *Analyze.* Given: heat and work. Find: magnitude and sign of ΔE.

Plan. In each case, evaluate q and w in the expression $\Delta E = q + w$. For an exothermic process, q is negative; for an endothermic process, q is positive. *Solve.*

(a) $q = 0.763$ kJ, $w = -840$ J $= -0.840$ kJ. $\Delta E = 0.763$ kJ $- 0.840$ kJ $= -0.077$ kJ. The process is endothermic.

(b) q is negative because the system releases heat, and w is positive because work is done on the system. $\Delta E = -66.1$ kJ $+ 44.0$ kJ $= -22.1$ kJ. The process is exothermic.

5.27 *Analyze.* How do the different physical situations (cases) affect the changes to heat and work of the system upon addition of 100 J of energy?

Plan. Use the definitions of heat and work and the First Law to answer the questions.

Solve. If the piston is allowed to move, case (1), the heated gas will expand and push the piston up, doing work on the surroundings. If the piston is fixed, case (2), most of the electrical energy will be manifested as an increase in heat of the system.

(a) Because little or no work is done by the system in case (2), the gas will absorb most of the energy as heat; the case (2) gas will have the higher temperature.

(b) In Case 1, w is negative because work is done on the surroundings by expansion. Because the transfer of electrical energy is never completely efficient and some energy will be transferred as heat, q is positive. In Case 2, w is zero because no work (expansion) is done. The value of q is positive because all energy is transferred as heat.

(c) ΔE is greater for case (2) because the entire 100 J increases the internal energy of the system, rather than a part of the energy doing work on the surroundings.

5.29 (a) A *state function* is a property of a system that depends only on the physical state (pressure, temperature, etc.) of the system, not on the route used by the system to get to the current state.

(b) Internal energy and enthalpy **are** state functions; heat **is not** a state function.

(c) Volume **is** a state function. The volume of a system depends only on conditions (pressure, temperature, amount of substance), not the route or method used to establish that volume.

Enthalpy (Sections 5.3 and 5.4)

5.31 *Analyze.* Given, $P = 1.0$ atm, $\Delta V = +0.50$ L. Find work involved, in J.

Plan. This change is P – V work done at constant P. $w = -P\Delta V$. 1 L-atm = 101.3 J

Solve. $w = -1.0$ atm(0.50 L) $= -0.50$ L-atm; 0.50 L-atm × 101.3 J/L-atm $= -50.65 = -51$ J

The negative sign indicates that work is done by the system on the surroundings.

5.33 (a) Change in enthalpy (ΔH) is usually easier to measure than change in internal energy (ΔE) because, at constant pressure, $\Delta H = q_p$. The heat flow associated with a process at constant pressure can easily be measured as a change in temperature. Measuring ΔE requires a means to measure both q and w.

(b) H describes the enthalpy of a system at a certain set of conditions; the value of H depends only on these conditions. q describes energy transferred as heat, an energy *change*, which, in the general case, does depend on how the change occurs. We can equate change in enthalpy, ΔH, with heat, q_p, only for the specific conditions of constant pressure and exclusively P-V work.

(c) If ΔH is positive, the enthalpy of the system increases, and the process is endothermic.

5.35 (a) At constant pressure, ΔE = ΔH – PΔV. To calculate ΔE, more information about the conditions of the reaction must be known. For an ideal gas at constant pressure and temperature, PΔV = RTΔn. We know the value of Δn = –3 from the chemical reaction. We must know either the temperature, T, or the values of P and ΔV to calculate ΔE from ΔH.

(b) ΔE is larger than ΔH.

(c) Because the value of Δn is negative, the quantity (–PΔV) is positive. We add a positive quantity to ΔH to calculate ΔE, so ΔE must be larger.

5.37 *Analyze/Plan.* q = 824 J = 0.824 kJ (heat is absorbed by the system), w = 0.65 kJ (work is done on the system). *Solve.*

ΔE = q + w = 0.824 kJ + 0.65 kJ = 1.47 kJ. ΔH = q = 0.824 kJ (at constant pressure).

Check. The reaction is endothermic.

5.39 (a) $C_2H_5OH(l) + 3\,O_2(g) \rightarrow 3\,H_2O(g) + 2\,CO_2(g) \qquad \Delta H = -1235 \text{ kJ}$

(b) *Analyze.* How are reactants and products arranged on an enthalpy diagram?

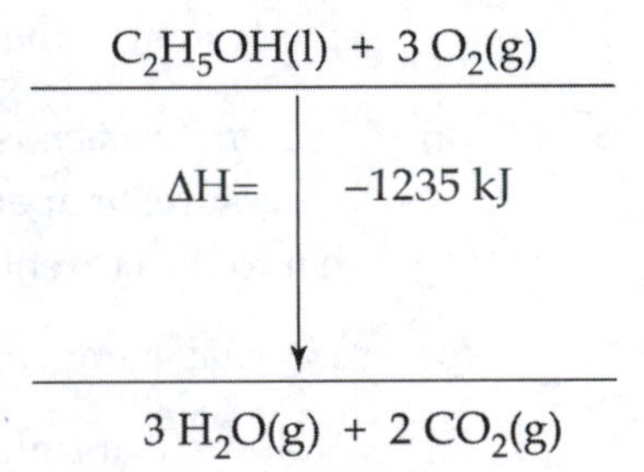

Plan. The substances (reactants or products, collectively) with higher enthalpy are shown on the upper level, and those with lower enthalpy are shown on the lower level.

Solve. For this reaction, ΔH is negative, so the products have lower enthalpy and are shown on the lower level; reactants are on the upper level. The arrow points in the direction of reactants to products and is labeled with the value of ΔH.

5.41 *Analyze/Plan.* Consider ΔH for the exothermic reaction as written. *Solve.*

(a) $\Delta H = -284.6 \text{ kJ}/2 \text{ mol } O_3(g) = -142.3 \text{ kJ/mol } O_3(g)$

(b) Because ΔH is negative, the reactants, 2 $O_3(g)$ has the higher enthalpy.

5.43 *Analyze/Plan.* Follow the strategy in Sample Exercise 5.4. *Solve.*

(a) Exothermic (ΔH is negative)

(b) $3.55 \text{ g Mg} \times \frac{1 \text{ mol Mg}}{24.305 \text{ g Mg}} \times \frac{-1204 \text{ kJ}}{2 \text{ mol Mg}} = -87.9 \text{ kJ heat transferred}$

Check. The units of kJ are correct for heat. The negative sign indicates heat is evolved.

(c) $-234 \text{ kJ} \times \frac{2 \text{ mol MgO}}{-1204 \text{ kJ}} \times \frac{40.30 \text{ g MgO}}{1 \text{ mol Mg}} = 15.7 \text{ g MgO produced}$

Check. Units are correct for mass. $(200 \times 2 \times 40/1200) \approx (16{,}000/1200) > 10$ g

(d) $2\,MgO(s) \rightarrow 2\,Mg(s) + O_2(g) \qquad \Delta H = +1204 \text{ kJ}$

This is the reverse of the reaction given above, so the sign of ΔH is reversed.

$$40.3 \text{ g MgO} \times \frac{1 \text{ mol MgO}}{40.30 \text{ g MgO}} \times \frac{1204 \text{ kJ}}{2 \text{ mol MgO}} = +602 \text{ kJ heat absorbed}$$

Check. 40.3 g MgO is just 1 mol MgO, so the calculated value is the heat absorbed per mol of MgO, 1204 kJ/2 mol MgO = 602 kJ.

5.45 *Analyze.* Given: balanced thermochemical equation, various quantities of substances and/or enthalpy. *Plan.* Enthalpy is an extensive property; it is "stoichiometric." Use the mole ratios implicit in the balanced thermochemical equation to solve for the desired quantity. Use molar masses to change mass to moles and vice versa where appropriate. *Solve.*

(a) $0.450 \text{ mol AgCl} \times \frac{-65.5 \text{ kJ}}{1 \text{ mol AgCl}} = -29.5 \text{ kJ}$

Check. Units are correct; sign indicates heat evolved.

(b) $9.00 \text{ g AgCl} \times \frac{1 \text{ mol AgCl}}{143.3 \text{ g AgCl}} \times \frac{-65.5 \text{ kJ}}{1 \text{ mol AgCl}} = -4.11 \text{ kJ}$

Check. Units correct; sign indicates heat evolved.

(c) $9.25 \times 10^{-4} \text{ mol AgCl} \times \frac{+65.5 \text{ kJ}}{1 \text{ mol AgCl}} = 0.0606 \text{ kJ} = 60.6 \text{ J}$

Check. Units correct; sign of ΔH reversed; sign indicates heat is absorbed during the reverse reaction.

5.47 *Analyze.* Given: balanced thermochemical equation. *Plan.* Follow the guidelines given in Section 5.4 for evaluating thermochemical equations. *Solve.*

(a) When a chemical equation is reversed, the sign of ΔH is reversed.

$CO_2(g) + 2\,H_2O(l) \rightarrow CH_3OH(l) + 3/2\,O_2(g) \qquad \Delta H = +726.5 \text{ kJ}$

(b) Enthalpy is extensive. If the coefficients in the chemical equation are multiplied by 2 to obtain all integer coefficients, the enthalpy change is also multiplied by 2.

$2\,CH_3OH(l) + 3\,O_2(g) \rightarrow 2\,CO_2(g) + 4\,H_2O(l) \qquad \Delta H = 2(-726.5) \text{ kJ} = -1453 \text{ kJ}$

(c) The exothermic forward reaction is more likely to be thermodynamically favored.

(d) Decrease. Vaporization (liquid $\rightarrow$ gas) is endothermic. If the product were $H_2O(g)$, the reaction would be more endothermic and would have a smaller negative ΔH. (Depending on temperature, the enthalpy of vaporization for 2 mol H_2O is about +88 kJ, not large enough to cause the overall reaction to be endothermic.)

Calorimetry (Section 5.5)

The specific heat of water to four significant figures, **4.184 J/g-K,** will be used in many of the following exercises; temperature units of K and °C will be used interchangeably.

5.49 (a) J/mol-K or J/mol-°C. Heat capacity is the amount of heat in J required to raise the temperature of an object or a certain amount of substance 1 °C or 1 K. Molar heat capacity is the heat capacity of one mole of substance.

(b) $\frac{J}{g\text{-}°C}$ or $\frac{J}{g\text{-}K}$ Specific heat is a particular kind of heat capacity where the amount of substance is 1 g.

(c) To calculate heat capacity from specific heat, the **mass** of the particular piece of copper pipe must be known.

5.51 *Plan.* Manipulate the definition of specific heat to solve for the desired quantity, paying close attention to units. $C_s = q/(m \times \Delta T)$. *Solve.*

(a) $\frac{4.184\ J}{1\ g\text{-}K}$ or $\frac{4.184\ J}{1\ g\text{-}°C}$

(b) $$\frac{4.184\ J}{1\ g\text{-}°C} \times \frac{18.02\ g\ H_2O}{1\ mol\ H_2O} = \frac{75.40\ J}{mol\text{-}°C}$$

(c) $$\frac{185\ g\ H_2O \times 4.184\ J}{1\ g\text{-}°C} = 774\ J/°C$$

(d) $$10.00\ kg\ H_2O \times \frac{1000\ g}{1\ kg} \times \frac{4.184\ J}{1\ g\text{-}°C} \times \frac{1\ kJ}{1000\ J} \times (46.2°C - 24.6\ °C) = 904\ kJ$$

Check. $(10 \times 4 \times 20) \approx 800$ kJ; the units are correct. Note that the conversion factors for kg $\rightarrow$ g and J $\rightarrow$ kJ cancel. An equally correct form of specific heat would be kJ/kg-°C

5.53 *Analyze/Plan.* Follow the logic in Sample Exercise 5.5. *Solve.*

(a) $$80.0\ g\ C_8H_{18} \times \frac{2.22\ J}{g\text{-}K} \times (25.0\ °C - 10.0\ °C) = 2.66 \times 10^3\ J\ (\text{or } 2.66\ kJ)$$

(b) *Plan.* Calculate the molar heat capacity of octane and compare it with the molar heat capacity of water, 75.40 J/mol-°C, as calculated in Exercise 5.51(b). *Solve.*

$$\frac{2.22\ J}{g\text{-}K} \times \frac{114.2\ g\ C_8H_{18}}{1\ mol\ C_8H_{18}} = \frac{253.58\ J}{mol\text{-}K} = \frac{254\ J}{mol\text{-}K}$$

The molar heat capacity of $C_8H_{18}(l)$, 254 J/mol-K, is greater than that of $H_2O(l)$, so it will require more heat to increase the temperature of octane than to increase the temperature of water.

5.55 *Analyze.* Because the temperature of the water increases, the dissolving process is exothermic and the sign of ΔH is negative. The heat lost by the NaOH(s) dissolving equals the heat gained by the solution.

Plan/Solve. Calculate the heat gained by the solution. The temperature change is $37.8 - 21.6 = 16.2$ °C. The total mass of solution is (100.0 g H_2O + 6.50 g NaOH) = 106.5 g.

$$106.5\ g\ solution \times \frac{4.184\ J}{1\ g\text{-}°C} \times 16.2\ °C \times \frac{1\ kJ}{1000\ J} = 7.2187 = 7.22\ kJ$$

This is the amount of heat lost when 6.50 g of NaOH dissolves.

The heat loss per mole NaOH is

$$\frac{-7.2187 \text{ kJ}}{6.50 \text{ g NaOH}} \times \frac{40.00 \text{ g NaOH}}{1 \text{ mol NaOH}} = -44.4 \text{ kJ/mol} \qquad \Delta H = q_p = -44.4 \text{ kJ/mol NaOH}$$

Check. $(-7/7 \times 40) \approx -40$ kJ; the units and sign are correct.

5.57 *Analyze/Plan.* Follow the logic in Sample Exercise 5.7. *Solve.*

$q_{bomb} = -q_{rxn}$; $\Delta T = 30.57\ °C - 23.44\ °C = 7.13\ °C$

$$q_{bomb} = \frac{7.854 \text{ kJ}}{1\ °C} \times 7.13\ °C = 56.00 = 56.0 \text{ kJ}$$

At constant volume, $q_v = \Delta E$. ΔE and ΔH are very similar.

$$\Delta H_{rxn} \approx \Delta E_{rxn} = q_{rxn} = -q_{bomb} = \frac{-56.0 \text{ kJ}}{2.200 \text{ g } C_6H_4O_2} = -25.454 = -25.5 \text{ kJ/g } C_6H_4O_2$$

$$\Delta H_{rxn} = \frac{-25.454 \text{ kJ}}{1 \text{ g } C_6H_4O_2} \times \frac{108.1 \text{ g } C_6H_4O_2}{1 \text{ mol } C_6H_4O_2} = -2.75 \times 10^3 \text{ kJ/mol } C_6H_4O_2$$

5.59 *Analyze.* Given: specific heat and mass of glucose, ΔT for calorimeter. Find: heat capacity, C, of calorimeter. *Plan.* All heat from the combustion raises the temperature of the calorimeter. Calculate heat from combustion of glucose, divide by ΔT for calorimeter to get kJ/°C. $\Delta T = 24.72\ °C - 20.94\ °C = 3.78\ °C$ *Solve.*

(a) $$C_{total} = 3.500 \text{ g glucose} \times \frac{15.57 \text{ kJ}}{1 \text{ g glucose}} \times \frac{1}{3.78\ °C} = 14.42 = 14.4 \text{ kJ/°C}$$

(b) Qualitatively, assuming the same exact initial conditions in the calorimeter, twice as much glucose produces twice as much heat, which raises the calorimeter temperature by twice as many °C. Quantitatively,

$$7.000 \text{ g glucose} \times \frac{15.57 \text{ kJ}}{1 \text{ g glucose}} \times \frac{1\ °C}{14.42 \text{ kJ}} = 7.56\ °C$$

Check. Units are correct. ΔT is twice as large as in part (a). The result has 3 sig figs, because the heat capacity of the calorimeter is known to 3 sig figs.

Hess's Law (Section 5.6)

5.61 Yes, because internal energy is a state function. Hess's Law works for any state function.

5.63 *Analyze/Plan.* Follow the logic in Sample Exercise 5.8. Manipulate the equations so that "unwanted" substances can be canceled from reactants and products. Adjust the corresponding sign and magnitude of ΔH. *Solve.*

$$\begin{array}{ll} P_4O_6(s) \rightarrow P_4(s) + 3\,O_2(g) & \Delta H = 1640.1 \text{ kJ} \\ P_4(s) + 5\,O_2(g) \rightarrow P_4O_{10}(s) & \Delta H = -2940.1 \text{ kJ} \\ \hline P_4O_6(s) + 2\,O_2(g) \rightarrow P_4O_{10}(s) & \Delta H = -1300.0 \text{ kJ} \end{array}$$

Check. We have obtained the desired reaction.

5.65 *Analyze/Plan.* Follow the logic in Sample Exercise 5.9. Manipulate the equations so that "unwanted" substances can be canceled from reactants and products. Adjust the corresponding sign and magnitude of ΔH. *Solve.*

$C_2H_4(g)$	→	$2\ H_2(g) + 2\ C(s)$	$\Delta H = -52.3$ kJ
$2\ C(s) + 4\ F_2(g)$	→	$2\ CF_4(g)$	$\Delta H = 2(-680$ kJ)
$2\ H_2(g) + 2\ F_2(g)$	→	$4\ HF(g)$	$\Delta H = 2(-537$ kJ)
$C_2H_4(g) + 6\ F_2(g)$	→	$2\ CF_4(g) + 4\ HF(g)$	$\Delta H = -2.49\times10^3$ kJ

Check. We have obtained the desired reaction.

Enthalpies of Formation (Section 5.7)

5.67 (a) *Standard conditions* for enthalpy changes are usually P = 1 atm and T = 298 K. For the purpose of comparison, standard enthalpy changes, ΔH°, are tabulated for reactions at these conditions.

(b) *Enthalpy of formation,* ΔH_f, is the enthalpy change that occurs when a compound is formed from its component elements.

(c) *Standard enthalpy of formation,* ΔH_f°, is the enthalpy change that accompanies formation of 1 mole of a substance from elements in their standard states.

5.69 (a) $1/2\ N_2(g) + O_2(g) \rightarrow NO_2(g)$ $\qquad \Delta H_f^\circ = \ \ 33.84$ kJ

(b) $S(s) + 3/2\ O_2(g) \rightarrow SO_3(g)$ $\qquad \Delta H_f^\circ = -395.2$ kJ

(c) $Na(s) + 1/2\ Br_2(l) \rightarrow NaBr(s)$ $\qquad \Delta H_f^\circ = -361.4$ kJ

(d) $Pb(s) + N_2(g) + 3\ O_2(g) \rightarrow Pb(NO_3)_2(s)$ $\qquad \Delta H_f^\circ = -451.9$ kJ

5.71 *Plan.* $\Delta H^\circ_{rxn} = \Sigma n\Delta H_f^\circ$ (products) $- \Sigma n\Delta H_f^\circ$ (reactants). Be careful with coefficients, states, and signs. *Solve.*

$\Delta H^\circ_{rxn} = \Delta H_f^\circ\ Al_2O_3(s) + 2\ \Delta H_f^\circ\ Fe(s) - \Delta H_f^\circ\ Fe_2O_3(s) - 2\ \Delta H_f^\circ\ Al(s)$

$\Delta H^\circ_{rxn} = (-1669.8\text{ kJ}) + 2(0) - (-822.16\text{ kJ}) - 2(0) = -847.6\text{ kJ}$

5.73 *Plan.* $\Delta H^\circ_{rxn} = \Sigma n\Delta H_f^\circ$ (products) $- \Sigma n\Delta H_f^\circ$ (reactants). Be careful with coefficients, states, and signs. *Solve.*

(a) $\Delta H^\circ_{rxn} = 2\ \Delta H_f^\circ\ SO_3(g) - 2\ \Delta H_f^\circ\ SO_2(g) - \Delta H_f^\circ\ O_2(g)$

$= 2(-395.2\text{ kJ}) - 2(-296.9\text{ kJ}) - 0 = -196.6\text{ kJ}$

(b) $\Delta H^\circ_{rxn} = \Delta H_f^\circ\ MgO(s) + \Delta H_f^\circ\ H_2O(l) - \Delta H_f^\circ\ Mg(OH)_2(s)$

$= -601.8\text{ kJ} + (-285.83\text{ kJ}) - (-924.7\text{ kJ}) = 37.1\text{ kJ}$

(c) $\Delta H^\circ_{rxn} = 4\ \Delta H_f^\circ\ H_2O(g) + \Delta H_f^\circ\ N_2(g) - \Delta H_f^\circ\ N_2O_4(g) - 4\ \Delta H_f^\circ\ H_2(g)$

$= 4(-241.82\text{ kJ}) + 0 - (9.66\text{ kJ}) - 4(0) = -976.94\text{ kJ}$

(d) $\Delta H^\circ_{rxn} = \Delta H_f^\circ\ SiO_2(s) + 4\ \Delta H_f^\circ\ HCl(g) - \Delta H_f^\circ\ SiCl_4(l) - 2\ \Delta H_f^\circ\ H_2O(l)$

$= -910.9\text{ kJ} + 4(-92.30\text{ kJ}) - (-640.1\text{ kJ}) - 2(-285.83\text{ kJ}) = -68.3\text{ kJ}$

5.75 *Analyze.* Given: combustion reaction, enthalpy of combustion, enthalpies of formation for most reactants and products. Find: enthalpy of formation for acetone.

Plan. Rearrange the expression for enthalpy of reaction to calculate the desired enthalpy of formation. *Solve.*

$$\Delta H^\circ_{rxn} = 3\ \Delta H^\circ_f\ CO_2(g) + 3\ \Delta H^\circ_f\ H_2O(l) - \Delta H^\circ_f\ C_3H_6O(l) - 4\ \Delta H^\circ_f\ O_2(g)$$

$$-1790\text{ kJ} = 3(-393.5\text{ kJ}) + 3(-285.83\text{ kJ}) - \Delta H^\circ_f\ C_3H_6O(l) - 4(0)$$

$$\Delta H^\circ_f\ C_3H_6O(l) = 3(-393.5\text{ kJ}) + 3(-285.83\text{ kJ}) + 1790\text{ kJ} = -248\text{ kJ}$$

5.77 (a) $C_8H_{18}(l) + 25/2\ O_2(g) \rightarrow 8\ CO_2(g) + 9\ H_2O(g) \quad \Delta H^\circ = -5064.9\text{ kJ}$

(b) *Plan.* Follow the logic in Solution 5.75 and 5.76. *Solve.*

$$\Delta H^\circ_{rxn} = 8\ \Delta H^\circ_f\ CO_2(g) + 9\ \Delta H^\circ_f\ H_2O(g) - \Delta H^\circ_f\ C_8H_{18}(l) - 25/2\ \Delta H^\circ_f\ O_2(g)$$

$$-5064.9\text{ kJ} = 8(-393.5\text{ kJ}) + 9(-241.82\text{ kJ}) - \Delta H^\circ_f\ C_8H_{18}(l) - 25/2(0)$$

$$\Delta H^\circ_f\ C_8H_{18}(l) = 8(-393.5\text{ kJ}) + 9(-241.82\text{ kJ}) + 5064.9\text{ kJ} = -259.5\text{ kJ}$$

5.79 (a) $C_2H_5OH(l) + 3\ O_2(g) \rightarrow 2\ CO_2(g) + 3\ H_2O(g)$

(b) $$\Delta H^\circ_{rxn} = 2\ \Delta H^\circ_f\ CO_2(g) + 3\ \Delta H^\circ_f\ H_2O(g) - \Delta H^\circ_f\ C_2H_5OH(l) - 3\ \Delta H^\circ_f\ O_2(g)$$

$$= 2(-393.5\text{ kJ}) + 3(-241.82\text{ kJ}) - (-277.7\text{ kJ}) - 3(0) = -1234.76 = -1234.8\text{ kJ}$$

(c) *Plan.* The enthalpy of combustion of ethanol [from part (b)] is −1234.8 kJ/mol. Change mol to mass using molar mass, then mass to volume using density. *Solve.*

$$\frac{-1234.76\text{ kJ}}{\text{mol } C_2H_5OH} \times \frac{1\text{ mol } C_2H_5OH}{46.06844\text{ g}} \times \frac{0.789\text{ g}}{\text{mL}} \times \frac{1000\text{ mL}}{\text{L}} = -21{,}147 = -2.11 \times 10^4\text{ kJ/L}$$

Check. $(1200/50) \approx 25;\ 25 \times 800 \approx 20{,}000$

(d) *Plan.* The enthalpy of combustion corresponds to any of the molar amounts in the equation as written. Production of −1234.76 kJ also produces 2 mol CO_2. Use this relationship to calculate mass CO_2/kJ.

$$\frac{2\text{ mol } CO_2}{-1234.76\text{ kJ}} \times \frac{44.0095\text{ g } CO_2}{\text{mol}} = 0.071284\text{ g } CO_2\text{ / kJ emitted}$$

Check. The negative sign associated with enthalpy indicates that energy is emitted.

Bond Enthalpies (Section 5.8)

5.81 (a) + ΔH; energy must be supplied to separate oppositely charged ions.

(b) – ΔH; energy is released when a chemical bond is formed.

(c) + ΔH; energy must be supplied to separate a negatively charged electron from a neutral atom.

(d) + ΔH; energy must be supplied to melt a solid.

5.83 *Analyze.* Given: structural formulas. Find: enthalpy of reaction.

Plan. Count the number and kinds of bonds that are broken and formed by the reaction. Use bond enthalpies from Table 5.4 and Equation 5.32 to calculate the overall enthalpy of reaction, ΔH. *Solve.*

(a) $\Delta H = D(H–H) + D(Br–Br) – 2\,D(H–Br)$

$= 436\text{ kJ} + 193\text{ kJ} – 2(366\text{ kJ}) = –103\text{ kJ}$

(b) $\Delta H = 6\,D(C–H) + 2\,D(C–O) + 2\,D(O–H) + 3\,D(O{=}O)$

$– 4\,D(C{=}O) – 8\,D(O–H)$

$= 6\,D(C–H) + 2\,D(C–O) + 3\,D(O{=}O) – 4\,D(C{=}O) – 6\,D(O–H)$

$\Delta H = 6(413) + 2(358) + 3(495) – 4(799) – 6(463) = –1295\text{ kJ}$

5.85 (a) *Plan.* $\Delta H^\circ_{rxn} = \Sigma n\Delta H^\circ_f$ (products) $- \Sigma n\Delta H^\circ_f$ (reactants). Be careful with coefficients, states, and signs. *Solve.*

$\Delta H^\circ_{rxn} = 2\,\Delta H^\circ_f\ Br(g) – \Delta H^\circ_f\ Br_2(g)$

$= 2(111.8) – 30.71 = 192.9\text{ kJ}$

This reaction is just the breaking of a Br–Br single bond to form Br atoms; reactants and products are all in the gas phase. The enthalpy of reaction represents the bond enthalpy D(Br–Br), 193 kJ.

(b) The value of D(Br–Br) in Table 5.4 is 193 kJ, the same as the enthalpy calculated in part (a). The difference between the two values is zero (to three significant figures).

5.87 (a) $\Delta H = 2\,D(H–H) + D(O{=}O) – 4\,D(O–H) = 2(436) + 495 – 4(463) = –485\text{ kJ}$

(b) The estimate from part (a) is less negative or larger than the true reaction enthalpy. When we use bond enthalpies to estimate reaction enthalpies, we assume all reactants and products are gases. We have estimated the enthalpy change for production of $H_2O(g)$. Because condensation, $[(g) \rightarrow (l)]$, is exothermic, we expect ΔH for production of liquid water to be more negative or smaller than the value we estimated in part (a).

(c) $\Delta H^\circ_{rxn} = 2\,\Delta H^\circ_f\ H_2O(l) – 2\,\Delta H^\circ_f\ H_2(g) – \Delta H^\circ_f\ O_2(g)$

$= 2(–285.83) – 2(0) – 0 = –571.66 = –572\text{ kJ}$

As predicted in part (b), the true enthalpy of reaction is more negative than the result calculated using bond enthalpies.

Foods and Fuels (Section 5.9)

5.89 (a) *Fuel value* is the amount of energy produced when 1 gram of a substance (fuel) is combusted.

(b) The fuel value of fats is 9 kcal/g and of carbohydrates is 4 kcal/g. Therefore, 5 g of fat produce 45 kcal, whereas 9 g of carbohydrates produce 36 kcal; 5 g of fat are a greater energy source.

(c) These products of metabolism are expelled as waste, $H_2O(l)$ primarily in urine and feces, and $CO_2(g)$ as gas when breathing.

5.91 (a) *Plan.* Calculate the Cal (kcal) from each nutritional component of the soup, then sum. *Solve.*

$$2.5 \text{ g fat} \times \frac{38 \text{ kJ}}{1 \text{ g fat}} = 95.0 \text{ or } 0.95 \times 10^2 \text{ kJ}$$

$$14 \text{ g carbohydrates} \times \frac{17 \text{ kJ}}{1 \text{ g carbohydrate}} = 238 \text{ or } 2.4 \times 10^2 \text{ kJ}$$

$$7 \text{ g protein} \times \frac{17 \text{ kJ}}{1 \text{ g protein}} = 119 \text{ or } 1 \times 10^2 \text{ kJ}$$

total energy = 95.0 kJ + 238 kJ + 119 kJ = 452 or 5×10^2 kJ

$$452 \text{ kJ} \times \frac{1 \text{ kcal}}{4.184 \text{ kJ}} \times \frac{1 \text{ Cal}}{1\text{kcal}} = 108.03 \text{ or } 1 \times 10^2 \text{ Cal/serving}$$

Check. 100 Cal/serving is a reasonable result; units are correct. The data and the result have 1 sig fig.

(b) Sodium does not contribute to the calorie content of the food, because it is not metabolized by the body; it enters and leaves as Na^+.

5.93 *Plan.* $g \rightarrow mol \rightarrow kJ \rightarrow Cal$ *Solve.*

$$16.0 \text{ g } C_6H_{12}O_6 \times \frac{1 \text{ mol } C_6H_{12}O_6}{180.2 \text{ g } C_6H_{12}O_6} \times \frac{2812 \text{ kJ}}{\text{mol } C_6H_{12}O_6} \times \frac{1 \text{ Cal}}{4.184 \text{ kJ}} = 59.7 \text{ Cal}$$

Check. 60 Cal is a reasonable result for most of the food value in an apple.

5.95 *Plan.* Use enthalpies of formation to calculate molar heat (enthalpy) of combustion using Hess's Law. Use molar mass to calculate heat of combustion per kg of hydrocarbon. *Solve.*

Propyne: $C_3H_4(g) + 4\,O_2(g) \rightarrow 3\,CO_2(g) + 2\,H_2O(g)$

(a) $\Delta H^\circ_{rxn} = 3(-393.5 \text{ kJ}) + 2(-241.82 \text{ kJ}) - (185.4 \text{ kJ}) - 4(0) = -1849.5$

$= -1850 \text{ kJ/mol } C_3H_4$

(b) $$\frac{-1849.5 \text{ kJ}}{1 \text{ mol } C_3H_4} \times \frac{1 \text{ mol } C_3H_4}{40.065 \text{ g } C_3H_4} \times \frac{1000 \text{ g } C_3H_4}{1 \text{ kg } C_3H_4} = -4.616 \times 10^4 \text{ kJ/kg } C_3H_4$$

Propylene: $C_3H_6(g) + 9/2\,O_2(g) \rightarrow 3\,CO_2(g) + 3\,H_2O(g)$

(a) $\Delta H^\circ_{rxn} = 3(-393.5 \text{ kJ}) + 3(-241.82 \text{ kJ}) - (20.4 \text{ kJ}) - 9/2(0) = -1926.4$

$= -1926 \text{ kJ/mol } C_3H_6$

(b) $$\frac{-1926.4 \text{ kJ}}{1 \text{ mol } C_3H_6} \times \frac{1 \text{ mol } C_3H_6}{42.080 \text{ g } C_3H_6} \times \frac{1000 \text{ g } C_3H_6}{1 \text{ kg } C_3H_6} = -4.578 \times 10^4 \text{ kJ/kg } C_3H_6$$

Propane: $C_3H_8(g) + 5\,O_2(g) \rightarrow 3\,CO_2(g) + 4\,H_2O(g)$

(a) $\Delta H^\circ_{rxn} = 3(-393.5 \text{ kJ}) + 4(-241.82 \text{ kJ}) - (-103.8 \text{ kJ}) - 5(0) = -2044.0$

$= -2044 \text{ kJ/mol } C_3H_8$

(b) $$\frac{-2044.0 \text{ kJ}}{1 \text{ mol } C_3H_8} \times \frac{1 \text{ mol } C_3H_8}{44.096 \text{ g } C_3H_8} \times \frac{1000 \text{ g } C_3H_8}{1 \text{ kg } C_3H_8} = -4.635 \times 10^4 \text{ kJ/kg } C_3H_8$$

(c) These three substances yield nearly identical quantities of heat per unit mass, but propane is marginally higher than the other two.

5.97 *Analyze/Plan.* Given population, Cal/person/day and kJ/mol glucose, calculate kg glucose/yr. Calculate kJ/yr, then kg/yr. 1 billion = 1×10^9. 365 day = 1 yr. 1 Cal = 1 kcal, 4.184 kJ = 1 kcal = 1 Cal. *Solve.*

$$7.0 \times 10^9 \text{ persons} \times \frac{1500 \text{ Cal}}{\text{person-day}} \times \frac{365 \text{ day}}{1 \text{ yr}} \times \frac{4.184 \text{ kJ}}{1 \text{ Cal}} = 1.6035 \times 10^{16} = 1.6 \times 10^{16} \text{ kJ/yr}$$

$$\frac{1.6035 \times 10^{16} \text{ kJ}}{\text{yr}} \times \frac{1 \text{ mol } C_6H_{12}O_6}{2803 \text{ kJ}} \times \frac{180.2 \text{ g } C_6H_{12}O_6}{1 \text{ mol } C_6H_{12}O_6} \times \frac{1 \text{ kg}}{1000 \text{ g}} = 1.0 \times 10^{12} \text{ kg } C_6H_{12}O_6/\text{yr}$$

Check. 1×10^{12} kg is 1 *trillion* kg of glucose.

Additional Exercises

5.99 Like the combustion of $H_2(g)$ and $O_2(g)$ described in Section 5.4, the reaction that inflates airbags is spontaneous after initiation. Spontaneous reactions are usually exothermic, $-\Delta H$. The airbag reaction occurs at constant atmospheric pressure, $\Delta H = q_p$; both are likely to be large and negative. When the bag inflates, work is done by the system on the surroundings, so the sign of w is negative.

5.103 $\Delta E = q + w = +38.95 \text{ kJ} - 2.47 \text{ kJ} = +36.48 \text{ kJ}$

$\Delta H = q_p = +38.95 \text{ kJ}$

5.105 Find the heat capacity of 1.7×10^3 gal H_2O.

$$C_{H_2O} = 1.7 \times 10^3 \text{ gal } H_2O \times \frac{4 \text{ qt}}{1 \text{ gal}} \times \frac{1 \text{ L}}{1.057 \text{ qt}} \times \frac{1 \times 10^3 \text{ cm}^3}{1 \text{ L}} \times \frac{1 \text{ g}}{1 \text{ cm}^3} \times \frac{4.184 \text{ J}}{1 \text{ g-°C}}$$

$= 2.692 \times 10^7 \text{ J/°C} = 2.7 \times 10^4 \text{ kJ/°C}$; then,

$$\frac{2.692 \times 10^7 \text{ J}}{1 \text{°C}} \times \frac{1 \text{ g-°C}}{0.85 \text{ J}} \times \frac{1 \text{ kg}}{1 \times 10^3 \text{g}} \times \frac{1 \text{ brick}}{1.8 \text{ kg}} = 1.8 \times 10^4 \text{ or 18,000 bricks}$$

Check. $(1.7 \times {\sim}16 \times 10^6)/({\sim}1.6 \times 10^3) \approx 17 \times 10^3$ bricks; the units are correct.

5.108 (a) $Mg(s) + 2\,H_2O(l) \rightarrow Mg(OH)_2(s) + H_2(g)$

$$\Delta H^\circ_{rxn} = \Delta H^\circ_f\, Mg(OH)_2(s) + \Delta H^\circ_f\, H_2(g) - 2\,\Delta H^\circ_f\, H_2O(l) - \Delta H^\circ_f\, Mg(s)$$

$= -924.7 \text{ kJ} + 0 - 2(-285.83 \text{ kJ}) - 0 = -353.04 = -353.0 \text{ kJ}$

(b) Use the specific heat of water, 4.184 J/g-°C, to calculate the energy required to heat the water. Use the density of water at 25 °C to calculate the mass of H_2O to be heated. (The change in density of H_2O going from 21 °C to 79 °C does not substantially affect the strategy of the exercise.) Then use the 'heat stoichiometry' in (a) to calculate mass of Mg(s) needed.

$$75 \text{ mL} \times \frac{0.997 \text{ g } H_2O}{\text{mL}} \times \frac{4.184 \text{ J}}{\text{g-°C}} \times 58 \text{ °C} \times \frac{1 \text{ kJ}}{1000 \text{ J}} = 18.146 \text{ kJ} = 18 \text{ kJ required}$$

$$18.146 \text{ kJ} \times \frac{1 \text{ mol Mg}}{353.04 \text{ kJ}} \times \frac{24.305 \text{ g Mg}}{1 \text{ mol Mg}} = 1.249 \text{ g} = 1.2 \text{ g Mg needed}$$

5.112 (a) $3\,C_2H_2(g) \rightarrow C_6H_6(l)$

$$\Delta H^\circ_{rxn} = \Delta H^\circ_f\ C_6H_6(l) - 3\ \Delta H^\circ_f\ C_2H_2(g) = 49.0\ kJ - 3(226.77\ kJ) = -631.31 = -631.3\ kJ$$

(b) Because the reaction is exothermic (ΔH is negative), the reactant, 3 moles of $C_2H_2(g)$, has more enthalpy than the product, 1 mole of $C_6H_6(l)$.

(c) The fuel value of a substance is the amount of heat (kJ) produced when 1 gram of the substance is burned. Calculate the molar heat of combustion (kJ/mol) and use this to find kJ/g of fuel.

$C_2H_2(g) + 5/2\,O_2(g) \rightarrow 2\,CO_2(g) + H_2O(l)$

$$\Delta H^\circ_{rxn} = 2\ \Delta H^\circ_f\ CO_2(g) + \Delta H^\circ_f\ H_2O(l) - \Delta H^\circ_f\ C_2H_2(g) - 5/2\ \Delta H^\circ_f\ O_2(g)$$

$$= 2(-393.5\ kJ) + (-285.83\ kJ) - 226.77\ kJ - 5/2\ (0) = -1299.6\ kJ/mol\ C_2H_2$$

$$\frac{-1299.6\ kJ}{1\ mol\ C_2H_2} \times \frac{1\ mol\ C_2H_2}{26.036\ g\ C_2H_2} = 49.916 = 50\ kJ/g\ C_2H_2$$

$C_6H_6(l) + 15/2\,O_2(g) \rightarrow 6\,CO_2(g) + 3\,H_2O(l)$

$$\Delta H^\circ_{rxn} = 6\ \Delta H^\circ_f\ CO_2(g) + 3\ \Delta H^\circ_f\ H_2O(l) - \Delta H^\circ_f\ C_6H_6(l) - 15/2\ \Delta H^\circ_f\ O_2(g)$$

$$= 6(-393.5\ kJ) + 3(-285.83\ kJ) - 49.0\ kJ - 15/2\ (0) = -3267.5\ kJ/mol\ C_6H_6$$

$$\frac{-3267.5\ kJ}{1\ mol\ C_6H_6} \times \frac{1\ mol\ C_6H_6}{78.114\ g\ C_6H_6} = 41.830 = 42\ kJ/g\ C_6H_6$$

5.115 $\Delta E_p = m \times g \times d$. Be careful with units. $1\ J = 1\ kg\text{-}m^2/s^2$

$$201\ lb \times \frac{1\ kg}{2.205\ lb} \times \frac{9.81\ m}{s^2} \times \frac{45\ ft}{time} \times \frac{1\ yd}{3\ ft} \times \frac{1\ m}{1.0936\ yd} \times 20\ times$$

$$= 2.453 \times 10^5\ kg\text{-}m^2/s^2 = 2.453 \times 10^5\ J = 2.5 \times 10^2\ kJ$$

1 Cal = 1 kcal = 4.184 kJ

$$2.453 \times 10^2\ kJ \times \frac{1\ Cal}{4.184\ kJ} = 58.63 = 59\ Cal$$

No, if all work is used to increase the man's potential energy, 20 rounds of stair-climbing will not compensate for one extra order of 245 Cal fries. In fact, more than 59 Cal of work will be required to climb the stairs, because some energy is required to move limbs and some energy will be lost as heat.

Integrative Exercises

5.120 (a) $CH_4(g) + 2\,O_2(g) \rightarrow CO_2(g) + 2\,H_2O(l)$

$$\Delta H^\circ = \Delta H^\circ_f\ CO_2(g) + 2\ \Delta H^\circ_f\ H_2O(l) - \Delta H^\circ_f\ CH_4(g) - 2\ \Delta H^\circ_f\ O_2(g)$$

$$= -393.5\ kJ + 2(-285.83\ kJ) - (-74.8\ kJ) - 2(0) = -890.36 = -890.4\ kJ/mol\ CH_4$$

The minus sign indicates that 890.4 kJ are produced per mole of CH_4 burned.

$$\frac{890.36\ kJ}{mol\ CH_4} \times \frac{1000\ J}{1\ kJ} \times \frac{1\ mol}{6.022 \times 10^{23}\ molecules\ CH_4} = 1.4785 \times 10^{-18}$$

$$= 1.479 \times 10^{-18}\ J/molecule$$

(b) $1\text{eV} = 96.485\ \text{kJ/mol}$

$$8\ \text{keV} \times \frac{1000\ \text{eV}}{1\ \text{keV}} \times \frac{96.485\ \text{kJ}}{\text{eV-mol}} \times \frac{1\ \text{mol}}{6.022 \times 10^{23}} \times \frac{1000\ \text{J}}{\text{kJ}} = 1.282 \times 10^{-15} = 1 \times 10^{-15}\ \text{J/X-ray}$$

The energy produced by the combustion of 1 molecule of $CH_4(g)$ is much smaller than the energy of the X-ray.

5.122 (a) $\Delta H^\circ = \Delta H_f^\circ\ NaNO_3(aq) + \Delta H_f^\circ\ H_2O(l) - \Delta H_f^\circ\ HNO_3(aq) - \Delta H_f^\circ\ NaOH(aq)$

$\Delta H^\circ = -446.2\ \text{kJ} - 285.83\ \text{kJ} - (-206.6\ \text{kJ}) - (-469.6\ \text{kJ}) = -55.8\ \text{kJ}$

$\Delta H^\circ = \Delta H_f^\circ\ NaCl(aq) + \Delta H_f^\circ\ H_2O(l) - \Delta H_f^\circ\ HCl(aq) - \Delta H_f^\circ\ NaOH(aq)$

$\Delta H^\circ = -407.1\ \text{kJ} - 285.83\ \text{kJ} - (-167.2\ \text{kJ}) - (-469.6\ \text{kJ}) = -56.1\ \text{kJ}$

$\Delta H^\circ = \Delta H_f^\circ\ NH_3(aq) + \Delta H_f^\circ\ Na^+(aq) + \Delta H_f^\circ\ H_2O(l) - \Delta H_f^\circ\ NH_4^+(aq) - \Delta H_f^\circ\ NaOH(aq)$

$= -80.29\ \text{kJ} - 240.1\ \text{kJ} - 285.83\ \text{kJ} - (-132.5\ \text{kJ}) - (-469.6\ \text{kJ}) = -4.1\ \text{kJ}$

(b) $H^+(aq) + OH^-(aq) \rightarrow H_2O(l)$

(c) The ΔH° values for the first two reactions are nearly identical, –55.8 kJ and –56.1 kJ. The spectator ions by definition do not change during the course of a reaction, so ΔH° is the enthalpy change for the net ionic equation. Because the first two reactions have the same net ionic equation, it is not surprising that they have the same ΔH°.

(d) Strong acids are more likely than weak acids to donate H^+. The neutralizations of the two strong acids are energetically favorable, whereas the neutralization of $NH_4^+(aq)$ is significantly less favorable. $NH_4^+(aq)$ is probably a weak acid.

5.124 (a) $AgNO_3(aq) + NaCl(aq) \rightarrow NaNO_3(aq) + AgCl(s)$

net ionic equation: $Ag^+(aq) + Cl^-(aq) \rightarrow AgCl(s)$

$\Delta H^o = \Delta H_f^\circ\ AgCl(s) - \Delta H_f^\circ\ Ag^+(aq) - \Delta H_f^\circ Cl^-(aq)$

$\Delta H^\circ = -127.0\ \text{kJ} - (105.90\ \text{kJ}) - (-167.2\ \text{kJ}) = -65.7\ \text{kJ}$

(b) ΔH° for the complete molecular equation will be the same as ΔH° for the net ionic equation. $Na^+(aq)$ and $NO_3^-(aq)$ are spectator ions; they appear on both sides of the chemical equation. Because the overall enthalpy change is the enthalpy of the products minus the enthalpy of the reactants, the contributions of the spectator ions cancel.

(c) $\Delta H^o = \Delta H_f^\circ\ NaNO_3(aq) + \Delta H_f^\circ\ AgCl(s) - \Delta H_f^\circ\ AgNO_3(aq) - \Delta H_f^\circ\ NaCl(aq)$

$\Delta H_f^\circ\ AgNO_3(aq) = \Delta H_f^\circ\ NaNO_3(aq) + \Delta H_f^\circ\ AgCl(s) - \Delta H_f^\circ\ NaCl(aq) - \Delta H^o$

$\Delta H_f^\circ\ AgNO_3(aq) = -446.2\ \text{kJ} + (-127.0\ \text{kJ}) - (-407.1\ \text{kJ}) - (-65.7\ \text{kJ})$

$\Delta H_f^\circ\ AgNO_3(aq) = -100.4\ \text{kJ/mol}$

6 Electronic Structure of Atoms

Visualizing Concepts

6.2 Given: 2450 MHz radiation. Hz = s^{-1}, unit of frequency. M = 1×10^6; 2450×10^6 Hz = 2.45×10^9 Hz = 2.45×10^9 s^{-1}.

(a) Find 2.45×10^9 s^{-1} on the frequency axis of Figure 6.4. The wavelength that corresponds to this frequency is approximately 1×10^{-1} = 0.1 m or 10 cm.

(b) No, visible radiation has wavelengths of 4×10^{-7} to 7×10^{-7} m, much shorter than 0.1 m.

(c) Energy and wavelength are inversely proportional. Photons of the longer 0.1 m radiation have less energy than visible photons.

(d) Radiation of 0.1 m is in the low energy portion of the microwave region. The appliance is probably a microwave oven. (Appliances with heating elements that glow red or orange give off wavelengths in the visible or the near visible portion of the infrared. The 0.1 m wavelength is too long to belong to these appliances.)

6.5 (a) Increase. The rainbow has shorter wavelength blue light on the inside and longer wavelength red light on the outside. (See Figure 6.4.)

(b) Decrease. Wavelength and frequency are inversely related. Wavelength increases so frequency decreases going from the inside to the outside of the rainbow.

6.9 (a) $l = 1$

(b) $3p_y$ (n = 3 shell, dumbbell shape, node at the nucleus, oriented along the y-axis)

(c) (iii) (m_l indicates the orientation of an orbital)

The Wave Nature of Light (Section 6.1)

6.13 (a) Meters (m) (b) 1/seconds (s^{-1}) (c) meters/second (m-s^{-1} or m/s)

6.15 (a) True.

(b) False. Ultraviolet light has shorter wavelengths than visible light. [See Solution 6.14(b).]

(c) False. X-rays travel at the same speed as microwaves. (X-rays and microwaves are both electromagnetic radiation.)

(d) False. Electromagnetic radiation and sound waves travel at different speeds. (Sound is not a form of electromagnetic radiation.)

6.17 *Analyze/Plan.* Use the electromagnetic spectrum in Figure 6.4 to determine the wavelength of each type of radiation; put them in order from shortest to longest wavelength. *Solve.*

Wavelength of X-rays < ultraviolet < green light < red light < infrared < radio waves

Check. These types of radiation should read from left to right on Figure 6.4.

6.19 *Analyze/Plan.* These questions involve relationships between wavelength, frequency, and the speed of light. Manipulate the equation $\nu = c/\lambda$ to obtain the desired quantities, paying attention to units. *Solve.*

(a) $\nu = c/\lambda;\ \frac{2.998 \times 10^8 \text{ m}}{\text{s}} \times \frac{1}{10\ \mu\text{m}} \times \frac{1\ \mu\text{m}}{1 \times 10^{-6} \text{ m}} = 3.0 \times 10^{13} \text{ s}^{-1}$

(b) $\lambda = c/\nu;\ \frac{2.998 \times 10^8 \text{ m}}{s} \times \frac{1 \text{ s}}{5.50 \times 10^{14}} = 5.45 \times 10^{-7} \text{ m} \quad (545 \text{ nm})$

(c) The radiation in (b) is in the visible region and is "visible" to humans.
The 10 μm (1×10^{-5} m) radiation in (a) is in the infrared region and is not visible.

(d) $50.0\ \mu\text{s} \times \frac{1 \text{ s}}{1 \times 10^6\ \mu\text{s}} \times \frac{2.998 \times 10^8 \text{ m}}{\text{s}} = 1.50 \times 10^4 \text{ m}$

Check. Confirm that powers of 10 make sense and units are correct.

6.21 *Analyze/Plan.* $\nu = c/\lambda$; change nm → m.

Solve. $\nu = c/\lambda;\ \frac{2.998 \times 10^8 \text{ m}}{1 \text{ s}} \times \frac{1}{650 \text{ nm}} \times \frac{1 \text{ nm}}{1 \times 10^{-9} \text{ m}} = 4.6 \times 10^{14} \text{ s}^{-1}$

The color is red.

Check. $(3000 \times 10^5 / 650 \times 10^{-9}) \approx 4.5 \times 10^{14} \text{ s}^{-1}$; units are correct.

Quantized Energy and Photons (Section 6.2)

6.23 (iii) Quantization means that energy changes can only happen in certain allowed increments. If the human growth quantum is one-foot, growth occurs instantaneously in one-foot increments.

6.25 *Analyze/Plan.* These questions deal with the relationships between energy, wavelength, and frequency. Use the relationships $E = h\nu = hc/\lambda$ to calculate the desired quantities. Pay attention to units. *Solve.*

(a) $E = h\nu = 6.626 \times 10^{-34} \text{ J-s} \times 2.94 \times 10^{14} \text{ s}^{-1} = 1.95 \times 10^{-19} \text{ J}$

(b) $E = h\nu = \frac{hc}{\lambda} = \frac{6.626 \times 10^{-34} \text{ J-s} \times 2.998 \times 10^8 \text{ m/s}}{413 \text{ nm}} \times \frac{1 \text{ nm}}{1 \times 10^{-9} \text{ m}} = 4.81 \times 10^{-19} \text{ J}$

(c) $\lambda = hc/\Delta E = \frac{6.626 \times 10^{-34} \text{ J-s}}{6.06 \times 10^{-19} \text{ J}} \times \frac{2.998 \times 10^8 \text{ m}}{1 \text{ s}} = 3.28 \times 10^{-7} \text{ m} = 328 \text{ nm}$

6.27 *Analyze/Plan*. Use E = hc/λ; pay close attention to units. *Solve*.

(a) $$E = hc/\lambda = 6.626 \times 10^{-34}\ \text{J-s} \times \frac{2.998 \times 10^8\ \text{m}}{1\ \text{s}} \times \frac{1}{3.3\ \mu\text{m}} \times \frac{1\ \mu\text{m}}{1 \times 10^{-6}\ \text{m}} = 6.0 \times 10^{-20}\ \text{J}$$

$$E = hc/\lambda = 6.626 \times 10^{-34}\ \text{J-s} \times \frac{2.998 \times 10^8\ \text{m}}{1\ \text{s}} \times \frac{1}{0.154\ \text{nm}} \times \frac{1\ \text{nm}}{1 \times 10^{-9}\ \text{m}} = 1.29 \times 10^{-15}\ \text{J}$$

Check. $(6.6 \times 3/3.3) \times (10^{-34} \times 10^{8}/10^{-6}) \approx 6 \times 10^{-20}$ J

$(6.6 \times 3/0.15) \times (10^{-34} \times 10^{8}/10^{-9}) \approx 120 \times 10^{-17} \approx 1.2 \times 10^{-15}$ J

The results are reasonable. We expect the longer wavelength 3.3 μm radiation to have the lower energy.

(b) The 3.3 μm photon is in the infrared and the 0.154 nm (1.54×10^{-10} m) photon is in the X-ray region; the X-ray photon has the greater energy.

6.29 *Analyze/Plan*. Use E = hc/λ to calculate J/photon; Avogadro's number to calculate J/mol; photon/J [the result from part (a)] to calculate photons in 1.00 mJ. Pay attention to units. *Solve*.

(a) $$E_{photon} = hc/\lambda = \frac{6.626 \times 10^{-34}\ \text{J-s}}{325 \times 10^{-9}\ \text{m}} \times \frac{2.998 \times 10^8\ \text{m}}{\text{s}} = 6.1122 \times 10^{-19} = 6.11 \times 10^{-19}\ \text{J/photon}$$

(b) $$\frac{6.1122 \times 10^{-19}\ \text{J}}{1\ \text{photon}} \times \frac{6.022 \times 10^{23}\ \text{photons}}{1\ \text{mol}} = 3.68 \times 10^5\ \text{J/mol} = 368\ \text{kJ/mol}$$

(c) $$\frac{1\ \text{photon}}{6.1122 \times 10^{-19}\ \text{J}} \times 1.00\ \text{mJ} \times \frac{1 \times 10^{-3}\ \text{J}}{1\ \text{mJ}} = 1.64 \times 10^{15}\ \text{photons}$$

Check. Powers of 10 (orders of magnitude) and units are correct.

(d) If the energy of one 325 nm photon breaks exactly one bond, 1 mol of photons break 1 mol of bonds. The average bond energy in kJ/mol is the energy of 1 mol of photons (from part b) 368 kJ/mol.

6.31 *Analyze/Plan*. E = hc/λ gives J/photon. Use this result with J/s (given) to calculate photons/s. *Solve*.

(a) The ~1×10^{-6} m radiation is infrared but very near the visible edge.

(b) $$E_{photon} = hc/\lambda = \frac{6.626 \times 10^{-34}\ \text{J-s}}{987 \times 10^{-9}\ \text{m}} \times \frac{2.998 \times 10^8\ \text{m}}{1\ \text{s}} = 2.0126 \times 10^{-19} = 2.01 \times 10^{-19}\ \text{J/photon}$$

$$\frac{0.52\ \text{J}}{32\ \text{s}} \times \frac{1\ \text{photon}}{2.0126 \times 10^{-19}\ \text{J}} = 8.1 \times 10^{16}\ \text{photons/s}$$

Check. $(7 \times 3/1000) \times (10^{-34} \times 10^{8}/10^{-9}) \approx 21 \times 10^{-20} \approx 2.1 \times 10^{-19}$ J/photon

$(0.5/30/2) \times (1/10^{-19}) = 0.008 \times 10^{19} = 8 \times 10^{16}$ photons/s

Units are correct; powers of 10 are reasonable.

6.33 *Analyze/Plan.* Use E = hν and ν = c/λ. Calculate the desired characteristics of the photons. Assume 1 photon interacts with 1 electron. Compare E_{min} and E_{120} to calculate maximum kinetic energy of the emitted electron. *Solve.*

(a) $E = h\nu = 6.626 \times 10^{-34}\ \text{J-s} \times 1.09 \times 10^{15}\ \text{s}^{-1} = 7.22 \times 10^{-19}\ \text{J}$

(b) $\lambda = c/\nu = \dfrac{2.998 \times 10^{8}\ \text{m}}{1\ \text{s}} \times \dfrac{1\ \text{s}}{1.09 \times 10^{15}} = 2.75 \times 10^{-7}\ \text{m} = 275\ \text{nm}$

(c) $E_{120} = hc/\lambda = 6.626 \times 10^{-34}\ \text{J-s} \times \dfrac{2.998 \times 10^{8}\ \text{m}}{1\ \text{s}} \times \dfrac{1}{120\ \text{nm}} \times \dfrac{1\ \text{nm}}{1 \times 10^{-9}\ \text{m}}$

$= 1.655 \times 10^{-18} = 1.66 \times 10^{-18}\ \text{J}$

The excess energy of the 120 nm photon is converted into the kinetic energy of the emitted electron.

$E_k = E_{120} - E_{min} = 16.55 \times 10^{-19}\ \text{J} - 7.22 \times 10^{-19}\ \text{J} = 9.3 \times 10^{-19}\ \text{J/electron}$

Check. E_{120} must be greater than E_{min} in order for the photon to impart kinetic energy to the emitted electron. Our calculations are consistent with this requirement.

Bohr's Model; Matter Waves (Sections 6.3 and 6.4)

6.35 When the electron in a hydrogen atom transitions from $n = 1$ to $n = 3$, the atom "expands." The average distance from the nucleus of an $n = 3$ electron is greater than that of an $n = 1$ electron. The volume where there is a significant probability of finding an electron increases.

6.37 *Analyze/Plan.* An isolated electron is assigned an energy of zero; the closer the electron comes to the nucleus, the more negative its energy. Thus, as an electron moves closer to the nucleus, the energy of the electron decreases and the excess energy is emitted. Conversely, as an electron moves further from the nucleus, the energy of the electron increases and energy must be absorbed. *Solve.*

(a) As the principle quantum number decreases, the electron moves toward the nucleus and energy is emitted.

(b) An increase in the radius of the orbit means the electron moves away from the nucleus; energy is absorbed.

(c) An isolated electron is assigned an energy of zero. As the electron moves to the $n = 3$ state closer to the H^+ nucleus, its energy becomes more negative (decreases) and energy is emitted.

6.39 *Analyze/Plan.* Equation 6.5: $E = (-2.18 \times 10^{-18}\ \text{J})(1/n^2)$. *Solve.*

(a) $E_2 = -2.18 \times 10^{-18}\ \text{J}/(2)^2 = -5.45 \times 10^{-19}\ \text{J}$

$E_6 = -2.18 \times 10^{-18}\ \text{J}/(6)^2 = -6.0556 \times 10^{-20} = -0.606 \times 10^{-19}\ \text{J}$

$\Delta E = E_2 - E_6 = (-5.45 \times 10^{-19}\ \text{J}) - (-0.606 \times 10^{-19}\ \text{J})$

$= -4.844 \times 10^{-19}\ \text{J} = -4.84 \times 10^{-19}\ \text{J}$

$\lambda = hc/\Delta E = \dfrac{6.626 \times 10^{-34}\ \text{J-s}}{4.844 \times 10^{-19}\ \text{J}} \times \dfrac{2.998 \times 10^{8}\ \text{m}}{\text{s}} = 4.10 \times 10^{-7}\ \text{m} = 410\ \text{nm}$

(b) The visible range is 400–700 nm, so this line is visible; the observed color is violet.

Check. We expect E_6 to be a more positive (or less negative) than E_2, and it is. ΔE is negative, which indicates emission. The orders of magnitude make sense and units are correct.

6.41 (a) Statement (ii) is the best explanation. Only lines with $n_f = 2$ represent ΔE values and wavelengths that lie in the visible portion of the spectrum. Lines with $n_f = 3$ have smaller ΔE values and lie in the lower energy, longer wavelength infrared portion of the electromagnetic spectrum.

(b) *Analyze/Plan.* Use Equation 6.6 to calculate ΔE, then $\lambda = hc/\Delta E$. *Solve.*

$$n_i = 3, n_f = 2; \quad \Delta E = -2.18 \times 10^{-18}\ \text{J}\left[\frac{1}{n_f^2} - \frac{1}{n_i^2}\right] = -2.18 \times 10^{-18}\ \text{J}\,(1/4 - 1/9)$$

$$\lambda = hc/E = \frac{6.626 \times 10^{-34}\ \text{J-s} \times 2.998 \times 10^8\ \text{m/s}}{-2.18 \times 10^{-18}\ \text{J}\,(1/4 - 1/9)} = 6.56 \times 10^{-7}\ \text{m}$$

This is the red line at 656 nm.

$$n_i = 4, n_f = 2; \quad \lambda = hc/E = \frac{6.626 \times 10^{-34}\ \text{J-s} \times 2.998 \times 10^8\ \text{m/s}}{-2.18 \times 10^{-18}\ \text{J}\,(1/4 - 1/16)} = 4.86 \times 10^{-7}\ \text{m}$$

This is the blue-green line at 486 nm.

$$n_i = 5, n_f = 2; \quad \lambda = hc/E = \frac{6.626 \times 10^{-34}\ \text{J-s} \times 2.998 \times 10^8\ \text{m/s}}{-2.18 \times 10^{-18}\ \text{J}\,(1/4 - 1/25)} = 4.34 \times 10^{-7}\ \text{m}$$

This is the blue-violet line at 434 nm.

Check. The calculated wavelengths correspond well to three lines in the H emission spectrum in Figure 6.11, so the results are sensible.

6.43 (a) $93.07\ \text{nm} \times \dfrac{1 \times 10^{-9}\ \text{m}}{1\ \text{nm}} = 9.307 \times 10^{-8}\ \text{m}$; this line is in the ultraviolet region.

(b) *Analyze/Plan.* Only lines with $n_f = 1$ have a large enough ΔE to lie in the ultraviolet region (see Solutions 6.41 and 6.42). Solve Equation 6.6 for n_i, recalling that ΔE is negative for emission. *Solve.*

$$\frac{-hc}{\lambda} = -2.18 \times 10^{-18}\ \text{J}\left[\frac{1}{n_f^2} - \frac{1}{n_i^2}\right]; \quad \frac{hc}{\lambda(2.18 \times 10^{-18}\ \text{J})} = \left[1 - \frac{1}{n_i^2}\right]$$

$$-\frac{1}{n_i^2} = \left[\frac{hc}{\lambda(2.18 \times 10^{-18}\ \text{J})} - 1\right]; \frac{1}{n_i^2} = \left[1 - \frac{hc}{\lambda(2.18 \times 10^{-18}\ \text{J})}\right]$$

$$n_i^2 = \left[1 - \frac{hc}{\lambda(2.18 \times 10^{-18}\ \text{J})}\right]^{-1}; n_i = \left[1 - \frac{hc}{\lambda(2.18 \times 10^{-18}\ \text{J})}\right]^{-1/2}$$

$$n_i = \left(1 - \frac{6.626 \times 10^{-34}\ \text{J-s} \times 2.998 \times 10^8\ \text{m/s}}{9.307 \times 10^{-8}\ \text{m} \times 2.18 \times 10^{-18}\ \text{J}}\right)^{-1/2} = 7\ (n \text{ values must be integers})$$

$n_i = 7, n_f = 1$

Check. From Solution 6.42, we know that $n_i > 4$ for $\lambda = 93.07$ nm. The calculated result is close to 7, so the answer is reasonable.

6.45 *Analyze/Plan.* According to Equation 6.6 and several preceding solutions, the greater the energy of light absorbed, the lower the value of n_i. The greater the energy of light, the greater its frequency, $\Delta E = h\nu$.

Solve. The order of increasing frequency (and energy) of light absorbed is:

$n = 4$ to $n = 9 < n = 3$ to $n = 6 < n = 2$ to $n = 3 < n = 1$ to $n = 2$

6.47 *Analyze/Plan.* $\lambda = \frac{h}{mv}$; $1\ J = \frac{1\ kg\text{-}m^2}{s^2}$;

Change mass to kg and velocity to m/s in each case. *Solve.*

(a) $$\frac{50\ km}{1\ hr} \times \frac{1000\ m}{1\ km} \times \frac{1\ hr}{60\ min} \times \frac{1\ min}{60\ s} = 13.89 = 14\ m/s$$

$$\lambda = \frac{6.626 \times 10^{-34}\ kg\text{-}m^2\text{-}s}{1\ s^2} \times \frac{1}{85\ kg} \times \frac{1\ s}{13.89\ m} = 5.6 \times 10^{-37}\ m$$

(b) $$10.0\ g \times \frac{1\ kg}{1000\ g} = 0.0100\ kg$$

$$\lambda = \frac{6.626 \times 10^{-34}\ kg\text{-}m^2\text{-}s}{1\ s^2} \times \frac{1}{0.0100\ kg} \times \frac{1\ s}{250\ m} = 2.65 \times 10^{-34}\ m$$

(c) We need to calculate the mass of a single Li atom in kg.

$$\frac{6.94\ g\ Li}{1\ mol\ Li} \times \frac{1\ kg}{1000\ g} \times \frac{1\ mol}{6.022 \times 10^{23}\ Li\ atoms} = 1.152 \times 10^{-26} = 1.15 \times 10^{-26}\ kg$$

$$\lambda = \frac{6.626 \times 10^{-34}\ kg\text{-}m^2\text{-}s}{1\ s^2} \times \frac{1}{1.152 \times 10^{-26}\ kg} \times \frac{1\ s}{2.5 \times 10^{5}\ m} = 2.3 \times 10^{-13}\ m$$

(d) Calculate the mass of a single O_3 molecule in kg.

$$\frac{48.00\ g\ O_3}{1\ mol\ O_3} \times \frac{1\ kg}{1000\ g} \times \frac{1\ mol}{6.022 \times 10^{23}\ O_3\ molecules} = 7.971 \times 10^{-26}$$

$$= 7.97 \times 10^{-26}\ kg$$

$$\lambda = \frac{6.626 \times 10^{-34}\ kg\text{-}m^2\text{-}s}{1\ s^2} \times \frac{1}{7.971 \times 10^{-26}\ kg} \times \frac{1\ s}{550\ m}$$

$$= 1.51 \times 10^{-11}\ m\ (15\ pm)$$

6.49 *Analyze/Plan.* Use $v = h/m\lambda$; change wavelength to meters and mass of neutron (back inside cover) to kg. *Solve.*

$$\lambda = 1.25\ \text{Å} \times \frac{1 \times 10^{-10}\ m}{1\ \text{Å}} = 1.25 \times 10^{-10}\ m;\ mass = 1.6749 \times 10^{-27}\ kg$$

$$v = \frac{6.626 \times 10^{-34}\ kg\text{-}m^2\text{-}s}{1\ s^2} \times \frac{1}{1.6749 \times 10^{-27}\ kg} \times \frac{1}{1.25 \times 10^{-10}\ m} = 3.16 \times 10^{3}\ m/s$$

Check. $(6.6/1.6/1.25) \times (10^{-34}/10^{-27}/10^{-10}) \approx 3 \times 10^{3}\ m/s$

6.51 *Analyze/Plan.* Use $\Delta x \geq h/4\pi m \Delta v$, paying attention to appropriate units. Note that the uncertainty in speed of the particle (Δv) is important, rather than the speed itself. *Solve.*

(a) $m = 1.50 \text{ mg} \times \frac{1 \text{ g}}{1000 \text{ mg}} \times \frac{1 \text{ kg}}{1000 \text{ g}} = 1.50 \times 10^{-6} \text{ kg}; \Delta v = 0.01 \text{ m/s}$

$$\Delta x \geq = \frac{6.626 \times 10^{-34} \text{ J-s}}{4\pi(1.50 \times 10^{-6} \text{ kg})(0.01 \text{ m/s})} \geq 3.52 \times 10^{-27} = 4 \times 10^{-27} \text{ m}$$

(b) $m = 1.673 \times 10^{-24} \text{ g} = 1.673 \times 10^{-27} \text{ kg}; \Delta v = 0.01 \times 10^{4} \text{ m/s}$

$$\Delta x \geq = \frac{6.626 \times 10^{-34} \text{ J-s}}{4\pi(1.673 \times 10^{-27} \text{ kg})(0.01 \times 10^{4} \text{ m/s})} \geq 3 \times 10^{-10} \text{ m}$$

Check. The more massive particle in (a) has a much smaller uncertainty in position.

Quantum Mechanics and Atomic Orbitals (Sections 6.5 and 6.6)

6.53 (a) False. The contour representation shows a volume where there is significant electron density. The electron can be moving anywhere within this volume.

(b) False. The probability of finding an electron at a given distance from the nucleus is the radial probability function. This is the probability density summed over all points that lie a distance r from the nucleus.

6.55 (a) The possible values of l are $(n - 1)$ to 0. $n = 4$, $l = 3, 2, 1, 0$

(b) The possible values of m_l are $-l$ to $+l$. $l = 2$, $m_l = -2, -1, 0, 1, 2$

(c) Because the value of m_l is less than or equal to the value of l, $m_l = 2$ must have an l-value greater than or equal to 2. In terms of elements that have been observed, the possibilities are 2, 3, and 4.

6.57 (a) 3p: $n = 3, l = 1$ (b) 2s: $n = 2, l = 0$

(c) 4f: $n = 4, l = 3$ (d) 5d: $n = 5, l = 2$

6.59 (a) 2, 1 ,0, –1, –2 (b) ½, –½

6.61 Impossible: (a) 1p, only $l = 0$ is possible for $n = 1$; (d) 2d, for $n = 2$, $l = 1$ or 0, but not 2

6.63

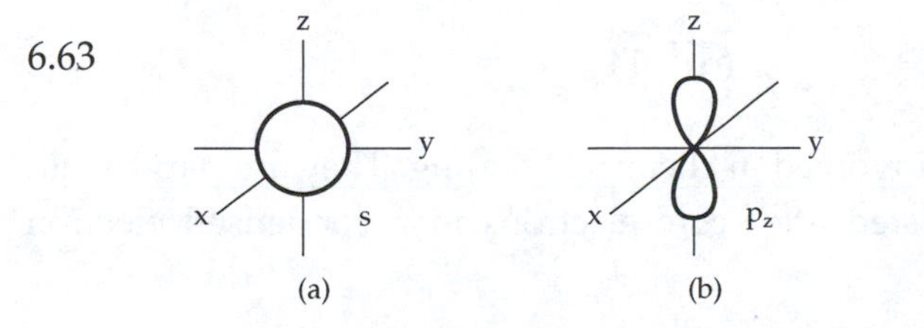

Note that the lobes of the d_{xy} orbital lie in the xy-plane and point between the x-axes and y-axes.

6.65 (a) The 1s and 2s orbitals of an H atom have the same overall spherical shape. The 2s orbital has a larger radial extension and one node, whereas the 1s orbital has continuous electron density. Because the 2s orbital is "larger," there is greater probability of finding an electron farther from the nucleus in the 2s orbital.

(b) A single 2p orbital is directional in that its electron density is concentrated along one of the three Cartesian axes of the atom. The $d_{x^2-y^2}$ orbital has electron density along both the x- and y-axes, whereas the p_x orbital has density only along the x-axis.

(c) The average distance of an electron from the nucleus in a 3s orbital is greater than for an electron in a 2s orbital. In general, for the same kind of orbital, the larger the n value, the greater the average distance of an electron from the nucleus of the atom.

(d) 1s < 2p < 3d < 4f < 6s. In the H atom, orbitals with the same n value are degenerate and energy increases with increasing n value. Thus, the order of increasing energy is given above.

Many-Electron Atoms and Electron Configurations (Sections 6.7 – 6.9)

6.67 (a) Yes. An He^+ ion has only one electron. In a one-electron particle, the energy of an orbital depends only on the value of n. Therefore, the 2s and 2p orbitals of He^+ have the same energy.

(b) Yes. In an He atom, the 2s orbital is lower in energy than the 2p orbitals. A helium atom is a two-electron particle. In multi-electron particles, electron-electron repulsions cause orbitals with the same n-value but different l-values to have different energies.

6.69 (a) No, both configurations follow the Pauli exclusion principle.

(b) No, both configurations obey Hund's rule.

(c) No. In the absence of a magnetic field, we cannot say which configuration has lower energy. The difference in the two configurations is the m_s value of the electron in the 2s orbital. In the absence of an external magnetic field, electrons that only differ in m_s have the same energy.

6.71 *Analyze/Plan.* Each subshell has an l-value associated with it. For a particular l-value, permissible m_l-values are $-l$ to $+l$. Each m_l-value represents an orbital, which can hold two electrons. *Solve.*

(a) 6 (b) 10 (c) 2 (d) 14

6.73 (a) "Valence electrons" are those involved in chemical bonding. They are part (or all) of the outer-shell electrons listed after core electrons in a condensed electron configuration.

(b) "Core electrons" are inner shell electrons that have the electron configuration of the nearest noble-gas element.

(c) Each box represents an orbital.

(d) Each half-arrow in an orbital diagram represents an electron. The direction of the half-arrow represents electron spin.

6.75 *Analyze/Plan*. Follow the logic in Sample Exercise 6.9. *Solve.*

(a) Cs: $[Xe]6s^1$

(b) Ni: $[Ar]4s^2 3d^8$

(c) Se: $[Ar]4s^2 3d^{10} 4p^4$

(d) Cd: $[Kr]5s^2 4d^{10}$

(e) U: $[Rn]5f^3 6d^1 7s^2$. (Note the U and several other *f*-block elements have irregular *d*- and *f*-electron orders.)

(f) Pb: $[Xe]6s^2 4f^{14} 5d^{10} 6p^2$

6.79 (a) The orbitals are not filled in order of increasing energy. The fifth electron would fill the 2p subshell (same *n*-value as 2s) before the 3s.

(b) The orbitals are not filled in order of increasing energy. After 6s, electrons fill the 4f subshell.

(c) The orbitals are not filled in order of increasing energy. The 3p subshell would fill before the 3d because it has the lower *l*-value and the same *n*-value. (If there were more electrons, 4s would also fill before 3d.)

Additional Exercises

6.81 (a) $\lambda_A = 1.6 \times 10^{-7}\text{ m} / 4.5 = 3.56 \times 10^{-8} = 3.6 \times 10^{-8}\text{ m}$

$\lambda_B = 1.6 \times 10^{-7}\text{ m} / 2 = 8.0 \times 10^{-8}\text{ m}$

(b) $\nu = c/\lambda;\ \nu_A = \dfrac{2.998 \times 10^8\text{ m}}{1\text{ s}} \times \dfrac{1}{3.56 \times 10^{-8}\text{ m}} = 8.4 \times 10^{15}\text{ s}^{-1}$

$$\nu_B = \frac{2.998 \times 10^8\text{ m}}{1\text{ s}} \times \frac{1}{8.0 \times 10^{-8}\text{ m}} = 3.7 \times 10^{15}\text{ s}^{-1}$$

(c) A: ultraviolet, B: ultraviolet

6.84 All electromagnetic radiation travels at the same speed, 2.998×10^8 m/s. Change miles to meters and seconds to some appropriate unit of time.

$$391 \times 10^6\text{ mi} \times \frac{1.6093\text{ km}}{1\text{ mi}} \times \frac{1000\text{ m}}{1\text{ km}} \times \frac{1\text{ s}}{2.998 \times 10^8\text{ m}} \times \frac{1\text{ min}}{60\text{ s}} = 35.0\text{ min}$$

6.86 $E = hc/\lambda \rightarrow$ J/photon; total energy = power × time; photons = total energy / J / photon

$$E = \frac{6.626 \times 10^{-34}\text{ J-s} \times 2.998 \times 10^8\text{ m/s}}{780 \times 10^{-9}\text{ m}} = 2.5468 \times 10^{-19} = 2.55 \times 10^{-19}\text{ J/photon}$$

$$0.10\text{ mW} = \frac{0.10 \times 10^{-3}\text{ J}}{1\text{ s}} \times 69\text{ min} \times \frac{60\text{ s}}{1\text{ min}} = 0.4140 = 0.41\text{ J}$$

$$0.4140\text{ J} \times \frac{1\text{ photon}}{2.5468 \times 10^{-19}\text{ J}} = 1.626 \times 10^{18} = 1.6 \times 10^{18}\text{ photons}$$

6.91 (a) Lines with $n_f = 1$ lie in the ultraviolet (see Solution 6.42) and with $n_f = 2$ lie in the visible (see Solution 6.41). Lines with $n_f = 3$ will have smaller ΔE and longer wavelengths and lie in the infrared.

(b) Use Equation 6.6 to calculate ΔE, then $\lambda = hc/\Delta E$.

$$n_i = 4, n_f = 3; \Delta E = -2.18 \times 10^{-18} \text{ J}\left[\frac{1}{n_f^2} - \frac{1}{n_i^2}\right] = -2.18 \times 10^{-18} \text{ J} (1/9 - 1/16)$$

$$\lambda = hc/E = \frac{6.626 \times 10^{-34} \text{ J-s} \times 2.998 \times 10^8 \text{ m/s}}{-2.18 \times 10^{-18} (1/9 - 1/16)} = 1.87 \times 10^{-6} \text{ m}$$

$$n_i = 5, n_f = 3; \lambda = hc/E = \frac{6.626 \times 10^{-34} \text{ J-s} \times 2.998 \times 10^8 \text{ m/s}}{-2.18 \times 10^{-18} (1/9 - 1/25)} = 1.28 \times 10^{-6} \text{ m}$$

$$n_i = 6, n_f = 3; \lambda = hc/E = \frac{6.626 \times 10^{-34} \text{ J-s} \times 2.998 \times 10^8 \text{ m/s}}{-2.18 \times 10^{-18} (1/9 - 1/36)} = 1.09 \times 10^{-6} \text{ m}$$

These three wavelengths are all greater than 1 μm or 1×10^{-6} m. They are in the infrared, close to the visible edge (0.7×10^{-6} m).

6.95 *Plan.* Calculate v from kinetic energy. $\lambda = h/mv$. *Solve.*

$$E_k = mv^2/2; v^2 = 2E_k/m; v = \sqrt{2E_k/m}$$

$$v = \left(\frac{2 \times 2.147 \times 10^{-15} \text{ kg-m}^2/\text{s}^2}{9.1094 \times 10^{-31} \text{ kg}}\right)^{1/2} = 6.866 \times 10^7 = 6.87 \times 10^7 \text{ m/s}$$

$$\lambda = h/mv = \frac{6.626 \times 10^{-34} \text{ J-s}}{9.1094 \times 10^{-31} \text{ kg} \times 6.866 \times 10^7 \text{ m/s}} \times \frac{1 \text{ kg-m}^2/\text{s}^2}{1 \text{ J}} = 1.06 \times 10^{-11} \text{ m} = 10.6 \text{ pm}$$

6.99 (a) The p_z orbital has a nodal plane where z = 0. This is the xy plane.

(b) The d_{xy} orbital has 4 lobes and 2 nodal planes, the 2 planes where x = 0 and y = 0. These are the yz and xz planes.

(c) The $d_{x^2-y^2}$ has 4 lobes and 2 nodal planes, the planes where $x^2 - y^2 = 0$. These are the planes that bisect the x and y axes and contain the z axis.

6.102 (a) Br: $[Ar]4s^2 3d^{10} 4p^5$, 1 unpaired electron

(b) Ga: $[Ar]4s^2 3d^{10} 4p^1$, 1 unpaired electron

(c) Hf: $[Xe]6s^2 4f^{14} 5d^2$, 2 unpaired electrons

(d) Sb: $[Kr]5s^2 4d^{10} 5p^3$, 3 unpaired electrons

(e) Bi: $[Xe]6s^2 4f^{14} 5d^{10} 6p^3$, 3 unpaired electrons

(f) Sg: $[Rn]7s^2 5f^{14} 6d^4$, 4 unpaired electrons

Integrative Exercises

6.105 (a) We know the wavelength of microwave radiation, the volume of coffee to be heated, and the desired temperature change. Assume the density and heat capacity of coffee are the same as pure water. We need to calculate: (i) the total energy required to heat the coffee and (ii) the energy of a single photon to find (iii) the number of photons required.

(i) From Chapter 5, the heat capacity of liquid water is 4.184 J/g-°C.

To find the mass of 200 mL of coffee at 23 °C, use the density of water given in Appendix B.

$$200 \text{ mL} \times \frac{0.997 \text{ g}}{1 \text{ mL}} = 199.4 = 199 \text{ g coffee}$$

$$\frac{4.184 \text{ J}}{1 \text{ g-°C}} \times 199.4 \text{ g} \times (60 \text{ °C} - 23 \text{ °C}) = 3.087 \times 10^4 \text{ J} = 31 \text{ kJ}$$

(ii) $$E = hc/\lambda = 6.626 \times 10^{-34} \text{ J-s} \times \frac{2.998 \times 10^8 \text{ m}}{1 \text{ s}} \times \frac{1}{0.112 \text{ m}} = \frac{1.77 \times 10^{-24} \text{ J}}{1 \text{ photon}}$$

(iii) $$3.087 \times 10^4 \text{ J} \times \frac{1 \text{ photon}}{1.774 \times 10^{-24} \text{ J}} = 1.7 \times 10^{28} \text{ photons}$$

(The answer has 2 sig figs because the temperature change, 37 °C, has 2 sig figs.)

(b) 1 W = 1 J/s. 900 W = 900 J/s. From part (a), 31 kJ are required to heat the coffee.

$$3.087 \times 10^4 \text{ J} \times \frac{1 \text{ s}}{900 \text{ J}} = 34.30 = 34 \text{ s}$$

6.109 (a) Bohr's theory was based on the Rutherford "nuclear" model of the atom. That is, Bohr theory assumed a dense positive charge at the center of the atom and a diffuse negative charge (electrons) surrounding it. Bohr's theory then specified the nature of the diffuse negative charge. The prevailing theory before the nuclear model was Thomson's plum pudding or watermelon model, with discrete electrons scattered about a diffuse positive charge cloud. Bohr's theory could not have been based on the Thomson model of the atom.

(b) De Broglie's hypothesis is that electrons exhibit both particle and wave properties. Thomson's conclusion that electrons have mass is a particle property, whereas the nature of cathode rays is a wave property. De Broglie's hypothesis actually rationalizes these two seemingly contradictory observations about the properties of electrons.

7 Periodic Properties of the Elements

Visualizing Concepts

7.2 The order of radii is $Br^- > Br > F$, so the largest brown sphere is Br^-, the intermediate blue one is Br, and the smallest red one is F.

7.5 (a) The bonding atomic radius of A, r_A, is $d_1/2$. The distance d_2 is the sum of the bonding atomic radii of A and X, $r_A + r_X$. We know that $r_A = d_1/2$, so $d_2 = r_X + d_1/2$, $r_X = d_2 - d_1/2$.

(b) The length of the X–X bond is $2r_X$.

$2r_X = 2\,(d_2 - d_1/2) = 2d_2 - d_1$.

7.8 (a) $X + 2\,F_2 \rightarrow XF_4$

(b) If X is a nonmetal, XF_4 is a molecular compound. If X is a metal, XF_4 is ionic. For an ionic compound with this formula, X would have a charge of 4+, and a much smaller bonding atomic radius than F^-. X in the diagram has about the same bonding radius as F, so it is likely to be a nonmetal.

Periodic Table; Effective Nuclear Charge (Sections 7.1 and 7.2)

7.9 (a) The results are 2, 8, 18, 32.

(b) The atomic numbers of the noble gases are 2, 10, 18, 36, 54, and 86. The differences between sequential pairs of these atomic numbers is 8, 8, 18, 18, and 32. These differences correspond to the results in (a). They represent the filling of new subshells when moving across the next row of the periodic chart.

(c) The Pauli exclusion principle is the source of the "2" in the expressions in part (a). The Pauli principle states that no two electrons can have the same four quantum numbers. Because m_s has only two possible values, the consequence is that an atomic orbital can hold a maximum of "2" electrons.

7.11 (a) According to Figure 7.1, of the elements listed, only Fe was known before 1700.

(b) The seven metals known in ancient times, Fe, Cu, Ag, Sn, Au, Hg, and Pb, are mostly near the bottom of the activity series, Table 4.5. These less active metals are present in nature in elemental form; they can be observed directly and their isolation does not require chemical processing.

7.13 *Analyze/Plan.* Z_{eff} values for elements 3–18 are shown as the green line in Figure 7.5. Use Equation 7.1 to calculate Z_{eff} values for elements 1 and 2, H and He. Compare the values. *Solve.*

$Z_{eff} = Z - S$, where Z is the atomic number and S is the number of core electrons. For H and He, the valence electrons have n = 1, and there are no core electrons; S = 0.

H: 1 – 0 = 1; He: 2 – 0 = 2.

The minimum value of Z_{eff} from Figure 7.5 is 1 for Li and Na. The maximum value is 8 for Ne and Ar.

For elements 1–18, H, Li and Na have minimum values of Z_{eff}; Ne and Ar have maximum values.

7.15 (a) *Analyze/Plan.* $Z_{eff} = Z - S$. Find the atomic number, Z, of Na and K. Write their electron configurations and count the number of core electrons. Assume S = number of core electrons.

Solve. Na: Z = 11; $[Ne]3s^1$. In the Ne core there are 10 electrons. $Z_{eff} = 11 - 10 = 1$. K: Z = 19; $[Ar]4s^1$. In the Ar core there are 18 electrons. $Z_{eff} = 19 - 18 = 1$.

(b) *Analyze/Plan.* $Z_{eff} = Z - S$. Write the complete electron configuration for each element to show counting for Slater's rules. S = 0.35 [# of electrons with same *n*] + 0.85 [# of electrons with (*n*–1)] + 1[# of electrons with (*n*–2)].

Solve. Na: $1s^2 2s^2 2p^6 3s^1$. S = 0.35(0) + 0.85(8) + 1(2) = 8.8. $Z_{eff} = 11 - 8.8 = 2.2$

K: $1s^2 2s^2 2p^6 3s^2 3p^6 4s^1$. S = 0.35(0) + 0.85(8) + 1(10) = 16.8. $Z_{eff} = 19 - 16.8 = 2.2$

(c) For both Na and K, the two values of Z_{eff} are 1.0 and 2.2. The Slater value of 2.2 is closer to the values of 2.51 (Na) and 3.49 (K) obtained from detailed calculations.

(d) Both approximations, "core electrons 100% effective" and Slater, yield the same value of Z_{eff} for Na and K. Neither approximation accounts for the gradual increase in Z_{eff} moving down a group.

(e) Following the trend from detailed calculations, we predict a Z_{eff} value of approximately 4.5 for Rb.

7.17 Krypton has a larger nuclear charge (Z = 36) than argon (Z = 18). The shielding of electrons in the *n* = 3 shell by the 1s, 2s, and 2p core electrons in the two atoms is approximately equal, so the *n* = 3 electrons in Kr experience a greater effective nuclear charge and are thus situated closer to the nucleus.

Atomic and Ionic Radii (Section 7.3)

7.19 The quantity described in (b) must be measured experimentally in order to determine the bonding atomic radius of an atom. Bonding atomic radius is a property of a bonded atom. The measurement must be done on an atom participating in a chemical bond.

7.21 (a) The atomic (*metallic*) radius of W is the interatomic W–W distance divided by two, 2.74 Å/2= 1.37 Å.

(b) Under high pressure, we expect atoms in a pure substance to move closer together. That is, the distance between W atoms will decrease.

7.23 From bonding atomic radii in Figure 7.7, As–I = 1.19 Å + 1.39 Å = 2.58 Å. This is very close to the experimental value of 2.55 Å in AsI_3.

7.25 *Plan.* Locate each element on the periodic charge and use trends in radii to predict their order. *Solve.*

(a) Cs > K > Li (b) Pb > Sn > Si (c) N > O > F

7.27 (a) False. Cations are smaller than their corresponding neutral atoms. Electrostatic repulsions are reduced by removing an electron from a neutral atom, Z_{eff} increases, and the cation is smaller.

(b) True. [See (a) above.]

(c) False. I^- is bigger than Cl^-. Going down a column, the *n* value of the valence electrons increases and they are farther from the nucleus. Valence electrons also experience greater shielding by core electrons. The greater radial extent of the valence electrons outweighs the increase in Z, and the size of particles with like charge increases.

7.29 Ga^{3+}: none; Zr^{4+}: Kr; Mn^{7+}: Ar; I^-: Xe; Pb^{2+}: Hg

7.31 (a) *Analyze/Plan.* Follow the logic in Sample Exercise 7.4.

Solve. Na^+ is smaller. Because F^- and Na^+ are isoelectronic, the ion with the larger nuclear charge, Na^+, has the smaller radius.

(b) *Analyze/Plan.* The electron configuration of the ions is [Ne] or [He]$2s^22p^6$. The ions have either 10 core electrons or 2 core electrons. Apply Equation 7.1 to both cases and check the result.

Solve. F^-: Z = 9. For 10 core electrons, $Z_{eff} = 9 - 10 = -1$. Although we might be able to interpret a negative value for Z_{eff}, positive values will be easier to compare; we will assume a He core of 2 electrons.

F^-, Z = 9. $Z_{eff} = 9 - 2 = 7$. Na^+: $Z_{eff} = 11 - 2 = 9$

(c) *Analyze/Plan.* The electron of interest has *n* = 2. There are seven other *n* = 2 electrons, and two *n* = 1 electrons.

Solve. S = 0.35(7) + 0.85(2) + 1(0) = 4.15

F^-: $Z_{eff} = 9 - 4.15 = 4.85$. Na^+: $Z_{eff} = 11 - 4.15 = 6.85$

(d) For isoelectronic ions (without d electrons), the electron configurations and therefore shielding values (S) are the same. Only the nuclear charge changes. So, as nuclear charge (Z) increases, effective nuclear charge (Z_{eff}) increases and ionic radius decreases.

7.33 *Analyze/Plan.* Use relative location on periodic chart and trends in atomic and ionic radii to establish the order.

(a) $Cl < S < K$ (b) $K^+ < Cl^- < S^{2-}$

(c) Even though K has the largest Z value, the *n*-value of the outer electron is larger than the *n*-value of valence electrons in S and Cl so K atoms are largest. When the 4s electron is removed, K^+ is isoelectronic with Cl^- and S^{2-}. The larger Z value causes the 3p electrons in K^+ to experience the largest effective nuclear charge and K^+ is the smallest ion.

7.35 (a) O^{2-} is larger than O because the increase in electron repulsions that accompany addition of an electron causes the electron cloud to expand.

(b) S^{2-} is larger than O^{2-} because for particles with like charges, size increases going down a family.

(c) S^{2-} is larger than K^+ because the two ions are isoelectronic and K^+ has the larger Z and Z_{eff}.

(d) K^+ is larger than Ca^{2+} because the two ions are isoelectronic and Ca^{2+} has the larger Z and Z_{eff}.

Ionization Energies; Electron Affinities (Sections 7.4 and 7.5)

7.37 $Al(g) \rightarrow Al^+(g) + e^-$; $Al^+(g) \rightarrow Al^{2+}(g) + e^-$; $Al^{2+}(g) \rightarrow Al^{3+}(g) + e^-$

The process for the first ionization energy requires the least amount of energy. When an electron is removed from an atom or ion, electrostatic repulsions are reduced, Z_{eff} increases, and the energy required to remove the next electron increases. This rationale is confirmed by the ionization energies listed in Table 7.2.

7.39 *Analyze/Plan.* We are asked about the second ionization energy of three elements. This involves removing an electron from the 1+ ion of each element. Write the electron configurations of each ion and consider the attraction for the nucleus of the valence electron to be lost. *Solve.*

The electron configurations are: Li^+, $1s^2$ or [He]; Be^+, [He]$2s^1$; K^+, [Ne] $3s^2 3p^6$ or [Ar]. Be has one more valence electron to lose whereas Li^+ has the stable noble gas configuration of He. It requires much more energy to remove a 1s core electron close to the nucleus of Li^+ than a 2s valence electron farther from the nucleus of Be^+. K^+ also has a stable noble gas configuration. The electron to be lost is a core 2p electron. This electron is farther from the nucleus than the 1s electron in Li^+ and will require less energy to remove. Of these three elements, Li has the highest second ionization energy.

7.41 (a) In general, the smaller the atom, the larger its first ionization energy.

(b) According to Figure 7.10, He has the largest and Cs has the smallest first ionization energy of the nonradioactive elements.

7.43 *Plan.* Use periodic trends in first ionization energy. *Solve.*

(a) Cl (b) Ca (c) K (d) Ge (e) Sn

7.45 *Plan.* Follow the logic of Sample Exercise 7.7. *Solve.*

(a) Co^{2+}: $[Ar]3d^7$ (b) Sn^{2+}: $[Kr]5s^2 4d^{10}$

(c) Zr^{4+}: [Kr], noble-gas configuration (d) Ag^{+}: $[Kr]4d^{10}$

(e) S^{2-}: $[Ne]3s^2 3p^6$, noble-gas configuration

7.47 *Plan.* Focus on transition metals, which have d electrons in their outer shell. Use Figure 7.15 to find representative oxidations states for transition metals. Note that, by definition, metals lose electrons to form positive ions.

Solve. Elements in group 10 and beyond have at least eight d electrons. Of these, the group 10 metals all form 2+ ions with the electron configuration nd^8. Of the elements in groups 11 and 12, only Au adopts a sufficiently high positive charge to form an ion with the configuration nd^8. (Other possibilities not listed on Figure 7.15 exist.)

Ni^{2+}: $[Ar]3d^8$; Pd^{2+}: $[Kr]4d^8$; Pt^{2+}: $[Xe]4f^{14}5d^8$; Au^{3+}: $[Xe]4f^{14}5d^8$

7.49 *Analyze/Plan.* The second electron affinity of Cl is addition of an electron to Cl^-. Write the chemical equation and electron configurations for the relevant ions. Consider the attraction of the added electron for the Cl nucleus. *Solve.*

Second Electron affinity: $Cl^-(g) + e^- \rightarrow Cl^{2-}(g)$

$[Ne]3s^2 3p^6$ $[Ne]3s^2 3p^6 4s^1$

The Cl^- ion has the stable noble gas electron configuration of Ar. According to Figure 7.12 the first electron affinity of Ar is positive. Because the Cl^- ion has the electron configuration of Ar and a smaller nuclear charge, we predict that the second electron affinity of Cl will be positive as well. A positive value means that Cl^{2-} ion is unstable and will not form. It is probably not possible to directly measure the second electron affinity of Cl. [Sometimes unstable species have a finite lifetime before decomposing, so a very fast measurement may be possible.]

7.51 *Analyze/Plan.* Write chemical equations for the electron affinity of K^+ and K. Consider the effective nuclear charge, Z – S, experienced by the added electron. *Solve.*

$K^+(g) + e^- \rightarrow K(g)$; $K(g) + e^- \rightarrow K^-(g)$

$[Ne]3s^2 3p^6$ or [Ar] $[Ne]3s^2 3p^6 4s^1$ or $[Ar]4s^1$ $[Ar]4s^1$ $[Ar]4s^2$

The electron affinity of K^+ is more negative. The full nuclear charge, Z, is the same for all K atoms and ions. In both cases, the added electron occupies a 4s subshell; the screening by the Ar core of electrons is the same. The single difference is the electron-electron repulsion experienced by the electron added to neutral K(g). This repulsion raises the energy of K^- relative to K and the added electron. According to Figure 7.12, the electron affinity of K is negative. The electron affinity of K^+ will also be negative and have a greater magnitude.

7.53 *Analyze/Plan.* Consider the definitions of ionization energy and electron affinity, along with the appropriate electron configurations. *Solve.*

(a) Ionization energy of Ne: $Ne(g) \rightarrow Ne^+(g) + e^-$

$[He]2s^22p^6 \quad\quad [He]2s^22p^5$

Electron affinity of F: $F(g) + e^- \rightarrow F^-(g)$

$[He]2s^22p^5 \quad\quad [He]2s^22p^6$

(b) The I_1 of Ne is positive, whereas E_1 of F is negative. All ionization energies are positive.

(c) One process is apparently the reverse of the other, with one important difference. The Z (and Z_{eff}) for Ne is greater than Z (and Z_{eff}) for F^-. We expect the magnitude of $I_1(Ne)$ to be somewhat greater than the magnitude of $E_1(F)$. [Repulsion effects approximately cancel; repulsion decrease upon I_1 causes smaller positive value; repulsion increase upon E_1 causes smaller negative value.]

Properties of Metals and Nonmetals (Section 7.6)

7.55 (a) Decrease

(b) Increase

(c) The smaller the first ionization energy of an element, the greater the metallic character of that element. The trends in (a) and (b) are the opposite of the trends in ionization energy.

7.57 *Analyze/Plan.* Use Figure 7.13, "Metals, metalloids, and nonmetals," and Figure 7.15, "Representative oxidation states of the elements," to inform our discussion.

Solve. Agree. An element that commonly forms a cation is a metal. The only exception to this statement shown on Figure 7.15 is antimony, Sb, a metalloid that commonly forms cations. Although Sb is a metalloid, it is far down (in the fifth row) on the chart and likely to have significant metallic character.

7.59 *Analyze/Plan.* Ionic compounds are formed by combining a metal and a nonmetal; molecular compounds are formed by two or more nonmetals. *Solve.*

Ionic: SnO_2, Al_2O_3, Li_2O, Fe_2O_3; molecular: CO_2, H_2O

7.61 *Analyze/Plan.* Decide whether MnO is an acidic or basic oxide.

Solve. MnO will react more readily with HCl(aq). Manganese is a metal. Oxides of metals usually act like bases, which react readily with the strong acid HCl(aq).

7.63 *Analyze/Plan.* Cl_2O_7 is a molecular compound formed by two nonmetallic elements. More specifically, it is a nonmetallic oxide and acidic. *Solve.*

(a) Dichlorine heptoxide

(b) Elemental chlorine and oxygen are diatomic gases.

$2\,Cl_2(g) + 7\,O_2(g) \rightarrow 2\,Cl_2O_7(l)$

(c) Cl_2O_7 is an acidic oxide, so it will be more reactive to base, OH^-.
$Cl_2O_7(l) + 2\,OH^-(aq) \rightarrow 2\,ClO_4^-(aq) + H_2O(l)$

(d) The oxidation state of Cl in Cl_2O_7 is +7. In this oxidation state, the electron configuration of Cl is $[He]2s^22p^6$ or [Ne].

7.65 (a) $BaO(s) + H_2O(l) \rightarrow Ba(OH)_2(aq)$

(b) $FeO(s) + 2\,HClO_4(aq) \rightarrow Fe(ClO_4)_2(aq) + H_2O(l)$

(c) $SO_3(g) + H_2O(l) \rightarrow H_2SO_4(aq)$

(d) $CO_2(g) + 2\,NaOH(aq) \rightarrow Na_2CO_3(aq) + H_2O(l)$

Group Trends in Metals and Nonmetals (Sections 7.7 and 7.8)

7.67 (a) Ca and Mg are both metals; they tend to lose electrons and form cations when they react. Ca is more reactive because it has a lower ionization energy than Mg. The Ca valence electrons in the 4s orbital are less tightly held because they are farther from the nucleus than the 3s valence electrons of Mg.

(b) K and Ca are both metals; they tend to lose electrons and form cations when they react. K is more reactive because it has the lower first ionization energy. The 4s valence electron in K is less tightly held because it has the same n value as the valence electrons of Ca and experiences smaller Z and Z_{eff}.

7.69 (a) $2\,K(s) + Cl_2(g) \rightarrow 2\,KCl(s)$

(b) $SrO(s) + H_2O(l) \rightarrow Sr(OH)_2(aq)$

(c) $4\,Li(s) + O_2(g) \rightarrow 2\,Li_2O(s)$

(d) $2\,Na(s) + S(l) \rightarrow Na_2S(s)$

7.71 (a) The reactions of the alkali metals with hydrogen and with a halogen are redox reactions. In both classes of reaction, the alkali metal loses electrons and is oxidized. Both hydrogen and the halogen gain electrons and are reduced. Hydrogen or the halogen act as oxidizing agents in these reactions.

$Ca(s) + F_2(g) \rightarrow CaF_2(s)$ $\qquad$ $Ca(s) + H_2(g) \rightarrow CaH_2(s)$

(b) The oxidation number of Ca in both products is +2. The electron configuration is that of Ar, $[Ne]3s^23p^6$.

7.73

	Br	Cl
(a)	$[Ar]4s^24p^5$	$[Ne]3s^23p^5$
(b)	–1	–1
(c)	1140 kJ/mol	1251 kJ/mol
(d)	reacts slowly to form HBr+HOBr	reacts slowly to form HCl+HOCl
(e)	–325 kJ/mol	–349 kJ/mol
(f)	1.20 Å	1.02 Å

The $n = 4$ valence electrons in Br are farther from the nucleus and less tightly held than the $n = 3$ valence electrons in Cl. Therefore, the ionization energy of Cl is greater, the electron affinity is more negative and the atomic radius is smaller.

7.75 (a) The term "inert" was dropped because it no longer described all the group 8A elements.

(b) In the 1960s, scientists discovered that Xe would react with substances such as F_2 and PtF_6 that have a strong tendency to remove electrons. Thus, Xe could not be categorized as an "inert" gas.

(c) The group is now called the noble gases.

7.77 (a) $2\ O_3(g) \rightarrow 3\ O_2(g)$

(b) $Xe(g) + F_2(g) \rightarrow XeF_2(g)$

$Xe(g) + 2\ F_2(g) \rightarrow XeF_4(s)$

$Xe(g) + 3\ F_2(g) \rightarrow XeF_6(s)$

(c) $S(s) + H_2(g) \rightarrow H_2S(g)$

(d) $2\ F_2(g) + 2\ H_2O(l) \rightarrow 4\ HF(aq) + O_2(g)$

Additional Exercises

7.79 Up to Z = 82, there are three instances where atomic weights are reversed relative to atomic numbers: Ar and K; Co and Ni; Te and I.

7.81 (a) P: $[Ne]3s^2 3p^3$. $Z_{eff} = Z - S = 15 - 10 = 5$.

(b) Four other $n = 3$ electrons, eight $n = 2$ electrons, two $n = 1$ electron. $S = 0.35(4) + 0.85(8) + 1(2) = 10.2$. $Z_{eff} = Z - S = 15 - 10.2 = 4.8$.

(c) The 3s electrons penetrate the [Ne] core electrons (by analogy to Figure 7.4) and experience less shielding than the 3p electrons. That is, S is greater for 3p electrons, owing to the penetration of the 3s electrons, so Z – S (3p) is less than Z – S (3s).

(d) The 3p electrons are the outermost electrons; they experience a smaller Z_{eff} than 3s electrons and thus a smaller attraction for the nucleus, given equal *n*-values. The first electron lost is a 3p electron. Each 3p orbital holds one electron, so there is no preference as to which 3p electron will be lost.

7.85 (a) Assume that the bonding atomic radius of the element will be one-half of the bond distance in the element.

$r_{As} = (2.48\ \text{Å})/2 = 1.24\ \text{Å}$; $r_{Cl} = (1.99\ \text{Å})/2 = 0.995\ \text{Å}$

The As–Cl distance is the sum of these radii: 1.24 Å + 0.995 Å = 2.235 = 2.24 Å.

(b) From Figure 7.7, the predicted As–Cl distance = 1.19 Å + 1.02 Å = 2.21 Å

7.87 (a) The most common oxidation state of the chalcogens is –2, whereas that of the halogens is –1.

(b) The family listed has the larger value of the stated property.
atomic radii, chalcogens
ionic radii of the most common oxidation state, chalcogens
first ionization energy, halogens
second ionization energy, halogens

7.91 C: $1s^2 2s^2 2p^2$. I_1 through I_4 represent loss of the 2p and 2s electrons in the outer shell of the atom. The values of I_1–I_4 increase as expected. The nuclear charge is constant, but removing each electron reduces repulsive interactions between the remaining electrons, so effective nuclear charge increases and ionization energy increases. I_5 and I_6 represent loss of the 1s core electrons. These 1s electrons are much closer to the nucleus and experience the full nuclear charge (they are not shielded), so the values of I_5 and I_6 are significantly greater than I_1–I_4. I_6 is larger than I_5 because all repulsive interactions have been eliminated.

7.96 (a) Cl^-, K^+ (b) Mn^{2+}, Fe^{3+} (c) Sn^{2+}, Sb^{3+}

7.99 (a) For both H and the alkali metals, the added electron will complete an *n*s subshell (1s for H and *n*s for the alkali metals) so shielding and repulsion effects will be similar. For the halogens, the electron is added to an *n*p subshell, so the energy change is likely to be quite different.

(b) True. Only He has a smaller estimated "bonding" atomic radius, and no known compounds of He exist. The electron configuration of H is $1s^1$. The single 1s electron experiences no repulsion from other electrons and feels the full unshielded nuclear charge. It is held very close to the nucleus. The outer electrons of all other elements that form compounds are shielded by a spherical inner core of electrons and are less strongly attracted to the nucleus, resulting in larger bonding atomic radii.

(c) Ionization is the process of removing an electron from an atom. For the alkali metals, the *n*s electron being removed is effectively shielded by the core electrons, so ionization energies are low. For the halogens, a significant increase in nuclear charge occurs as the *n*p orbitals fill, and this is not offset by an increase in shielding. The relatively large effective nuclear charge experienced by *n*p electrons of the halogens is similar to the unshielded nuclear charge experienced by the H 1s electron. Both H and the halogens have large ionization energies.

(d) Ionization energy of hydride: $H^-(g) \rightarrow H(g) + e^-$

(e) Electron affinity of hydrogen: $H(g) + e^- \rightarrow H^-(g)$

The two processes in parts (d) and (e) are the exact reverse of one another. The value for the ionization energy of hydride is equal in magnitude but opposite in sign to the electron affinity of hydrogen.

7.102 The most likely product is (i). The ionization energy of K is less than that of H, so K will lose an electron and H will gain one; the (mostly) ionic solid KH is the product. This is one instance when H acts like a nonmetal, even though it appears with the metals on the periodic chart.

7.107 *Plan.* According to the periodic table on the inside cover of your text, element 116 is in group 6A, so element 117 will be in group 7A, the halogens. Write the electron configuration and use information from Figures 7.7, 7.10, 7.12, 7.15, and Table 7.7 along with periodic trends to estimate values for properties. Remember that element 117 is two rows below iodine, and that the increase in Z and Z_{eff} that accompanies filling of the f orbitals will decrease the size of the changes in ionization energy, electron affinity and atomic size. *Solve.*

Electron configuration:	$[Rn]7s^2 5f^{14} 6d^{10} 7p^5$
First ionization energy:	805 kJ/mol
Electron affinity:	–235 kJ/mol
Atomic size:	1.65 Å
Common oxidation state:	–1

Integrative Exercises

7.110 (a) Li: $[He]2s^1$. Assume that the [He] core is 100% effective at shielding the 2s valence electron $Z_{eff} = Z - S \approx 3 - 2 = 1+$.

(b) The first ionization energy represents loss of the 2s electron.

ΔE = energy of free electron (n = ∞) – energy of electron in ground state (n = 2)

$\Delta E = I_1 = [-2.18 \times 10^{-18}\text{ J }(Z^2/\infty^2)] - [-2/18 \times 10^{-18}\text{ J }(Z^2/2^2)]$

$\Delta E = I_1 = 0 + 2.18 \times 10^{-18}\text{ J }(Z^2/2^2)$

For Li, which is not a one-electron particle, let $Z = Z_{eff}$.

$\Delta E \approx 2.18 \times 10^{-18}\text{ J }(+1^2/4) \approx 5.45 \times 10^{-19}\text{ J/atom}$

(c) Change the result from part (b) to kJ/mol so it can be compared to the value in Table 7.4.

$$5.45 \times 10^{-19}\ \frac{\text{J}}{\text{atom}} \times \frac{6.022 \times 10^{23}\text{ atom}}{\text{mol}} \times \frac{1\text{ kJ}}{1000\text{ J}} = 328\text{ kJ/mol}$$

The value in Table 7.4 is 520 kJ/mol. This means that our estimate for Z_{eff} was a lower limit, that the [He] core electrons do not perfectly shield the 2s electron from the nuclear charge.

(d) From Table 7.4, I_1 = 520 kJ/mol.

$$\frac{520\text{ kJ}}{\text{mol}} \times \frac{1000\text{ J}}{\text{kJ}} \times \frac{1\text{ mol}}{6.022 \times 10^{23}\text{ atoms}} = 8.6350 \times 10^{-19}\text{ J/atom}$$

Use the relationship for I_1 and Z_{eff} developed in part (b).

$$Z_{eff}^2 = \frac{4(8.6350 \times 10^{-19}\text{ J})}{2.18 \times 10^{-18}\text{ J}} = 1.5844 = 1.58;\ Z_{eff} = 1.26$$

This value, Z_{eff} = 1.26, based on the experimental ionization energy, is greater than our estimate from part (a), which is consistent with the explanation in part (c).

7.113 (a) Mg_3N_2

(b) $Mg_3N_2(s) + 3\,H_2O(l) \rightarrow 3\,MgO(s) + 2\,NH_3(g)$

The driving force is the production of $NH_3(g)$.

(c) After the second heating, all the Mg is converted to MgO.

Calculate the initial mass Mg.

$$0.486 \text{ g MgO} \times \frac{24.305 \text{ g Mg}}{40.305 \text{ g MgO}} = 0.293 \text{ g Mg}$$

x = g Mg converted to MgO; y = g Mg converted to Mg_3N_2; x = 0.293 – y

$$\text{g MgO} = x\left(\frac{40.305 \text{ g MgO}}{24.305 \text{ g Mg}}\right); \text{ g } Mg_3N_2 = y\left(\frac{100.929 \text{ g } Mg_3N_2}{72.915 \text{ g Mg}}\right)$$

$$\text{g MgO} + \text{g } Mg_3N_2 = 0.470$$

$$(0.293 - y)\left(\frac{40.305}{24.305}\right) + y\left(\frac{100.929}{72.915}\right) = 0.470$$

$$(0.293 - y)(1.6583) + y(1.3842) = 0.470$$

$$-1.6583\,y + 1.3842\,y = 0.470 - 0.48588$$

$$-0.2741\,y = -0.01588 = -0.016$$

$$y = 0.05794 = 0.058 \text{ g Mg in } Mg_3N_2$$

$$\text{g } Mg_3N_2 = 0.05794 \text{ g Mg} \times \frac{100.929 \text{ g } Mg_3N_2}{72.915 \text{ g Mg}} = 0.0802 = 0.080 \text{ g } Mg_3N_2$$

$$\text{mass \% } Mg_3N_2 = \frac{0.0802 \text{ g } Mg_3N_2}{0.470 \text{ g } (MgO + Mg_3N_2)} \times 100 = 17\%$$

(The final mass % has 2 sig figs because the mass of Mg obtained from solving simultaneous equations has 2 sig figs.)

(d) $3\,Mg(s) + 2\,NH_3(g) \rightarrow Mg_3N_2(s) + 3\,H_2(g)$

$$6.3 \text{ g Mg} \times \frac{1 \text{ mol Mg}}{24.305 \text{ g Mg}} = 0.2592 = 0.26 \text{ mol Mg}$$

$$2.57 \text{ g } NH_3 \times \frac{1 \text{ mol } NH_3}{17.031 \text{ g } NH_3} = 0.1509 = 0.15 \text{ mol } NH_3$$

$$0.2592 \text{ mol Mg} \times \frac{2 \text{ mol } NH_3}{3 \text{ mol Mg}} = 0.1728 = 0.17 \text{ mol } NH_3$$

0.26 mol Mg requires more than the available NH_3 so NH_3 is the limiting reactant.

$$0.1509 \text{ mol } NH_3 \times \frac{3 \text{ mol } H_2}{2 \text{ mol } NH_3} \times \frac{2.016 \text{ g } H_2}{\text{mol } H_2} = 0.4563 = 0.46 \text{ g } H_2$$

(e) $\Delta H^\circ_{rxn} = \Delta H^\circ_f\ Mg_3N_2(s) + 3\,\Delta H^\circ_f\ H_2(g) - 3\,\Delta H^\circ_f\ Mg(s) - 2\,\Delta H^\circ_f\ NH_3(g)$

$= -461.08 \text{ kJ} + 3(0) - 3(0) - 2(-46.19) = -368.70 \text{ kJ}$

8 Basic Concepts of Chemical Bonding

Visualizing Concepts

8.1 *Analyze/Plan*. Count the number of electrons in the Lewis symbol. This corresponds to the 'A'-group number of the family. *Solve.*

- (a) Group 4A or 14
- (b) Group 2A or 2
- (c) Group 5A or 15

(These are the appropriate groups in the s and p blocks, where Lewis symbols are most useful.)

8.4 *Analyze/Plan*. Count the valence electrons in the orbital diagram, take ion charge into account, and find the element with this orbital electron count on the periodic table. Write the complete electron configuration for the ion. *Solve.*

- (a) This ion has six 4d-electrons. Transition metals, or d-block elements, have valence electrons in d-orbitals. Transition metal ions first lose electrons from the 5s orbital, then from 4d if required by the charge. This 2+ ion has lost two electrons from 5s, none from 4d. The transition metal with six 4d-electrons is ruthenium, Ru.
- (b) The electron configuration of Ru is $[Kr]5s^2 4d^6$. (The configuration of Ru^{2+} is $[Kr]4d^6$).

8.7 *Analyze/Plan*. Because there are no unshared pairs in the molecule, we use single bonds to H to complete the octet of each C atom. For the same pair of bonded atoms, the greater the bond order, the shorter and stronger the bond. *Solve.*

- (a) Four. Moving from left to right along the molecule, the first C needs two H atoms, the second needs one, the third needs none, and the fourth needs one. The complete molecule is:

```
    H   H
    | 1 |  2    3
H — C ═ C — C ≡ C — H
```

- (b) In order of increasing bond length: 3 < 1 < 2
- (c) Bond 3 is strongest. For the same pair of bonded atoms, the shorter the bond length, the stronger the bond.

Lewis Symbols (Section 8.1)

8.9 (a) False. The total number of electrons in an atom is the same as its atomic number. Valence electrons are those that take part in chemical bonding, those in the outermost shell of the atom.

(b) N: [He]$2s^2 2p^3$ (Valence electrons: $2s^2 2p^3$) A nitrogen atom has 5 valence electrons.

(c) $1s^2 2s^2 2p^6$ [Ne] $3s^2 3p^2$ valence electrons. The atom (Si) has 4 valence electrons.

8.11 (a) Si: $1s^2 2s^2 2p^6 3s^2 3p^2$.

(b) Four.

(c) The 3s and 3p electrons are valence electrons.

8.13 (a) $\cdot\dot{Al}\cdot$ (b) $:\dot{\underset{..}{Br}}:$ (c) $:\overset{..}{\underset{..}{Ar}}:$ (d) $\dot{\underset{\cdot}{Sr}}$

Ionic Bonding (Section 8.2)

8.15 (a) $\dot{Mg}\cdot + \cdot\ddot{\underset{..}{O}}: \longrightarrow Mg^{2+} + [:\ddot{\underset{..}{O}}:]^{2-}$

(b) 2 electrons are transferred.

(c) Mg loses electrons.

8.17 (a) AlF_3 (b) K_2S (c) Y_2O_3 (d) Mg_3N_2

8.19 (a) Sr^{2+}: $[Ar]4s^2 3d^{10} 4p^6$ = [Kr], noble-gas configuration

(b) Ti^{2+}: $[Ar]3d^2$

(c) Se^{2-}: $[Ar]4s^2 3d^{10} 4p^6$ = [Kr], noble-gas configuration

(d) Ni^{2+}: $[Ar]3d^8$

(e) Br^-: $[Ar]4s^2 3d^{10} 4p^6$ = [Kr], noble-gas configuration

(f) Mn^{3+}: $[Ar]3d^4$

8.21 (a) Endothermic. *Lattice energy* is the energy required to totally separate 1 mole of solid ionic compound into its gaseous ions. E_{el} for attractive interactions among ions is negative, so the energy required to overcome these attractions and separate the ions is positive.

(b) $NaCl(s) \rightarrow Na^+(g) + Cl^-(g)$

(c) Salts like NaCl that have singly charged ions will have smaller lattice energies compared with salts that have doubly charged ions. The magnitude of lattice energy depends on the magnitudes of the charges of the two ions, their radii, and the arrangement of ions in the lattice. The main factor is the charges because the radii of ions do not vary over a wide range.

8.23 *Analyze/Plan.* Assign ion charges by the position of the elements in the periodic table. Lattice energy is directly related to the product of ion charges and inversely related to the ion separation. The dominant factor is ion charges, because the difference between ion separations from one compound to another is not as large as the possible difference between the products of ion charges. *Solve.*

(a) Na^+, 1+; Ca^{2+}, 2+

(b) F^-, 1– ; O^{2-}, 2–

(c) CaO will have the larger lattice energy. Lattice energy is directly related to the magnitudes of ion charges. CaO has larger cation and anion charges.

(d) Consider the relationship between the lattice energies of CaO and NaF. (Assume the lattice energy of NaF, 910 kJ, has 3 significant figures, similar to other values in Table 8.1.)

$$\frac{\text{CaO}}{\text{NaF}} = \frac{3414\text{ kJ}}{910\text{ kJ}} = 3.75$$

The ratio of the lattice energies is approximately 4 and the ratio of the two products of cation and anion charges is [(2)(2)/(1)(1)] = 4. If the charges in ScN are 3+ and 3–, respectively, the lattice enthalpy of ScN will be approximately (3)(3)(910) = 8190 kJ.

Since the calculated ratio is less than the integer value of 4, we expect 8190 kJ to be slightly greater than the measured lattice energy of ScN. From Table 8.1, the measured lattice energy of ScN is 7547, slightly less than our estimate. Recall that lattice energy is also inversely related to ion separation, which is probably greater for ScN than NaF. This also predicts that the measured lattice energy of ScN will be less than our estimate, which is based only on the differences in ion charges.

8.25 (a) K–F, 1.52 Å + 1.19 Å = 2.71 Å

Na–Cl, 1.16 Å + 1.67 Å = 2.83 Å

Na–Br, 1.16 Å + 1.82 Å = 2.98 Å

Li–Cl, 0.90 Å + 1.67 Å = 2.57 Å

(b) The order of decreasing lattice energy should be the order of increasing ion separation: LiCl > KF > NaCl > NaBr

(c) From Table 8.1: LiCl, 1030 kJ; KF, 808 kJ; NaCl, 788 kJ; NaBr, 732 kJ The predictions from ionic radii are correct.

8.27 Statement (a) is the best explanation. Equation 8.4 predicts that as the oppositely charged ions approach each other, the energy of interaction will be large and negative. This more than compensates for the energy required to form Ca^{2+} and O^{2-} from the neutral atoms.

8.29 $RbCl(s) \rightarrow Rb^+(g) + Cl^-(g)$ ΔH (lattice energy) = ?

By analogy to NaCl, Figure 8.6, the lattice energy is

$$\begin{aligned}\Delta H_{latt} &= -\Delta H_f^\circ\ RbCl(s) + \Delta H_f^\circ\ Rb(g) + \Delta H_f^\circ\ Cl(g) + I_1\ (Rb) + E\ (Cl)\\ &= -(-430.5\text{ kJ}) + 85.8\text{ kJ} + 121.7\text{ kJ} + 403\text{ kJ} + (-349\text{ kJ}) = +692\text{ kJ/mol}\end{aligned}$$

Covalent Bonding, Electronegativity, and Bond Polarity (Sections 8.3 and 8.4)

8.31 (a) The bonding in (iii) water and (iv) oxygen is likely to be covalent. (i) iron and (ii) sodium chloride contain metal atoms, which are unlikely to participate in covalent bonding in simple substances. (v) argon is a monatomic gas whose atoms do not strongly interact.

(b) Substance XY is likely to be covalent because it is a gas even below room temperature.

8.33 *Analyze/Plan.* Follow the logic in Sample Exercise 8.3. *Solve.*

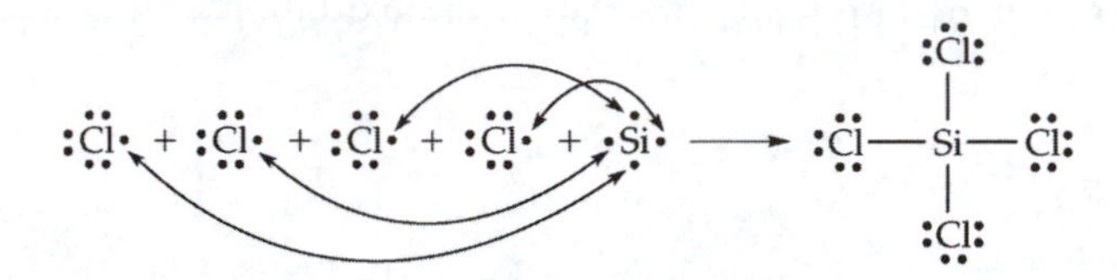

Check. Each pair of shared electrons in $SiCl_4$ is shown as a line; each atom is surrounded by an octet of electrons.

(a) 4 (b) 7 (c) 8 (d) 8 (e) 4

8.35 (a) :Ö=O:

(b) There are four bonding electrons (two bonding electron pairs) in the structure of O_2.

(c) The greater the number of shared electron pairs between two atoms, the shorter the distance between the atoms. If O_2 has two bonding electron pairs, the O–O distance will be shorter than the O–O single bond distance.

8.37 Statement (b) is false. Electron affinity is a property of gas phase atoms or ions, whereas electronegativity is a property of bonded atoms in a molecule.

8.39 *Plan.* Electronegativity increases going up and to the right in the periodic table. *Solve.*

(a) Mg (b) S (c) C (d) As

Check. The electronegativity values in Figure 8.8 confirm these selections.

8.41 The bonds in (a), (c), and (d) are polar because the atoms involved differ in electronegativity. The more electronegative element in each polar bond is:

(a) F (c) O (d) I

8.43 (a) *Analyze/Plan.* Q is the charge at either end of the dipole. Q = μ/r. The values for HBr are μ = 0.82 D and r = 1.41 Å. Change Å to m; use the definition of debyes and the charge of an electron to calculate effective charge in units of *e*. *Solve.*

$$Q = \frac{\mu}{r} = \frac{0.82\,\text{D}}{1.41\,\text{Å}} \times \frac{1\,\text{Å}}{1\times10^{-10}\,\text{m}} \times \frac{3.34\times10^{-30}\,\text{C-m}}{1\,\text{D}} \times \frac{1\,\text{e}}{1.60\times10^{-19}\,\text{C}} = 0.12\,e$$

(b) Decrease. Q = μ/r, μ = Q × r. If r decreases and Q remains the same, μ decreases.

8.45 *Analyze/Plan*. Generally, compounds formed by a metal and a nonmetal are described as ionic, whereas compounds formed from two or more nonmetals are covalent. However, substances with metals in a high oxidation state often have properties of molecular compounds. In this exercise we know that one substance in each pair is molecular and one is ionic; we may need to distinguish by comparison. *Solve.*

(a) SiF_4, metalloid and nonmetal, molecular, silicon tetrafluoride

LaF_3, metal and nonmetal, ionic, lanthanum(III) fluoride

(b) $FeCl_2$, metal and nonmetal, ionic, iron(II) chloride

$ReCl_6$, metal in high oxidation state, Re(VI), molecular, rhenium hexachloride

(c) $PbCl_4$, metal and nonmetal, Pb(IV) is relatively high oxidation state, molecular (by contrast with RbCl, which is definitely ionic), lead tetrachloride

RbCl, metal and nonmetal, ionic, rubidium chloride

Lewis Structures; Resonance Structures (Sections 8.5 and 8.6)

8.47 *Analyze.* Counting the **correct number of valence electrons** is the foundation of every Lewis structure. *Plan/Solve.*

(a) Count valence electrons: $4 + (4 \times 1) = 8\ e^-$, $4\ e^-$ pairs. Follow the procedure in Sample Exercise 8.6.

```
      H
      |
  H—Si—H
      |
      H
```

(b) Valence electrons: $4 + 6 = 10\ e^-$, $5\ e^-$ pairs

$:C\equiv O:$

(c) Valence electrons: $[6 + (2 \times 7)] = 20\ e^-$, $10\ e^-$ pairs

```
 ..   ..   ..
:F — S — F:
 ..   ..   ..
```

i. Place the S atom in the middle and connect each F atom with a single bond; this requires $2\ e^-$ pairs.

ii. Complete the octets of the F atoms with nonbonded pairs of electrons; this requires an additional $6\ e^-$ pairs.

iii. The remaining $2\ e^-$ pairs complete the octet of the central S atom.

(d) (Draw the structure that obeys the octet rule, for now.) 32 valence e^-, $16\ e^-$ pairs

```
       ..
      :O:
        |
 ..          ..
:O — S — O — H
 ..          ..
        |
      :O:
       ..
        |
        H
```

(e) Follow Sample Exercise 8.8. 20 valence e^-, 10 e^- pairs

$[:\ddot{\underset{..}{O}}-\ddot{\underset{..}{Cl}}-\ddot{\underset{..}{O}}:]^-$

(f) 14 valence e^-, 7 e^- pairs

H—N̈—Ö—H (with H bonded below N)

Check. In each molecule, bonding e^- pairs are shown as lines, and each atom is surrounded by an octet of electrons (duet for H).

8.49 Statement (b) is most true. (The other four are clearly false.) Keep in mind that when it is necessary to place more than an octet of electrons around an atom to minimize formal charge, there may not be a "best" Lewis structure.

8.51 *Analyze/Plan.* Draw the correct Lewis structure: count valence electrons in each atom, total valence electrons and electron pairs in the molecule or ion; connect bonded atoms with a line, place the remaining e^- pairs as needed, in nonbonded pairs or multiple bonds, so that each atom is surrounded by an octet (or duet for H). Calculate formal charges: assign electrons to individual atoms [nonbonding e^- + 1/2 (bonding e^-)]; formal charge = valence electrons – assigned electrons. Assign oxidation numbers, assuming that the more electronegative element holds all electrons in a bond.

Solve. Formal charges are shown near the atoms, oxidation numbers (ox. #) are listed below the structures.

(a) 16 e^-, 8 e^- pairs

Ö=C=S̈ (formal charges: O 0, C 0, S 0)

ox. #: O, –2; C, +4; S, –2

(b) 26 valence e^-, 13 e^- pairs

:Ö: (−1) bonded to S; 0 :C̈l—S̈—C̈l: 0 (S +1)

ox #: S, +4; Cl, –1; O, –2

(c) 26 valence e^-, 13 e^- pairs

[:Ö: (−1) bonded to Br; −1 :Ö—Br—Ö: −1 (Br +2)]$^{1-}$

ox. #: Br, +5; O, –2

(d) 20 valence e^-, 10 e^- pairs

0 H—Ö—C̈l—Ö: −1 (O 0, Cl +1)

ox. #: Cl, +3; H, +1; O, –2

Check. Each atom is surrounded by an octet (or duet) and the sum of the formal charges and oxidation numbers is the charge on the particle.

8.53 (a) *Plan.* Count valence electrons, draw all possible correct Lewis structures, taking note of alternate placements for multiple bonds. *Solve.*

$18\ e^-$, $9\ e^-$ pairs

$$[\mathrm{O{=}N{-}O}]^- \longleftrightarrow [\mathrm{O{-}N{=}O}]^-$$

Check. The octet rule is satisfied.

(b) *Plan.* Isoelectronic species have the same number of valence electrons and the same electron configuration. *Solve.*

A single O atom has 6 valence electrons, so the neutral ozone molecule O_3 is isoelectronic with NO_2^-.

$$\mathrm{O{=}O{-}O} \longleftrightarrow \mathrm{O{-}O{=}O}$$

Check. The octet rule is satisfied.

(c) Because each N–O bond has partial double bond character, the N–O bond length in NO_2^- should be shorter than N–O single bonds but longer than N=O double bonds.

8.55 *Plan/Solve.* The Lewis structures are as follows:

$5\ e^-$ pairs

$$:\mathrm{C{\equiv}O}:$$

$8\ e^-$ pairs

$$\mathrm{O{=}C{=}O}$$

$12\ e^-$ pairs

$$[\mathrm{CO_3}]^{2-} \longleftrightarrow [\mathrm{CO_3}]^{2-} \longleftrightarrow [\mathrm{CO_3}]^{2-}$$

The more pairs of electrons shared by two atoms, the shorter the bond between the atoms. The average number of electron pairs shared by C and O in the three species is 3 for CO, 2 for CO_2, and 1.33 for CO_3^{2-}. The order of bond lengths, from shortest to longest, is: $CO < CO_2 < CO_3^{2-}$.

8.57 (a) False. Because of resonance (see Figure 8.15), the C–C bonds in benzene are the same length, but they are shorter than a typical single C–C bond and longer than a typical double C–C bond.

(b) HCCH, $10\ e^-$, $5\ e^-$ pr

$$\mathrm{H{-}C{\equiv}C{-}H}$$

False. The C–C bond in acetylene is an isolated triple bond; it is shorter than an isolated double bond and therefore shorter than the average C–C bond in benzene.

Exceptions to the Octet Rule (Section 8.7)

8.59 *Analyze/Plan*. In order to decide whether a molecule is an exception to the octet rule, examine the Lewis structure. Does the Lewis structure have an odd number of electrons, an atom with less than eight electrons, or an atom with more than eight electrons? *Solve.*

(a) CO_2, 16 e^-, 8 e^-pr

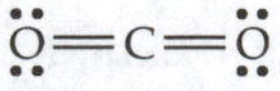

H_2O, 8 e^-, 4 e^-pr

NH_3, 8 e^-, 4 e^-pr

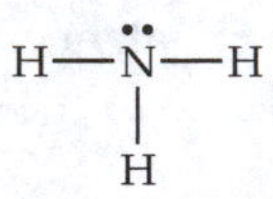

PF_3, 26 e^-, 13 e^-pr

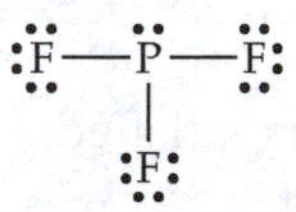

AsF_5, 40 e^-, 20 e^-pr

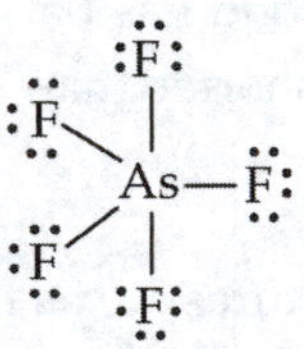

None of the molecules have an odd number of electrons. CO_2, H_2O, NH_3, and PF_3 obey the octet rule. In AsF_5, the central As atom is bound to five F atoms, so it has 10 electrons around it. AsF_5 is an exception to the octet rule.

(b) BH_4^-, 8 e^-, 4 e^- pr

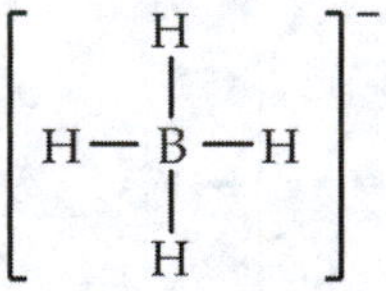

$B_3N_3H_6$, 30 e^-, 15 e^-pr

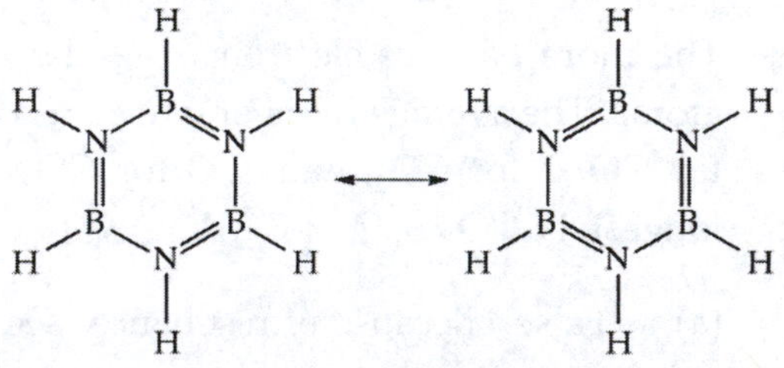

BCl_3, 24 e^-, 12 e^-pr

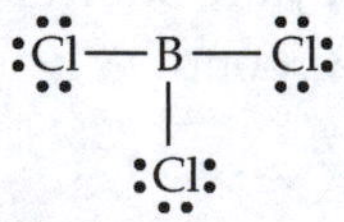

[The structure shown minimizes formal charges and is the dominant form (see Section 8.7)]

BCl_3 is an exception to the octet rule; the B atom has an incomplete octet.

8.61 *Analyze/Plan.* For each species, count the number of valence electrons and electron pairs. Draw the dominant Lewis structure. For the purpose of this exercise, assume that the dominant Lewis structure is the one that minimizes formal charge.

ClO, 13 e^-, 6.5 e^- pairs

$\cdot\ddot{\underset{\cdot\cdot}{Cl}}=\ddot{\underset{\cdot\cdot}{O}}$

Odd number of electrons
Does not obey the octet rule

ClO^-, 14 e^-, 7 e^- pairs

$\left[:\ddot{\underset{\cdot\cdot}{Cl}}-\ddot{\underset{\cdot\cdot}{O}}:\right]^-$

Obeys the octet rule

ClO_2^-, 20 e^-, 10 e^- pairs

$\left[\ddot{\underset{\cdot\cdot}{O}}=\ddot{\underset{\cdot\cdot}{Cl}}-\ddot{\underset{\cdot\cdot}{O}}:\right]^-$

Cl has expanded octet
Does not obey the octet rule

ClO_3^-, 26 e^-, 13 e^- pairs

$\left[\ddot{\underset{\cdot\cdot}{O}}=\ddot{Cl}(=\ddot{O}:)-\ddot{\underset{\cdot\cdot}{O}}:\right]^-$

Cl has expanded octet
Does not obey the octet rule

ClO_4^-, 32 e^-, 16 e^- pairs

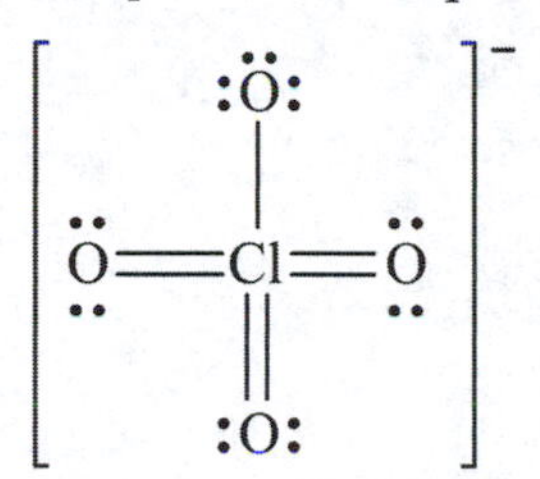

Cl has expanded octet
Does not obey the octet rule

In each species, Cl has a zero formal charge and O atoms that form double bonds have zero formal charge. The O atoms that from single bonds have –1 formal charge. For ClO_2^-, ClO_3^-, and ClO_4^-, structures that do not minimize formal charge but obey the octet rule can be drawn. The octet rule vs minimum formal charge debate is ongoing.

8.63 (a) 8 e^-, 4 e^- pairs [PH_3]

$H-\ddot{P}(-H)-H$

Obeys octet rule.

(b) 6 e^-, 3 e^- pairs

$H-Al(-H)-H$

Does not obey the octete rule. Central Al has 6 electrons (impossible to satisfy octet rule with only 6 valence electrons).

(c) 16 e^-, 8 e^- pairs

$[:N\equiv N-\ddot{\underset{..}{N}}:]^- \longleftrightarrow [:\ddot{\underset{..}{N}}-N\equiv N:]^- \longleftrightarrow [:\ddot{N}=N=\ddot{N}:]^-$

3 resonance structures; all obey octet rule.

(d) 20 e^-, 10 e^- pairs

$$\begin{array}{c} :\ddot{\underset{..}{Cl}}: \\ | \\ :\ddot{\underset{..}{Cl}}-C-H \\ | \\ H \end{array}$$

Obeys octet rule.

(e) 48 e^-, 24 e^- pairs [SnF_6^{2-}]

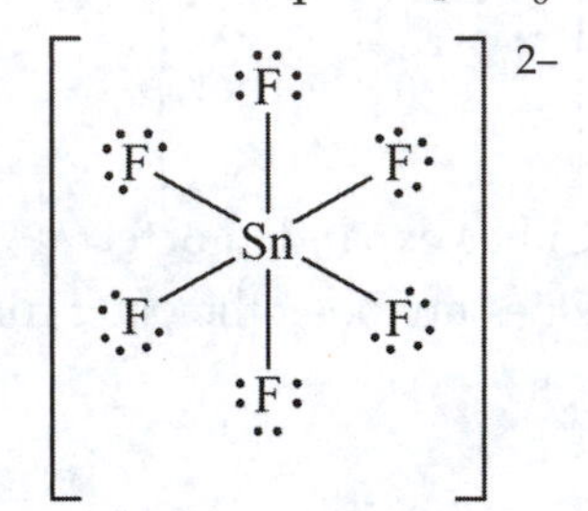

Does not obey octet rule. Central Sn has 12 electrons.

8.65 (a) 16 e^-, 8 e^- pairs

$:\ddot{\underset{..}{Cl}}-Be-\ddot{\underset{..}{Cl}}:$

This structure violates the octet rule; Be has only 4 e^- around it.

(b) $\ddot{\underset{..}{Cl}}=Be=\ddot{\underset{..}{Cl}} \longleftrightarrow :\ddot{\underset{..}{Cl}}-Be\equiv Cl: \longleftrightarrow :Cl\equiv Be-\ddot{\underset{..}{Cl}}:$

(c) The formal charges on each of the atoms in the four resonance structures are:

$:\ddot{\underset{..}{Cl}}-Be-\ddot{\underset{..}{Cl}}:$	$\ddot{\underset{..}{Cl}}=Be=\ddot{\underset{..}{Cl}}$	$:\ddot{\underset{..}{Cl}}-Be\equiv Cl:$	$:Cl\equiv Be-\ddot{\underset{..}{Cl}}:$
0 0 0	+1 −2 +1	0 −2 +2	+2 −2 0

Formal charges are minimized on the structure that violates the octet rule; this form is probably dominant.

8.67 (a) *Analyze/Plan.* Given H_2SO_4 with H attached to O, assume S is central and bound to the four O atoms. Draw a Lewis structure where S and O obey the octet rule and H atoms have two electrons and are terminal. *Solve.*

32 e^-, 16 e^- pr

$$\begin{array}{c} :\ddot{O}: \\ | \\ :\ddot{\underset{..}{O}}-S-\ddot{\underset{..}{O}}-H \\ | \\ :\underset{..}{O}: \\ | \\ H \end{array}$$

(b) *Analyze/Plan.* Starting with the Lewis structure from part (a), rearrange electrons to minimize formal charge. Formal charge = valence e^- – assigned e^-. To have a formal charge of zero, both S and O should have 6 assigned electrons.

Assigned e^- = [nonbonding e^- + ½(bonding e^-)] *Solve.*

For the structure in part (a),

S: assigned electrons = 0 + ½(8) = 4;
FC = 6 valence e^- – 4 assigned e^- = +2

O (terminal): assigned electrons = 6 + ½(2) = 7;
FC = 6 valence e^- – 7 assigned e^- = –1

O (bound to H): assigned electrons = 4 + ½(4) = 6;
FC = 6 valence e^- – 6 assigned e^- = 0

To minimize formal charges, change one nonbonding e^- pair on each terminal O atom into a bonding e^- pair between that atom and S. That is, the bonds between the terminal O atoms and S become double bonds.

:O:
‖
O═S—O—H
|
:O:
|
H

Strengths and Lengths of Covalent Bonds (Section 8.8)

8.69 *Analyze.* Given: structural formulas. Find: enthalpy of reaction.

Plan. Count the number and kinds of bonds that are broken and formed by the reaction. Use bond enthalpies from Table 8.3 and Equation 5.32 to calculate the overall enthalpy of reaction, ΔH. *Solve.*

(a) $\Delta H = 2\,D(O–H) + D(O–O) + 4\,D(C–H) + D(C{=}C)$
$-2\,D(O–H) - 2\,D(O–C) - 4\,D(C–H) - D(C–C)$

$\Delta H = D(O–O) + D(C{=}C) - 2\,D(O–C) - D(C–C)$

$= 146 + 614 - 2(358) - 348 = -304$ kJ

(b) $\Delta H = 5\,D(C–H) + D(C \equiv N) + D(C{=}C) - 5\,D(C–H) - D(C \equiv N) - 2\,D(C–C)$
$= D(C{=}C) - 2\,D(C–C) = 614 - 2(348) = -82$ kJ

(c) $\Delta H = 6\,D(N–Cl) - 3\,D(Cl–Cl) - D(N \equiv N)$
$= 6(200) - 3(242) - 941 = -467$ kJ

8.71 (a) False. For the same atom pair, the longer the bond the smaller the bond enthalpy.

(b) False. See average bond enthalpies in Table 8.3.

(c) False. The single bond lengths in Table 8.4 are all less than 5 Å.

(d) False. Breaking a chemical bond requires an input of energy.

(e) True.

8.73 Ionic bond enthalpies depend on the charge and size of the participating ions. The Ca–O bond will be stronger than the Na–Cl bond, because the ion charges are greater.

8.75 (a) CH_2, 6 e^-, 3 e^- pr H—C̈—H

The C atom in carbene is extremely electron deficient, which causes carbene to be very reactive.

(b) $2\ CH_2, \rightarrow C_2H_4$

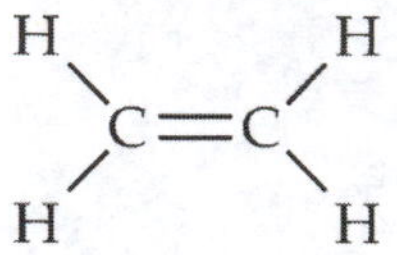

$H_2C{=}CH_2$

The product of this reaction contains a C–C double bond, with a typical bond length of 1.34 Å. (See Table 8.4.)

Additional Exercises

8.77 A triple C–C bond. A 1.15 Å bond length is relatively short for any atom pair. Table 8.4 indicates than an average triple C–C bond length is 1.20 Å.

8.85 (a) B–O. The most polar bond will be formed by the two elements with the greatest difference in electronegativity. Because electronegativity increases moving right and up on the periodic table, the possibilities are B–O and Te–O. These two bonds are likely to have similar electronegativity differences (3 columns apart vs. 3 rows apart). Values from Figure 8.8 confirm the similarity, and show that B–O is slightly more polar.

(b) Te–I. Both are in the fifth row of the periodic table and have the two largest covalent radii among this group of elements.

(c) TeI_2. Te needs to participate in two covalent bonds to satisfy the octet rule, and each I atom needs to participate in one bond, so by forming a TeI_2 molecule, the octet rule can be satisfied for all three atoms.

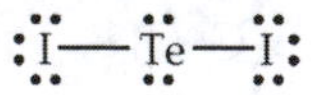

:Ï—T̈e—Ï:

(d) B_2O_3. Although this is probably not a purely ionic compound, it can be understood in terms of gaining and losing electrons to achieve a noble-gas configuration. If each B atom were to lose 3 e^- and each O atom were to gain 2 e^-, charge balance and the octet rule would be satisfied.

P_2O_3. Each P atom needs to share 3 e^- and each O atom 2 e^- to achieve an octet. Although the correct number of electrons seem to be available, a correct Lewis structure is difficult to imagine. In fact, phosphorus (III) oxide exists as P_4O_6 rather than P_2O_3 (Chapter 22).

8.90 Formal charge (FC) = # valence e^- – (# nonbonding e^- + 1/2 # bonding e^-)

(a) 18 e^-, 9 e^- pairs

O—O=O ⟷ O=O—O

FC for the central O = 6 – [2 + 1/2 (6)] = +1

(b) 48 e^-, 24 e^- pairs

$[PF_6]^-$ (P bonded to six F)

FC for P = 5 – [0 + 1/2 (12)] = –1

The three nonbonded pairs on each F have been omitted.

(c) 17 e^-; 8 e^- pairs, 1 odd e^-

O=N—O ⟷ O—N=O

The odd electron is probably on N because it is less electronegative than O. Assuming the odd electron is on N, FC for N = 5 – [1 + 1/2 (6)] = +1. If the odd electron is on O, FC for N = 5 – [2 + 1/2 (6)] = 0.

(d) 28 e^-, 14 e^- pairs

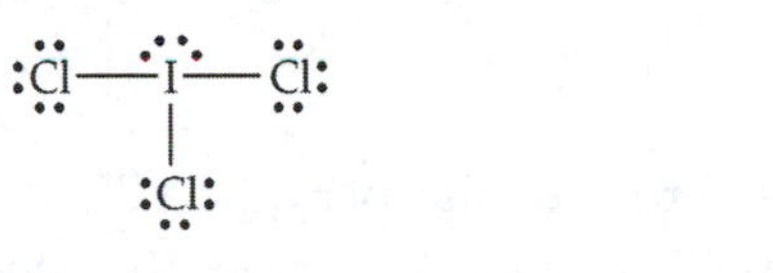

FC for I = 7 – [4 + 1/2 (6)] = 0

(e) 32 e^-, 16 e^- pairs

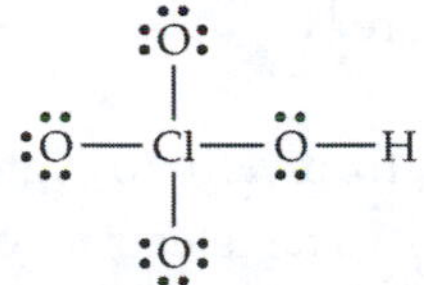

FC for Cl = 7 – [0 + 1/2 (8)] = +3

Integrative Exercises

8.95 (a) False. The B–A=B structure says nothing about the nonbonding electrons in the molecule. One possible example is NO_2, which has an odd electron.

(b) True.

8.98 (a) Ti^{2+} : [Ar]$3d^2$; Ca : [Ar]$4s^2$.

(b) Ca has no unpaired electrons and Ti^{2+} has two. The two valence electrons in Ca are paired in the 4s orbital. Each of the two valence electrons in Ti^{2+} occupies its own 3d orbital (Hund's rule).

(c) To be isoelectronic with Ca^{2+}, Ti would have a 4+ charge.

8.104 (a) Assume 100 g.

$$62.04 \text{ g Ba} \times \frac{1 \text{ mol}}{137.33 \text{ g Ba}} = 0.4518 \text{ mol Ba}; 0.4518/0.4518 = 1.0$$

$$37.96 \text{ g N} \times \frac{1 \text{ mol}}{14.007 \text{ g N}} = 2.710 \text{ mol N}; 2.710/0.4518 = 6.0$$

The empirical formula is BaN_6. Ba has an ionic charge of 2+, so there must be two 1– azide ions to balance the charge. The formula of each azide ion is N_3^-.

(b) 16 e^-, 8 e^- pairs

$[:\ddot{N}=N=\ddot{N}:]^-$ ⟷ $[:N\equiv N-\ddot{N}:]^-$ ⟷ $[:\ddot{N}-N\equiv N:]^-$

Formal charges: –1 +1 –1; 0 +1 –2; –2 +1 0

(c) The structure with two double bonds minimizes formal charges and is probably the main contributor.

(d) The two N–N bond lengths will be equal. The two minor contributors would individually cause unequal N–N distances, but collectively they contribute equally to the lengthening and shortening of each bond. The N–N distance will be approximately 1.24 Å, the average N=N distance.

8.109 (a) NH_3BF_3, 32 e^-, 16 e^- pr

```
     H   :F:
     |    |
  H—N—B—F:
     |    |
     H   :F:
```

(b) Moving from left to right across a row of the periodic table, electronegativity increases. The electron density will be greater around the atom with greater electronegativity, in this case N.

$$\overset{\delta+}{B}-\overset{\delta-}{N}$$

(c) The difference between NH_3BCl_3 and NH_3BF_3 is that Cl has replaced F as the element bound to B. Chlorine is less electronegative and less electron withdrawing than fluorine. This increases the electron density at B and renders the B–N bond less polar in NH_3BCl_3, than NH_3BF_3.

9 Molecular Geometry and Bonding Theories

Visualizing Concepts

9.1 Removing an atom from the equatorial plane of trigonal bipyramid in Figure 9.3 creates a seesaw shape. It might appear that you could also obtain a seesaw by removing two atoms from the square plane of the octahedron. However, one of the B–A–B angles in the seesaw is 120°, so it must be derived from a trigonal bipyramid.

9.3 *Analyze/Plan.* Visualize the molecular geometry and the electron-domain geometries that could produce it. Confirm your choices with Tables 9.2 and 9.3. In Table 9.3, note that octahedral electron-domain geometry results in only 3 possible molecular geometries: octahedral, square pyramidal, and square planar (not T-shaped, bent, or linear). *Solve.*

(a) 2. Molecular geometry: linear. Possible electron-domain geometries: linear, trigonal bipyramidal

(b) 1. Molecular geometry, T-shaped. Possible electron-domain geometries: trigonal bipyramidal

(c) 1. Molecular geometry, octahedral. Possible electron-domain geometries: octahedral

(d) 1. Molecular geometry, square-pyramidal. Possible electron-domain geometries: octahedral

(e) 1. Molecular geometry, square planar. Possible electron-domain geometries: octahedral

(f) 1. Molecular geometry, triangular pyramid. Possible electron-domain geometries: trigonal bipyramidal. This is an unusual molecular geometry that is not listed in Table 9.3. It could occur if the equatorial substituents on the trigonal bipyramid were extremely bulky, causing the nonbonding electron pair to occupy an axial position.

9.5 (a) Zero. Moving from left to right along the x-axis of the plot, the distance between the Cl atoms increases. At very large separation, the potential energy of interaction approaches zero.

(b) The Cl–Cl bond distance is approximately 2.0 Å. The Cl–Cl bond energy is approximately 240 kJ/mol. The minimum energy for the two atoms represents the stabilization obtained by bringing two Cl atoms together at the optimum (bond) distance. The x-coordinate of the minimum point on the plot is the Cl–Cl bond length; the y-coordinate is the bond strength or enthalpy.

(c) Weaker. Under extreme pressure, assume the Cl–Cl separation gets shorter. According to the plot, the potential energy of the atom pair increases and the bond gets weaker as the separation becomes shorter than the optimum bond distance.

9.11 *Analyze/Plan.* σ molecular orbitals (MOs) are symmetric about the internuclear axis, π MOs are not. Bonding MOs have most of their electron density in the area between the nuclei, antibonding MOs have a node between the nuclei.

(a) (i) Two s atomic orbitals (electron density at each nucleus).

(ii) Two p atomic orbitals overlapping end to end (node near each nucleus).

(iii) Two p atomic orbitals overlapping side to side (node near each nucleus).

(b) (i) σ-type (symmetric about the internuclear axis, s orbitals can produce only σ overlap).

(ii) σ-type (symmetric about internuclear axis)

(iii) π-type (not symmetric about internuclear axis, side-to-side overlap)

(c) (i) antibonding (node between nuclei)

(ii) bonding (concentration of electron density between nuclei)

(iii) antibonding (node between nuclei)

(d) (i) The nodal plane is between the atom centers, perpendicular to the interatomic axis and equidistant from each atom.

(ii) There are two nodal planes; both are perpendicular to the interatomic axis. One is left of the left atom and the second is right of the right atom.

(iii) There are two nodal planes; one is between the atom centers, perpendicular to the interatomic axis and equidistant from each atom. The second contains the interatomic axis and is perpendicular to the first.

Molecular Shapes; the VSEPR Model (Sections 9.1 and 9.2)

9.13 (a) It is not possible to tell the number of nonbonding electron pairs about the A atom from this information. If AB_2 obeys the octet rule, A would have 0 nonbonding pairs around it, as in CO_2. If AB_2 does not obey the octet rule, there could be 0 or 3 nonbonding pairs around A. Examples are BeH_2 and XeF_2.

(b) Three. XeF_2 has 11 electron pairs. Of these, 2 are bonding pairs between Xe and F and 6 are nonbonding pairs around the two F atoms. This leaves three nonbonding electron pairs around Xe.

(c) Yes. The electron domain geometry of XeF_2 is trigonal bipyramidal; the 3 nonbonding pairs are equatorial, the 2 bonding pairs are axial, and the molecular geometry is linear.

9.15 A molecule with tetrahedral molecular geometry has an atom at each vertex of the tetrahedron. A trigonal pyramidal molecule has one vertex of the tetrahedron occupied by a nonbonding electron pair rather than an atom. That is, a trigonal pyramid is a tetrahedron with one vacant vertex.

9.17 (a) Octahedral. There are 6 electron domains around A in an AB_6 molecule. Because none of the 6 electron domains are nonbonding, the electron domain geometry and molecular geometry are octahedral.

(b) Octahedral. An AB_4 molecule has 4 bonding electron domains around A. Additionally, this molecules has two nonbonding domains, for a total of 6 electron domains around A. A total of six electron pairs dictates octahedral electron domain geometry.

(c) Square planar. For an octahedral electron domain geometry that includes two nonbonding domains, the nonbonding domains are opposite each other to minimize repulsions. The four bonding domains occupy the remaining positions in the octahedron, forming a square plane.

9.19 *Analyze/Plan*. Draw the Lewis structure of each molecule and take note of nonbonding (lone) electron pairs about the central atom. *Solve.*

(a) SiH_4, 8 valence e^-, 4 e^- pr, 0 nonbonding pairs, no effect on molecular shape

```
     H
     |
H—Si—H
     |
     H
```

(b) PF_3, 26 valence e^-, 13 e^- pr, 1 nonbonding pair on P, influences molecular shape

```
 ..   ..   ..
:F—P—F:
 ..   |   ..
     :F:
      ..
```

(c) HBr, 8 valence e^-, 4 e^- pr, 3 nonbonding pairs on Br, no effect on molecular shape because Br is not "central"

```
    ..
H—Br:
    ..
```

(d) HCN, 10 valence e^-, 5 e^- pr, 0 nonbonding pairs on C, no effect on molecular shape

H—C≡N:

(e) SO_2, 18 valence e^-, 9 e^- pr, 1 nonbonding pair on S, influences molecular shape

```
 ..   ..   ..         ..   ..   ..          ..   ..   ..
:O—S=O  ←→  O=S—O:  ←→  O=S=O
 ..        ..         ..        ..          ..        ..
```

9.21 *Analyze/Plan*. Draw the Lewis structure of each molecule and count the number of nonbonding (lone) electron pairs. Note that the question asks 'in the molecule' rather than just around the central atom. *Solve.*

(a) $(CH_3)_2S$, 20 valence e^-, 10 e^- pr, 2 nonbonding pairs

```
   H       H
   |   ..  |
H—C—S—C—H
   |   ..  |
   H       H
```

(b) HCN, 10 valence e^-, 5 e^- pr, 1 nonbonding pair

H—C≡N:

(c) H_2C_2, 10 valence e^-, 5 e^- pr, 0 nonbonding pairs

H—C≡C—H

(d) CH_3F, 14 valence e^-, 7 e^- pr, 3 nonbonding pairs

$$\begin{array}{c} \text{H} \\ | \\ \text{H}-\text{C}-\ddot{\underset{\cdot\cdot}{\text{F}}}\!: \\ | \\ \text{H} \end{array}$$

9.23 *Analyze/Plan*. See Tables 9.2 and 9.3. *Solve*.

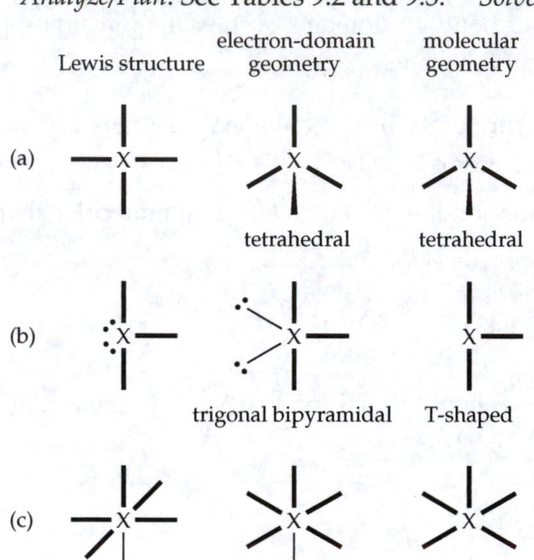

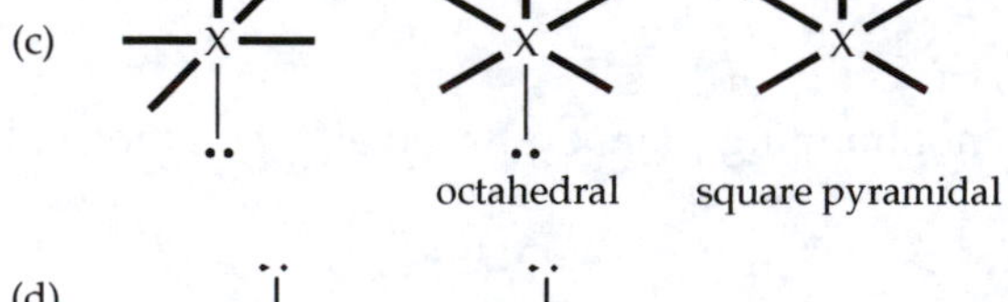

9.25 *Analyze/Plan*. Follow the logic in Sample Exercises 9.1 and 9.2. *Solve*.
bent (b), linear (l), octahedral (oh), seesaw (ss), square pyramidal (sp), square planar (spl), tetrahedral (td), trigonal bipyramidal (tbp), trigonal planar (tr), trigonal pyramidal (tp), T-shaped (T)

	Molecule or ion	Valence electrons	Lewis structure		Electron-domain geometry	Molecular geometry
(a)	HCN	10	:N≡C—H	N≡C—H	l	l
(b)	SO_3^{2-}	26	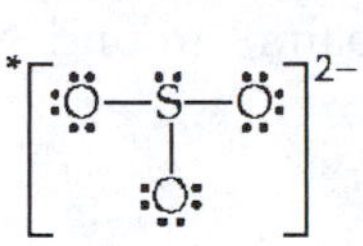 *$[$:Ö—S̈—Ö:$]^{2-}$ with :Ö: below S	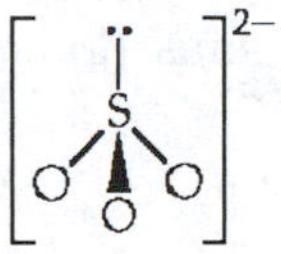 $[S(O)_3]^{2-}$ (pyramidal)	td	tp
(c)	SF_4	34	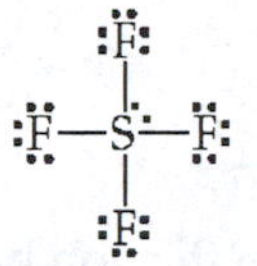 :F̤—S̤—F̤: with :F: above and below	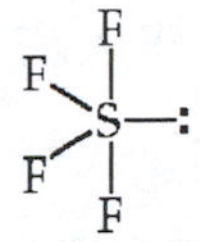 SF_4 (seesaw)	tbp	ss
(d)	PF_6^-	48	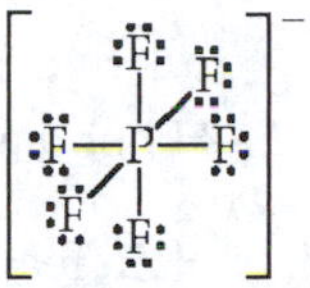 $[PF_6]^-$	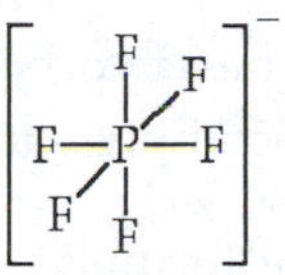 $[PF_6]^-$ (octahedral)	oh	oh
(e)	NH_3Cl^+	14	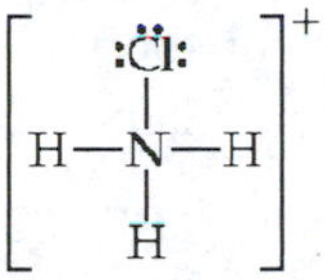 $[H—N(Cl)(H)—H]^+$	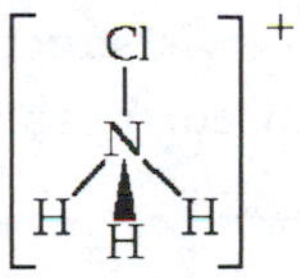 $[NH_3Cl]^+$ (tetrahedral)	td	td
(f)	N_3^-	16	*$[$N̤—N≡N:$]^-$	$[N—N≡N]^-$	l	l

*More than one resonance structure is possible. All equivalent resonance structures predict the same molecular geometry.

9.27 *Analyze/Plan*. Work backward from molecular geometry, using Tables 9.2 and 9.3. *Solve*.

(a) Electron-domain geometries: (i), trigonal planar; (ii), tetrahedral; (iii), trigonal bipyramidal

(b) nonbonding electron domains: (i), 0; (ii), 1; (iii), 2

(c) N and P. Shape (ii) has three bonding and one nonbonding electron domains. Li and Al would form ionic compounds with F, so there would be no nonbonding electron domains. Assuming that F always has three nonbonding domains, BF_3 and ClF_3 would have the wrong number of nonbonding domains to produce shape ii.

(d) Cl (also Br and I, because they have seven valence electrons). This T-shaped molecular geometry arises from a trigonal bipyramidal electron-domain geometry with two nonbonding domains (Table 9.3). Assuming each F atom has three nonbonding domains and forms only single bonds with A, A must have seven valence electrons to produce these electron-domain and molecular geometries. It must be in or below the third row of the periodic table, so that it can accommodate more than four electron domains.

9.29 *Analyze/Plan.* Follow the logic in Sample Exercise 9.3. *Solve.*

(a) 1, less than 109.5°; 2, less than 109.5°

(b) 3, different than 109.5°; 4, less than 109.5°

(c) 5 – 180°

(d) 6, slightly more than 120°; 7, less than 109.5°; 8, slightly different than 109.5°

9.31 *Analyze/Plan.* Draw correct Lewis structures for NH_2^-, NH_3, and NH_4^+. The more nonbonding electron domains (lone pairs) around N, the smaller the H–N–H bond angles. *Solve.*

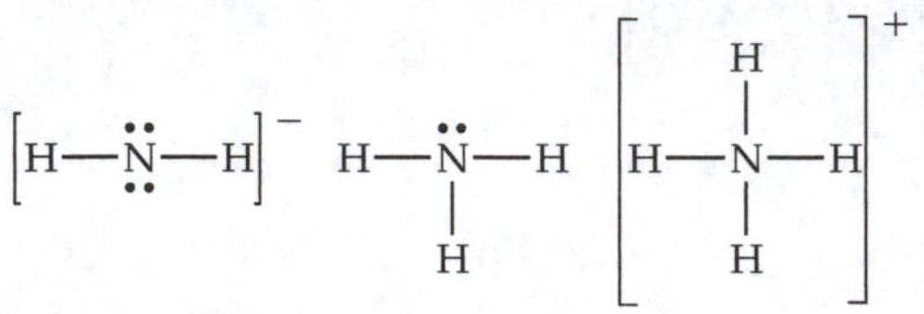

(a) NH_4^+. There are no lone pairs on N, so this ion has the largest bond angles.

(b) NH_2^-. Amide ion has two bonding and two nonbonding domains around N. The two lone pairs compress the H–N–H bond angle to its smallest value.

9.33 *Analyze.* Given: molecular formulas. Find: explain features of molecular geometries.

Plan. Draw the correct Lewis structures for the molecules and use VSEPR to predict and explain observed molecular geometry. *Solve.*

(a) BrF_4^- 36 e^-, 18 e^- pr — BF_4^- 32 e^-, 16 e^- pr

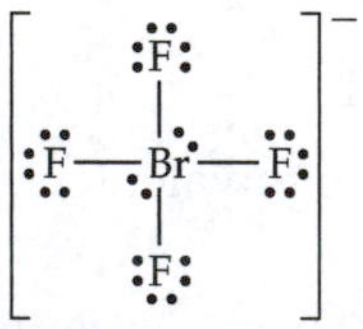

6 e^- pairs around Br
octahedral e^- domain geometry
square planar molecular geometry

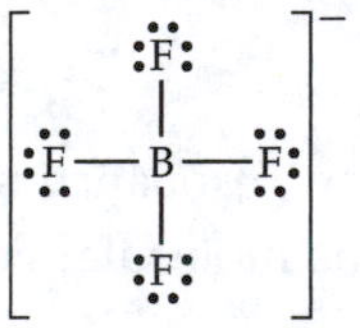

4 e^- pairs around B,
tetrahedral e^- domain geometry
tetrahedral molecular geometry

The fundamental feature that determines molecular geometry is the number of electron domains around the central atom, and the number of these that are bonding domains. Although BrF_4^- and BF_4^- are both of the form AX_4^-, the central atoms, and thus the number of valence electrons in the two ions, are different. This leads to different numbers of e^- domains about the two central atoms. Even though both ions have four bonding electron domains, the six total domains around Br require octahedral domain geometry and square planar molecular geometry, whereas the four total domains about B lead to tetrahedral domain and molecular geometry.

(b) H_2X, 8 e^-, 4 e^- pr

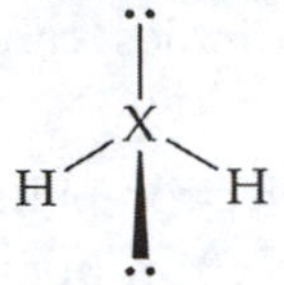

All molecules in the series have tetrahedral electron-domain geometry and bent molecular structure. To a first approximation, the H–X–H angles will be less than 109.5°. Any variation will be because of differences in repulsion among the nonbonding and bonding electron domains. The less electronegative the central atom, the larger the nonbonding electron domain, the greater the effect of repulsive forces on adjacent bonding domains. The less electronegative the central atom, the larger the deviation from ideal tetrahedral angles. The angles will vary as $H_2O > H_2S > H_2Se$.

Shapes and Polarity of Polyatomic Molecules (Section 9.3)

9.35 A bond dipole is the asymmetric charge distribution between two bonded atoms with unequal electronegativities. A molecular dipole moment is the three-dimensional sum of all the bond dipoles in a molecule. (A molecular dipole moment is a measurable physical property; a bond dipole is not measurable, unless the molecule is diatomic.)

9.37 *Analyze/plan*. Follow the logic in Sample Exercise 9.4. *Solve*.

(a) SCl_2, 20 e^-, 10 e^- pr

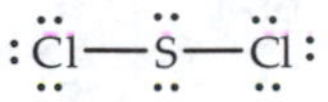

:C̤̈l—S̤̈—C̤̈l:

tetrahedral e^- domain geometry
bent molecular geometry

S and Cl have different electronegativities; the S–Cl bonds are polar. The bond dipoles are not opposite each other, so the molecule is polar. The dipole moment vector bisects the Cl–S–Cl bond angle. (A more difficult question is which end of the dipole moment vector is negative. The resultant of the two bond dipoles has its negative end toward the Cl atoms. However, the partial negative charge because of the lone pairs on S points opposite to the negative end of the resultant. A reasonable guess is that the negative end of the dipole moment vector is in the direction of the lone pairs.)

(b) $BeCl_2$, 16 e^-, 8 e^- pr

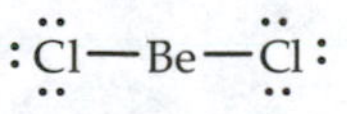

:C̤̈l—Be—C̤̈l:

linear electron-domain and molecular geometry
(Resonance structures with Be=Cl can be drawn, but electronegativity arguments predict that most electron density will reside on Cl and that the structure above is the main resonance contributor.) Be and Cl have very different electronegativities, so the Be–Cl bonds are polar. The individual bond dipoles are equal and opposite, so the net molecular dipole moment is zero.

9.39 (a) Nonpolar. The BF_3 molecule has polar B–F bonds, but they are arranged in a symmetrical trigonal plane. The individual bond dipoles cancel, leaving the molecule with zero net dipole moment.

(b) No. The BF_3^{2-} ion has 3 bonding and 1 nonbonding electron pairs around B. The added nonbonding electron pair requires that the electron domain geometry is tetrahedral and the shape is a trigonal pyramid.

(c) Yes. In BF_2Cl the B–F bond dipoles do not exactly cancel with the B–Cl bond dipole, resulting in a net dipole moment.

9.41 *Analyze/Plan*. Given molecular formulas, draw correct Lewis structures, determine molecular structure and polarity. *Solve.*

(a) Polar, $\Delta EN > 0$

I–F

(b) Nonpolar, the molecule is linear and the bond dipoles cancel.

S=C=S

(c) Nonpolar, in a symmetrical trigonal planar structure, the bond dipoles cancel.

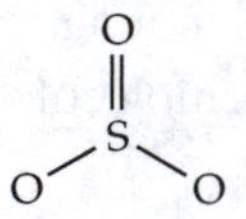

(d) Polar, although the bond dipoles are essentially zero, there is an unequal charge distribution because of the nonbonded electron pair on P.

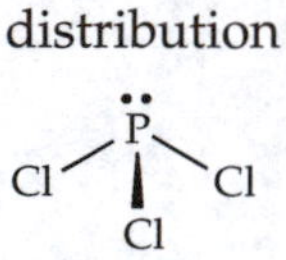

(e) Nonpolar, symmetrical octahedron

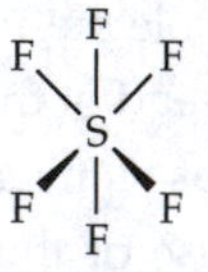

(f) Polar, square pyramidal molecular geometry, bond dipoles do not cancel.

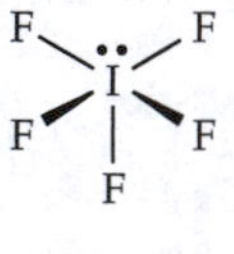

9.43 *Analyze/Plan*. Given molecular formulas, draw correct Lewis structures, analyze molecular structure and determine polarity. *Solve.*

(a) $C_2H_2Cl_2$, each isomer has 24 e^-, 12 e^- pr. Lewis structures:

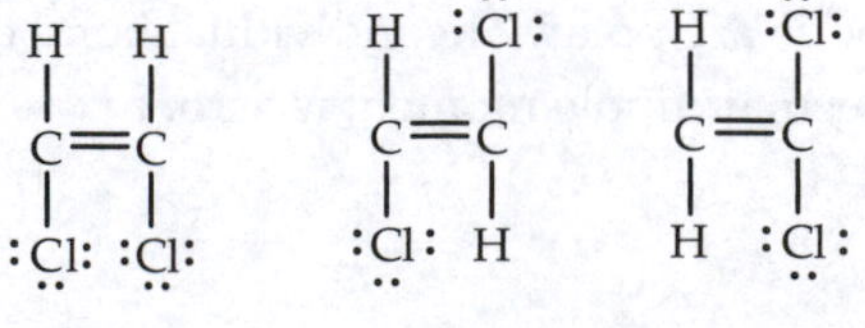

Molecular geometries:

polar nonpolar polar

(b) All three isomers are planar. The molecules on the left and right are polar because the C–Cl bond dipoles do not point in opposite directions. In the middle isomer, the C–Cl bonds and dipoles are pointing in opposite directions (as are the C–H bonds), the molecule is nonpolar and has a measured dipole moment of zero.

(c) C_2H_3Cl (lone pairs on Cl omitted for clarity)

There are four possible placements for Cl:

By rotating each of these structures in various directions, it becomes clear that the four structures are equivalent; C_2H_3Cl has only one isomer. Because C_2H_3Cl has only one C–Cl bond, the bond dipoles do not cancel, and the molecule has a dipole moment.

Orbital Overlap; Hybrid Orbitals (Sections 9.4 and 9.5)

9.45 (a) True.

(b) False. Examples of bonds that could involve an s orbital on one atom and a p orbital on another are H–F, H–Cl, etc.

(c) True. See Tables 9.2 and 9.3.

(d) False. A 1s orbital is shaped like a sphere; there are no nodal planes.

(e) True. All p orbitals have a nodal plane.

9.47 (a) False. The more orbital overlap in a bond, the stronger the bond.

(b) True.

(c) False. Hybrid orbitals are combinations of atomic orbitals on the same atom.

(d) False. Nonbonding electron pairs (lone pairs) can occupy hybrid orbitals.

9.49 (a) B: [He]$2s^2 2p^1$

(b) F, [He]$2s^2 2p^5$

(c) BF_3, 24 e^-, 12 e^- pairs

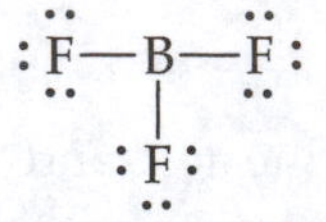

:F—B—F:
|
:F:

sp^2. The three electron domains around B require sp^2 hybrid orbitals.

(d) A single 2p orbital is unhybridized. It lies perpendicular to the trigonal plane of the sp^2 hybrid orbitals.

9.51 *Analyze/Plan.* Given the molecular (or ionic) formula, draw the correct Lewis structure and determine the electron-domain geometry, which determines hybridization. *Solve.*

(a) 24 e^-, 12 e^- pairs

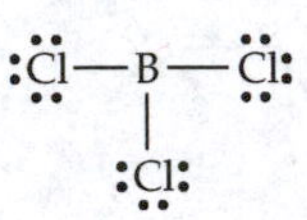

:Cl—B—Cl:
|
:Cl:

3 e^- pairs around B, trigonal planar e^- domain geometry, sp^2 hybridization

(b) 32 e^-, 16 e^- pairs

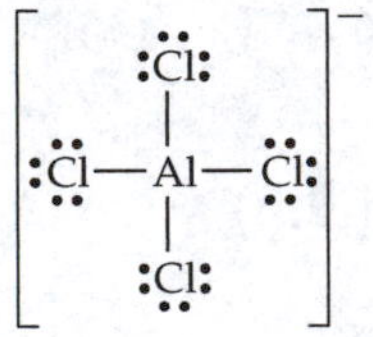

[:Cl: | :Cl—Al—Cl: | :Cl:]$^-$

4 e^- domains around Al, tetrahedral e^- domain geometry, sp^3 hybridization

(c) 16 e^-, 8 e^- pairs

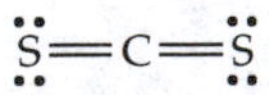

S=C=S

2 e^- domains around C, linear e^- domain geometry, sp hybridization

(d) 8 e^-, 4 e^- pairs

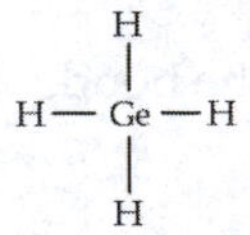

H
|
H— Ge —H
|
H

4 e^- pairs around Ge, tetrahedral e^- domain geometry, sp^3 hybridization

9.53 Left: No hybrid orbitals discussed in this chapter have angles of 90°; p atomic orbitals are perpendicular to each other.

Center: Angles of 109.5° are characteristic of sp^3 hybrid orbitals.

Right: Angles of 120° can be formed by sp^2 hybrids.

Multiple Bonds (Section 9.6)

9.55 (a)

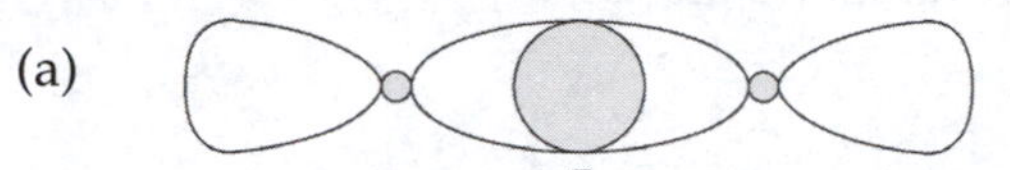

(b)

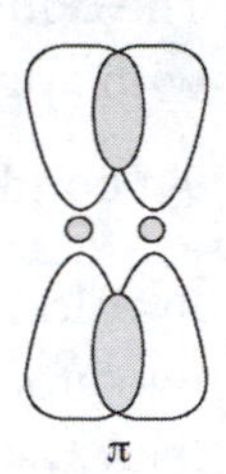

(c) A σ bond is generally stronger than a π bond, because there is more extensive orbital overlap.

(d) Two s orbitals cannot form a π bond. A π bond has no electron density along the internuclear axis. Overlap of s orbitals results in electron density along the internuclear axis. (Another way to say this is that *s* orbitals have the wrong symmetry to form a π bond.)

9.57 *Analyze/Plan.* Draw the correct Lewis structures, count electron domains and decide hybridization. Molecules with π bonds that require all bonded atoms to be in the same plane are planar. For bond-type counting, single bonds are σ bonds, double bonds consist of one σ and one π bond, triple bonds consist of one σ and two π bonds. *Solve.*

(a)

```
    H   H           H       H
    |   |            \     /
H — C — C — H         C = C          H — C ≡ C — H
    |   |            /     \
    H   H           H       H
```

(b)	sp^3	sp^2	sp
(c)	nonplanar	planar	planar
(d)	7 σ, 0 π	5 σ, 1 π	3 σ, 2 π

9.59 *Analyze/Plan.* Single bonds are σ bonds, double bonds consist of 1 σ and 1 π bond. Each bond is formed by a pair of valence electrons. *Solve.*

(a) C_3H_6 has 3(4) + 6(1) = 18 valence electrons

(b) 8 pairs or 16 total valence electrons form σ bonds

(c) 1 pair or 2 total valence electrons form π bonds

(d) no valence electrons are nonbonding

(e) The left and central C atoms are sp^2 hybridized; the right C atom is sp^3 hybridized.

9.61 *Analyze/Plan.* Given the correct Lewis structure, analyze the electron domain geometry at each central atom. This determines the hybridization and bond angles at that atom. *Solve.*

(a) ~109° bond angles about the left most C, sp^3; ~120° bond angles about the right-hand C, sp^2

(b) The doubly bonded O can be viewed as sp^2, the other as sp^3; the nitrogen is sp^3 with approximately less than 109.5° bond angles.

(c) 9 σ bonds, 1 π bond

9.63 (a) In a localized π bond, the electron density is concentrated strictly between the two atoms forming the bond. In a delocalized π bond, parallel p orbitals on more than two adjacent atoms overlap and the electron density is spread over all the atoms that contribute p orbitals to the network. There are still two regions of overlap, above and below the σ framework of the molecule.

(b) The existence of more than one resonance form is a good indication that a molecule will have delocalized π bonding.

(c) [O=N–O]⁻ ⟷ [O–N=O]⁻

The existence of more than one resonance form for NO_2 indicates that the π bond is delocalized. From an orbital perspective, the electron-domain geometry around N is trigonal planar, so the hybridization at N is sp^2. This leaves a p orbital on N and one on each O atom perpendicular to the trigonal plane of the molecule, in the correct orientation for delocalized π overlap. Physically, the two N–O bond lengths are equal, indicating that the two N–O bonds are equivalent, rather than one longer single bond and one shorter double bond.

9.65 *Analzye/Plan.* Follow the logic in Sample Exercise 9.7.

(a) 18 e^-, 9 e^- pairs

[H—C(—O:)=O]¹⁻

(b) sp^2

(c) Yes, there is one other resonance structure.

[H—C(=O:)—O:]¹⁻

(d) There are four electrons in the π system of the molecule. If the C and both O atoms are sp^2 hybridized, there are three bonding electron pairs and four nonbonding electron pairs in the σ system. This leaves two electron pairs or four electrons in the π system.

9.67 *Analyze/Plan.* Count valence e^- and e^- pairs in each molecule. Complete the Lewis structure by placing nonbonding electron pairs. Analyze the electron-domain geometry at each central atom; visualize and describe the molecular structure. *Solve.*

(a) 26 e^-, 13 e^- pairs

H—C≡C—C≡C—C≡N:

The molecule is linear. Each C atom has 2 bonding e^- domains, linear geometry, and sp hybridization. This requires that all atoms not only lie in the same plane, but in a line.

(b) 34 e^-, 17 e^- pairs

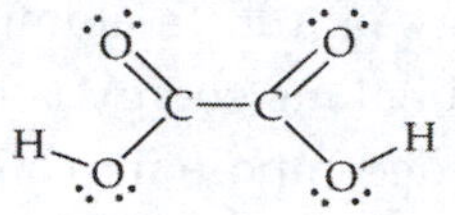

The two central C atoms each have 3 bonding e^- domains, trigonal planar geometry, and sp^2 hybridization. Each O–C–O group is planar, whereas the terminal H atoms can rotate out of these planes. In principle, there is free rotation about the C–C σ bond, but delocalization of the π electrons is possible if the two planes are coincident. It is possible to put all 8 atoms in the same plane.

(c) 12 e^-, 6 e^- pairs

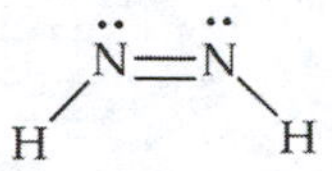

The molecule is planar. Each N atom has 3 bonding e^- domains, trigonal planar geometry, and sp^2 hybridization. Because the N atoms share a π bond, the planes must be coincident and all 4 atoms are required to lie in this plane. [The structure shown here has H atoms on the same side of the double bond. There is another isomer that has H atoms on opposite sides of the double bond. Both compounds are planar with 120 degree bond angles.]

Molecular Orbitals and Period 2 Diatomic Molecules (Sections 9.7 and 9.8)

9.69 (a) Hybrid orbitals are mixtures (linear combinations) of atomic orbitals from a single atom; the hybrid orbitals remain localized on that atom. Molecular orbitals are combinations of atomic orbitals from two or more atoms. They are associated with the entire molecule, not a single atom.

(b) Each MO, like each AO or hybrid, can hold a maximum of two electrons.

(c) Yes, antibonding MOs can have electrons in them.

9.71 (a) σ^*_{1s} σ^*_{1s} 1s 1s σ_{1s} σ_{1s} H_2^+

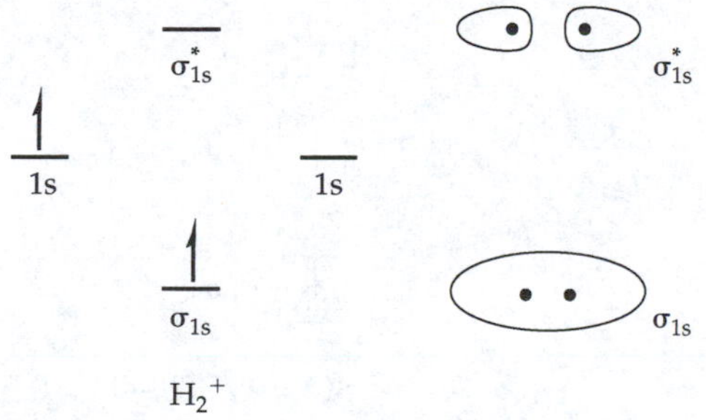

(b) There is one electron in H_2^+.

(c) σ^*_{1s} σ_{1s}

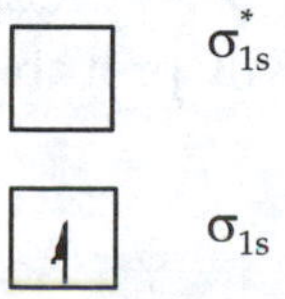

(d) Bond order = 1/2 (1 – 0) = ½

(e) Fall apart. The stability of H_2^+ is because of the lower-energy state of the σ bonding molecular orbital relative to the energy of a H 1s atomic orbital. If the single electron in H_2^+ is excited to the σ^*_{1s} orbital, its energy is higher than the energy of an H 1s atomic orbital and H_2^+ will decompose into a hydrogen atom and a hydrogen ion.

$$H_2^+ \xrightarrow{h\nu} H + H^+.$$

(f) Statement (i) is true.

9.73

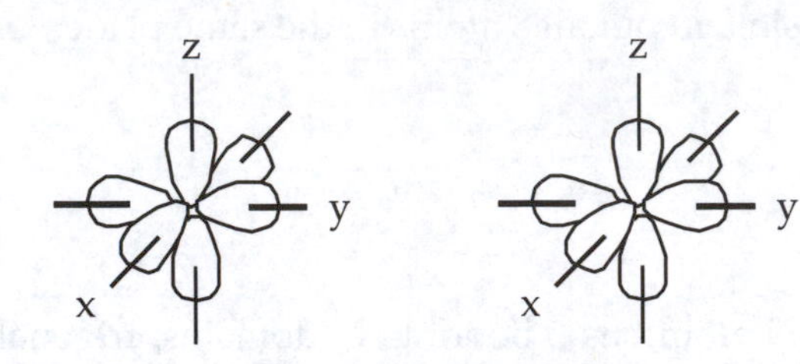

(a) One. With three mutually perpendicular p orbitals on each atom, only one set can be oriented for end-to-end sigma overlap.

(b) Two. The 2 *p* orbitals on each atom not involved in σ bonding can be aligned for side-to-side π overlap.

(c) Three, 1 σ* and 2 π*. There are a total of 6 *p* orbitals on the two atoms. When combining AOs to form MOs, total number of orbitals is conserved. If 3 of the 6 MOs are bonding MOs, as described in (a) and (b), then the remaining 3 MOs must be antibonding. They will have the same symmetry as the bonding MOs, 1 σ* and 2 π*.

9.75 (a) When comparing the same two bonded atoms, the greater the bond order, the shorter the bond length and the greater the bond energy. That is, bond order and bond energy are directly related, whereas bond order and bond length are inversely related. When comparing different bonded nuclei, there are no simple relationships (see Solution 8.100).

(b) Be_2, 4 e^- | Be_2^+, 3 e^-

Be_2, 4 e^-		Be_2^+, 3 e^-	
↑↓	σ^*_{2s}	↑	σ^*_{2s}
↑↓	σ_{2s}	↑↓	σ_{2s}
BO = 1/2(2 – 2) = 0		BO = 1/2(2 – 1) = 0.5	

Be_2 has a bond order of zero and is not energetically favored over isolated Be atoms; it is not expected to exist. Be_2^+ has a bond order of 0.5 and is slightly lower in energy than isolated Be atoms. It will probably exist under special experimental conditions, but be unstable.

9.77 (a,b) Substances with no unpaired electrons are weakly repelled by a magnetic field. This property is called *diamagnetism*.

(c) O_2^{2-}, Be_2^{2+} [see Figure 9.43 and Solution 9.75(b)]

9.79

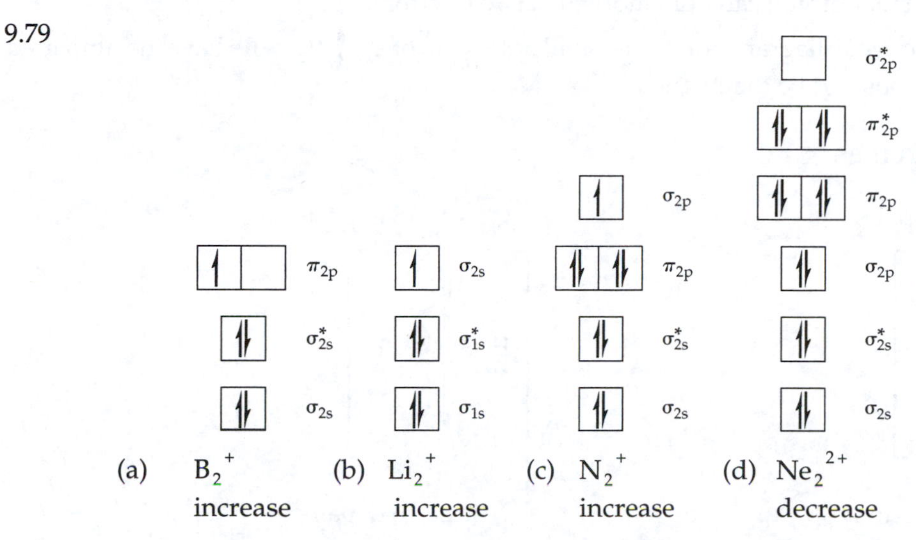

(a) B_2^+ increase (b) Li_2^+ increase (c) N_2^+ increase (d) Ne_2^{2+} decrease

Addition of an electron increases bond order if it occupies a bonding MO and decreases stability if it occupies an antibonding MO.

9.81 *Analyze/Plan*. Determine the number of "valence" (non-core) electrons in each molecule or ion. Use the homonuclear diatomic MO diagram from Figure 9.43 (shown below) to calculate bond order and magnetic properties of each species. The electronegativity difference between heteroatomics increases the energy difference between the 2s AO on one atom and the 2p AO on the other, rendering the "no interaction" MO diagram in Figure 9.43 appropriate. *Solve*.

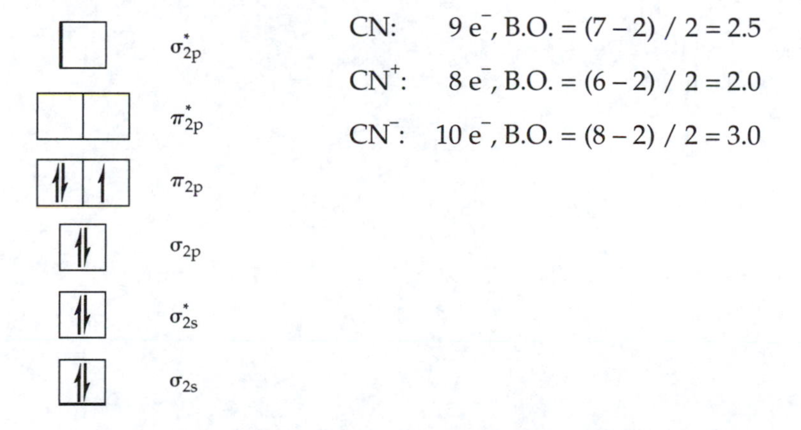

CN: 9 e^-, B.O. = (7 – 2) / 2 = 2.5

CN^+: 8 e^-, B.O. = (6 – 2) / 2 = 2.0

CN^-: 10 e^-, B.O. = (8 – 2) / 2 = 3.0

(a) CN^- has the highest bond order and therefore the strongest C–N bond.

(b) CN and CN^+. CN has an odd number of valence electrons, so it must have an unpaired electron. The electron configuration for CN is shown in the diagram. Removing one electron from the π_{2p} MOs to form CN^+ produces an ion with two unpaired electrons. Adding one electron to the π_{2p} MOs of CN to form CN^- produces an ion with all electrons paired.

9.83 (a) 3s, $3p_x$, $3p_y$, $3p_z$ (b) π_{3p}

(c) Two. Note that there are two degenerate π_{3p} bonding molecular orbitals; each holds two electrons. A total of 4 electrons can be designated as π_{3p}, but no single molecular orbital can hold more than two electrons.

(d) If the MO diagram for P_2 is similar to that of N_2, P_2 will have no unpaired electrons and be diamagnetic.

Additional Exercises

9.87 34 e^-, 17 e^- pairs 36 e^-, 18 e^- pairs

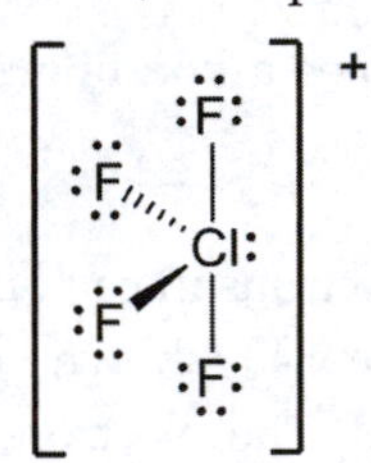

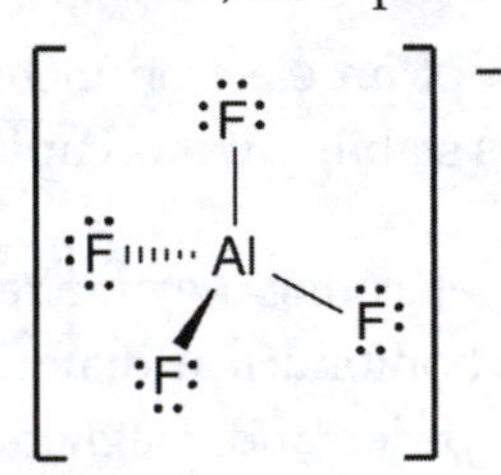

34 e^-, 17 e^- pairs 32 e^-, 16 e^- pairs

(a) Three. The P, Br, and Cl central atoms have more than an octet of electrons about them.

(b) One, AlF_4^-

(c) BrF_4^-

(d) PF_4^- and ClF_4^+

9.90 *Analyze/Plan.* For entries where the molecule is listed, follow the logic in Sample Exercises 9.4 and 9.5. For entries where no molecule is listed, decide electron-domain geometry from hybridization (or vice versa). If the molecule is nonpolar, the terminal atoms will be identical. If the molecule is polar, the terminal atoms will be different, or the central atom will have one or more lone pairs, or both. *Solve.*

Molecule	Molecular Structures	Electron Domain Geometry	Hybrdization of CentralAtom	Dipole Moment Yes or No
CO_2	O=C=O	linear	sp	no
NH_3	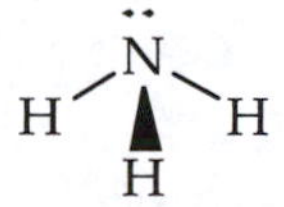	tetrahedral	sp^3	yes
CH_4	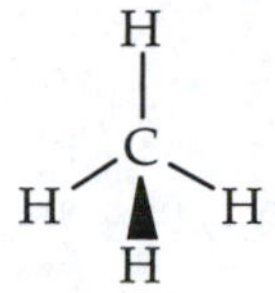	tetrahedral	sp^3	no
BH_3	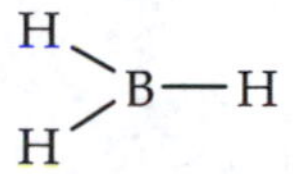	trigonal planar	sp^2	no
SF_4	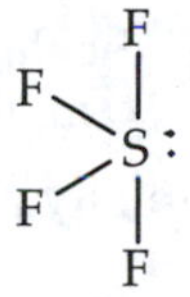	trigonal bipyramidal	not applicable	yes
SF_6		octahedral	not applicable	no
H_2CO	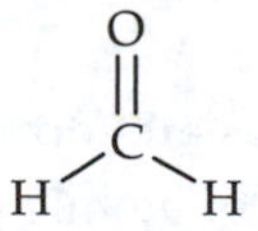	trigonal planar	sp^2	yes
PF_5	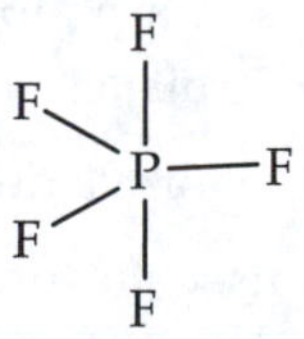	trigonal bipyramidal	not applicable	no
XeF_2	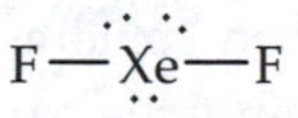 F—Xe—F	trigonal bipyramidal	not applicable	no

9.91 (a) CO_2, 16 valence e^-

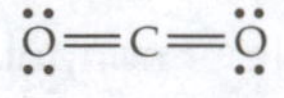

$\ddot{O}=C=\ddot{O}$

2σ 2π

(b) $(CN)_2$, 18 valence e^-

:N≡C—C≡N:

3σ, 3π

(c) H_2CO, 12 valence e^-

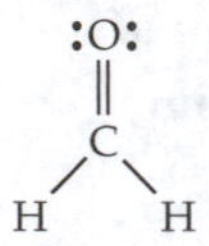

3σ, 1π

(d) HCOOH, 18 valence e^-

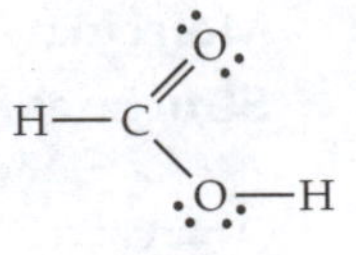

4σ, 1π

9.98

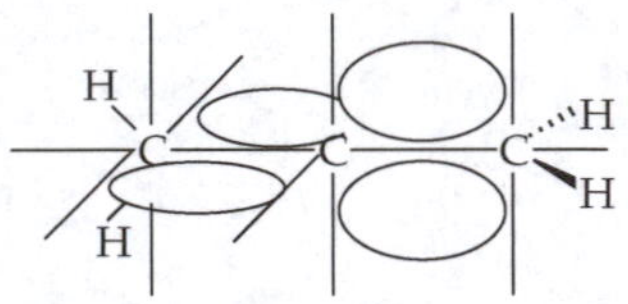

(a) The molecule is not planar. The CH_2 planes at each end are twisted 90° from one another.

(b) Allene has no dipole moment.

(c) The bonding in allene would not be described as delocalized. The π electron clouds of the two adjacent C=C are mutually perpendicular. The mechanism for delocalization of π electrons is mutual overlap of parallel p atomic orbitals on adjacent atoms. If adjacent π electron clouds are mutually perpendicular, there is no overlap and no delocalization of π electrons.

9.101 (a) $\ddot{O}=\ddot{O}-\ddot{O}: \longleftrightarrow :\ddot{O}-\ddot{O}=\ddot{O}$

To accommodate the π bonding by all 3 O atoms indicated in the resonance structures above, all O atoms are sp^2 hybridized.

(b) For the first resonance structure, both sigma bonds are formed by overlap of sp^2 hybrid orbitals, the π bond is formed by overlap of atomic p orbitals, one of the nonbonded pairs on the right terminal O atom is in a p atomic orbital, and the remaining five nonbonded pairs are in sp^2 hybrid orbitals.

(c) Only unhybridized p atomic orbitals can be used to form a delocalized π system.

(d) The unhybridized p orbital on each O atom is used to form the delocalized π system, and in both resonance structures one nonbonded electron pair resides in a p atomic orbital. The delocalized π system then contains four electrons, two from the π bond and two from the nonbonded pair in the p orbital.

9.108 The white solid has the larger HOMO-LUMO gap. The green solid absorbs visible red light and appears green. The white solid does not absorb visible light, but light of higher energy in the ultraviolet region of the spectrum. The white solid has the larger HOMO-LUMO gap.

Integrative Exercises

9.114 (a) $2SF_4(g) + O_2(g) \rightarrow 2OSF_4(g)$

(b) 40 e^-, 20 e^- pairs

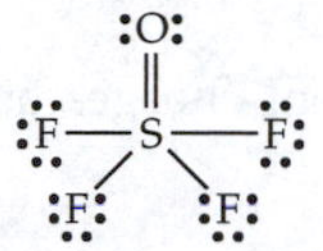

There must be a double bond drawn between O and S in order for their formal charges to be zero.

(c) $\Delta H = 8D(S–F) + D(O=O) – 8D(S–F) – 2D(S=O)$

$\Delta H = D(O=O) – 2D(S=O) = 495 – 2(523) = –551$ kJ, exothermic

(d) trigonal-bipyramidal electron-domain geometry

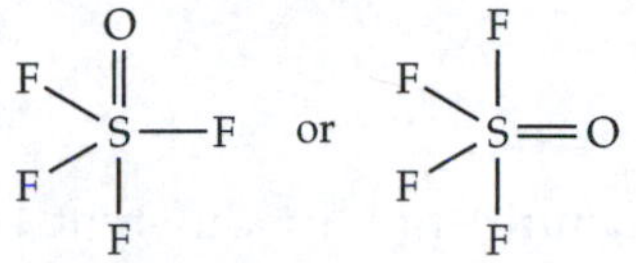

(e) In the structure on the left, there are 3 equatorial and 1 axial fluorine atoms. In the structure on the right, there are 2 equatorial and 2 axial fluorine atoms.

9.118

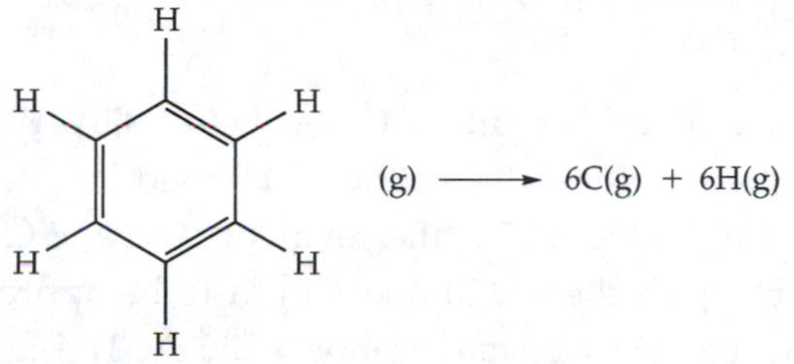

(g) ⟶ 6C(g) + 6H(g)

$\Delta H = 6D(C–H) + 3D(C–C) + 3D(C=C) – 0$

$= 6(413 \text{ kJ}) + 3(348 \text{ kJ}) + 3(614 \text{ kJ})$

$= 5364$ kJ

(The products are isolated atoms; there is no bond making.)

According to Hess' law:

$$\Delta H° = 6\Delta H_f^\circ\ C(g) + 6\Delta H_f^\circ\ H(g) - \Delta H_f^\circ\ C_6H_6(g)$$

$= 6(718.4 \text{ kJ}) + 6(217.94 \text{ kJ}) – (82.9 \text{ kJ})$

$= 5535$ kJ

The difference in the two results, 171 kJ/mol C_6H_6 is because of the resonance stabilization in benzene. That is, because the π electrons are delocalized, the molecule has a lower overall energy than that predicted for the presence of 3 localized C–C and C=C bonds. Thus, the amount of energy actually required to decompose 1 mole of $C_6H_6(g)$, represented by the Hess' law calculation, is greater than the sum of the localized bond enthalpies (not taking resonance into account) from the first calculation above.

9.122 (a) 22 valence e^-, 11 e^- pairs

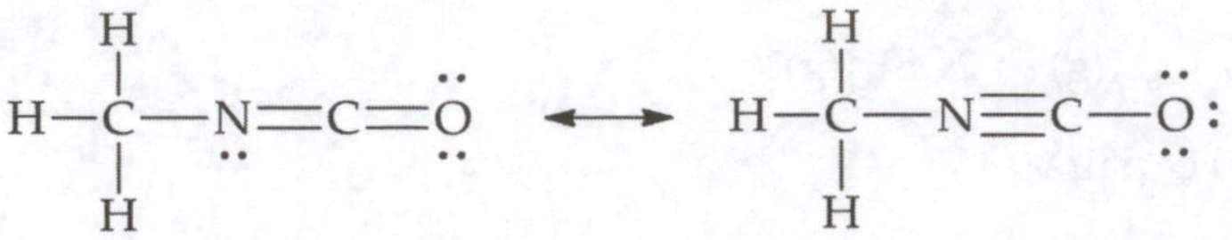

```
     H                              H
     |          ..                  |           ..
 H—C—N=C=O         ←→      H—C—N≡C—O:
     |    ..    ..                  |           ..
     H                              H
```

(The structure on the right does not minimize formal charges and will make a minor contribution to the true structure.)

(b) Both resonance structures predict the same bond angles. We expect H–C–N angles close to 109.5°.

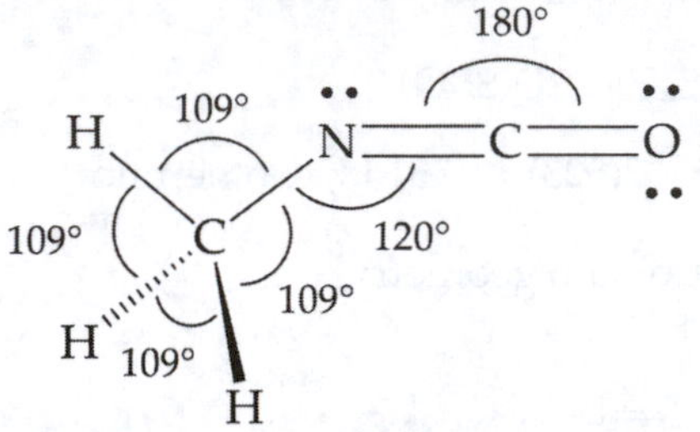

```
                         180°
          109°     ..          ..
   H              N═══C═══O
  109°      C            ..
                   120°
           109°
  H  109°
           H
```

(c) The two extreme Lewis structures predict different bond lengths. As the true bonding model is some blend of the extreme Lewis structures, the true bond lengths are a blend of the extreme values. Our bond length estimates take into account that the structure minimizing formal charge makes a larger contribution to the true structure.

From Figure 7.7, the sum of the covalent radii for C and H is 1.07Å. We list this value in the table. As corroboration, the value of the O–H bond length given in Exercise 9.95 is 0.96 Å. According to Figure 7.7, the covalent radius of C is 0.10 Å greater than that of O, so we expect the C–H bond length to be approximately 0.10 Å greater than the measured O–H distance, about 1.06 Å. The lengths for isolated C–N, N–C, and C–O bonds are taken from Table 8.4.

Bond	Length (Å) N=C=O	Length (Å) N≡C–O	Length (Å) estimated
C–H	1.07	1.07	1.07
C–N	1.43	1.43	1.43
N–C	1.38	1.16	1.33
C–O	1.23	1.43	1.28

(d) The molecule will have a dipole moment. The N–C and C–O bond dipoles are opposite each other, but they are not equal. And, there are nonbonding electron pairs that are not directly opposite each other (in either structure) and will not cancel. There will be a resulting dipole.

10 Gases

Visualizing Concepts

10.1 It would be much easier to drink from a straw on Mars. When a straw is placed in a glass of liquid, the liquid level in the straw equals the liquid level in the glass. The atmosphere exerts equal pressure inside and outside the straw. When we drink through a straw, we withdraw air, thereby reducing the pressure on the liquid inside. If only 0.007 atm is exerted on the liquid in the glass, a very small reduction in pressure inside the straw will cause the liquid to rise.

Another approach is to consider the gravitational force on Mars. Because the pull of gravity causes atmospheric pressure, the gravity on Mars must be much smaller than that on Earth. With a very small Martian gravity holding liquid in a glass, it would be very easy to raise the liquid through a straw.

10.3 Statement (b) is correct. At constant volume and temperature, pressure depends on total number of particles. To reduce the pressure by a factor of 2, the number of particles must be reduced by a factor of 2. At the lower pressure, the container would have half as many particles as at the higher pressure.

Compare the two situations using the ideal-gas law, $PV = nRT$. $P = nRT/V$.

$P_2 = P_1/2$. V, R, and T are the same for the two states.

$n_2RT/V = (n_1RT/V)/2$; $n_2 = n_1/2$

10.5 $PV = nRT$ (ideal-gas equation). In the ideal-gas equation, R is a constant. Given constant V and n (fixed amount of ideal gas), P and T are directly proportional. If P is doubled, T is also doubled. That is, if P is doubled, T increases by a factor of 2.

10.7 (a) Partial pressure depends on the number of particles of each gas present. Red has the fewest particles, then yellow, then blue. $P_{red} < P_{yellow} < P_{blue}$

(b) $P_{gas} = \chi_{gas} P_t$. Calculate the mole fraction, χ_{gas} = [mol gas / total moles] or [particles gas / total particles]. This is true because Avogadro's number is a counting number, and mole ratios are also particle ratios.

χ_{red} = 2 red atoms / 10 total atoms = 0.2; P_{red} = 0.2(1.40 atm) = 0.28 atm

χ_{yellow} = 3 yellow atoms / 10 total atoms = 0.3; P_{yellow} = 0.3(1.40 atm) = 0.42 atm

χ_{blue} = 5 blue atoms / 10 total atoms = 0.5; P_{blue} = 0.5(1.40 atm) = 0.70 atm

Check. (0.28 atm + 0.42 atm + 0.70 atm) = 1.40 atm. The sum of the calculated partial pressures equals the given total pressure.

10.9 (a) Curve B. At constant temperature, the root-mean-square (rms) speed (as well as the average speed) of a collection of gas particles is inversely related to molar mass; the lighter the particle, the faster it moves. Therefore, curve B represents He and curve A represents O_2. Curve B has the higher rms speed and He is the lighter gas. Curve A has the lower rms speed and O_2 is the heavier gas.

(b) B. For the same gas, average kinetic energy $(1/2\, mu_{rms}^2)$, and therefore root-mean-square speed (u_{rms}) is directly related to Kelvin temperature. Curve A is the lower temperature and curve B is the higher temperature.

(c) The root-mean-square speed. According to Figure 10.12(b), the most probable speed is lowest, then the average speed, and the rms speed is highest. This is true for the distribution of speeds for any gas, including the two in this exercise.

10.11 The $NH_4Cl(s)$ ring will form at location a. The process described in this exercise is diffusion, rather than simple effusion. According to Section 10.8, Graham's law approximates (but does not exactly describe) the diffusion rates of two gases under identical conditions. According to Graham's law, the ratio of rates is inversely related to the ratio of molar masses of the two gases. That is, the lighter gas moves faster than the heavier gas. When introduced into the tube, NH_3, MM = 17, moves faster and therefore farther than HCl, MM = 36. If NH_3 moves farther than HCl, the two gases meet and form $NH_4Cl(s)$ nearer the end where HCl was introduced; this is in the vicinity of location a.

Gas Characteristics; Pressure (Sections 10.1 and 10.2)

10.13 Statement (c) is false. Gaseous molecules are so far apart that there is no barrier to mixing, regardless of the identity of the molecules.

10.15 *Analyze.* Given: mass, area. Find: pressure. *Plan.* P = F/A. Data is given in terms of lb and in.^{2}; calculate pressure in lb/in.^{2} (or psi) for part (c), then use this value to find pressure in and atm for part (b) and kPa for part (a). *Solve.*

(a) $$P = \frac{F}{A} = \frac{130 \text{ lb}}{0.50 \text{ in.}^2} = 260 = 2.6 \times 10^2 \text{ lb/in.}^2$$

(b) 1 atm = 101.325 kPa

$$17.687 \text{ atm} \times \frac{101.325 \text{ kPa}}{1 \text{ atm}} = 1792 = 1.8 \times 10^3 \text{ kPa}$$

(c) 14.70 lb/in.^{2} = 1 atm

$$260 \text{ lb/in.}^2 \times \frac{1 \text{ atm}}{14.70 \text{ lb/in.}^2} = 17.687 = 18 \text{ atm}$$

10.17 *Analyze.* Given: 760 mm column of Hg, densities of Hg and glycerol. Find: height of a column of glycerol at the same pressure.

Plan. Develop a relationship between pressure, height of a column of liquid, and density of the liquid. Relationships that might prove useful: P = F/A; F = m × a; m = d × V(density)(volume); V = A × height *Solve.*

$$P = \frac{F}{A} = \frac{m \times a}{A} = \frac{d \times V \times a}{A} = \frac{d \times A \times h \times a}{A} = d \times h \times a$$

(a) $P_{Hg} = P_{glycerol}$; Using the relationship derived above: $(d \times h \times a)_{glycerol} = (d \times h \times a)_{Hg}$

Because a, the acceleration due to gravity, is equal in both liquids,
$(d \times h)_{glycerol} = (d \times h)_{Hg}$

$1.26\ g/mL \times h_{glycerol} = 13.6\ g/mL \times 760\ mm$

$$h_{glycerol} = \frac{13.6\ g/mL \times 760\ mm}{1.26\ g/mL} = 8.203 \times 10^3 = 8.20 \times 10^3\ mm = 8.20\ m$$

(b) *Plan.* Calculate the height of a column of Hg that exerts the same pressure as 15 ft of water. The density of water is 1.00 g/mL. Then calculate the total pressure, atmospheric plus water, in atmospheres.

$$15\ ft\ H_2O \times \frac{12\ in.}{1\ ft} \times \frac{2.54\ cm}{1\ in.} \times \frac{10\ mm}{1\ cm} = 4572 = 4.6 \times 10^3\ mm\ H_2O$$

$(d \times h)_{H_2O} = (d \times h)_{Hg}$

$1.00\ g/mL \times 4572\ mm\ H_2O = 13.6\ g/mL \times ?\ mm\ Hg$

$$h_{Hg} = \frac{1.00\ g/mL \times 4572\ mm}{13.6\ g/mL} = 336.2 = 3.4 \times 10^2\ mm\ Hg$$

$P_{total} = P_{atm} + P_{H_2O} = 750\ mm\ Hg + 336\ mm\ Hg = 1086\ mm\ Hg = 1.1 \times 10^2\ mm\ Hg$

$$P_{total} = 1086\ mm\ Hg \times \frac{1\ atm}{760\ mm\ Hg} = 1.429 = 1.4\ atm$$

10.19 *Analyze/Plan.* We are given pressure in one unit and asked to change it to another unit. Select appropriate conversion factors and use dimensional analysis to find the desired quantities. *Solve.*

(a) $265\ torr \times \frac{1\ atm}{760\ torr} = 0.349\ atm$

(b) $265\ torr \times \frac{1\ mm\ Hg}{1\ torr} = 265\ mm\ Hg$

(c) $265\ torr \times \frac{1.01325 \times 10^5\ Pa}{760\ torr} = 3.53 \times 10^4\ Pa$

(d) $265\ torr \times \frac{1.01325 \times 10^5\ Pa}{760\ torr} \times \frac{1\ bar}{1 \times 10^5\ Pa} = 0.353\ bar$

(e) $265\ torr \times \frac{1\ atm}{760\ torr} \times \frac{14.70\ psi}{1\ atm} = 5.13\ psi$

10.21 *Analyze/Plan.* Follow the logic in Solution 10.19. *Solve.*

(a) $30.45\ in.\ Hg \times \frac{25.4\ mm}{1\ in.} \times \frac{1\ torr}{1\ mm\ Hg} = 773.4\ torr$

[The result has 4 sig figs because 25.4 mm/in. is considered to be an exact number.]

(b) $30.45\ in\ Hg = 773.4\ torr;\ \ 773.4\ torr \times \frac{1\ atm}{760\ torr} = 1.018\ atm$

10.23 *Analyze/Plan*. Follow the logic in Sample Exercise 10.2. *Solve.*

(i) The Hg level is lower in the open end than the closed end, so the gas pressure is less than atmospheric pressure.

$$P_{gas} = 0.995 \text{ atm} - \left(52 \text{ cm} \times \frac{1 \text{ atm}}{76.0 \text{ cm}}\right) = 0.31 \text{ atm}$$

(ii) The Hg level is higher in the open end, so the gas pressure is greater than atmospheric pressure.

$$P_{gas} = 0.995 \text{ atm} + \left(67 \text{ cm Hg} \times \frac{1 \text{ atm}}{76.0 \text{ cm Hg}}\right) = 1.8766 = 1.88 \text{ atm}$$

(iii) This is a closed-end manometer, so $P_{gas} = h$.

$$P_{gas} = 10.3 \text{ cm} \times \frac{1 \text{ atm}}{76.0 \text{ cm}} = 0.136 \text{ atm}$$

The Gas Laws (Section 10.3)

10.25 *Analyze/Plan*. Given certain changes in volume and temperature of a gas contained in a cylinder with a moveable piston, predict which will double the gas pressure. Consider the gas law relationships in Section 10.3. *Solve.*

(a) No. P and V are inversely proportional at constant T. If the volume increases by a factor of 2, the pressure *decreases* by a factor of 2.

(b) No. P and *Kelvin* T are directly proportional at constant V. Doubling °C does increase V, but it does not double it.

(c) Yes. P and V are inversely proportional at constant T. If the volume decreases by a factor of 2, the pressure increases by a factor of 2.

10.27 (a) *Analyze/Plan*. Use Boyle's law, PV = constant or $P_1V_1 = P_2V_2$, and Charles' law, V/T = constant or $V_1/T_1 = V_2/T_2$, to derive Amonton's law for the relationship between P and T at constant V. *Solve.*

Boyle's law: $P_1V_1 = P_2V_2$ or $\frac{P_1}{P_2} = \frac{V_2}{V_1}$. At constant V, $\frac{V_2}{V_1} = 1$ and $\frac{P_1}{P_2} = 1$.

Charles' law: $\frac{V_1}{T_1} = \frac{V_2}{T_2}$ or $\frac{V_1}{V_2} = \frac{T_1}{T_2}$. At constant V, $\frac{V_1}{V_2} = 1$ and $\frac{T_1}{T_2} = 1$.

$\frac{P_1}{P_2} = 1$ and $\frac{T_1}{T_2} = 1$, then $\frac{P_1}{P_2} = \frac{T_1}{T_2}$ or $\frac{P_1}{T_1} = \frac{P_2}{T_2}$ or P/T = constant.

Amonton's law is that pressure and temperature are directly proportional at constant volume.

(b) *Analyze/Plan*. Recall that the proportional relationships for gases are true for any units of pressure (or volume) but only for Kelvin temperatures. Change °F to K, then use the relationship derived in (a) to calculate the new tire pressure. °C = 5/9(°F – 32); K = °C + 273. *Solve.*

T_1: °C = 5/9(75 – 32) = 23.89 = 24 °C. K = 24 °C + 273 = 297 K. T_1 = 32.0 psi

T_2: °C = 5/9(120 – 32) = 48.89 = 49 °C. K = 49 °C + 273 = 322 K. P_2 = ?

$$\frac{P_1}{T_1} = \frac{P_2}{T_2} \text{ or } P_2 = \frac{P_1T_2}{T_1} = \frac{32.0 \text{ psi} \times 322 \text{ K}}{297 \text{ K}} = 34.694 = 34.7 \text{ psi}$$

The Ideal-Gas Equation (Section 10.4)

(In *Solutions to Exercises*, the symbol for molar mass is MM.)

10.29 (a) STP stands for standard temperature, 0 °C (or 273 K), and standard pressure, 1 atm.

(b) $V = \frac{nRT}{P}$; $V = 1 \text{ mol} \times \frac{0.08206 \text{ L-atm}}{\text{mol-K}} \times \frac{273 \text{ K}}{1 \text{ atm}}$

V = 22.4 L for 1 mole of gas at STP

(c) 25 °C + 273 = 298 K

$V = \frac{nRT}{P}$; $V = 1 \text{ mol} \times \frac{0.08206 \text{ L-atm}}{\text{mol-K}} \times \frac{298 \text{ K}}{1 \text{ atm}}$

V = 24.5 L for 1 mol of gas at 1 atm and 25 °C

(d) $\frac{0.08206 \text{ L-atm}}{\text{mol-K}} \times \frac{1.01325 \times 10^5 \text{ Pa}}{1 \text{ atm}} \times \frac{1 \text{ bar}}{10^5 \text{ Pa}} = \frac{0.08315 \text{ L-bar}}{\text{mol-K}}$

10.31 *Analyze/Plan.* PV = nRT. At constant volume and temperature, P is directly proportional to n.

Solve. For samples with equal masses of gas, the gas with MM = 30 will have twice as many moles of particles and twice the pressure. Thus, flask A contains the gas with MM = 30 and flask B contains the gas with MM = 60.

10.33 *Analyze/Plan.* Follow the strategy for calculations involving many variables given in Section 10.4. *Solve.*

$T = \frac{PV}{nR} = 2.00 \text{ atm} \times \frac{1.00 \text{ L}}{0.500 \text{ mol}} \times \frac{\text{mol-K}}{0.08206 \text{ L-atm}} = 48.7 \text{ K}$

K = 27 °C + 273 = 300 K

$n = \frac{PV}{RT} = 0.300 \text{ atm} \times \frac{0.250 \text{ L}}{300 \text{ K}} \times \frac{\text{mol-K}}{0.08206\text{-atm}} = 3.05 \times 10^{-3} \text{ mol}$

$650 \text{ torr} \times \frac{1 \text{ atm}}{760 \text{ torr}} = 0.85526 = 0.855 \text{ atm}$

$V = \frac{nRT}{P} = 0.333 \text{ mol} \times \frac{350 \text{ K}}{0.85526 \text{ atm}} \times \frac{0.08206 \text{ L-atm}}{\text{mol-K}} = 11.2 \text{ L}$

585 mL = 0.585 L

$P = \frac{nRT}{V} = 0.250 \text{ mol} \times \frac{295 \text{ K}}{0.585 \text{ L}} \times \frac{0.08206 \text{ L-atm}}{\text{mol-K}} = 10.3 \text{ atm}$

P	V	n	T
2.00 atm	1.00 L	0.500 mol	**48.7 K**
0.300 atm	0.250 L	**3.05×10^{-3} mol**	27 °C
650 torr	**11.2 L**	0.333 mol	350 K
10.3 atm	585 mL	0.250 mol	295 K

10.35 *Analyze/Plan.* Follow the strategy for calculations involving many variables. *Solve.*

$n = g/MM$; $PV = nRT$; $PV = gRT/MM$; $g = MM \times PV/RT$

$P = 1.0$ atm, $T = 23$ °C $= 296$ K, $V = 1.75 \times 10^5$ ft^3. Change ft^3 to L, then calculate grams (or kg).

$$1.75\times10^5 \text{ft}^3 \times \frac{(12)^3 \text{ in}^3}{\text{ft}^3} \times \frac{(2.54)^3 \text{ cm}^3}{\text{in}^3} \times \frac{1 \text{ L}}{1\times10^3 \text{ cm}^3} = 4.9554\times10^6 = 4.96\times10^6 \text{ L}$$

$$g = \frac{4.003 \text{ g He}}{1 \text{ mol He}} \times \frac{\text{mol-K}}{0.08206 \text{ L-atm}} \times \frac{1.0 \text{ atm} \times 4.955\times10^6 \text{ L}}{296 \text{ K}} = 8.2\times10^5 \text{g} = 820 \text{ kg He}$$

10.37 *Analyze/Plan.* Follow the strategy for calculations involving many variables. *Solve.*

(a) $V = 2.25$ L; $T = 273 + 37$ °C $= 310$ K; $P = 735 \text{ torr} \times \frac{1 \text{ atm}}{760 \text{ torr}} = 0.96710 = 0.967$ atm

$PV = nRT$, $n = PV/RT$, number of molecules (#) $= n \times 6.022 \times 10^{23}$

$$\# = \frac{0.9671 \text{ atm} \times 2.25 \text{ L}}{310 \text{ K}} \times \frac{\text{mol-K}}{0.08206 \text{ L-atm}} \times \frac{6.022\times10^{23} \text{ molecules}}{\text{mol}}$$
$$= 5.15 \times 10^{22} \text{ molecules}$$

(b) $V = 5.0 \times 10^3$ L; $T = 273 + 0$ °C $= 273$ K; $P = 1.00$ atm; $MM = 28.98$ g/mol

$$PV = \frac{g}{MM}RT; \quad g = \frac{MM \times PV}{RT}$$

$$g = \frac{28.98 \text{ g air}}{1 \text{ mol air}} \times \frac{\text{mol-K}}{0.08206 \text{ L-atm}} \times \frac{1.00 \text{ atm} \times 5.0\times10^3 \text{ L}}{273 \text{ K}}$$
$$= 6468 \text{ g} = 6.5 \times 10^3 \text{ g air} = 6.5 \text{ kg air}$$

10.39 *Analyze/Plan.* Follow the strategy for calculations involving many variables. *Solve.*

(a) $P = \frac{nRT}{V}$; $n = 0.29 \text{ kg } O_2 \times \frac{1000 \text{ g}}{1 \text{ kg}} \times \frac{1 \text{ mol } O_2}{32.00 \text{ g } O_2} = 9.0625 = 9.1$ mol; $V = 2.3$ L;

$T = 273 + 9$ °C $= 282$ K

$$P = \frac{9.0625 \text{ mol}}{2.3 \text{ L}} \times \frac{0.08206 \text{ L-atm}}{\text{mol-K}} \times 282 \text{ K} = 91 \text{ atm}$$

(b) $$V = \frac{nRT}{P}; = \frac{9.0625 \text{ mol}}{0.95 \text{ atm}} \times \frac{0.08206 \text{ L-atm}}{\text{mol-K}} \times 299 \text{ K} = 2.3\times10^2 \text{ L}$$

10.41 *Analyze/Plan.* Follow the strategy for calculations involving many variables. *Solve.*

$P = \frac{nRT}{V}$; $n = 35.1 \text{ g } CO_2 \times \frac{1 \text{ mol } CO_2}{44.01 \text{ g } CO_2} = 0.7975 = 0.798$ mol; $V = 4.0$ L; $T = 298$ K

$$P = \frac{0.7975 \text{ mol}}{4.0 \text{ L}} \times \frac{0.08206 \text{ L-atm}}{\text{mol-K}} \times 298 \text{ K} = 4.876 = 4.9 \text{ atm}$$

10.43 *Analyze/Plan.* Follow the strategy for calculations involving many variables. *Solve.*

$$V = 8.70\ L,\ T = 24\ °C = 297\ K,\ P = 895\ torr \times \frac{1\ atm}{760\ torr} = 1.1776 = 1.18\ atm$$

(a) $$g = \frac{MM \times PV}{RT};\ g = \frac{70.91g\ Cl_2}{1\ mol\ Cl_2} \times \frac{mol\text{-}K}{0.08206\ L\text{-}atm} \times \frac{1.1776\ atm}{297\ K} \times 8.70\ L = 29.8g\ Cl_2$$

(b) $$V_2 = \frac{P_1V_1T_2}{T_1P_2} = \frac{895\ torr \times 8.70\ L \times 273\ K}{297\ K \times 760\ torr} = 9.42\ L$$

(c) $$T_2 = \frac{P_2V_2T_1}{P_1V_1} = \frac{876\ torr \times 15.00\ L \times 297\ K}{895\ torr \times 8.70\ L} = 501\ K$$

(d) $$P_2 = \frac{P_1V_1T_2}{V_2T_1} = \frac{895\ torr \times 8.70\ L \times 331\ K}{5.00\ L \times 297\ K} = 1.73 \times 10^3\ torr = 2.28\ atm$$

10.45 *Analyze.* Given: mass of cockroach, rate of O_2 consumption, temperature, percent O_2 in air, volume of air. Find: mol O_2 consumed per hour; mol O_2 in 1 quart of air; mol O_2 consumed in 48 hr.

(a) *Plan/Solve.* V of O_2 consumed = rate of consumption × mass × time. n = PV/RT.

$$5.2\ g \times 1\ hr \times \frac{0.8\ mL\ O_2}{1\ g\text{-}hr} = 4.16 = 4\ mL\ O_2\ consumed$$

$$n = \frac{PV}{RT} = 1\ atm \times \frac{mol\text{-}K}{0.08206\ L\text{-}atm} \times \frac{0.00416\ L}{297\ K} = 1.71 \times 10^{-4} = 2 \times 10^{-4}\ mol\ O_2$$

(b) *Plan/Solve.* qt air → L air → L O_2 available. mol O_2 available = PV/RT. mol O_2/hr (from part (a)) → total mol O_2 consumed. Compare O_2 available and O_2 consumed.

$$1\ qt\ air \times \frac{0.946\ L}{1\ qt} \times 0.21\ O_2\ in\ air = 0.199\ L\ O_2\ available$$

$$n = 1\ atm \times \frac{mol\text{-}K}{0.08206\ L\text{-}atm} \times \frac{0.199\ L}{297\ K} = 8.17 \times 10^{-3} = 8 \times 10^{-3}\ mol\ O_2\ available$$

$$cockroach\ uses\ \frac{1.71 \times 10^{-4}\ mol}{1\ hr} \times 48\ hr = 8.21 \times 10^{-3} = 8 \times 10^{-3}\ mol\ O_2\ consumed$$

Not only does the cockroach use 20% of the available O_2, it needs all the O_2 in the jar.

Further Applications of the Ideal-Gas Equation (Section 10.5)

10.47 *Analyze/Plan.* At the same temperature and pressure, the density of a gas increases with increasing molar mass.

HF (20 g/mol) < CO (28 g/mol) < N_2O (44 g/mol) < Cl_2 (71 g/mol)

10.49 (c) Because the helium atoms are of lower mass than the average air molecule, the helium gas is less dense than air. The balloon thus weighs less than the air displaced by its volume.

10.51 *Analyze/Plan*. Conditions (P, V, T) and amounts of gases are given. Rearrange the relationship $PV \times MM = gRT$ to obtain the desired quantity, paying attention (as always!) to units. *Solve.*

(a) $d = \frac{MM \times P}{RT}$; $MM = 46.0\ g/mol$; $P = 0.970\ atm$, $T = 35\ °C = 308\ K$

$$d = \frac{46.0\ g\ NO_2}{1\ mol} \times \frac{mol\text{-}K}{0.08206\ L\text{-}atm} \times \frac{0.970\ atm}{308\ K} = 1.77\ g/L$$

(b) $MM = \frac{gRT}{PV} = \frac{2.50\ g}{0.875\ L} \times \frac{0.08206\ L\text{-}atm}{mol\text{-}K} \times \frac{308\ K}{685\ torr} \times \frac{760\ torr}{1\ atm} = 80.1\ g/mol$

10.53 *Analyze/Plan*. Given: mass, conditions (P, V, T) of unknown gas. Find: molar mass. $MM = gRT/PV$. *Solve.*

$$MM = \frac{gRT}{PV} = \frac{1.012\ g}{0.354\ L} \times \frac{0.08206\ L\text{-}atm}{mol\text{-}K} \times \frac{372\ K}{742\ torr} \times \frac{760\ torr}{1\ atm} = 89.4\ g/mol$$

10.55 *Analyze/Plan*. Follow the logic in Sample Exercise 10.9. *Solve.*

$$mol\ O_2 = \frac{PV}{RT} = 3.5 \times 10^{-6}\ torr \times \frac{1\ atm}{760\ torr} \times \frac{mol\text{-}K}{0.08206\ L\text{-}atm} \times \frac{0.452\ L}{300\ K} = 8.456 \times 10^{-11}$$
$$= 8.5 \times 10^{-11}\ mol\ O_2$$

$$8.456 \times 10^{-11}\ mol\ O_2 \times \frac{2\ mol\ Mg}{1\ mol\ O_2} \times \frac{24.3\ g\ Mg}{1\ mol\ Mg} = 4.1 \times 10^{-9}\ g\ Mg\ (4.1\ ng\ Mg)$$

10.57 (a) *Analyze/Plan*. g glucose $\rightarrow$ mol glucose $\rightarrow$ mol $CO_2 \rightarrow$ V CO_2 *Solve.*

$$24.5\ g \times \frac{1\ mol\ glucose}{180.1\ g} \times \frac{6\ mol\ CO_2}{1\ mol\ glucose} = 0.8162 = 0.816\ mol\ CO_2$$

$$V = \frac{nRT}{P} = 0.8162\ mol \times \frac{0.08206\ L\text{-}atm}{mol\text{-}K} \times \frac{310\ K}{0.970\ atm} = 21.4\ L\ CO_2$$

(b) *Analyze/Plan*. g glucose $\rightarrow$ mol glucose $\rightarrow$ mol $O_2 \rightarrow$ V O_2 *Solve.*

$$50.0\ g \times \frac{1\ mol\ glucose}{180.1\ g} \times \frac{6\ mol\ O_2}{1\ mol\ glucose} = 1.6657 = 1.67\ mol\ O_2$$

$$V = \frac{nRT}{P} = 1.6657\ mol \times \frac{0.08206\ L\text{-}atm}{mol\text{-}K} \times \frac{298\ K}{1\ atm} = 40.7\ L\ O_2$$

10.59 *Analyze/Plan*. The gas sample is a mixture of $H_2(g)$ and $H_2O(g)$. Find the partial pressure of $H_2(g)$ and then the moles of $H_2(g)$ and Zn(s). *Solve.*

$P_t = 738\ torr = P_{H_2} + P_{H_2O}$

From Appendix B, the vapor pressure of H_2O at 24 °C = 22.38 torr

$$P_{H_2} = (738\ torr - 22.38\ torr) \times \frac{1\ atm}{760\ torr} = 0.9416 = 0.942\ atm$$

$$n_{H_2} = \frac{P_{H_2}V}{RT} = 0.9416\ atm \times \frac{mol\text{-}K}{0.08206\ L\text{-}atm} \times \frac{0.159\ L}{297\ K} = 0.006143 = 0.00614\ mol\ H_2$$

$$0.006143\ mol\ H_2 \times \frac{1\ mol\ Zn}{1\ mol\ H_2} \times \frac{65.39\ g\ Zn}{1\ mol\ Zn} = 0.402\ g\ Zn$$

Partial Pressures (Section 10.6)

10.61 (a) When the stopcock is opened, the volume occupied by $N_2(g)$ increases from 2.0 to 5.0 L. At constant T, $P_1V_1 = P_2V_2$. $1.0 \text{ atm} \times 2.0 \text{ L} = P_2 \times 5.0 \text{ L}$; $P_2 = 0.40 \text{ atm}$

(b) When the gases mix, the volume of $O_2(g)$ increases from 3.0 to 5.0 L. At constant T, $P_1V_1 = P_2V_2$. $2.0 \text{ atm} \times 3.0 \text{ L} = P_2 \times 5.0 \text{ L}$; $P_2 = 1.2 \text{ atm}$

(c) $P_t = P_{N_2} + P_{O_2} = 0.40 \text{ atm} + 1.2 \text{ atm} = 1.6 \text{ atm}$

10.63 *Analyze.* Given: amount, V, T of three gases. Find: P of each gas, total P.

Plan. $P = nRT/V$; $P_t = P_1 + P_2 + P_3 +$ *Solve.*

(a) $$P_{He} = \frac{nRT}{V} = 0.765 \text{ mol} \times \frac{0.08206 \text{ L-atm}}{\text{mol-K}} \times \frac{298 \text{ K}}{10.00 \text{ L}} = 1.871 = 1.87 \text{ atm}$$

$$P_{Ne} = \frac{nRT}{V} = 0.330 \text{ mol} \times \frac{0.08206 \text{ L-atm}}{\text{mol-K}} \times \frac{298 \text{ K}}{10.00 \text{ L}} = 0.8070 = 0.807 \text{ atm}$$

$$P_{Ar} = \frac{nRT}{V} = 0.110 \text{ mol} \times \frac{0.08206 \text{ L-atm}}{\text{mol-K}} \times \frac{298 \text{ K}}{10.00 \text{ L}} = 0.2690 = 0.269 \text{ atm}$$

(b) $P_t = 1.871 \text{ atm} + 0.8070 \text{ atm} + 0.2690 \text{ atm} = 2.9470 = 2.95 \text{ atm}$

10.65 *Analyze.* Given 407 ppm CO_2 in the atmosphere; 407 L CO_2 in 10^6 total L air. Find: the mole fraction of CO_2 in the atmosphere. *Plan.* Avogadro's law deals with the relationship between volume and moles of a gas.

Solve. Avogadro's law states that volume of a gas at constant temperature and pressure is directly proportional to moles of the gas. Using volume fraction to express concentration assumes that the 407 L CO_2 and 10^6 total L air are at the same temperature and pressure. That is, 407 L is the volume that the number of moles of CO_2 present in 10^6 L air would occupy at atmospheric temperature and pressure. The mole fraction of CO_2 in the atmosphere is then just the volume fraction from the concentration by volume.

$$X_{CO_2} = \frac{407 \text{ L } CO_2}{10^6 \text{ L air}} = 0.000407$$

10.67 *Analyze.* Given: mass CO_2 at V, T; pressure of air at same V, T. Find: partial pressure of CO_2 at these conditions, total pressure of gases at V, T.

Plan. $\text{g } CO_2 \rightarrow \text{mol } CO_2 \rightarrow P_{CO_2}$ (via $P = nRT/V$); $P_t = P_{CO_2} + P_{air}$ *Solve.*

$$5.50 \text{ g } CO_2 \times \frac{1 \text{ mol } CO_2}{44.01 \text{ g } CO_2} = 0.12497 = 0.125 \text{ mol } CO_2;\ T = 273 + 24 \text{ °C} = 297 \text{ K}$$

$$P_{CO_2} = 0.12497 \text{ mol} \times \frac{297 \text{ K}}{10.0 \text{ L}} \times \frac{0.08206 \text{ L-atm}}{\text{mol-K}} = 0.30458 = 0.305 \text{ atm}$$

$$P_{air} = 705 \text{ torr} \times \frac{1 \text{ atm}}{760 \text{ torr}} = 0.92763 = 0.928 \text{ atm}$$

$$P_t = P_{CO_2} + P_{air} = 0.30458 + 0.92763 = 1.23221 = 1.232 \text{ atm}$$

(Result has 3 decimal places and 4 sig figs.)

10.69 *Analyze/Plan.* When the sample is cooled, the water vapor condenses and all the gas pressure is because of $CO_2(g)$. The partial pressure CO_2 at 200 °C is equal to the mole fraction of CO_2 times the total pressure of the mixture. Apply Amonton's law (see Solution 10.27) to the CO_2 pressures at the two temperatures. *Solve.*

For a 3:1 mole ratio of CO_2 to H_2O

$$X_{CO_2} = \frac{3}{4} = 0.75; \quad P_{CO_2} = 0.75 \times 2.00 \text{ atm} = 1.50 \text{ atm}$$

$T_1 = 200\text{ °C} + 273 = 473\text{ K}$; $T_2 = 10\text{ °C} + 273 = 283\text{ K}$

$$\frac{P_1}{T_1} = \frac{P_2}{T_2} \text{ or } P_2 = \frac{P_1T_2}{T_1} = \frac{1.50 \text{ atm} \times 283 \text{ K}}{473 \text{ K}} = 0.8975 = 0.9 \text{ atm}$$

[The result has 1 sig fig if you consider the mole ratio to have 1 sig fig. If you think of the mole ratio as exact (experimentally unlikely), the result will have 3 sig figs.]

10.71 *Analyze/Plan.* Mole fraction = pressure fraction. Find the desired mole fraction of O_2 and change to mole percent. *Solve.*

$$X_{O_2} = \frac{P_{O_2}}{P_t} = \frac{0.21 \text{ atm}}{8.38 \text{ atm}} = 0.025; \text{ mole \%} = 0.025 \times 100 = 2.5\%$$

10.73 *Analyze/Plan.* $N_2(g)$ and $O_2(g)$ undergo changes of conditions and are mixed. Calculate the new pressure of each gas and add them to obtain the total pressure of the mixture. $P_2 = P_1V_1T_2/V_2T_1$; $P_t = P_{N_2} + P_{O_2}$. *Solve.*

$$P_{N_2} = \frac{P_1V_1T_2}{V_2T_1} = \frac{5.25 \text{ atm} \times 1.00 \text{ L} \times 293 \text{ K}}{12.5 \text{ L} \times 299 \text{ K}} = 0.41157 = 0.412 \text{ atm}$$

$$P_{O_2} = \frac{P_1V_1T_2}{V_2T_1} = \frac{5.25 \text{ atm} \times 5.00 \text{ L} \times 293 \text{ K}}{12.5 \text{ L} \times 299 \text{ K}} = 2.05786 = 2.06 \text{ atm}$$

$P_t = 0.41157 \text{ atm} + 2.05786 \text{ atm} = 2.46943 = 2.47 \text{ atm}$

Kinetic-Molecular Theory of Gases; Effusion and Diffusion (Sections 10.7 and 10.8)

10.75 (a) Decrease. Increasing the container volume increases the distance between collisions and decreases the number of collisions per unit time.

(b) Increase. Increasing temperature increases the rms speed of the gas molecules, which increases the number of collisions per unit time.

(c) Decrease. Increasing the molar mass of a gas decreases the rms speed of the molecules, which decreases the number of collisions per unit time.

10.77 *Analyze/Plan.* Given two gases, compare their rms speeds. Use Equation 10.23. *Solve.*

$$\frac{u_{rms}(\text{He})}{u_{rms}(\text{WF}_6)} = \sqrt{\frac{\text{MM}(\text{WF}_6)}{\text{MM}(\text{He})}} = \sqrt{\frac{297.9 \text{ g/mol}}{4.003 \text{ g/mol}}} = 8.63$$

The rms speed of WF_6 is approximately 9 times slower than that of He.

10.79 *Analyze/Plan.* Apply the concepts of the Kinetic-Molecular Theory (KMT) to the situation where a gas is heated at constant volume. Determine how the quantities in (a)–(d) are affected by this change. *Solve.*

(a) Average kinetic energy is proportional to temperature (K), so average kinetic energy of the molecules increases.

(b) The average kinetic energy of a gas is $1/2\ mu_{rms}^2$. Molecular mass doesn't change as T increases; average kinetic energy increases so rms speed (u) increases. (Also, $u_{rms} = (3RT/MM)^{1/2}$, so u_{rms} is directly related to T.)

(c) As T and thus rms molecular speed increase, molecular momentum (mu) increases and the strength of an average impact with the container wall increases.

(d) As T and rms molecular speed increase, the molecules collide more frequently with the container walls, and the total number of collisions per second increases.

10.81 (a) *Plan.* The larger the molar mass, the slower the average speed (at constant temperature).

Solve. In order of increasing speed (and decreasing molar mass):
$HBr < NF_3 < SO_2 < CO < Ne$

(b) *Plan.* Follow the logic of Sample Exercise 10.13. *Solve.*

$$u_{rms} = \sqrt{\frac{3\,RT}{MM}} = \left(\frac{3 \times 8.314\ \text{kg-m}^2/\text{s}^2\text{-mol-K} \times 298\ \text{K}}{71.0 \times 10^{-3}\ \text{kg/mol}}\right)^{1/2} = 324\ \text{m/s}$$

(c) *Plan.* Use Equation 10.21 to calculate the most probable speed, u_{mp}. MM of O_3 = 48.0 g/mol; T = 270 K. *Solve.*

$$u_{mp} = \sqrt{\frac{2\,RT}{MM}} = \left(\frac{2 \times 8.314\ \text{kg-m}^2/\text{s}^2\text{-mol-K} \times 270\ \text{K}}{48.0 \times 10^{-3}\ \text{kg/mol}}\right)^{1/2} = 306\ \text{m/s}$$

10.83 Statements (a) and (d) are true. Statement (b) is false because effusion is the escape of gas molecules through a tiny hole, while diffusion is the distribution of a gas throughout space or throughout another substance. Statement (c) is false because perfume molecules travel to your nose by the process of diffusion, not effusion.

10.85 *Plan.* The heavier the molecule, the slower the rate of effusion. Thus, the order for increasing rate of effusion is in the order of decreasing mass. *Solve.*
rate $^2H^{37}Cl$ < rate $^1H^{37}Cl$ < rate $^2H^{35}Cl$ < rate $^1H^{35}Cl$

10.87 *Analyze.* Given: relative effusion rates of two gases at same temperature. Find: molecular formula of one of the gases. *Plan.* Use Graham's law to calculate the formula weight of arsenic(III) sulfide, and thus the molecular formula. *Solve.*

$$\frac{\text{rate (sulfide)}}{\text{rate (Ar)}} = \left[\frac{39.9}{\text{MM (sulfide)}}\right]^{1/2} = 0.28$$

MM (sulfide) = $39.9 / 0.28^2$ = 509 g/mol (two significant figures)
The empirical formula of arsenic(III) sulfide is As_2S_3, which has a formula mass of 246.1. Twice this is 490 g/mol, close to the value estimated from the effusion experiment. Thus, the formula of the gas phase molecule is As_4S_6.

Nonideal-Gas Behavior (Section 10.9)

10.89 (a) Nonideal-gas behavior is observed at very high pressures and/or low temperatures.

(b) The real volumes of gas molecules and attractive intermolecular forces between molecules cause gases to behave nonideally.

10.91 Statement (b) is true. The constants *a* and *b* are characteristic of a particular gas and are independent of pressure and temperature.

10.93 *Analyze/Plan*. Follow the logic in Sample Exercise 10.15. Use the ideal-gas equation to calculate pressure in (a), the van der Waals equation in (b). n = 1.00 mol, V = 5.00 L, T = 25 °C = 298 K; a = 6.49 L^2-atm/mol^2, b = 0.0562 L/mol.

(a) $$P = \frac{nRT}{V} = 1.00 \text{ mol} \times \frac{298 \text{ K}}{5.00 \text{ L}} \times \frac{0.08206 \text{ L-atm}}{\text{mol-K}} = 4.89 \text{ atm}$$

(b) $$P = \frac{nRT}{V - nb} - \frac{n^2 a}{V^2};$$

$$P = \frac{(1.00 \text{ mol})(298 \text{ K})(0.08206 \text{ L-atm/mol-K})}{5.00 \text{ L} - (1.00 \text{ mol})(0.0562 \text{ L/mol})} - \frac{(1.00 \text{ mol})^2(6.49 \text{ L}^2\text{-atm/mol}^2)}{(5.00 \text{ L})^2}$$

$$P = 4.9463 \text{ atm} - 0.2596 \text{ atm} = 4.6868 = 4.69 \text{ atm}$$

(c) From Sample Exercise 10.15, the difference at 22.41 L between the ideal and van der Waals results is (1.00 – 0.990) = 0.010 atm. At 5.00 L, the difference is (4.89 – 4.69) = 0.20 atm. The effects of both molecular attractions, the *a* correction, and molecular volume, the *b* correction, increase with decreasing volume. For the *a* correction, V^2 appears in the denominator, so the correction increases exponentially as V decreases. For the *b* correction, n*b* is a larger portion of the total volume as V decreases. That is, 0.0562 L is 1.1% of 5.0 L, but only 0.25% of 22.41 L. Qualitatively, molecular attractions are more important as the amount of free space decreases and the number of molecular collisions increase. Molecular volume is a larger part of the total volume as the container volume decreases.

10.95 *Analyze*. Given the *b* value of Xe, 0.0510 L/mol, calculate the radius of a Xe atom.

Plan. Use Avogadro's number to change L/mol to L/atom. Use the volume formula, $V = 4/3\ \pi r^3$ and units conversion to obtain the radius in Å. 1 L = 1 dm^3. *Solve*.

$$\frac{0.0510 \text{ L}}{1 \text{ mol Xe}} \times \frac{1 \text{ mol Xe}}{6.022 \times 10^{23} \text{ Xe atoms}} \times \frac{1 \text{ dm}^3}{1 \text{ L}} = 8.4689 \times 10^{-26} = 8.47 \times 10^{-26} \text{ dm}^3$$

$$V = 4/3\ \pi r^3; \quad r^3 = 3V/4\pi; \quad r = (3V/4\pi)^{1/3}$$

$$r = \left(\frac{3 \times 8.4689 \times 10^{-26} \text{ dm}^3}{4 \times 3.14159}\right)^{1/3} = 2.7243 \times 10^{-9} = 2.72 \times 10^{-9} \text{ dm}$$

$$2.72 \times 10^{-9} \text{ dm} \times \frac{1 \text{ m}}{10 \text{ dm}} \times \frac{1 \text{ Å}}{1 \times 10^{-10} \text{ m}} = 2.72 \text{ Å}$$

The calculated value is the nonbonding radius. From Figure 7.7 in Section 7.3, the bonding atomic radius of Xe is 1.40 Å. We expect the nonbonding radius of an atom to be larger than the bonding radius, but our calculated value is nearly twice as large.

Additional Exercises

10.98 $P_1V_1 = P_2V_2$; $V_2 = P_1V_1/P_2$

$$V_2 = \frac{3.0\text{ atm} \times 1.0\text{ mm}^3}{730\text{ torr}} \times \frac{760\text{ torr}}{1\text{ atm}} = 3.1\text{ mm}^3$$

10.100 $P = \frac{nRT}{V}$; $n = 1.4 \times 10^{-5}$ mol, $V = 0.600$ L, $T = 23\ °C = 296$ K

$$P = 1.4 \times 10^{-5}\text{ mol} \times \frac{0.08206\text{ L-atm}}{\text{mol-K}} \times \frac{296\text{ K}}{0.600\text{ L}} = 5.7 \times 10^{-4}\text{ atm} = 0.43\text{ mm Hg}$$

10.102 (a) $n = \frac{PV}{RT} = 3.00\text{ atm} \times \frac{\text{mol-K}}{0.08206\text{ L-atm}} \times \frac{110\text{ L}}{300\text{ K}} = 13.4\text{ mol } C_3H_8(g)$

(b) $\frac{0.590\text{ g } C_3H_8\text{ (l)}}{1\text{ mL}} \times 110 \times 10^3\text{ mL} \times \frac{1\text{ mol } C_3H_8}{44.094\text{ g}} = 1.47 \times 10^3\text{ mol } C_3H_8\text{ (l)}$

(c) Using C_3H_8 in a 110-L container as an example, the ratio of moles liquid to moles gas that can be stored in a certain volume is $\frac{1.47 \times 10^3\text{ mol liquid}}{13.4\text{ mol gas}} = 110$.

A container with a fixed volume holds many more moles (molecules) of $C_3H_8(l)$ because in the liquid phase the molecules are touching. In the gas phase, the molecules are far apart (statement 2, Section 10.7), and many fewer molecules will fit in the container.

10.104 (a) mol = g/MM; assume mol Ar = mol X;

$$\frac{\text{g Ar}}{39.948\text{ g/mol}} = \frac{\text{g X}}{\text{MM X}};\ \frac{3.224\text{ g Ar}}{39.948\text{ g/mol}} = \frac{8.102\text{ g X}}{\text{MM X}}$$

$$\text{MM X} = \frac{(8.102\text{ g X})(39.948\text{ g/mol})}{3.224\text{ g Ar}} = 100.39 = 100.4\text{ g/mol}$$

(b) Assume mol Ar = mol X. For gases, PV= nRT and n = PV/RT. For moles of the two gases to be equal, the implied assumption is that P, V, and T are constant. Because we use the same container for both gas samples, constant V is a good assumption. The values of P and T are not explicitly stated.

We also assume that the gases behave ideally. At ambient conditions, this is a reasonable assumption.

10.106 (a) $n = \frac{PV}{RT} = 0.980\text{ atm} \times \frac{\text{mol-K}}{0.08206\text{ L-atm}} \times \frac{0.524\text{ L}}{347\text{ K}} = 0.018034 = 0.0180\text{ mol air}$

$$\text{mol } O_2 = 0.018034\text{ mol air} \times \frac{0.2095\text{ mol } O_2}{1\text{ mol air}} = 0.003778 = 0.00378\text{ mol } O_2$$

(b) $C_8H_{18}(l) + 25/2\ O_2(g) \rightarrow 8\ CO_2(g) + 9\ H_2O(g)$

(The H_2O produced in an automobile engine is in the gaseous state.)

$$0.003778\text{ mol } O_2 \times \frac{1\text{ mol } C_8H_{18}}{12.5\text{ mol } O_2} \times \frac{114.2\text{ g } C_8H_{18}}{1\text{ mol } C_8H_{18}} = 0.0345\text{ g } C_8H_{18}$$

10.108 V and T are the same for He and O_2.

$P_{He}V = n_{He}RT, \quad P_{He}/n_{He} = RT/V; \quad P_{O_2}/n_{O_2} = RT/V$

$$\frac{P_{He}}{n_{He}} = \frac{P_{O_2}}{n_{O_2}} = n_{O_2} \times \frac{P_{O_2} \times n_{He}}{P_{He}}; \; n_{He} = 1.42 \text{ g He} \times \frac{1 \text{ mol He}}{4.003 \text{ g He}} = 0.3547 = 0.355 \text{ mol He}$$

$$n_{O_2} = \frac{158 \text{ torr}}{42.5 \text{ torr}} \times 0.355 \text{ mol} = 1.3188 = 1.32 \text{ mol } O_2; 1.3188 \text{ mol } O_2 \times \frac{32.00 \text{ g } O_2}{1 \text{ mol } O_2} = 42.2 \text{ g } O_2$$

10.110 (a) The quantity $d/P = MM/RT$ should be a constant at all pressures for an ideal gas. It is not, however, because of nonideal behavior. If we graph d/P versus P, the ratio should approach ideal behavior at low P. At P = 0, $d/P = 2.2525$. Using this value in the formula $MM = d/P \times RT$, MM = 50.46 g/mol.

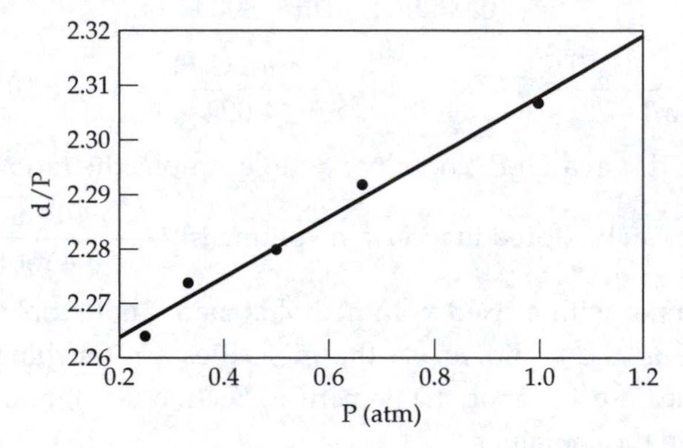

(b) The ratio d/P varies with pressure because of the finite volumes of gas molecules and attractive intermolecular forces.

10.112 $u_{rms} = (3RT/MM)^{1/2}$; $u_{rms2} = 2\,u_{rms1}$; $T_1 = -33\ °C = 240$ K

$u_{rms1} = (3RT_1/MM)^{1/2}$; $u_{rms1}{}^2 = 3(240)R/MM = 720R/MM$

$u_{rms1} = (720R/MM)^{1/2}$; $u_{rms2} = 2u_{rms1} = (2)(720R/MM)^{1/2}$

$(2)(720R/MM)^{1/2} = (3RT_2/MM)^{1/2}$

$(2)^2(720R/MM) = 3RT_2/MM$; $(2)^2(720) = 3T_2$

$T_2 = (4)(720)/3 = 960 \text{ K} = 687\ °C$

Increasing the rms speed (u) by a factor of 2 requires heating to 960 K (or 687 °C), increasing the temperature by a factor of 4.

10.114 (a) The effect of intermolecular attraction becomes more significant as a gas is compressed to a smaller volume at constant temperature. This compression causes the pressure, and thus the number of intermolecular collisions, to increase. Intermolecular attraction causes some of these collisions to be inelastic, which amplifies the deviation from ideal behavior.

(b) The effect of intermolecular attraction becomes less significant as the temperature of a gas is increased at constant volume. When the temperature of a gas is increased at constant volume, the pressure of the gas, the number of intermolecular collisions, and the average kinetic energy of the gas particles increase. This higher average kinetic energy means that a larger fraction of the molecules has sufficient kinetic energy to overcome intermolecular attractions, even though there are more total collisions. This increases the fraction of elastic collisions, and the gas more closely obeys the ideal-gas equation.

10.116 (a) At STP, the molar volume $= 1\ \text{mol} \times \frac{0.08206\ \text{L-atm}}{\text{mol-K}} \times \frac{273\ \text{K}}{1\ \text{atm}} = 22.4\ \text{L}$

Dividing the value for *b*, 0.0322 L/mol, by 4, we obtain 0.00805 L. Thus, the volume of the Ar atoms is (0.00805/22.4)100 = 0.0359% of the total volume.

(b) At 200 atm pressure (and 0 °C, standard temperature) the molar volume is 0.112 L, and the volume of the Ar atoms is 7.19% of the total volume.

Integrative Exercises

10.118 (a) $MM = \frac{gRT}{VP} = \frac{1.56\ \text{g}}{1.00\ \text{L}} \times \frac{0.08206\ \text{L-atm}}{\text{mol-K}} \times \frac{323\ \text{K}}{0.984\ \text{atm}} = 42.0\ \text{g/mol}$

Assume 100 g cyclopropane

$$100\ \text{g} \times 0.857\ \text{C} = 85.7\ \text{g C} \times \frac{1\ \text{mol C}}{12.01\ \text{g}} = \frac{7.136\ \text{mol C}}{7.136} = 1\ \text{mol C}$$

$$100\ \text{g} \times 0.143\ \text{H} = 14.3\ \text{g H} \times \frac{1\ \text{mol H}}{1.008\ \text{g}} = \frac{14.19\ \text{mol H}}{7.136} = 2\ \text{mol H}$$

The empirical formula of cyclopropane is CH_2 and the empirical formula weight is 12 + 2 = 14 g. The ratio of molar mass to empirical formula weight, 42.0 g/14 g, is 3; therefore, there are three empirical formula units in one cyclopropane molecule. The molecular formula is $3 \times (CH_2) = C_3H_6$.

(b) Ar is a monoatomic gas. Cyclopropane molecules are larger and more structurally complex, even though the molar masses of Ar and C_3H_6 are similar. If both gases are at the same relatively low temperature, they have approximately the same average kinetic energy, and the same ability to overcome intermolecular attractions. We expect intermolecular attractions to be more significant for the more complex C_3H_6 molecules, and that C_3H_6 will deviate more from ideal behavior at the conditions listed. This conclusion is supported by the *a* values in Table 10.3. The *a* values for CH_4 and CO_2, more complex molecules than Ar atoms, are larger than the value for Ar. If the pressure is high enough for the volume correction in the van der Waals equation to dominate behavior, the larger C_3H_6 molecules definitely deviate more than Ar atoms from ideal behavior.

(c) Cyclopropane, C_3H_6, MM = 42.0 g/mol; methane, CH_4 MM = 16.0. Rate of effusion through a pinhole is inversely related to molar mass. Cyclopropane would effuse through a pinhole slower than methane, because it has the greater molar mass.

10.120 (a) Get g C from mL CO_2; get g H from mL H_2O. Also calculate mol C and H, to use in part (b). Get g N by subtraction. Calculate % composition.

n = PV/RT. At STP, P = 1 atm, T = 273 K. (STP implies an infinite number of sig figs.)

$$n_{CO_2} = 0.08316\ \text{L} \times \frac{1\ \text{atm}}{273\ \text{K}} \times \frac{\text{mol-K}}{0.08206\ \text{L-atm}} = 0.003712\ \text{mol}\ CO_2$$

$$0.003712\ \text{mol}\ CO_2 \times \frac{1\ \text{mol C}}{1\ \text{mol}\ CO_2} \times \frac{12.0107\ \text{g C}}{\text{mol C}} = 0.044585 = 0.04458\ \text{g}\ CO_2$$

$$n_{H_2O} = 0.07330\ \text{L} \times \frac{1\ \text{atm}}{273\ \text{K}} \times \frac{\text{mol-K}}{0.08206\ \text{L-atm}} = 3.2720 \times 10^{-3}$$

$$= 3.272 \times 10^{-3}\ \text{mol}\ H_2O$$

$$3.2720 \times 10^{-3}\ \text{mol}\ H_2O \times \frac{2\ \text{mol H}}{1\ \text{mol}\ H_2O} \times \frac{1.00794\ \text{g H}}{\text{mol H}} = 6.5959 \times 10^{-3}$$

$$= 6.596 \times 10^{-3}\ \text{g H}$$

$$\text{mass \% X} = \frac{\text{mass X}}{\text{sample mass}} \times 100;\ \text{sample mass} = 100.0\ \text{mg} = 0.1000\ \text{g}$$

$$\text{\% C} = \frac{0.044585\ \text{g}}{0.1000\ \text{g}} \times 100 = 44.585 = 44.58\text{\% C}$$

$$\text{\% H} = \frac{6.5959 \times 10^{-3}\ \text{g H}}{0.1000\ \text{g}} \times 100 = 6.5959 = 6.596\text{\% H}$$

$$\text{\% Cl} = \frac{0.01644\ \text{g Cl}}{0.1000\ \text{g}} \times 100 = 16.44\text{\% Cl}$$

% N = 100 – 44.58 – 6.596 – 16.44 = 32.38% N

(b) 0.003712 mol C; $2(3.272 \times 10^{-3}) = 6.544 \times 10^{-3}$ mol H

$$0.01644\ \text{g Cl} \times \frac{1\ \text{mol C}}{35.453\ \text{g Cl}} = 4.637 \times 10^{-4}\ \text{mol Cl}$$

0.1000 g sample $\times$ 0.3238 mass fraction N = 0.03238 g N

$$0.03238\ \text{g N} \times \frac{1\ \text{mol N}}{14.0067\ \text{g N}} = 0.0023118 = 0.002312\ \text{mol N}$$

Divide by the smallest number of mol to find the simplest ratio of moles.

$$\frac{0.003712\ \text{mol C}}{4.637 \times 10^{-4}} = 8.005\ \text{C}$$

$$\frac{6.544 \times 10^{-3}\ \text{mol H}}{4.637 \times 10^{-4}} = 14.11\ \text{H}$$

$$\frac{4.637 \times 10^{-4}\ \text{mol Cl}}{4.637 \times 10^{-4}} = 1.000\ \text{Cl}$$

$$\frac{0.002312\ \text{mol N}}{4.637 \times 10^{-4}} = 4.985\ \text{N}$$

If we assume 14.11 is "close" to 14 (a reasonable assumption), the empirical formula is $C_8H_{14}N_5Cl$.

(c) Molar mass of the compound is required to determine molecular formula when the empirical formula is known.

10.122 (a) $5.00 \text{ g HCl} \times \frac{1 \text{ mol HCl}}{36.46 \text{ g HCl}} = 0.1371 = 0.137 \text{ mol HCl}$

$5.00 \text{ g } NH_3 \times \frac{1 \text{ mol } NH_3}{17.03 \text{ g } NH_3} = 0.2936 = 0.294 \text{ mol } NH_3$

The gases react in a 1:1 mole ratio, HCl is the limiting reactant and is completely consumed. (0.2936 mol – 0.1371 mol) = 0.1565 = 0.157 mol NH_3 remain in the system. $NH_3(g)$ is the only gas remaining after reaction.

(b) $V_t = 4.00 \text{ L}. \quad P = \frac{nRT}{V} = 0.1565 \text{ mol} \times \frac{0.08206 \text{ L-atm}}{\text{mol-K}} \times \frac{298 \text{ K}}{4.00 \text{ L}} = 0.957 \text{ atm}$

(c) $0.137 \text{ mol HCl} \times \frac{1 \text{ mol } NH_4Cl}{1 \text{ mol HCl}} \times \frac{53.49 \text{ g } NH_4Cl}{1 \text{ mol } NH_4Cl} = 7.3284 = 7.33 \text{ g } NH_4Cl$

10.124 (a) 19 e^-, 9.5 e^- pairs

:Ö—Ċl—Ö:

Resonance structures can be drawn with the odd electron on O, but electronegativity considerations predict that it will be on Cl for most of the time.

(b) ClO_2 is very reactive because it is an odd-electron molecule. Adding an electron (reduction) both pairs the odd electron and completes the octet of Cl. Thus, ClO_2 has a strong tendency to gain an electron and be reduced.

(c) ClO_2^-, 20 e^-, 10 e^- pairs

[:Ö—C̈l—Ö:]$^-$

(d) 4 e^- domains around Cl, O–Cl–O bond angle ~107° (<109° owing to repulsion by nonbonding domains)

(e) Calculate mol Cl_2 from ideal-gas equation; determine limiting reactant; mass ClO_2 via mol ratios.

$\text{mol } Cl_2 = \frac{PV}{RT} = 1.50 \text{ atm} \times \frac{2.00 \text{ L}}{294 \text{ K}} \times \frac{\text{mol-K}}{0.08206 \text{ L-atm}} = 0.1243 = 0.124 \text{ mol } Cl_2$

$15.0 \text{ g } NaClO_2 \times \frac{1 \text{ mol } NaClO_2}{90.44 \text{ g}} = 0.1659 = 0.166 \text{ mol } NaClO_2$

2 mol $NaClO_2$ are required for 1 mol Cl_2, so $NaClO_2$ is the limiting reactant. For every 2 mol $NaClO_2$ reacted, 2 mol ClO_2 are produced, so mol ClO_2 = mol $NaClO_2$.

$0.1659 \text{ mol } ClO_2 \times \frac{67.45 \text{ g } ClO_2}{\text{mol}} = 11.2 \text{ g } ClO_2$

10.126 After reaction, the flask contains $IF_5(g)$ and whichever reactant is in excess. Determine the limiting reactant, which regulates the moles of IF_5 produced and moles of excess reactant.

$I_2(s) + 5F_2(g) \rightarrow 2\,IF_5(g)$

$$10.0\text{ g } I_2 \times \frac{1\text{ mol } I_2}{253.8\text{ g } I_2} \times \frac{5\text{ mol } F_2}{1\text{ mol } I_2} = 0.1970 = 0.197\text{ mol } F_2$$

$$10.0\text{ g } F_2 \times \frac{1\text{ mol } F_2}{38.00\text{ g } F_2} = 0.2632 = 0.263\text{ mol } F_2\text{ available}$$

I_2 is the limiting reactant; F_2 is in excess.

0.263 mol F_2 available – 0.197 mol F_2 reacted = 0.066 mol F_2 remain.

$$10.0\text{ g } I_2 \times \frac{1\text{ mol } I_2}{253.8\text{ g } I_2} \times \frac{2\text{ mol } IF_5}{1\text{ mol } I_2} = 0.0788\text{ mol } IF_5\text{ produced}$$

(a) $$P_{IF_5} = \frac{nRT}{V} = 0.0788\text{ mol} \times \frac{0.08206\text{ L-atm}}{\text{mol-K}} \times \frac{398\text{ K}}{5.00\text{ L}} = 0.515\text{ atm}$$

(b) $$\chi_{IF_5} = \frac{\text{mol } IF_5}{\text{mol } IF_5 + \text{mol } F_2} = \frac{0.0788}{0.0788 + 0.066} = 0.544$$

(c) 42 valence e^-, 21 e^- pairs

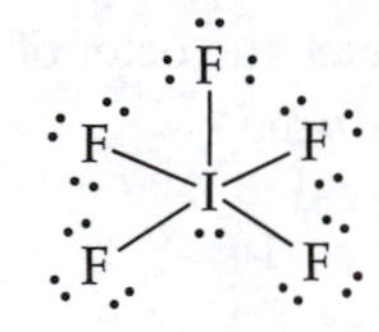

(d) $$0.0788\text{ mol } IF_5 \times \frac{221.90\text{ g } IF_5}{\text{mol } IF_5} = 17.4857 = 17.5\text{ g } IF_5\text{ produced}$$

$$0.066\text{ mol } F_2 \times \frac{38.00\text{ g } F_2}{\text{mol } F_2} = 2.508 = 2.5\text{g } F_2\text{ remain}$$

Total mass in flask = 17.5 g IF_5 + 2.5 g F_2 = 20.00 g; mass is conserved.

11 Liquids and Intermolecular Forces

Visualizing Concepts

In this chapter we will use the temperature units °C and K interchangeably when designating specific heats and *changes* in temperature.

11.1 (a) The diagram best describes a liquid.

(b) In the diagram, the particles are close together, mostly touching but there is no regular arrangement or order. This rules out a gaseous sample, where the particles are far apart, and a crystalline solid, which has a regular repeating structure in all three directions.

11.4 *Analyze.* When heat is added to a liquid, the temperature of the liquid rises. If enough heat is added to reach the boiling point (bp), any excess heat is used to vaporize the liquid. If heat is still available when all the liquid is converted to gas, the temperature of the gas rises.

Plan. Use the specific heat of $CH_4(l)$ to calculate the amount of heat required to raise the temperature of 32.0 g of $CH_4(l)$ from –170 °C to –161.5 °C. If this is less than 42 kJ, use ΔH_{vap} to calculate the energy required to vaporize the liquid, and so on, until exactly 42.0 kJ has been used to increase the temperature and/or change the state of CH_4.

Solve. Heat the liquid to its boiling point: $\Delta T = [-161.5\ °C - (-170\ °C)] = 8.5\ °C = 8.5\ K$ (Note that the magnitude of one degree is the same in Kelvins and Celsius.)

$$\frac{3.48\ J}{g\text{-}K} \times 32.0\ g\ CH_4 \times 8.5\ K = 946.56 = 9.5 \times 10^2\ J = 0.95\ kJ$$

Heating $CH_4(l)$ to its boiling point requires only 0.95 kJ. We have added 42 kJ, so there is definitely enough heat to vaporize the liquid. ΔH_{vap} for $CH_4(l)$ is 8.20 kJ/mol. The 32.0 g sample is 2.00 mol $CH_4(l)$, so the energy required to vaporize the sample at –161.5 °C is (2 × 8.20 kJ/mol =) 16.4 kJ. The energy used to heat the sample to –161.5 °C and vaporize it at this temperature is (0.947 kJ + 16.4 kJ) = 17.347 = 17.3 kJ. We have (42.0 kJ – 17.347 kJ) = 24.653 = 24.7 kJ left to heat the gas.

$$\Delta T = 24.65\ kJ \times \frac{g\text{-}K}{2.22\ J} \times \frac{1000\ J}{kJ} \times \frac{1}{32.0\ g\ CH_4} = 346.99 = 347\ K = 347\ °C$$

The final temperature of the methane gas, $CH_4(g)$, is then (–161.5 °C + 346.99 °C) = 185.49 = 185 °C.

11.6 *Analyze/Plan.* We are given the structures of two molecules with the same molecular formula and asked questions about their physical properties. Because the molecules have the same molecular formula, their van der Waals forces are similar. Consider any differences in intermolecular forces experienced by the two molecules. *Solve.*

(a) Propanol, the molecule on the left, has an O–H bond and experiences hydrogen bonding, while ethyl methyl ether does not.

(b) We expect propanol to have a larger dipole moment. Both molecules are somewhat polar, but propanol has hydrogen bonding.

(c) Propanol boils at 97.2 °C, while ethyl methyl ether boils at 10.8 °C. Molecules in liquid propanol are attracted to each other by hydrogen bonding. More kinetic energy and thus a higher temperature is required to separate the molecules and produce a gas.

Molecular Comparisons of Gases, Liquids, and Solids (Section 11.1)

11.9 (a) solid < liquid < gas

(b) gas < liquid < solid

(c) Matter in the gaseous state is most easily compressed, because particles are far apart and there is much empty space.

11.11 (a) It increases. Kinetic energy is the energy of motion. As melting occurs, the motion of atoms relative to each other increases, which increases kinetic energy. As the kinetic energy of individual atoms increases, the overall average kinetic energy of the sample increases.

(b) It increases. As the atoms move relative to one another, the average distance between them increases. The physical property that corroborates this is density. The density of a solid is usually greater than the density of its liquid, indicating a greater sample volume for the liquid. This greater sample volume is the result of greater distance between atoms in three dimensions.

11.13 (a) At standard temperature and pressure, the molar volumes of Cl_2 and NH_3 are nearly the same because they are both gases. In the gas phase, molecules are far apart and most of the volume occupied by the substance is empty space. Differences in molecular characteristics such as weight, shape, and dipole moment have little bearing on the molar volume of a gas.

The ideal-gas law states that one mole of any gas at STP will occupy a fixed volume. The slight difference in molar volumes of the two gases is predicted by the van der Waals correction, which quantifies deviation from ideal behavior.

(b) On cooling to 160 K, both compounds condense from the gas phase to the solid state. Condensation, as the word implies, eliminates most of the empty space between molecules, so we expect a significant decrease in the molar volume.

(c) $$\frac{1\text{ cm}^3}{2.02\text{ g Cl}_2} \times \frac{70.096\text{ g Cl}_2}{1\text{ mol Cl}_2} = 35.1\text{ cm}^3/\text{mol Cl}_2 = 0.0351\text{ L/mol Cl}_2$$

$$\frac{1\text{ cm}^3}{0.84\text{ g NH}_3} \times \frac{17.031\text{ g NH}_3}{1\text{ mol NH}_3} = 20.3\text{ cm}^3/\text{mol NH}_3 = 0.0203\text{ L/mol NH}_3$$

(d) Solid state molar volumes are not as similar as those in the gaseous state. In the solid state, most of the empty space is gone, so molecular characteristics do influence molar volumes. Cl_2 is heavier than NH_3 and the Cl–Cl bond distance is almost double the N–H bond distance (Figure 7.7). Intermolecular attractive forces among polar NH_3 molecules bind them more tightly than forces among nonpolar Cl_2 molecules. These factors all contribute to a molar volume for $Cl_2(s)$ that is almost twice that of $NH_3(s)$.

(e) Like solids, liquids are condensed phases. That is, there is little empty space between molecules in the liquid state. We expect the molar volumes of the liquids to be closer to those in the solid state than those in the gaseous state.

Intermolecular Forces (Section 11.2)

11.15 (a) London dispersion forces

(b) dipole-dipole forces

(c) hydrogen-bonding forces

11.17 (a) SO_2 is a polar covalent molecule, so dipole-dipole and London dispersion forces must be overcome to convert the liquid to a gas.

(b) CH_3COOH is a polar covalent molecule that experiences London dispersion, dipole-dipole, and hydrogen-bonding (O–H bonds) forces. All of these forces must be overcome to convert the liquid to a gas.

(c) H_2S is a polar covalent molecule that experiences London dispersion and dipole-dipole forces, so these must be overcome to change the liquid into a gas. (H–S bonds do not lead to hydrogen-bonding interactions.)

11.19 (a) *Polarizability* increases as molecular size (and thus molecular weight) increases. In order of increasing polarizability: $CH_4 < SiH_4 < SiCl_4 < GeCl_4 < GeBr_4$.

(b) The magnitude of London dispersion forces and thus the boiling points of molecules increase as polarizability increases. The order of increasing boiling points is the order of increasing polarizability:

$CH_4 < SiH_4 < SiCl_4 < GeCl_4 < GeBr_4$

11.21 *Analyze/Plan*. For molecules with similar structures, the strength of dispersion forces increases with molecular size (molecular weight and number of electrons in the molecule).

Solve: (a) H_2S (b) CO_2 (c) GeH_4

11.23 Both hydrocarbons experience dispersion forces. Rod-like butane molecules can contact each other over the length of the molecule, whereas spherical 2-methylpropane molecules can only touch tangentially. The larger contact surface of butane facilitates stronger forces and produces a higher boiling point.

11.25 (a) A molecule must contain H atoms bound to either N, O, or F atoms to participate in hydrogen bonding with like molecules.

(b) $\mathbf{CH_3NH_2}$ and $\mathbf{CH_3OH}$ have N–H and O–H bonds, respectively; they will form hydrogen bonds with other molecules of the same kind. (CH_3F has C–F and C–H bonds, but no H–F bonds.)

11.27 (a) Replacing a hydroxyl hydrogen with a CH_3 group eliminates hydrogen bonding in that part of the molecule. This reduces the strength of intermolecular forces and leads to a (much) lower boiling point.

(b) $CH_3OCH_2CH_2OCH_3$ is a larger, more polarizable molecule with stronger London dispersion forces and thus a higher boiling point.

11.29

Physical Property	H_2O	H_2S
Normal Boiling Point, °C	100.00	–60.7
Normal Melting Point, °C	0.00	–85.5

Based on its much higher normal melting and boiling point, H_2O has stronger intermolecular forces.

H_2O, with H bound to O, has hydrogen bonding. H_2S, with H bound to S, has dipole-dipole forces. (The electronegativities of H and S, 2.1 and 2.5, respectively, are similar. The H–S bond dipoles in H_2S are not large, but S does have two nonbonded electron pairs. The molecule has medium polarity.) Both molecules have London dispersion forces.

11.31 SO_4^{2-} has a greater negative charge than BF_4^-, so ion-ion electrostatic attractions are greater in sulfate salts. These strong forces limit the ion mobility required for the formation of an ionic liquid. (This is called an electronic effect.)

Select Properties of Liquids (Section 11.3)

11.33 (a) As temperature increases, surface tension decreases; they are inversely related.

(b) As temperature increases, viscosity decreases; they are inversely related.

(c) Surface tension and viscosity are both directly related to the strength of intermolecular attractive forces. The same attractive forces that cause surface molecules to be difficult to separate cause molecules elsewhere in the sample to resist movement relative to one another. Liquids with high surface tension have intermolecular attractive forces sufficient to produce a high viscosity as well.

11.35 (a) Diagram (ii) shows stronger adhesive forces between the surface and the liquid. These adhesive forces attract the liquid to the surface and flatten the drop.

(b) Diagram (i) represents water on a nonpolar surface. The stronger hydrogen bonding cohesive forces among water molecules in the liquid prevent the drop from spreading.

(c) Diagram (ii) represents water on a polar surface. Adhesive dipole-dipole interactions between water molecules and the surface compete successfully with cohesive hydrogen bonding forces in the liquid and the drop spreads.

11.37 (a) The three molecules have similar structures and all experience hydrogen-bonding, dipole-dipole, and dispersion forces. The main difference in the series is the increase in the number of carbon atoms in the alkyl chain, with a corresponding increase in chain length, molecular weight, and strength of dispersion forces. The boiling points, surface tension, and viscosities all increase because the strength of dispersion forces increases.

(b) Ethylene glycol has an –OH group at both ends of the molecule. This greatly increases the possibilities for hydrogen bonding, so the overall intermolecular attractive forces are greater and the viscosity of ethylene glycol is much greater.

(c) Water has the highest surface tension but lowest viscosity because it is the smallest molecule in the series. Because water molecules are small, they approach each other closely and form many strong hydrogen bonds. There is no hydrocarbon chain to disrupt hydrogen bond formation or to inhibit their attraction to molecules in the interior of the drop. Water molecules at the surface of a drop are missing a few hydrogen bonds and are strongly pulled into the center of the drop, resulting in high surface tension. The absence of an alkyl chain also means the molecules can move around each other easily, resulting in the low viscosity.

Phase Changes (Section 11.4)

11.39 (a) melting, endothermic

(b) evaporation or vaporization, endothermic

(c) deposition, exothermic

(d) condensation, exothermic

11.41 (a) Melting, (s) → (l)

(b) Endothermic. Energy is always required to overcome intermolecular forces that organize molecules into a solid.

(c) Heat of vaporization is usually larger than heat of fusion. Vaporization requires enough energy to separate molecules by large distances. Melting or fusion only requires that molecules move relative to each other, not that they are separated.

11.43 *Analyze.* The heat required to vaporize 60 g of H_2O equals the heat lost by the cooled water.

Plan. Using the enthalpy of vaporization, calculate the heat required to vaporize 60 g of H_2O in this temperature range. Using the specific heat capacity of water, calculate the mass of water than can be cooled 15 °C if this much heat is lost.

Solve. Evaporation of 60 g of water requires:

$$60 \text{ g } H_2O \times \frac{2.4 \text{ kJ}}{1 \text{ g } H_2O} = 1.44 \times 10^2 \text{ kJ} = 1.4 \times 10^5 \text{ J}$$

Cooling a certain amount of water by 15 °C:

$$1.44 \times 10^5 \text{ J} \times \frac{1 \text{ g-K}}{4.184 \text{ J}} \times \frac{1}{15 \text{ °C}} = 2294 = 2.3 \times 10^3 \text{ g } H_2O$$

Check. The units are correct. A surprisingly large mass of water (2300 g ≈ 2.3 L) can be cooled by this method.

11.45 *Analyze/Plan*. Follow the logic in Sample Exercise 11.3. *Solve*. Physical data for ethanol, C_2H_5OH, is: mp = –114 °C; ΔH_{fus} = 5.02 kJ/mol; $C_{s(solid)}$ = 0.97 J/g-K; bp = 78 °C; ΔH_{vap} = 38.56 kJ/mol; $C_{s(liquid)}$ = 2.3 J/g-K. *Solve*.

(a) Heat the liquid from 35 °C to 78 °C, ΔT = 43 °C = 43 K.

$$42.0 \text{ g } C_2H_5OH \times \frac{2.3 \text{ J}}{\text{g-K}} \times 43 \text{ K} \times \frac{1 \text{ kJ}}{1000 \text{ J}} = 4.1538 = 4.2 \text{ kJ}$$

Vaporize (boil) the liquid at 78 °C, using ΔH_{vap}.

$$42.0 \text{ g } C_2H_5OH \times \frac{1 \text{ mol } C_2H_5OH}{46.07 \text{ g}} \times \frac{38.56 \text{ kJ}}{\text{mol}} = 35.1535 = 35.2 \text{ kJ}$$

Total energy required is 4.1538 kJ + 35.1535 kJ = 39.3073 = 39.3 kJ.

(b) Heat the solid from –155 °C to –114 °C, ΔT = 41 °C = 41 K.

$$42.0 \text{ g } C_2H_5OH \times \frac{0.97 \text{ J}}{\text{g-K}} \times 41 \text{ K} \times \frac{1 \text{ kJ}}{1000 \text{ J}} = 1.6703 = 1.7 \text{ kJ}$$

Melt the solid at –114 °C, using ΔH_{fus}.

$$42.0 \text{ g } C_2H_5OH \times \frac{1 \text{ mol } C_2H_5OH}{46.07 \text{ g}} \times \frac{5.02 \text{ kJ}}{\text{mol}} = 4.5765 = 4.58 \text{ kJ}$$

Heat the liquid from –114 °C to 78 °C, ΔT = 192 °C = 192 K.

$$42.0 \text{ g } C_2H_5OH \times \frac{2.3 \text{ J}}{\text{g-K}} \times 192 \text{ K} \times \frac{1 \text{ kJ}}{1000 \text{ J}} = 18.5472 = 19 \text{ kJ}$$

From part (a), vaporizing (boiling) 42.0 g of C_2H_5OH liquid at 78 °C requires 35.1535 kJ = 35.2 kJ.

Total energy required = 1.6703 kJ + 4.5765 kJ + 18.5472 kJ + 35.1535 kJ = 59.9476 = 60 kJ.

Check. The relative energies of the various steps are reasonable; vaporization is the largest. The sum has no decimal places because (19 kJ) has no decimal places.

11.47 (a) False. The critical pressure is the pressure required to cause liquefaction at the critical temperature.

(b) True.

(c) False. In general, the higher the critical temperature, the higher the critical pressure.

(d) True. The more intermolecular forces in a substance, the greater the kinetic energy required to overcome them. This greater kinetic energy translates to higher critical temperatures and pressures (as well as melting and boiling points).

Vapor Pressure (Section 11.5)

11.49 Properties (c) intermolecular attractive forces, (d) temperature, and (e) density of the liquid affect vapor pressure of a liquid.

11.51 (a) *Analyze/Plan.* Given the molecular formulae of several substances, determine the kind of intermolecular forces present, and rank the strength of these forces. The weaker the forces, the more volatile the substance. *Solve.*

$CBr_4 < CHBr_3 < CH_2Br_2 < CH_2Cl_2 < CH_3Cl < CH_4$

(b) $CH_4 < CH_3Cl < CH_2Cl_2 < CH_2Br_2 < CHBr_3 < CBr_4$

(c) Boiling point increases as the strength of intermolecular forces increases. By analogy to attractive forces in HCl (Section 11.2), the trend will be dominated by dispersion forces, even though four of the molecules ($CHBr_3$, CH_2Br_2, CH_2Cl_2, and CH_3Cl) are polar. Thus, the order of increasing boiling point is the order of increasing molar mass and increasing strength of dispersion forces.

11.53 (a) The water in the two pans is at the same temperature, the boiling point of water at the atmospheric pressure of the room. During a phase change, the temperature of a system is constant. All energy gained from the surroundings is used to accomplish the transition, in this case to vaporize the liquid water. The pan of water that is boiling vigorously is gaining more energy and the liquid is being vaporized more quickly than in the other pan, but the temperature of the phase change is the same.

(b) Vapor pressure does not depend on either volume or surface area of the liquid. As long as the containers are at the same temperature, the vapor pressures of water in the two containers are the same.

11.55 *Analyze/Plan.* Follow the logic in Sample Exercise 11.4. The boiling point is the temperature at which the vapor pressure of a liquid equals atmospheric pressure. *Solve.*

(a) The boiling point of ethanol at 200 torr is ~48 °C.

(b) The vapor pressure of ethanol at 60 °C is approximately 340 torr. Thus, at 60 °C ethyl alcohol would boil at an external pressure of 340 torr.

(c) The boiling point of diethyl ether at 400 torr is ~17 °C.

(d) 40 °C is above the normal boiling point of diethyl ether, so the pressure at which 40 °C is the boiling point is greater than 760 torr. According to Figure 11.25, a boiling point of 40 °C requires an external pressure of 1000 torr. (At these conditions, the vapor pressure of diethyl ether is 1000 torr.)

Phase Diagrams (Section 11.6)

11.57 (a) The *critical point* is the temperature and pressure beyond which the gas and liquid phases are indistinguishable.

(b) The gas/liquid line ends at the critical point because at conditions beyond the critical temperature and pressure, there is no distinction between gas and liquid. In experimental terms, a gas cannot be liquefied at temperatures higher than the critical temperature, regardless of pressure.

(At conditions beyond the critical point, a substance is known as a supercritical fluid. Supercritical fluids have many practical applications, such as decaffeination of green coffee beans and dry cleaning.)

11.59 (a) The water vapor would deposit to form a solid at a pressure of around 4 torr. At higher pressure, perhaps 5 atm or so, the solid would melt to form liquid water. This occurs because the melting point of ice, which is 0 °C at 1 atm, decreases with increasing pressure.

(b) In thinking about this exercise, keep in mind that the **total** pressure is being maintained at a constant 0.50 atm. That pressure is composed of water vapor pressure and some other pressure, which could come from an inert gas. At 100 °C and 0.50 atm, water is in the vapor phase. As it cools, the water vapor will condense to the liquid at the temperature where the vapor pressure of liquid water is 0.50 atm. From Appendix B, we see that condensation occurs at approximately 82 °C. Further cooling of the liquid water results in freezing to the solid at approximately 0 °C. The freezing point of water increases with decreasing pressure, so at 0.50 atm, the freezing temperature is very slightly above 0 °C.

11.61 *Analyze/Plan.* Follow the logic in Sample Exercise 11.5, using the phase diagram for neon. *Solve.*

(a) The normal melting point is the temperature where solid becomes liquid at 1 atm pressure. Following a horizontal line at 1 atm to the solid-liquid line, the normal melting point is approximately 24 K.

(b) Neon sublimes, changes directly from solid to gas, at pressures less than the triple point pressure, approximately 0.5 atm.

(c) Room temperature is 298 K, in the region where neon is a supercritical fluid. Neon cannot be liquefied at any temperature above the critical temperature, approximately 45 K, regardless of pressure.

11.63 *Analyze/Plan.* Follow the logic in Sample Exercise 11.5, using the phase diagram for methane in Figure 11.30. *Solve.*

(a) According to Sample Exercise 11.5, the triple point of methane (CH_4) is approximately (–180 °C, 0.1 atm). The solid-liquid line in the phase diagram is essentially vertical in the pressure range 0.1-100 atm. This means that conditions at the surface of Titan (–178 °C, 1.6 atm) are very close to the solid-liquid line. Methane on the surface of titan is likely to exist in both solid and liquid forms.

(b) Methane is a liquid at –178 °C and 1.6 atm. Moving upward through Titan's atmosphere at a constant temperature of –178 °C, pressure decreases. At a pressure slightly greater than 0.1 atm, we expect to see vaporization to gaseous methane. If we begin with solid methane at 1.6 atm and a temperature slightly below –180 °C, we expect sublimation to gaseous methane at a pressure slightly less than 0.1 atm.

Liquid Crystals (Section 11.7)

11.65 In a nematic liquid crystalline phase, molecules are aligned along their long axes, but the molecular ends are not aligned. In an ordinary liquid, molecules have no orderly arrangement; they are randomly oriented, or amorphous. Both an ordinary liquid and a nematic liquid crystal phase are fluids; molecules are free to move relative to one another. In an ordinary liquid, molecules can move in any direction. In a nematic phase, molecules are free to translate in all dimensions. Molecules cannot tumble or rotate out of the molecular plane, or the order of the nematic phase is lost and the sample becomes an ordinary liquid.

11.67 (a) True.

(b) False. Liquid crystalline molecules are often rod-like with some rigidity in the long direction.

(c) True. Liquid crystalline is a phase of matter with distinct phase-change temperatures.

(d) False. If no significant intermolecular forces were present, there would be no driving force for the intermediate ordering typical of liquid crystals.

(e) False. Molecules containing only carbon and hydrogen do no exhibit the significant intermolecular forces required for liquid crystal formation.

(f) True.

11.69 Because order is maintained in at least one dimension, the molecules in a liquid-crystalline phase are not totally free to change orientation. This makes the liquid-crystalline phase more resistant to flow, more viscous, than the isotropic liquid.

11.71 As the temperature of a substance increases, the average kinetic energy of the molecules increases. More molecules have sufficient kinetic energy to overcome intermolecular attractive forces, so overall ordering of the molecules decreases as temperature increases. Melting provides kinetic energy sufficient to disrupt alignment in one dimension in the solid, producing a smectic phase with ordering in two dimensions. Additional heating of the smectic phase provides kinetic energy sufficient to disrupt alignment in another dimension, producing a nematic phase with one-dimensional order.

Additional Exercises

11.73 (a) decrease (b) increase (c) increase (d) increase

(e) increase (f) increase (g) increase

11.76 (a) The cis isomer has stronger dipole-dipole forces; the trans isomer is nonpolar. (Because both molecules have the same molecular weight, we can say that the dipole-dipole and dispersion forces of the cis isomer are stronger than the dispersion-only forces of the trans isomer.)

(b) The molecule with the stronger intermolecular interactions will have the higher boiling point. The cis isomer boils at 60.3 °C and the trans isomer boils at 47.5 °C.

11.78 (a) Four, all of them. All covalent compounds exhibit dispersion interactions.

(b) Three. Benzene is nonpolar and exhibits only dispersion forces. When Cl, Br, or OH is substituted for H in benzene, the molecule becomes polar and the molecules exhibit dipole-dipole interactions.

(c) One. The O–H bond in phenol participates in hydrogen bonding.

(d) Bromine is larger and more polarizable than chlorine, so the dispersion forces in bromobenzene are stronger than those in chlorobenzene.

(e) Phenol exhibits hydrogen bonding, which is the strongest intermolecular interaction among covalent molecules.

11.81 A plot of the number of carbon atoms versus boiling point follows. For eight C atoms, C_8H_{18}, the boiling point is approximately 130 °C. The more carbon atoms in the hydrocarbon, the longer the chain, the more polarizable the electron cloud, the stronger the London dispersion forces, and the higher the boiling point.

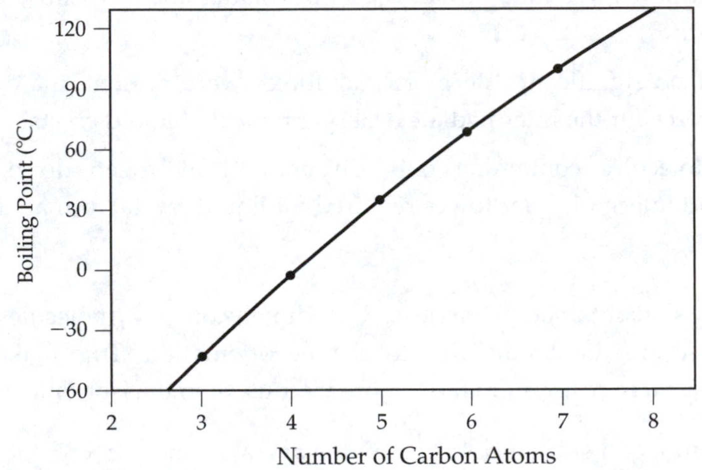

11.83 (a) Sweat, or salt water, on the surface of the body vaporizes to establish its typical vapor pressure at atmospheric pressure. Because the atmosphere is a totally open system, saturated vapor pressure is never reached, and the sweat evaporates continuously. Evaporation is an endothermic process. The heat required to vaporize sweat is absorbed from your body, helping to keep it cool.

(b) The vacuum pump reduces the pressure of the atmosphere (air + water vapor) above the water. Eventually, atmospheric pressure equals the vapor pressure of water and the water boils. Boiling is an endothermic process, and the temperature drops if the system is not able to absorb heat from the surroundings fast enough. As the temperature of the water decreases, the water freezes. (On a molecular level, the evaporation of water removes the molecules with the highest kinetic energies from the liquid. This decrease in average kinetic energy is what we experience as a temperature decrease.)

11.88 When voltage is applied to a liquid crystal display, the molecules align with the voltage and the appearance of the display changes. At low Antarctic temperatures, the liquid crystalline phase is closer to its freezing point. The molecules have less kinetic energy due to temperature and the applied voltage may not be sufficient to overcome orienting forces among the molecules. If some or all of the molecules do not rotate when the voltage is applied, the display will not function properly.

Integrative Exercises

11.92

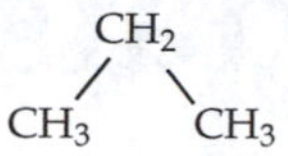

(i) $CH_3-CH_2-CH_3$ MM = 44

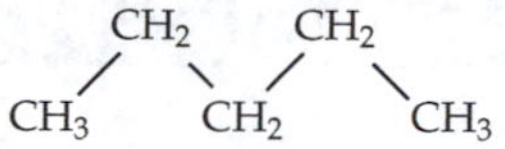

(ii) $CH_3-CH_2-CH_2-CH_2-CH_3$ MM = 72

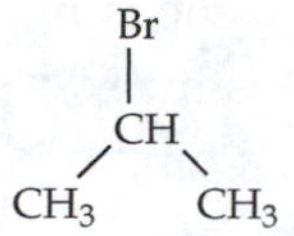

(iii) $CH_3-CH(Br)-CH_3$ MM = 123

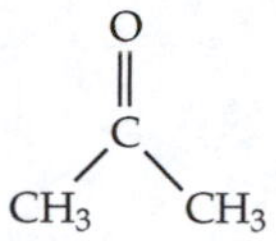

(iv) $CH_3-C(=O)-CH_3$ MM = 58

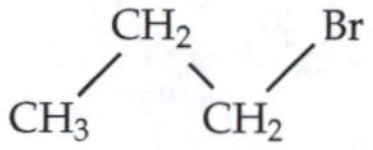

(v) $CH_3-CH_2-CH_2-Br$ MM = 123

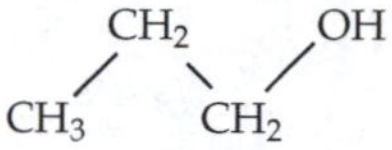

(vi) $CH_3-CH_2-CH_2-OH$ MM = 60

It is useful to draw the structural formulas because intermolecular forces are determined by the size and shape (structure) of molecules.

(a) *Molar mass*: compounds (i) and (ii) have similar rod-like structures; (ii) has a longer rod. The longer chain leads to greater molar mass, stronger London dispersion forces and higher heat of vaporization.

(b) *Molecular shape*: compounds (iii) and (v) have the same chemical formula and molar mass but different molecular shapes (they are structural isomers). The more rod-like shape of (v) leads to more contact between molecules, stronger dispersion forces, and higher heat of vaporization.

(c) *Molecular polarity*: rod-like hydrocarbons (i) and (ii) are essentially nonpolar, owing to free rotation about C–C σ bonds, whereas (iv) is quite polar, owing to the C=O group. (iv) has a smaller molar mass than (ii) but a larger heat of vaporization, which must be due to the presence of dipole-dipole forces in (iv). [Note that (iii) and (iv), with similar shape and molecular polarity, have very similar heats of vaporization.]

(d) *Hydrogen-bonding interactions*: molecules (v) and (vi) have similar structures, but (vi) has hydrogen bonding and (v) does not. Even though molar mass and thus dispersion forces are larger for (v), (vi) has the higher heat of vaporization. This must be due to hydrogen-bonding interactions.

11.96 $P = \frac{nRT}{V} = \frac{g\,RT}{M\,V}$; $T = 273.15 + 26.0\ ^\circ C = 299.15 = 299.2\ K$; $V = 5.00\ L$

$g\ C_6H_6(g) = 7.2146 - 5.1493 = 2.0653\ g\ C_6H_6(g)$

$$P\ (vapor) = \frac{2.0653\ g}{78.11\ g/mol} \times \frac{299.15\ K}{5.00\ L} \times \frac{0.08206\ L\text{-}atm}{mol\text{-}K} \times \frac{760\ torr}{1\ atm} = 98.660 = 98.7\ torr$$

12 Solids and Modern Materials

Visualizing Concepts

12.1 The red-orange compound is more likely to be a semiconductor and the white one an insulator. The red-orange compound absorbs light in the visible spectrum (red-orange is reflected, so blue-green is absorbed), whereas the white compound does not. This indicates that the red-orange compound has a lower energy electron transition than the white one. Semiconductors have lower energy electron transitions than insulators.

12.5 (a) Clearly, the structure is close packed. The question is: cubic or hexagonal? This is a side view of a close-packed array, like the one in Figure 12.13. The key is the arrangement of the third row relative to the first. Looking at any three rows of cannon balls, there is a ball in the third row directly above (at the same horizontal position as) one in the first row. This is an ABABAB pattern and the structure is hexagonal close packed.

(b) CN = 12, regardless of whether the structure is hexagonal or cubic close packed.

(c) CN(1) = 9, CN(2) = 6. The coordination numbers of these two balls are less than 12 because they are on the "surface" of the structure. Of the 12 maximum closest positions, several are unoccupied. Ball 1 is missing two balls that would be in front of it in its own layer, and one ball of the triad in the layer above, for a total of 3 missing and 9 occupied positions. Ball 2 is missing these same three balls, along will all 3 balls in the triad of the layer that would be below it. Ball 2 has six missing and six occupied nearest neighbors.

12.7 Fragment (b) is more likely to give rise to electrical conductivity. Arrangement (b) has a delocalized system of π electrons, in which electrons are free to move. Mobile electrons are required for electrical conductivity.

12.9 Polymer (a) is more crystalline and has the higher melting point. The polymer chains in cartoon (a) are linear and have two ordered regions shown in the upper left and lower right. The greater the degree of order, the more crystalline the material. The ordered regions of polymer (a) indicate that there are stronger intermolecular forces attracting the chains to each other. Stronger intermolecular forces mean that polymer (a) will have the higher melting point.

Classification of Solids (Section 12.1)

12.11 Statement (b) best explains the difference. In molecular solids, relatively weak intermolecular forces (hydrogen-bonding, dipole-dipole, dispersion) bind the molecules in the lattice, so relatively little energy is required to disrupt these forces. In covalent-network solids, covalent bonds join atoms into an extended network. Melting or deforming a covalent-network solid means breaking these covalent bonds, which requires a large amount of energy.

12.13 (a) hydrogen-bonding forces, dipole-dipole forces, London dispersion forces
(b) covalent chemical bonds (mainly)
(c) ionic bonds, the Coulombic forces between anions and cations (mainly)
(d) metallic bonds

12.15 (a) ionic (b) metallic
(c) covalent-network (Based on melting point, it is probably a network solid. Transition metals in high oxidation states often form bonds with nonmetals that have significant covalent character. It could also be characterized as ionic with some covalent character to the bonds.)
(d) molecular (e) molecular (f) molecular

12.17 Metallic. The melting point eliminates molecular and covalent-network solids. Because the solid conducts electricity and is insoluble in water, it is metallic. Ionic solids are insulators that are often soluble in water.

Structures of Solids (Section 12.2)

12.19 *Analyze/Plan.* Crystalline solids have a regular repeat in all three directions. Amorphous solids have no regular repeating structure. Draw diagrams that reflect these definitions. *Solve.*

(a)

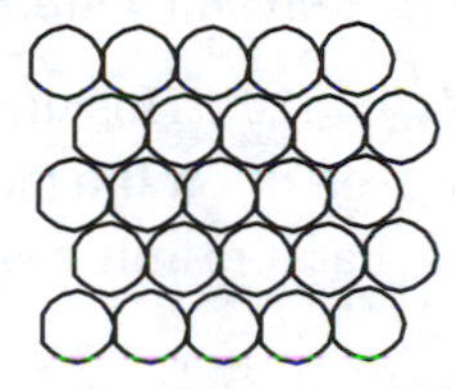

crystalline

(b)

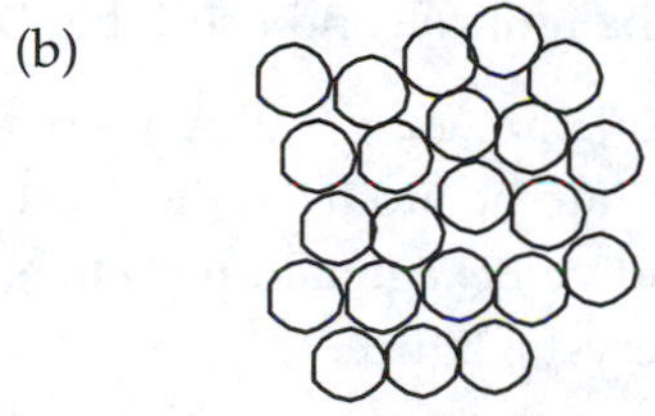

amorphous

12.21 *Analyze.* Given two two-dimensional structures, draw and describe the unit cells and lattice vectors. *Plan.* When choosing a unit cell, the environment of each lattice point must be identical and the unit cells must *tile* to generate the complete two-dimensional lattice. For a given structure, there are often several ways to draw a unit cell. The radii of A and B are equal. *Solve.*

Two-dimensional structure	(i)	(ii)
(a) unit cell		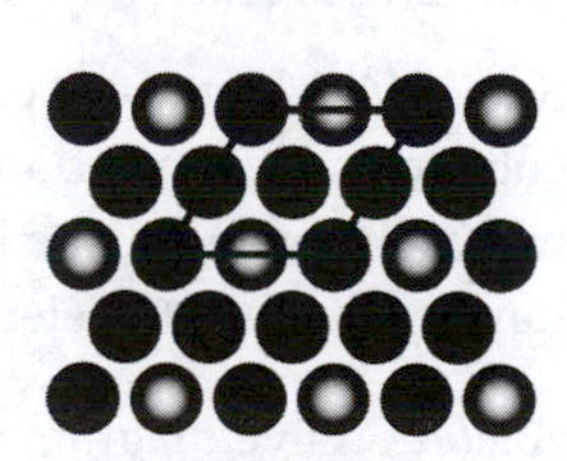
(b) γ, a, b	$\gamma = 90^\circ$, $a = b$	$\gamma = 120^\circ$, $a = b$
(c) lattice type	square	hexagonal

12.23 *Plan.* Refer to Figure 12.6 to find geometric characteristics of the seven three-dimensional primitive lattices.

Solve. Tetragonal, $a = b \neq c$, $\alpha = \beta = \gamma = 90°$. In the new lattice, one of the edge lengths is longer than (not the same as) the other two, and all angles remain 90°.

12.25 Choice (e), both rhombohedral and triclinic. According to Figure 12.6, if no lattice vectors are perpendicular to each other, none of the unit cell angles (α, β, γ) are 90°. This is characteristic of two of the three- dimensional primitive lattices: triclinic and rhombohedral.

12.27 Choice (b), 2. A body-centered cubic lattice is composed of body-centered cubic unit cells. A unit cell contains the minimum number of atoms when it has atoms only at the lattice points. A body-centered cubic unit cell like this is shown in Figure 12.12(b). There is one atom totally inside the cell (1×1) and one at each corner ($8 \times 1/8$) for a total of 2 atoms in the unit cell. (Only metallic elements have body-centered cubic lattices and unit cells.)

12.29 *Analyze.* Given a diagram of the unit cell dimensions and contents of nickel arsenide, determine what kind of lattice this crystal possess, and the empirical formula of the compound. *Plan.* Refer to Figure 12.6 to find geometric characteristics of the seven three-dimensional primitive lattices. Decide where atoms of the two elements are located in the unit cell and use Table 12.1 to help determine the empirical formula.

(a) $a = b = 3.57$ Å. $c = 5.10$ Å $\neq a$ or b. $\alpha = \beta = 90°$, $\gamma = 120°$. This unit cell is hexagonal. There are no atoms in the exact middle of the cell or on the face centers, so it is a primitive hexagonal unit cell. Nickel arsenide has a primitive hexagonal unit cell and crystal lattice.

(b) There are Ni atoms at each corner of the cell ($8 \times 1/8$) and centered on four of the unit cell edges ($4 \times 1/4$) for a total of 2 Ni atoms. There are 2 As atoms totally inside the cell. The unit cell contains 2 Ni and 2 As atoms; the empirical formula is NiAs.

Metallic Solids (Section 12.3)

12.31 *Analyze/Plan.* Consider body-centered and face-centered structures shown in Figures 12.11 and 12.12 to relate the structures and densities of the metals.

Solve. A body-centered cubic structure has more empty space than a face-centered cubic one. (A faced-centered cubic structure is one of the close-packed structures.) The more empty space, the less dense the solid. We expect potassium, which has the lowest density of the listed elements, to adopt the body-centered cubic structure.

12.33 *Analyze.* Give diagrams of three structure types, find which is most densely packed and which is least densely packed. *Plan.* Assume that the same element packs in each of the three structures, so that atomic mass and volume are constant. Then we are analyzing the packing efficiency or relative amount of empty space in each structure. *Solve.*

Structure type A has a face-centered cubic unit cell with metal atoms only at the lattice points; this corresponds to a cubic close-packed structure. Structure type B has a body-centered cubic unit cell with metal atoms at the lattice points; this is also a body-centered cubic structure. Structure type C has a hexagonal unit cell with two atoms totally inside the cell. Building up many unit cells into a lattice (Figure 12.14) leads to a hexagonal close-packed structure.

(a) In both cubic and hexagonal-close packed structures, any individual atom has twelve nearest neighbor atoms. Both structures are close packed and have equal amounts of empty space. Structure types A and C have equally dense packing and are more densely packed than structure type B.

(b) Structure type B, which is not close packed, has the least dense atom packing.

12.35 *Analyze.* Given the cubic unit cell edge length and arrangement of Ir atoms, calculate the atomic radius and the density of the metal. *Plan.* In a face-centered cubic metal structure, there is space between the atoms along the unit cell edge, but they touch along the face diagonal. See Figure 12.12. Use the geometry of the right equilateral triangle to calculate the atomic radius. From the definition of density and paying attention to units, calculate the density of Ir(s). *Solve.*

(a) The length of the face diagonal of a face-centered cubic unit cell is four times the radius of the atom and $\sqrt{2}$ times the unit cell dimension or edge length, a for cubic unit cells.

$$4\,r = \sqrt{2}\,a;\; r = \sqrt{2}\,a/4 = \frac{\sqrt{2} \times 3.833\ \text{Å}}{4} = 1.3552 = 1.355\ \text{Å}$$

(b) The density of iridium is the mass of the unit cell contents divided by the unit cell volume. There are 4 Ir atoms in a face-centered cubic unit cell.

$$\rho = \frac{4\ \text{Ir atoms}}{(3.833 \times 10^{-8}\ \text{cm})^3} \times \frac{192.22\ \text{g Ir}}{6.022 \times 10^{23}\ \text{Ir atoms}} = 22.67\ \text{g/cm}^3$$

Check. The units of density are correct. Note that Ir is quite dense.

12.37 (a) *Analyze.* We are given a face-centered cubic unit cell with edge length 5.588 Å. *Plan.* In a face-centered cubic metal structure, the atoms touch along the face diagonal of the unit cell. The length of the face diagonal of a face-centered cubic unit cell is four times the radius of the atom and $\sqrt{2}$ times the unit cell dimension or edge length, a for cubic unit cells. *Solve.*

$$4\,r = \sqrt{2}\,a;\; r = \sqrt{2}\,a/4 = \frac{\sqrt{2} \times 5.588\ \text{Å}}{4} = 1.9757 = 1.976\ \text{Å}$$

(b) *Plan.* The density of calcium is the mass of the unit cell contents divided by the unit cell volume. There are 4 Ca atoms in a face-centered cubic unit cell. *Solve.*

$$\rho = \frac{4\ \text{Ca atoms}}{(5.588 \times 10^{-8}\ \text{cm})^3} \times \frac{40.078\ \text{g Ca}}{6.022 \times 10^{23}\ \text{Ca atoms}} = 1.526\ \text{g/cm}^3$$

Check. The units of density are correct.

12.39 *Analyze.* Given the structure of aluminum metal and the atomic radius of an Al atom, find the number of Al atoms in each unit cell and the coordination number of each Al atom. Calculate (estimate) the length of the unit cell edge and the density of aluminum metal.

Plan. Use Figure 12.12(c) to count the number of Al atoms in one unit cell and use Figure 12.13 to visualize the coordination number of each Al atom. According to Figure 12.12(c), there is space between the atoms along the unit cell edge, but they touch along the face diagonal. Use the geometry of the right equilateral triangle and the atomic radius to calculate the unit cell edge length. From the definition of density and paying attention to units, calculate the density of aluminum metal. *Solve.*

(a) 8 corners × 1/8 atom/corner + 6 faces × ½ atom/face = 4 atoms

(b) Each aluminum atom is in contact with 12 nearest neighbors, 6 in one plane, 3 above that plane, and 3 below. Its coordination number is thus 12.

(c) The length of the face diagonal of a face-centered cubic unit cell is four times the radius of the metal and $\sqrt{2}$ times the unit cell dimension (usually designated *a* for cubic cells).

$$4 \times 1.43 \text{ Å} = \sqrt{2} \times a; \quad a = \frac{4 \times 1.43 \text{ Å}}{\sqrt{2}} = 4.0447 = 4.04 \text{ Å} = 4.04 \times 10^{-8} \text{ cm}$$

(d) The density of the metal is the mass of the unit cell contents divided by the volume of the unit cell.

$$\text{density} = \frac{4 \text{ Al atoms}}{(4.0447 \times 10^{-8} \text{ cm})^3} \times \frac{26.98 \text{ g Al}}{6.022 \times 10^{23} \text{ Al atoms}} = 2.71 \text{ g/cm}^3$$

12.41 Statement (b) is false. Alloys are mixtures, not compounds, that vary in composition. One or more of the components of the alloy can be a nonmetal.

12.43 *Analyze/Plan.* Consider the descriptions of various alloy types in Section 12.3. *Solve.*

(a) $Fe_{0.97}Si_{0.03}$; interstitial alloy. The radii of Fe and Si are substantially different, so Si could fit in "holes" in the Fe lattice. Also, the small amount of Si relative to Fe is characteristic of an interstitial alloy.

(b) $Fe_{0.60}Ni_{0.40}$, substitutional alloy. The two metals have very similar atomic radii and are present in similar amounts.

(c) $SmCo_5$, intermetallic compound. The two elements are present in stoichiometric amounts.

12.45 (a) True

(b) False. Interstitial alloys form between elements with very different bonding atomic radii.

(c) False. Nonmetallic elements are typically found in interstitial alloys.

12.47 *Analyze.* Given the color of a gold alloy, find the other element(s) in the alloy and the type of alloy formed. *Plan.* Refer to "Chemistry Put to Work: ALLOYS OF GOLD". *Solve.*

(a) White gold, nickel or palladium, substitutional alloy

(b) Rose gold, copper, substitutional alloy

(c) Green gold, silver, substitutional alloy

Metallic Bonding (Section 12.4)

12.49 (a) True

(b) False. See statement (a).

(c) False. Delocalized electrons in metals facilitate the transfer of kinetic energy, which is the basis of thermal conductivity.

(d) False. Metals have large thermal conductivities.

12.51 *Plan.* By analogy to Figure 12.22, the most bonding, lowest energy MOs have the fewest nodes. As energy increases, the number of nodes increases. When constructing an MO diagram from AOs, total number of orbitals is conserved. The MO diagram for a linear chain of six Li atoms will have six MOs, starting with zero nodes and maximum overlap, and ending with five nodes and minimum overlap. *Solve.*

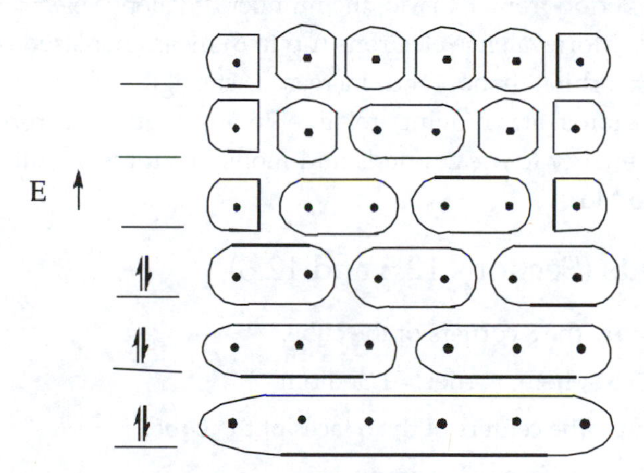

(a) Six. Six AOs require six MOs.

(b) Zero nodes in lowest energy orbital

(c) Five nodes in highest energy orbital

(d) Two nodes in the HOMO

(e) Three nodes in the LUMO

(f) The HOMO-LUMO energy gap for the six-atom diagram is smaller than the one for the four-atom diagram. In general, the more atoms in the chain, the smaller the HOMO-LUMO energy gap.

12.53 *Analyze/Plan*. Consider the definition of ductility, as well as the discussion of metallic bonding in Section 12.4. *Solve*.

Ductility is the property related to the ease with which a solid can be drawn into a wire. Basically, the softer the solid the more ductile it is. The more rigid the solid, the less ductile it is. For metals, ductility decreases as the strength of metal-metal bonding increases, producing a stiffer lattice less susceptible to distortion.

(a) Ag is more ductile. Mo, with 6 valence electrons, has a filled bonding band, strong metal-metal interactions, and a rigid lattice. This predicts high hardness and low ductility. Ag, with 11 valence electrons, has a nearly filled antibonding band as well as a filled bonding band. Bonding is weaker than in Mo, and Ag is more ductile.

(b) Zn is more ductile. Si is a covalent-network solid with all valence electrons localized in bonds between Si atoms. Covalent-network substances are high-melting, hard, and not particularly ductile.

12.55 The relevant electron configurations are:

Y: $[Kr]5s^2 4d^1$; Zr: $[Kr]5s^2 4d^2$; Nb: $[Kr]5s^2 4d^3$; Mo: $[Kr]5s^1 4d^5$;

The order of increasing melting points is Y < Zr < Nb < Mo.

Moving across the fifth period from Y to Mo, the number of valence electrons increases, from 3 for Y to 6 for Mo. More valence electrons (up to 6) mean increased occupancy of the bonding molecular orbital band, and increased strength of metallic bonding. Melting requires that atoms are moving relative to each other. Stronger metallic bonding requires more energy to break bonds and mobilize atoms, resulting in higher melting points from Y to Mo.

Ionic and Molecular Solids (Sections 12.5 and 12.6)

12.57 (a) Sr: Sr atoms occupy the 8 corners of the cube.

8 corners × 1/8 sphere/corner = 1 Sr atom

O: O atoms occupy the centers of the 6 faces of the cube.

6 faces × 1/2 atom/face = 3 O atoms

Ti: There is 1 Ti atom at the body center of the cube.

Empirical Formula: $SrTiO_3$

(b) Six. The Ti atom at the center of the cube is coordinated to the six O atoms at the face centers.

(c) Twelve. Each Sr atom occupies one corner of 8 unit cells. Sr is coordinated to 3 oxygen positions in each unit cell for a total of 24 oxygen positions. However, each O position is in the center of a cell face, with half-occupancy in each cell. 24 oxygen positions × ½ occupancy = 12 oxygen atoms. Each Ti atom is coordinated to 12 oxygen atoms.

12.59 *Analyze/Plan.* The density of MnS is the mass of the unit cell contents divided by the volume of the unit cell. MnS has the same structure as rock salt, or NaCl. Refer to the NaCl structure in Figures 12.25 and 12.26 to determine the unit cell contents. Use the unit cell edge length to find unit cell volume. Calculate density. *Solve.*

By analogy to the NaCl structure, with Mn replacing Na and S replacing Cl, there are 8 sulfide ions (green spheres) on the corners of the unit cell, and 6 sulfide ions in the middle of the faces. The number of sulfide ions per unit cell is then [8(1/8) + 6(1/2)] = 4. There is 1 manganese ion (purple sphere almost hidden in figure) completely inside the unit cell and 12 manganese ions (purple spheres) along the unit cell edges. The number of manganese ions is then [1 + 12(1/4)] = 4. This result satisfies charge balance requirements.

4 Mn^{2+}, 4 S^{2-}, 4 MnS formula units. The mass of 1 MnS formula unit is 87.003 g/ 6.022×10^{23} NaF units.

$$d = \frac{4 \text{ MnS units}}{(5.223 \text{ Å})^3} \times \frac{87.003 \text{ g}}{6.0221 \times 10^{23} \text{ MnS units}} \times \left(\frac{1 \text{ Å}}{1 \times 10^{-8} \text{ cm}}\right)^3 = 4.056 \text{ g/cm}^3$$

Check. The value for the density of alabandite (MnS) reported in the *CRC Handbook of Chemistry and Physics*, 74th Ed., is 3.99 g/cm^3. The calculated density is within 2% of the reported value.

12.61 *Analyze.* Given the atomic arrangement and length of the unit cell side, calculate the density of HgS and HgSe. Qualitatively and quantitatively compare the densities of the two solids. *Plan.* Calculate the mass and volume of a single unit cell and then use them to find density. The unit cell volume is the cube of edge length. *Solve.*

(a) According to Figures 12.25 and 12.26, sulfide ions (yellow spheres) occupy the corners and faces of the unit cell (in a face-centered cubic arrangement), for a total of 6(1/2) + 8(1/8) = 4 S^{2-} ions per unit cell. There are 4 mercury ions (gray spheres) totally in the interior of the cell. This means there are 4 HgS units in a unit cell with the zinc blende structure.

$$\text{density} = \frac{4 \text{ HgS units}}{(5.852 \text{ Å})^3} \times \frac{232.655 \text{ g}}{6.022 \times 10^{23} \text{ HgS units}} \times \left(\frac{1 \text{ Å}}{1 \times 10^{-8} \text{ cm}}\right)^3 = 7.711 \text{ g/cm}^3$$

(b) We expect Se^{2-} to have a larger ionic radius than S^{2-}, because Se is below S in the chalcogen family and both ions have the same charge. Thus, HgSe will occupy a larger volume and the unit cell edge will be longer.

(c) For HgSe, also with the zinc blende structure:

$$\text{density} = \frac{4 \text{ HgSe units}}{(6.085 \text{ Å})^3} \times \frac{279.55 \text{ g HgSe}}{6.022 \times 10^{23} \text{ HgSe units}} \times \left(\frac{1 \text{ Å}}{1 \times 10^{-8} \text{ cm}}\right)^3 = 8.241 \text{ g/cm}^3$$

Even though HgSe has a larger unit cell volume than HgS, it also has a larger molar mass. The mass of Se is more than twice that of S, whereas the radius of Se^{2-} is only slightly larger than that of S^{2-} (Figure 7.8). The greater mass of Se accounts for the greater density of HgSe.

12.63 *Analyze.* Given that CuI, CsI, and NaI uniquely adopt one of the structure types pictured in Figure 12.26, match the ionic compound with its structure. Use ionic radii to inform your decision. *Plan.* Note the relative cation and anion radii in the three structures in Figure 12.26. Match these ratios with those of CuI, CsI, and NaI. *Solve.*

(a) In the CsCl structure, the anion and cation have about the same radius; in the NaCl structure, the anion is somewhat larger than the cation; in the ZnS structure the anion is much larger than the cation.

In the three compounds given, Cs^+(r = 1.81 Å) and I^- (r = 2.06 Å) have the most similar radii; CsI will adopt the CsCl-type structure. The radii of Na^+ (r = 1.16 Å) and I^- (r = 2.06 Å) are somewhat different; NaI will adopt the NaCl-type structure. The radii of Cu^+ (r = 0.74 Å) and I^- (r = 2.06 Å) are very different; CuI has the ZnS-type structure.

(b) As cation size decreases, coordination number of the anion decreases. In CsI, I^- has a coordination number of eight; in NaI, I^- has a coordination number of six; in CuI, I^- has a coordination number of four.

12.65 *Analyze.* Given three magnesium compounds in which the coordination number (CN) of Mg^{2+} is six, determine the coordination number of the anion. *Plan.* Use Equation 12.1 with the cation/anion ratio of each compound and the Mg^{2+} coordination number to calculate the anion coordination number in each compound. *Solve.*

(a) MgS: 1 cation, 1 anion, cation CN = 6

$$\frac{\text{number of cations per formula unit}}{\text{number of anions per formula unit}} = \frac{\text{anion coordination number}}{\text{cation coordination number}}$$

$$\text{anion CN} = \frac{\text{cation CN} \times \text{\# of cations per formula unit}}{\text{\# of anions per formula unit}} = \frac{6 \times 1}{1} = 6$$

(b) MgF_2: 1 cation, 2 anions, cation CN = 6

$$\text{anion CN} = \frac{\text{cation CN} \times \text{\# of cations per formula unit}}{\text{\# of anions per formula unit}} = \frac{6 \times 1}{2} = 3$$

(c) MgO: 1 cation, 1 anion, cation CN = 6, anion CN = 6. The cation/anion ratio and cation CN are the same as in part (a), so the anion CN is the same (6).

12.67 (a) False. Although both molecular solids and covalent-network solids have covalent bonds, the melting points of molecular solids are much lower because intermolecular forces among their molecules are much weaker than covalent bonds among atoms in a covalent-network solid.

(b) True. The statement is true if there are no significant differences in polarity or molar mass of the molecules being compared.

Covalent-Network Solids (Section 12.7)

12.69 (a) Ionic solids are much more likely to dissolve in water. Polar water molecules can disrupt ionic bonds to surround and separate ions and form a solution, but they cannot break the covalent bonds of a covalent-network solid.

(b) Covalent-network solids can become better conductors of electricity via chemical substitution. Most semiconductors are covalent-network solids, and doping or chemical substitution changes their electrical properties.

12.71 *Analyze/Plan.* Follow the logic in Sample Exercise 12.3. *Solve.*

(a) CdS. Both semiconductors contain Cd; S and Te are in the same family, and S is higher.

(b) GaN. Ga is in the same family and higher than In; N is in the same family and higher than P.

(c) GaAs. Both semiconductors contain As; Ga and In are in the same family and Ga is higher.

12.73 *Analyze.* Given: GaAs. Find: dopant to make n-type semiconductor.

Plan. An n-type semiconductor has extra negative charges. If the dopant replaces a few Ga atoms, it should have more valence electrons than Ga, Group 3A.

Solve. The obvious choice is a Group 4A element, either Ge or Si. Ge would be closer to Ga in bonding atomic radius (Figure 7.7).

12.75 (a) *Analyze.* Given: 1.1 eV. Find: wavelength in meters that corresponds to the energy 1.1 eV. *Plan.* Use dimensional analysis to find wavelength.

Solve. $1 \text{ eV} = 1.602 \times 10^{-19}$ J (inside back cover of text); $\lambda = hc/E$

$$\lambda = 6.626 \times 10^{-34} \text{ J-s} \times \frac{3.00 \times 10^{8} \text{ m}}{\text{s}} \times \frac{1}{1.1 \text{ eV}} \times \frac{1 \text{ eV}}{1.602 \times 10^{-19} \text{ J}} = 1.128 \times 10^{-6} = 1.1 \times 10^{-6} \text{ m}$$

(b) Si can absorb energies ≥ 1.1 eV, or wavelengths $\leq 1.1 \times 10^{-6}$ m. A wavelength of 1.1×10^{-6} m corresponds to 1100 nm. The range of wavelengths in the visible portion of the spectrum is 300 to 750 nm. Silicon absorbs all wavelengths in the visible portion of the solar spectrum.

(c) Si absorbs wavelengths less than 1100 nm. This corresponds to approximately 80–90% of the total area under the curve.

12.77 *Plan/Solve.* Follow the logic in Solution 12.75.

$$\lambda = hc/E = 6.626 \times 10^{-34} \text{ J-s} \times \frac{3.00 \times 10^{8} \text{ m}}{\text{s}} \times \frac{1}{1.74 \text{ eV}} \times \frac{1 \text{ eV}}{1.602 \times 10^{-19} \text{ J}} = 7.131 \times 10^{-7}$$

$$= 7.13 \times 10^{-7} \text{ m} = 713 \text{ nm}$$

The emitted light with a wavelength of 713 nm is red light in the visible region of the electromagnetic spectrum.

12.79 *Analyze/Plan.* From Table 12.4, E_g for GaAs (x = 0) is 1.43 eV and for GaP (x = 1) is 2.26 eV. If E_g varies linearly with x, the band gap for x = 0.5 should be approximately the average of the two extreme values.

Solve. (1.43 + 2.26)/2 = 1.845 = 1.85 eV.

$$\lambda = hc/E = \frac{6.626\times10^{-34}\text{ J-s}\times2.998\times10^{8}\text{ m}}{\text{s}}\times\frac{1}{1.845\text{ eV}}\times\frac{1\text{ eV}}{1.602\times10^{-19}\text{ J}}$$

$$= 6.721\times10^{-7}\text{ m} = 672\text{ nm}$$

Polymers (Section 12.8)

12.81 (a) A monomer is a small molecule with low molecular mass that can be joined with other monomers to form a polymer. It is the repeating unit of a polymer.

(b) Ethene (ethylene) is commonly used as a monomer, while ethanol and methane are not. Methane does not have reactive groups that can be chemically linked into polymers. Ethanol is reactive because of its one alcohol functional group, but it cannot form long polymer chains. Alcohol monomers used to form polyesters have two alcohol groups.

12.83 A value of 100 amu is too low to be the molecular weight of a polymer; it would not represent enough monomer units. Often a single monomer has a molecular weight greater than 100 amu. Reasonable values for a polymer's molecular weight are 10,000 amu (typical for low density polyethylene), 100,000 amu, and 1,000,000 amu (typical for high density polyethylene).

12.85 *Analyze.* Given two types of reactant molecules, we are asked to write a condensation reaction with an ester product. *Plan.* A condensation reaction occurs when two smaller molecules combine to form a larger molecule and a small molecule, often water. Consider the structures of the two reactants and how they could combine to join the larger fragments and split water. *Solve.*

A carboxylic acid contains the $-\overset{\overset{\large O}{\|}}{C}-OH$ functional group; an alcohol contains the –OH functional group. These can be arranged to form the $-\overset{\overset{\large O}{\|}}{C}-O-C$ ester functional group and H_2O. Condensation reaction to form an ester:

$$CH_3-\overset{\overset{\large O}{\|}}{C}-\boxed{O-H + H}-O-CH_2-CH_3 \rightarrow CH_3-\overset{\overset{\large O}{\|}}{C}-O-CH_2CH_3 + H_2O$$

acetic acid — ethanol — ethyl acetate

If a dicarboxylic acid (two –COOH groups, usually at opposite ends of the molecule) and a dialcohol (two –OH groups, usually at opposite ends of the molecule) are combined, there is the potential for propagation of the polymer chain at both ends of both monomers. Polyethylene terephthalate (Table 12.6) is an example of a polyester formed from the monomers ethylene glycol and terephthalic acid.

12.87 *Analyze/Plan.* In an addition polymerization, the monomer contains a multiple bond. We expect the monomer associated with this polymer to have either a double or triple bond.

Solve. This polymer is similar to polyethylene (Figure 12.35), with two hydrogen atoms replaced by chlorine atoms. The monomer in polyethylene is ethylene or ethene. The monomer for this polymer is dichloroethylene or dichloroethene.

Cl H
C=C
Cl H

12.89 *Analyze/Plan.* Given the structure of a polymer, identify the monomers that react to form the polymer. Compare the polymer structure to other condensation polymers shown in Table 12.6. *Solve.*

Kevlar is a polymer similar to nylon. The connecting unit is —C(=O)—N(H)—

According to Figure 12.37, the monomers involved in this type polymer are a diacid, a molecule with two carboxylic acid groups, and a diamine, a molecule with two $-NH_2$ groups. The spacers in Kevlar are both benzene rings. The monomers used to produce Kevlar are:

HOOC—C₆H₄—COOH H₂N—C₆H₄—NH₂

12.91 (a) Most of a polymer backbone is composed of σ bonds. The geometry around individual atoms is tetrahedral with bond angles of 109°, so the polymer is not flat, and there is relatively free rotation around the σ bonds. The flexibility of the molecular chains causes flexibility of the bulk material. Flexibility is enhanced by molecular features that inhibit order, such as branching, and diminished by features that encourage order, such as cross-linking or delocalized π electron density.

(b) Less flexible. Cross-linking is the formation of chemical bonds between polymer chains. It reduces flexibility of the molecular chains and increases the hardness of the material. Cross-linked polymers are less chemically reactive because of the links.

12.93 Low degree of crystallinity. A good plastic wrap is extremely flexible; the lower the degree of crystallinity, the less rigid and more flexible the polymer.

Nanomaterials (Section 12.9)

12.95 Continuous energy bands of molecular orbitals require a large number of atoms contributing a large number of atomic orbitals to the molecular orbital scheme. If a solid has dimensions 1-10 nm, nanoscale dimensions, there may not be enough contributing atomic orbitals to produce continuous energy bands of molecular orbitals.

12.97 (a) False. As particle size decreases, the band gap increases. The smaller the particle, the fewer AOs that contribute to the MO scheme, the more localized the bonding and the larger the band gap.

(b) False. The wavelength of emitted light corresponds to the energy of the band gap. As particle size decreases, band gap increases and wavelength decreases ($E = hc/\lambda$).

12.99 *Analyze.* Given: Au, 4 atoms per unit cell, 4.08 Å cell edge, volume of sphere = $4/3\,\pi\, r^3$. Find: Au atoms in 20 nm diameter sphere.

Plan. Relate the number of Au atoms in the volume of 1 cubic unit cell to the number of Au atoms in a 20 nm diameter sphere. Change units to Å (you could just as well have chosen nm as the common unit), calculate the volumes of the unit cell and sphere, and use a ratio to calculate atoms in the sphere.

Solve. vol. of unit cell = $(4.08\ \text{Å})^3 = 67.9173 = 67.9\ \text{Å}^3$

$$20\ \text{nm diameter} = 10\ \text{nm radius};\ 10\ \text{nm} \times \frac{1\times10^{-9}\ \text{m}}{1\ \text{nm}} \times \frac{1\ \text{Å}}{1\times10^{-10}\ \text{m}} = 100\ \text{Å radius}$$

(Note that 1 nm = 10 Å.)

vol. of sphere = $4/3 \times 3.14159 \times (100\ \text{Å})^3 = 4.18879 \times 10^6 = 4.19 \times 10^6\ \text{Å}^3$

$$\frac{4\ \text{Au atoms}}{67.9173\ \text{Å}^3} = \frac{x\ \text{Au atoms}}{4.18879\times10^6\ \text{Å}^3};\ x = 2.46699\times10^5 = 2.47\times10^5\ \text{Au atoms}$$

12.101 Statement (b) is correct.

(a) Neither graphite nor graphene is molecular. Both are covalent-network solids with covalent-network bonding.

(c) Neither graphite nor graphene is an insulator or a metal.

(d) Both are pure carbon.

(e) The carbon atoms in both are sp^2 hybridized.

Additional Exercises

12.109 Ni_3Al: Ni is at the face centers ($6 \times 1/2 = 3$ Ni atoms) and Al is at the corners ($8 \times 1/8 =$ 1 Al atom). The atom ratio in the structure matches the empirical formula.

Nb_3Sn: The top, front, and side faces of the unit cell are clearly visible; each has 2 Nb atoms centered on it, as opposed to totally inside the cell. The three opposite faces are not completely visible in the diagram, but must be the same by translational symmetry. Two Nb atoms centered on each of 6 faces ($12 \times 1/2 = 6$ Nb atoms). One Sn atom completely inside the unit cell and one at each corner ($1 + 8 \times 1/8 = 2$ Sn atoms). The atom ratio is 6 Nb : 2 Sn or 3 Nb : 1 Sn; the atom ratio in the structure matches the empirical formula.

$SmCo_5$: One Sm atom at each corner ($8 \times 1/8 = 1$ Sm atom). One Co atom totally inside the unit cell (1 Co atom), two Co atoms centered on the top and bottom faces ($4 \times 1/2 =$ 2 Co atoms), four Co atoms centered on 4 side faces ($4 \times 1/2 = 2$ Co atoms), for a total of ($1 + 2 + 2 = 5$ Co atoms). The ratio is 1 Sm : 5 Co, which matches the empirical formula.

12.111 Calculate the wavelength of light that corresponds to an energy of 2.20 eV.

$$\lambda = \frac{hc}{E} = \frac{6.626 \times 10^{-34} \text{ J-s} \times 2.998 \times 10^{8} \text{ m}}{\text{s}} \times \frac{1}{2.20 \text{ eV}} \times \frac{1 \text{ eV}}{1.602 \times 10^{-19} \text{ J}}$$

$$= 5.636 \times 10^{-7} = 5.64 \times 10^{-7} \text{ m (564 nm)}$$

[The absorbed 564-nm light is yellow-green and cinnabar appears the complementary color, red-violet or vermillion.]

12.113 (a) Zinc sulfide, ZnS (Figure 12.26).

(b) Covalent. Silicon and carbon are both nonmetals, and their electronegativities are similar.

(c) Silicon carbide is hard and high-melting because it is a covalent-network solid. In SiC, the C atoms form a face-centered cubic array with Si atoms occupying alternate tetrahedral holes in the lattice. This means that the coordination numbers of both Si and C are 4; each Si is bound to four C atoms in a tetrahedral arrangement, and each C is bound to four Si atoms in a tetrahedral arrangement, producing an extended three-dimensional network. SiC is high-melting because a great deal of chemical energy is stored in the covalent Si–C bonds. Melting requires breaking covalent Si–C bonds, which takes a huge amount of thermal energy. It is hard because the three-dimensional lattice resists any change that would weaken the Si–C bonding network.

12.120 $n\lambda = 2d \sin\theta$; $n = 1$, $\lambda = 1.54$ Å, $\theta = 14.22$ °; calculate d.

$$d = \frac{n\lambda}{2 \sin\theta} = \frac{1 \times 1.54 \text{ Å}}{2 \sin(14.22)} = 3.1346 = 3.13 \text{ Å}$$

Integrative Exercises

12.126 Refer to Section 12.7 and Figure 12.29.

(a) In diamond, each C atom is bound to 4 other C atoms. According to VSEPR, the geometry around a central atom with 4 bonding electron pairs is tetrahedral and the C–C–C bond angles are 109°.

(b) Within a graphite sheet, each C atom is bound to 3 other C atoms in a trigonal planar arrangement. The C–C–C bond angles are 120°.)

(c) Within a sheet, sp^2 hybrid orbitals are involved in the sigma bonding network framework. This leaves atomic p orbitals to be involved in interactions between sheets. These interactions are responsible for stacking.

12.130 (a) Follow the logic outlined in Solution 12.99.

vol. of unit cell = $(5.43\ \text{Å})^3 = 160.1030 = 1.60 \times 10^2\ \text{Å}^3$

$$1\ \text{cm}^3 \times \frac{(1)^3\ \text{Å}^3}{(1\times10^{-8})^3\ \text{cm}^3} = 1\times10^{24}\ \text{Å}^3\ \text{(volume of material)}$$

$$\frac{4\ \text{Si atoms}}{160.103\ \text{Å}^3} = \frac{\text{x Si atoms}}{1\times10^{24}\ \text{Å}^3};\ x = 2.4984\times10^{22} = 2.50\times10^{22}\ \text{Si atoms}$$

(To 1 sig fig, the result is 2×10^{22} Si atoms.)

(b) 1 ppm phosphorus = 1 P atom per 1×10^6 Si atoms

$$\frac{1\ \text{P atom}}{1\times10^6\ \text{Si atoms}} = \frac{\text{x P atoms}}{2.4984\times10^{22}\ \text{Si Atoms}};\ x = 2.4984\times10^{22}$$

$$= 2.50\times10^{16}\ \text{P atoms}$$

$$2.4984\times10^{16}\ \text{P atom} \times \frac{1\ \text{mol}}{6.022\times10^{23}\ \text{atoms}} \times \frac{30.97376\ \text{g P}}{\text{mol}} \times \frac{1\ \text{mg}}{1\times10^{-3}\ \text{g}}$$

$$= 1.29\times10^{-3}\ \text{mg P (1.29 μg)}$$

12.132 The bonding atomic radius of a Si atom is 1.11 Å (Figure 7.7), so the diameter is 2.22 Å.

$$14\ \text{nm} \times \frac{1\times10^{-9}\ \text{m}}{1\ \text{nm}} \times \frac{1\ \text{Å}}{1\times10^{-10}\ \text{m}} \times \frac{1\ \text{Si atom}}{2.22\ \text{Å}} = 63\ \text{Si atoms.}$$

13 Properties of Solutions

Visualizing Concepts

13.1 (a) < (b) < (c). In Section 13.1, *entropy* is qualitatively defined as randomness or dispersal in space. In container (a), the two kinds of particles are not mixed and the particles are close together, so (a) has the least entropy. In container (b), the particles occupy approximately the same volume as container (a) but the two kinds of particles are homogeneously mixed, so the degree of dispersal and randomness is greater than in (a). In container (c), the two kinds of particles are homogeneously mixed and they occupy a larger volume than in (b), so (c) has the greatest entropy.

13.3 (a) No.

(b) The ionic solid with the smaller lattice energy will be more soluble in water. Lattice energy is the main component of ΔH_{solute}, the enthalpy required to separate solute particles. The solid with the smaller lattice energy will have the less endothermic ΔH_{solute}. Assuming $\Delta H_{solvent}$ and ΔH_{mix} are the same for the two solids, the one with the smaller lattice energy will have a less endothermic ΔH_{soln} and be more soluble in water.

13.7 Vitamin B_6 is more water soluble and vitamin E is more fat soluble. The B_6 molecule contains three –OH groups and the $-\ddot{N}-$ that can enter into many hydrogen-bonding interactions with water. Its relatively small molecular size indicates that dispersion forces will not play a large role in intermolecular interactions and that hydrogen bonding will dominate. On the other hand, the long, rod-like hydrocarbon chain of vitamin E will lead to stronger dispersion forces among it and the mostly nonpolar fats. Although vitamin E has one –OH and one $-\ddot{O}-$ group, the long hydrocarbon chain prevents water from surrounding and separating the vitamin E molecules, reducing its water solubility.

13.9 (a) Yes, the *molarity* changes with a change in temperature. Molarity is defined as moles solute per unit volume of solution. If solution volume is different, molarity is different.

(b) No, *molality* does not change with change in temperature. Molality is defined as moles solute per kilogram of solvent. Even though the volume of solution has changed because of increased kinetic energy, the mass of solute and solvent have not changed, and the molality stays the same.

13.11 Ideally, 0.50 L. If the volume outside the balloon is very large compared to 0.25 L, solvent will flow across the semipermeable membrane until the molarities of the inner and outer solutions are equal, 0.1 *M*. This requires an "inner" solution volume twice as large as the initial volume, or 0.50 L. (In reality, osmosis across the balloon membrane is not perfect. The solution concentration inside the balloon will be slightly greater than 0.1 *M* and the volume of the balloon will be slightly less than 0.50 L.)

The Solution Process (Section 13.1)

13.13 (a) False. It is generally true that solute-solvent interactions must balance the sum of solute-solute and solvent-solvent interactions. If solute-solute interactions are significantly stronger than solute-solvent interactions, dissolving will probably not occur. Entropy of mixing can compensate for a slightly endothermic enthalpy of solution, but not a large one.

(b) False. If a solution forms, enthalpy of mixing is exothermic, a negative number.

(c) True.

13.15 *Analyze/Plan.* Decide whether the solute and solvent in question are ionic, polar covalent, or nonpolar covalent. Draw Lewis structures as needed. Then state the appropriate type of solute-solvent interaction. *Solve.*

(a) CCl_4, nonpolar; benzene, nonpolar; dispersion forces

(b) methanol, polar with hydrogen bonding; water, polar with hydrogen bonding; hydrogen bonding

(c) KBr, ionic; water, polar; ion-dipole forces

(d) HCl, polar; CH_3CN, polar; dipole-dipole forces

13.17 (a) Very soluble.

(b) ΔH_{mix} will be the largest negative number. ΔH_{solute} and $\Delta H_{solvent}$ are both positive (endothermic); they represent the energy required to overcome attractive interactions in the solute and solvent, respectively. ΔH_{mix} represents the attractive interactions between solute and solvent particles. In order for ΔH_{soln} to be negative (exothermic), ΔH_{mix} must have a greater magnitude than (ΔH_{solute} + $\Delta H_{solvent}$).

13.19 (a) Lattice energy is the amount of energy required to completely separate a mole of solid ionic compound into its gaseous ions (Section 8.2). For ionic solutes, this corresponds to ΔH_{solute} (solute-solute interactions) in Equation 13.1.

(b) In Equation 13.1, ΔH_{mix} is always exothermic. Formation of attractive interactions, no matter how weak, always lowers the energy of the system, relative to the energy of the isolated particles.

13.21 (a) ΔH_{soln} is nearly zero. ΔH_{soln} is determined by the relative magnitudes of the "old" solute-solute (ΔH_{solute}) and solvent-solvent ($\Delta H_{solvent}$) interactions and the new solute-solvent interactions (ΔH_{mix}); $\Delta H_{soln} = \Delta H_{solute} + \Delta H_{solvent} + \Delta H_{mix}$. Because the solute and solvent in this case experience very similar London dispersion forces, the energy required to separate them individually and the energy released when they are mixed are approximately equal. $\Delta H_{solute} + \Delta H_{solvent} \approx -\Delta H_{mix}$ and ΔH_{soln} is nearly zero.

(b) The entropy of the system increases when heptane and hexane form a solution. The change involves taking two pure liquids and forming a (homogeneous) mixture. The result is an increase in randomness or disorder. From part (a), the enthalpy of mixing is nearly zero, so the increase in entropy is the driving force for mixing in all proportions.

Saturated Solutions; Factors Affecting Solubility (Sections 13.2 and 13.3)

13.23 (a) Supersaturated, because the solution contains more solute than a saturated solution at this temperature.

(b) The bits of glass scraped from the vessel act as a seed crystal, a place where solute molecules can align to form a crystal. The excess chromium nitrate is crystallizing out.

(c) 324 g dissolved at 35 °C – 208 g soluble at 15 °C = 116 g crystals form

13.25 *Analyze/Plan*. On Figure 13.15, find the solubility curve for the appropriate solute. Find the intersection of 40 °C and 40 g solute on the graph. If this point is below the solubility curve, more solute can dissolve and the solution is unsaturated. If the intersection is on or above the curve, the solution is saturated. *Solve*.

(a) unsaturated (b) saturated (c) saturated (d) unsaturated

13.27 (a) We expect the liquids water and glycerol to be miscible in all proportions. Glycerol has an –OH group on each C atom in the molecule. This structure facilitates strong hydrogen bonding similar to that in water. When two liquids have very similar intermolecular interactions, entropy of mixing drives the solution process and the two liquids are usually miscible in all proportions (Solution 13.21). Like dissolves like.

(b) Hydrogen bonding, dipole-dipole forces, London dispersion forces

13.29 *Analyze/Plan*. Evaluate molecules in the four common laboratory solvents for strength of intermolecular interactions with nonpolar solutes. *Solve*. Toluene, $C_6H_5CH_3$, is the best solvent for nonpolar solutes. Without polar groups or nonbonding electron pairs, it forms only dispersion interactions with itself and other molecules. The enthalpy of solution, ΔH_{soln}, is essentially zero (as in Solution 13.21) and solution occurs because of the favorable entropy of mixing.

13.31 (a) Despite the presence of the –COOH group, stearic acid is more soluble in nonpolar CCl_4 than in polar (hydrogen bonding) water. Dispersion interactions among nonpolar $CH_3(CH_2)_{16}$– chains dominate the properties of stearic acid.

(b) Dioxane will be more soluble in water than cyclohexane will, because dioxane can act as a hydrogen bond acceptor.

13.33 *Analyze/Plan*. Hexane is a nonpolar hydrocarbon that experiences dispersion forces with other nonpolar molecules. Solutes that primarily experience dispersion forces will be more soluble in hexane. *Solve*.

(a) CCl_4 is more soluble because dispersion forces among nonpolar CCl_4 molecules are similar to dispersion forces in hexane. Ionic bonds in $CaCl_2$ are unlikely to be broken by weak solute-solvent interactions. For $CaCl_2$, ΔH_{solute} is large, relative to ΔH_{mix}.

(b) Benzene, C_6H_6, is also a nonpolar hydrocarbon and will be more soluble in hexane. Glycerol experiences hydrogen bonding with itself; these solute-solute interactions are less likely to be overcome by weak solute-solvent interactions.

(c) Octanoic acid, $CH_3(CH_2)_6COOH$, will be more soluble than acetic acid CH_3COOH. Both solutes experience hydrogen bonding by –COOH groups, but octanoic acid has a long, rod-like hydrocarbon chain with dispersion forces similar to those in hexane, facilitating solubility in hexane.

13.35 (a) False. The lower the temperature, the more soluble most gases are in water.

(b) True.

(c) False. In a supersaturated solution all solute remains dissolved.

(d) True. Solubility of liquids and solids usually increases with increasing temperature.

13.37 *Analyze/Plan.* Follow the logic in Sample Exercise 13.2. *Solve.*

$S_{He} = 3.7 \times 10^{-4}\ M/\text{atm} \times 1.5\ \text{atm} = 5.6 \times 10^{-4}\ M$

$S_{N_2} = 6.0 \times 10^{-4}\ M/\text{atm} \times 1.5\ \text{atm} = 9.0 \times 10^{-4}\ M$

Concentrations of Solutions (Section 13.4)

13.39 *Analyze/Plan.* Follow the logic in Sample Exercise 13.3. *Solve.*

(a) $$\text{mass \%} = \frac{\text{mass solute}}{\text{total mass solution}} \times 100 = \frac{10.6\ \text{g Na}_2\text{SO}_4}{10.6\ \text{g Na}_2\text{SO}_4 + 483\ \text{g H}_2\text{O}} \times 100 = 2.15\%$$

(b) $$\text{ppm} = \frac{\text{mass solute}}{\text{total mass solution}} \times 10^6;\ \frac{2.86\ \text{g Ag}}{1\ \text{ton ore}} \times \frac{1\ \text{ton}}{2000\ \text{lb}} \times \frac{1\ \text{lb}}{453.6\ \text{g}} \times 10^6 = 3.15\ \text{ppm}$$

13.41 *Analyze/Plan.* Given masses of CH_3OH and H_2O, calculate moles of each component.

(a) Mole fraction CH_3OH = (mol CH_3OH)/(total mol)

(b) mass % CH_3OH = [(g CH_3OH)/(total mass)] × 100

(c) molality CH_3OH = (mol CH_3OH)/(kg H_2O). *Solve.*

(a) $$14.6\ \text{g CH}_3\text{OH} \times \frac{1\ \text{mol CH}_3\text{OH}}{32.04\ \text{g CH}_3\text{OH}} = 0.4557 = 0.456\ \text{mol CH}_3\text{OH}$$

$$184\ \text{g H}_2\text{O} \times \frac{1\ \text{mol H}_2\text{O}}{18.02\ \text{g H}_2\text{O}} = 10.211 = 10.2\ \text{mol H}_2\text{O}$$

$$\chi_{CH_3OH} = \frac{0.4557}{0.4557 + 10.211} = 0.04272 = 0.0427$$

(b) $$\text{mass \% CH}_3\text{OH} = \frac{14.6\ \text{g CH}_3\text{OH}}{14.6\ \text{g CH}_3\text{OH} + 184\ \text{g H}_2\text{O}} \times 100 = 7.35\%\ \text{CH}_3\text{OH}$$

(c) $$m = \frac{0.4557\ \text{mol CH}_3\text{OH}}{0.184\ \text{kg H}_2\text{O}} = 2.477 = 2.48\ m\ \text{CH}_3\text{OH}$$

13.43 *Analyze/Plan*. Given mass solute and volume solution, calculate mol solute, then molarity = mol solute/L solution. Or, for dilution, $M_c \times L_c = M_d \times L_d$. *Solve*.

(a) $$M = \frac{\text{mol solute}}{\text{L soln}};\ \frac{0.540\ \text{g Mg(NO}_3)_2}{0.2500\ \text{L soln}} \times \frac{1\ \text{mol Mg(NO}_3)_2}{148.3\ \text{g Mg(NO}_3)_2} = 1.46 \times 10^{-2}\ M\ \text{Mg(NO}_3)_2$$

(b) $$\frac{22.4\ \text{g LiClO}_4 \square 3\text{H}_2\text{O}}{0.125\ \text{L soln}} \times \frac{1\ \text{mol LiClO}_4 \square 3\text{H}_2\text{O}}{160.4\ \text{g LiClO}_4 \square 3\text{H}_2\text{O}} = 1.12\ M\ \text{LiClO}_4 \square 3\text{H}_2\text{O}$$

(c) $M_c \times L_c = M_d \times L_d$; $3.50\ M\ HNO_3 \times 0.0250\ L = ?M\ HNO_3 \times 0.250\ L$
250 mL of 0.350 $M\ HNO_3$

13.45 *Analyze/Plan*. Follow the logic in Sample Exercise 13.4. *Solve*.

(a) $$m = \frac{\text{mol solute}}{\text{kg solvent}};\ \frac{8.66\ \text{g C}_6\text{H}_6}{23.6\ \text{g CCl}_4} \times \frac{1\ \text{mol C}_6\text{H}_6}{78.11\ \text{g C}_6\text{H}_6} \times \frac{1000\ \text{g CCl}_4}{1\ \text{kg CCl}_4} = 4.70\ m\ \text{C}_6\text{H}_6$$

(b) The density of H_2O = 0.997 g/mL = 0.997 kg/L.

$$\frac{4.80\ \text{g NaCl}}{0.350\ \text{L H}_2\text{O}} \times \frac{1\ \text{mol NaCl}}{58.44\ \text{g NaCl}} \times \frac{1\ \text{L H}_2\text{O}}{0.997\ \text{kg H}_2\text{O}} = 0.235\ m\ \text{NaCl}$$

13.47 *Analyze/Plan*. Assume 1 L of solution. Density gives the total mass of 1 L of solution. The g H_2SO_4/L are also given in the problem. Mass % = (mass solute/total mass solution) × 100. Calculate mass solvent from mass solution and mass solute. Calculate moles solute and solvent and use the appropriate definitions to calculate mole fraction, molality, and molarity. *Solve*.

(a) $$\frac{571.6\ \text{g H}_2\text{SO}_4}{1\ \text{L soln}} \times \frac{1\ \text{L soln}}{1329\ \text{g soln}} = 0.430098\ \text{g H}_2\text{SO}_4/\text{g soln}$$

mass % is thus 0.4301 × 100 = 43.01% H_2SO_4

(b) In a liter of solution there are 1329 – 571.6 = 757.4 = 757 g H_2O.

$$\frac{571.6\ \text{g H}_2\text{SO}_4}{98.09\ \text{g/mol}} = 5.827\ \text{mol H}_2\text{SO}_4;\ \frac{757.4\ \text{g H}_2\text{O}}{18.02\ \text{g/mol}} = 42.03 = 42.0\ \text{mol H}_2\text{O}$$

$$X_{\text{H}_2\text{SO}_4} = \frac{5.827}{42.03 + 5.827} = 0.122$$

(The result has 3 sig figs because (g H_2O) resulting from subtraction is limited to 3 sig figs.)

(c) $$\text{molality} = \frac{5.827\ \text{mol H}_2\text{SO}_4}{0.7574\ \text{kg H}_2\text{O}} = 7.693 = 7.69\ m\ \text{H}_2\text{SO}_4$$

(d) $$\text{molarity} = \frac{5.827\ \text{mol H}_2\text{SO}_4}{1\ \text{L soln}} = 5.827\ M\ \text{H}_2\text{SO}_4$$

13.49 *Analyze/Plan.* Given: 98.7 mL of $CH_3CN(l)$, 0.786 g/mL; 22.5 mL CH_3OH, 0.791 g/mL. Use the density and volume of each component to calculate mass and then moles of each component. Use the definitions to calculate mole fraction, molality, and molarity. *Solve.*

(a) $$\text{mol } CH_3CN = \frac{0.786 \text{ g}}{1 \text{ mL}} \times 98.7 \text{ mL} \times \frac{1 \text{ mol } CH_3CN}{41.05 \text{ g } CH_3CN} = 1.8898 = 1.89 \text{ mol}$$

$$\text{mol } CH_3OH = \frac{0.791 \text{ g}}{1 \text{ mL}} \times 22.5 \text{ mL} \times \frac{1 \text{ mol } CH_3OH}{32.04 \text{ g } CH_3OH} = 0.5555 = 0.556 \text{ mol}$$

$$X_{CH_3OH} = \frac{0.5555 \text{ mol } CH_3OH}{1.8898 \text{ mol } CH_3CN + 0.5555 \text{ mol } CH_3OH} = 0.227$$

(b) Assuming CH_3OH is the solute and CH_3CN is the solvent,

$$98.7 \text{ mL } CH_3CN \times \frac{0.786 \text{ g}}{1 \text{ mL}} \times \frac{1 \text{ kg}}{1000 \text{ g}} = 0.07758 = 0.0776 \text{ kg } CH_3CN$$

$$m_{CH_3OH} = \frac{0.5555 \text{ mol } CH_3OH}{0.07758 \text{ kg } CH_3CN} = 7.1604 = 7.16\ m\ CH_3OH$$

(c) The total volume of the solution is 121.2 mL, assuming volumes are additive.

$$M = \frac{0.5555 \text{ mol } CH_3OH}{0.1212 \text{ L solution}} = 4.58\ M\ CH_3OH$$

13.51 *Analyze/Plan.* Given concentration and volume of solution use definitions of the appropriate concentration units to calculate amount of solute; change amount to moles if needed. *Solve.*

(a) $$\text{mol} = M \times L;\ \frac{0.250 \text{ mol } SrBr_2}{1 \text{ L soln}} \times 0.600 \text{ L} = 0.150 \text{ mol } SrBr_2$$

(b) Assume that for dilute aqueous solutions, the mass of the solvent is the mass of solution. Use proportions to get mol KCl.

$$\frac{0.180 \text{ mol KCl}}{1 \text{ kg } H_2O} = \frac{x \text{ mol KCl}}{0.0864 \text{ kg } H_2O};\ x = 1.56 \times 10^{-2} \text{ mol KCl}$$

(c) Use proportions to get mass of glucose, then change to mol glucose.

$$\frac{6.45 \text{ g } C_6H_{12}O_6}{100 \text{ g soln}} = \frac{x \text{ g } C_6H_{12}O_6}{124.0 \text{ g soln}};\ x = 8.00 \text{ g } C_6H_{12}O_6$$

$$8.00 \text{ g } C_6H_{12}O_6 \times \frac{1 \text{ mol } C_6H_{12}O_6}{180.2 \text{ g } C_6H_{12}O_6} = 4.44 \times 10^{-2} \text{ mol } C_6H_{12}O_6$$

13.53 *Analyze/Plan.* When preparing solution, we must know amount of solute and solvent. Use the appropriate concentration definition to calculate amount of solute. If this amount is in moles, use molar mass to get grams; use mass in grams directly. Amount of solvent can be expressed as total volume or mass of solution. Combine mass solute and solvent to produce the required amount (mass or volume) of solution. *Solve.*

(a) $$\text{mol} = M \times L;\ \frac{1.50 \times 10^{-2} \text{ mol KBr}}{1 \text{ L soln}} \times 0.75 \text{ L} \times \frac{119.0 \text{ g KBr}}{1 \text{ mol KBr}} = 1.3 \text{ g KBr}$$

Weigh out 1.3 g KBr, dissolve in water, dilute with stirring to 0.75 L (750 mL).

(b) Mass of solution is required, but density is not specified. Use molality to calculate mass fraction, and then the masses of solute and solvent needed for 125 g of solution.

$$\frac{0.180 \text{ mol KBr}}{1000 \text{ g } H_2O} \times \frac{119.0 \text{ g KBr}}{1 \text{ mol KBr}} = 21.42 = 21.4 \text{ g KBr/kg } H_2O. \text{ Thus,}$$

$$\text{mass fraction} = \frac{21.42 \text{ g KBr}}{1000 + 21.42} = 0.02097 = 0.0210$$

In 125 g of the 0.180 *m* solution, there are

$$(125 \text{ g soln}) \times \frac{0.02097 \text{ g KBr}}{1 \text{ g soln}} = 2.621 = 2.62 \text{ g KBr}$$

Weigh out 2.62 g KBr, dissolve it in 125 – 2.62 = 122.38 = 122 g H_2O to make exactly 125 g of 0.180 *m* solution.

(c) Using solution density, calculate the total mass of 1.85 L of solution, and from the mass % of KBr, the mass of KBr required.

$$1.85 \text{ L soln} \times \frac{1000 \text{ mL}}{1 \text{ L}} \times \frac{1.10 \text{ g soln}}{1 \text{ mL}} = 2035 = 2.04 \times 10^3 \text{ g soln}$$

0.120 (2035 g soln) = 244.2 = 244 g KBr

Dissolve 244 g KBr in water, dilute with stirring to 1.85 L.

(d) Calculate moles KBr needed to precipitate 16.0 g AgBr. $AgNO_3$ is present in excess.

$$16.0 \text{ g AgBr} \times \frac{1 \text{ mol AgBr}}{187.8 \text{ g AgBr}} \times \frac{1 \text{ mol KBr}}{1 \text{ mol AgBr}} = 0.08520 = 0.0852 \text{ mol KBr}$$

$$0.0852 \text{ mol KBr} \times \frac{1 \text{ L soln}}{0.150 \text{ mol KBr}} = 0.568 \text{ L soln}$$

Weigh out 0.0852 mol KBr (10.1 g KBr), dissolve it in a small amount of water, and dilute to 0.568 L.

13.55 *Analyze/Plan*. Assume a solution volume of 1.00 L. Calculate the mass of 1.00 L of solution and the mass of HNO_3 in 1.00 L of solution. Mass % = (mass solute/mass solution) × 100. *Solve.*

$$1.00 \text{ L} \times \frac{1000 \text{ mL}}{1 \text{ L}} \times \frac{1.42 \text{ g soln}}{\text{mL soln}} = 1.42 \times 10^3 \text{ g soln}$$

$$16\ M = \frac{16 \text{ mol } HNO_3}{1 \text{ L soln}} \times \frac{63.02 \text{ g } HNO_3}{1 \text{ mol } HNO_3} = 1008 = 1.0 \times 10^3 \text{ g } HNO_3$$

$$\text{mass \%} = \frac{1008 \text{ g } HNO_3}{1.42 \times 10^3 \text{ g soln}} \times 100 = 71\% \text{ } HNO_3$$

13.57 *Analyze*. Given: 80.0% Cu, 20.0% Zn by mass; density = 8750 kg/m^3. Find: (a) *m* of Zn (b) *M* of Zn

(a) *Plan*. In the brass alloy, Zn is the solute (lesser component) and Cu is the solvent (greater component). *m* = mol Zn/kg Cu. 1 m^3 brass alloy weighs 8750 kg. 80.0% is Cu, 20.0% is Zn. Change g Zn → mol Zn and solve for *m*. *Solve.*

$$8750 \text{ kg brass} \times \frac{80 \text{ g Cu}}{100 \text{ g brass}} = 7.00 \times 10^3 \text{ kg Cu}$$

$$8750 \text{ kg brass} - 7000 \text{ kg Cu} = 1750 \text{ kg Zn}$$

$$1750 \text{ kg Zn} \times \frac{1000 \text{ g}}{\text{kg}} \times \frac{1 \text{ mol Zn}}{65.39 \text{ g Zn}} = 26{,}762.5 = 2.68 \times 10^4 \text{ mol Zn}$$

$$m = \frac{2.676 \times 10^4 \text{ mol Zn}}{7000 \text{ kg Cu}} = 3.82\ m \text{ Zn}$$

(b) *Plan.* M = mol Zn/L brass. Use mol Zn from part (a). Change 1 $m^3 \to$ L brass and calculate M. *Solve.*

$$1 \text{ m}^3 \times \frac{(10)^3 \text{ dm}^3}{\text{m}^3} \times \frac{1 \text{ L}}{1 \text{ dm}^3} = 1000 \text{ L}$$

$$M = \frac{2.676 \times 10^4 \text{ mol Zn}}{1000 \text{ L brass}} = 26.76 = 26.8\ M \text{ Zn}$$

13.59 *Analyze.* Given: 4.6% CO_2 by volume (in air), 1 atm total pressure. Find: partial pressure and molarity of CO_2 in air.

Plan. 4.6% CO_2 by volume means 4.6 mL of CO_2 could be isolated from 100 mL of air, at the same temperature and pressure. According to Avogadro's law, equal volumes of gases at the same temperature and pressure contain equal numbers of moles. By inference, the volume ratio of CO_2 to air, 4.6/100 or 0.046, is also the mole ratio. *Solve.*

(a) $P_{CO_2} = X_{CO_2} \times P_t = 0.046 \text{ (1 atm)} = 0.046 \text{ atm}$

(b) $M = \text{mol } CO_2/\text{L air} = n/V$. $PV = nRT$, $M = n/V = P/RT$

$$M_{CO_2} = \frac{P_{CO_2}}{RT} = \frac{0.046 \text{ atm}}{310 \text{ K}} \times \frac{\text{mol-K}}{0.08206 \text{ L-atm}} = 1.8 \times 10^{-3}\ M$$

Colligative Properties (Section 13.5)

13.61 (a) False (b) True (c) True (d) False

13.63 *Analyze/Plan.* H_2O vapor pressure will be determined by the mole fraction of H_2O in the solution. The vapor pressure of pure H_2O at 20 °C = 17.5 torr. *Solve.*

The density of water at 20 °C is not exactly 1 g/mL. From Appendix B, the density of water at 25 °C is 0.99707 g/mL. For the purpose of this calculation, assume this is the density at 20 °C.

$$\frac{10.0 \text{ g } C_6H_{12}O_6}{180.15 \text{ g/mol}} = 0.055509 = 0.0555 \text{ mol}; \quad \frac{997 \text{ g } H_2O}{18.02 \text{ g/mol}} = 55.327 = 55.3 \text{ mol}$$

$$P_{H_2O} = X_{H_2O}\ P^\circ_{H_2O} = \frac{55.3 \text{ mol } H_2O}{55.3 + 0.0555} \times 17.5 \text{ torr} = 17.48 = 17.5 \text{ torr}$$

$$\frac{10.0 \text{ g } C_{12}H_{22}O_{11}}{342.3 \text{ g/mol}} = 0.029214 = 0.0292 \text{ mol}; \quad \frac{997 \text{ g } H_2O}{18.02 \text{ g/mol}} = 55.327 = 55.3 \text{ mol}$$

$$P_{H_2O} = X_{H_2O}\ P^{\circ}_{H_2O} = \frac{55.3 \text{ mol } H_2O}{55.3 + 0.0292} \times 17.5 \text{ torr} = 17.49 = 17.5 \text{ torr}$$

Because these two solutions are so dilute, they have essentially the same vapor pressure. Generally, the less concentrated solution, the one with fewer moles of solute per kilogram of solvent, will have the higher vapor pressure.

13.65 (a) *Analyze/Plan.* H_2O vapor pressure will be determined by the mole fraction of H_2O in the solution. The vapor pressure of pure H_2O at 338 K (65 °C) = 187.5 torr.

Solve.

$$\frac{22.5 \text{ g } C_{12}H_{22}O_{11}}{342.3 \text{ g/mol}} = 0.06573 = 0.0657 \text{ mol}; \quad \frac{200.0 \text{ g } H_2O}{18.02 \text{ g/mol}} = 11.09878 = 11.10 \text{ mol}$$

$$P_{H_2O} = X_{H_2O}\ P^{\circ}_{H_2O} = \frac{11.09878 \text{ mol } H_2O}{11.09878 + 0.06573} \times 187.5 \text{ torr} = 186.4 \text{ torr}$$

(b) *Analyze/Plan.* For this problem, it will be convenient to express Raoult's law in terms of the lowering of the vapor pressure of the solvent, ΔP_A.

$\Delta P_A = P_A{}^{\circ} - X_A P_A{}^{\circ} = P_A{}^{\circ}(1 - X_A)$. $1 - X_A = X_B$, the mole fraction of the *solute* particles

$\Delta P_A = X_B P_A{}^{\circ}$; the vapor pressure of the solvent (A) is lowered according to the mole fraction of solute (B) particles present. *Solve.*

$$P_{H_2O} \text{ at } 40\ ^{\circ}C = 55.3 \text{ torr}; \quad \frac{340 \text{ g } H_2O}{18.02 \text{ g/mol}} = 18.868 = 18.9 \text{ mol } H_2O$$

$$X_{C_3H_8O_2} = \frac{2.88 \text{ torr}}{55.3 \text{ torr}} = \frac{y \text{ mol } C_3H_8O_2}{y \text{ mol } C_3H_8O_2 + 18.868 \text{ mol } H_2O} = 0.05208 = 0.0521$$

$$0.05208 = \frac{y}{y + 18.868}; \ 0.05208\ y + 0.98263 = y; \ 0.94792\ y = 0.98263,$$

$y = 1.0366 = 1.04 \text{ mol } C_3H_8O_2$

This result has 3 sig figs because (0.340 kg water) has 3 sig figs.

$$1.0366 \text{ mol } C_3H_8O_2 \times \frac{76.09 \text{ g } C_3H_8O_2}{\text{mol } C_3H_8O_2} = 78.88 = 78.9 \text{ g } C_3H_8O_2$$

13.67 *Analyze/Plan.* At 63.5 °C, $P^{\circ}_{H_2O} = 175$ torr, $P^{\circ}_{Eth} = 400$ torr. Let G = the mass of H_2O and/or C_2H_5OH. *Solve.*

(a)

$$X_{Eth} = \frac{\dfrac{G}{46.07 \text{ g/mol } C_2H_5OH}}{\dfrac{G}{46.07 \text{ g/mol } C_2H_5OH} + \dfrac{G}{18.02 \text{ g/mol } H_2O}}$$

Multiplying top and bottom of the right side of the equation by 1/G gives:

$$X_{Eth} = \frac{1/46.07}{1/46.07 + 1/18.02} = \frac{0.02171}{0.02171 + 0.05549} = 0.2812$$

(b) $P_t = P_{Eth} + P_{H_2O}$; $P_{Eth} = \chi_{Eth} \times P^\circ_{Eth}$; $P_{H_2O} = \chi_{H_2O} P^\circ_{H_2O}$

$\chi_{Eth} = 0.2812$, $P_{Eth} = 0.2812\ (400\ \text{torr}) = 112.48 = 112\ \text{torr}$

$\chi_{H_2O} = 1 - 0.2812 = 0.7188$; $P_{H_2O} = 0.7188(175\ \text{torr}) = 125.8 = 126\ \text{torr}$

$P_t = 112.5\ \text{torr} + 125.8\ \text{torr} = 238.3 = 238\ \text{torr}$

(c) $$\chi_{Eth}\ \text{in vapor} = \frac{P_{Eth}}{P_{total}} = \frac{112.5\ \text{torr}}{238.3\ \text{torr}} = 0.4721 = 0.472$$

13.69 (a) Because NaCl is a soluble ionic compound and a strong electrolyte, there are 2 mol of dissolved particles for every 1 mol of NaCl solute. $C_6H_{12}O_6$ is a molecular solute, so there is 1 mol of dissolved particles per mol solute. Boiling point elevation is directly related to total moles of dissolved particles; 0.10 *m* NaCl has more dissolved particles so its boiling point is higher than 0.10 *m* $C_6H_{12}O_6$.

(b) In solutions of strong electrolytes like NaCl, electrostatic attractions between ions lead to ion pairing. Ion pairing reduces the effective number of particles in solution, decreasing the **change** in boiling point. The actual boiling point is then lower than the calculated boiling point for a 0.10 *m* solution.

13.71 *Analyze/Plan.* Rank the solutions in order of increasing boiling point. All solutes are nonvolatile. The more nonvolatile solute particles, the higher the boiling point of the solution. *Solve.*

Solve. Because LiBr and $Zn(NO_3)_2$ are electrolytes, the particle concentrations in these solutions are 0.10 *m* and 0.15 *m*, respectively (although ion-ion attractive forces may decrease the effective concentrations somewhat). Thus, the order of increasing particle concentration and boiling point is:

0.050 *m* LiBr < 0.120 *m* glucose < 0.050 *m* $Zn(NO_3)_2$

13.73 *Analyze/Plan.* $\Delta T = K(m)$; first, calculate the **molality** of each solution. *Solve.*

(a) 0.22 *m*

(b) $$2.45\ \text{mol CHCl}_3 \times \frac{119.4\ \text{g CHCl}_3}{\text{mol CHCl}_3} = 292.53\ \text{g} = 0.293\ \text{kg};$$

$$\frac{0.240\ \text{mol C}_{10}\text{H}_8}{0.29253\ \text{kg CHCl}_3} = 0.8204 = 0.820\ m$$

(c) $$1.50\ \text{g NaCl} \times \frac{1\ \text{mol NaCl}}{58.44\ \text{g NaCl}} \times \frac{2\ \text{mol particles}}{1\ \text{mol NaCl}} = 0.05133 = 0.0513\ \text{mol particles}$$

$$m = \frac{0.05133\ \text{mol NaCl}}{0.250\ \text{kg H}_2\text{O}} = 0.20534 = 0.205\ m$$

(d) $$2.04\ \text{g KBr} \times \frac{1\ \text{mol KBr}}{119.0\ \text{g KBr}} \times \frac{2\ \text{mol particles}}{1\ \text{mol KBr}} = 0.03429 = 0.0343\ \text{mol particles}$$

$$4.82\ \text{g C}_6\text{H}_{12}\text{O}_6 \times \frac{1\ \text{mol C}_6\text{H}_{12}\text{O}_6}{180.2\ \text{g C}_6\text{H}_{12}\text{O}_6} = 0.02675 = 0.0268\ \text{mol particles}$$

$$m = \frac{(0.03429 + 0.02675)\ \text{mol particles}}{0.188\ \text{kg H}_2\text{O}} = 0.32465 = 0.325\ m$$

Solve. Then, fp = $T_f - K_f(m)$; bp = $T_b + K_b(m)$; T in °C

	m	T_f	$-K_f(m)$	fp	T_b	$+K_b(m)$	bp
(a)	0.22	−114.6	−1.99(0.22) = −0.44	−115.0	78.4	1.22(0.22) = 0.27	78.7
(b)	0.820	−63.5	−4.68(0.820) = −3.84	−67.3	61.2	3.63(0.820) = 2.98	64.2
(c)	0.205	0.0	−1.86(0.205) = −0.381	−0.4	100.0	0.51(0.205) = 0.10	100.1
(d)	0.325	0.0	−1.86(0.325) = −0.605	−0.6	100.0	0.51(0.325) = 0.17	100.2

13.75 *Analyze.* Given freezing point of solution and mass of solvent, calculate mass of solute.

Plan. Use $\Delta T_f = K_f(m)$ to calculate the required molality, and then apply the definition of molality to calculate moles and grams of $C_2H_6O_2$.

Solve. fp of solution = −5.00 °C; fp of solvent (H_2O) = 0.0 °C

$\Delta T_f = 5.00\,°C = K_f(m)$; $5.00\,°C = 1.86\,°C/m(m)$

$$m = \frac{5.00\ °C}{1.86\ °C/m} = 2.688 = 2.69\ m\ C_2H_6O_2$$

$$m = \frac{\text{mol } C_2H_6O_2}{\text{kg } H_2O} = C_2H_6O_2 = m \times \text{kg } H_2O$$

$$2.688\ m\ C_2H_6O_2 \times 1.00 \text{ kg } H_2O = 2.688 = 2.69 \text{ mol } C_2H_6O_2$$

$$2.688\ m\ C_2H_6O_2 \times \frac{62.07 \text{ g } C_2H_6O_2}{1 \text{ mol}} = 166.84 = 167 \text{ g } C_2H_6O_2$$

13.77 *Analyze/Plan.* $\Pi = MRT$; T = 25 °C + 273 = 298 K; M = mol $C_9H_8O_4$/L soln *Solve.*

$$M = \frac{44.2 \text{ mg } C_9H_8O_4}{0.358 \text{ L}} \times \frac{1 \text{ g}}{1000 \text{ mg}} \times \frac{1 \text{ mol } C_9H_8O_4}{180.2 \text{ g } C_9H_8O_4} = 6.851 \times 10^{-4} = 6.85 \times 10^{-4}\ M$$

$$\Pi = \frac{6.851 \times 10^{-4} \text{ mol}}{\text{L}} \times \frac{0.08206 \text{ L-atm}}{\text{mol-K}} \times 298 \text{ K} = 0.01675 = 0.0168 \text{ atm} = 12.7 \text{ torr}$$

13.79 *Analyze/Plan.* Follow the logic in Sample Exercise 13.10 to calculate the molar mass of adrenaline based on the boiling point data. Use the structure to obtain the molecular formula and molar mass. Compare the two values. *Solve.*

$$\Delta T_b = K_b\ m; \quad m = \frac{\Delta T_b}{K_b} = \frac{+0.49}{5.02} = 0.0976 = 0.098\ m \text{ adrenaline}$$

$$m = \frac{\text{mol adrenaline}}{\text{kg } CCl_4} = \frac{\text{g adrenaline}}{\text{MM adrenaline} \times \text{kg } CCl_4}$$

$$\text{MM adrenaline} = \frac{\text{g adrenaline}}{m \times \text{kg } CCl_4} = \frac{0.64 \text{ g adrenaline}}{0.0976\ m \times 0.0360 \text{ kg } CCl_4} = 1.8 \times 10^2 \text{ g/mol adrenaline}$$

Check. The molecular formula is $C_9H_{13}NO_3$, MM = 183 g/mol. The values agree to 2 sig figs, the precision of the experimental value.

13.81 *Analyze/Plan.* Follow the logic in Sample Exercise 13.11. *Solve.*

$$\Pi = MRT; \quad M = \frac{\Pi}{RT}; \quad T = 25\ °C + 273 = 298\ K$$

$$M = 0.953\ \text{torr} \times \frac{1\ \text{atm}}{760\ \text{torr}} \times \frac{\text{mol-K}}{0.08206\ \text{L-atm}} \times \frac{1}{298\ K} = 5.128 \times 10^{-5} = 5.13 \times 10^{-5}\ M$$

$$\text{mol} = M \times L = 5.128 \times 10^{-5} \times 0.210\ L = 1.077 \times 10^{-5} = 1.08 \times 10^{-5}\ \text{mol lysozyme}$$

$$MM = \frac{g}{mol} = \frac{0.150\ g}{1.077 \times 10^{-5}\ \text{mol}} = 1.39 \times 10^{4}\ \text{g/mol lysozyme}$$

13.83 *Analyze/Plan.* $i = \Pi$ (measured) / Π (calculated for a nonelectrolyte);

Π (calculated) = MRT. *Solve.*

$$\Pi\ (\text{calculated}) = \frac{0.010\ \text{mol}}{L} \times \frac{0.08206\ \text{L-atm}}{\text{mol-K}} \times 298\ K = 0.2445 = 0.24\ \text{atm}$$

$i = 0.674\ \text{atm}/0.2445\ \text{atm} = 2.756 = 2.8$

Colloids (Section 13.6)

13.85 (a) No. In the gaseous state, the particles are far apart and intermolecular attractive forces are small. When two gases combine, all terms in Equation 13.1 are essentially zero and the mixture is always homogeneous.

(b) The outline of a light beam passing through a colloid is visible, whereas light passing through a true solution is invisible unless collected on a screen. This is the Tyndall effect. To determine whether Faraday's (or anyone's) apparently homogeneous dispersion is a true solution or a colloid, shine a beam of light on it and see if the light is scattered.

13.87 The best emulsifying agent is (d) $CH_3(CH_2)_{11}COONa$. A good emulsifying agent has a polar (or ionic) end to interact with hydrophilic substances, and an nonpolar end to interact with hydrophobic substances. Choices (c) and (d) fit this description, but the ionic end of (d) will help stabilize the colloid.

13.89 (a) No. Adsorbed ions stabilize hydrophobic colloids in water. The hydrophobic/hydrophilic nature of the protein will determine which electrolyte at which concentration will be the most effective precipitating salt.

(b) Stronger. If a protein has been "salted out," protein-protein interactions are sufficiently strong so that the protein molecules "stick together" and form a solid. Before the electrolyte is added, protein-protein interactions are weaker than protein-water interactions and the protein molecules remained suspended in solution.

(c) The first hypothesis seems plausible, because ion-dipole interactions among electrolytes and water molecules are stronger than dipole-dipole and hydrogen-bonding interactions between water and protein molecules. However, this ignores the strength of ion-dipole interactions between the electrolyte and protein molecules. And, we know from Figure 13.27 that ions are adsorbed on the surface of hydrophobic colloids. With the right protein and electrolyte, the second hypothesis also seems plausible.

The van't Hoff effect is a result of ion pairing. We know from Table 13.4 that the effect of ion pairing increases with concentration and the charge on the ions of the electrolyte. If we could measure the charge and adsorbed water content of protein molecules as a function of salt concentration, then we could distinguish between these two hypotheses.

Additional Exercises

13.91 (a) Hydrochloride. The hydrochloride is a salt, an ionic compound, with the possibility of ion-dipole interactions in addition to hydrogen bonding and dipole-dipole interactions. Of the two forms, it will be more soluble in water.

(b) Free base. Both forms have several hydrogen bond receptors (O and N atoms with nonbonded electron pairs), but the free base is less soluble because it does not have the possibility of ion-dipole interactions.

(c) *Analyze/Plan.* Calculate the molar mass of the free base, then moles and molarity. mol = g/molar mass; M = mol/L. *Solve.*

The molar mass of the free base is 303.353 g/mol. 6.70 mL = 0.00670 L ethanol

$$1.00 \text{ g free base} \times \frac{1 \text{ mol free base}}{303.353 \text{ g free base}} \times \frac{1}{0.00670 \text{ L ethanol}} = 0.492\, M \text{ free base}$$

(d) Use the method from part (c) to calculate molarity of the hydrochloride.

The molar mass of the hydrochloride is 339.814 g/mol. 0.400 mL = 0.000400 L

$$1.00 \text{ g hydrochloride} \times \frac{1 \text{ mol hydrochloride}}{339.814 \text{ g free base}} \times \frac{1}{0.000400 \text{ L water}} = 7.36\, M \text{ hydrochloride}$$

(e) *Analyze/Plan.* According to the chemical reaction given in the exercise, the free base reacts in a 1:1 mole ratio with HCl(aq). Calculate moles of free base in 1.00 kg and then liters of 12.0 M HCl(aq) required. 1.00 kg = 1.00×10^3 g. *Solve.*

$$1.00 \times 10^3 \text{g free base} \times \frac{1 \text{ mol free base}}{303.353 \text{ g free base}} \times \frac{1 \text{ L}}{12.0 \text{ mol HCl}} = 0.275 \text{ L} = 275 \text{ mL}$$

13.94 (a) $S_{Rn} = kP_{Rn};\ k = S_{Rn}/P_{Rn} = 7.27 \times 10^{-3}\ M/1 \text{ atm} = 7.27 \times 10^{-3} \text{ mol/L-atm}$

(b) $P_{Rn} = \chi_{Rn} P_{total};\ P_{Rn} = 3.5 \times 10^{-6} (32 \text{ atm}) = 1.12 \times 10^{-4} = 1.1 \times 10^{-4}$ atm

$$S_{Rn} = k\ P_{Rn};\ S_{Rn} = \frac{7.27 \times 10^{-3} \text{ mol}}{\text{L-atm}} \times 1.12 \times 10^{-4} \text{ atm} = 8.1 \times 10^{-7}\ M$$

13.98 (a) $$\frac{1.80 \text{ mol LiBr}}{1 \text{ L soln}} \times \frac{86.85 \text{ g LiBr}}{1 \text{ mol LiBr}} = 156.3 = 156 \text{ g LiBr}$$

1 L soln = 826 g soln; g CH_3CN = 826 – 156.3 = 669.7 = 670 g CH_3CN

$$m \text{ LiBr} = \frac{1.80 \text{ mol LiBr}}{0.6697 \text{ kg } CH_3CN} = 2.69\ m$$

(b) $$\frac{669.7 \text{ g } CH_3CN}{41.05 \text{ g/mol}} = 16.31 = 16.3 \text{ mol } CH_3CN;\ X_{LiBr} = \frac{1.80}{1.80 + 16.31} = 0.0994$$

(c) $\text{mass \%} = \frac{669.7 \text{ g } CH_3CN}{826 \text{ g soln}} \times 100 = 81.1\% \ CH_3CN$

13.100 *Analyze.* Given vapor pressure of both pure water and the aqueous solution and moles H_2O find moles of solute in the solution.

Plan. Use vapor pressure lowering, $P_A = X_A P_A^\circ$, to calculate X_A, mole fraction solvent, and then use the definition of mole fraction to calculate moles solute particles. Because NaCl is a strong electrolyte, there is 1 mol NaCl for every 2 mol solute particles.

Solve.

$$X_{H_2O} = P_{soln}/P_{H_2O} = 25.7/31.8 = 0.80818 = 0.808$$

$$X_{H_2O} = \frac{\text{mol } H_2O}{\text{mol ions} + \text{mol } H_2O}; \quad 0.80818 = \frac{0.115}{(\text{mol ions} + 0.115)}$$

$0.80818\ (0.115 + \text{mol ions}) = 0.115;\ 0.80818\ (\text{mol ions}) = 0.115 - 0.092940$

$\text{mol ions} = 0.02206/0.80818 = 0.02730 = 0.0273;$

$\text{mol NaCl} = \text{mol ions}/2 = 0.02730/2 = 0.01365 = 0.0137 \text{ mol NaCl}$

$$0.01365 \text{ mol NaCl} \times \frac{58.443 \text{ g } C_2H_5OH}{1 \text{ mol}} = 0.7977 = 0.798 \text{ g NaCl}$$

13.103 (a) $0.100\ m\ K_2SO_4$ is $0.300\ m$ in particles. H_2O is the solvent.

$\Delta T_f = K_f m = -1.86(0.300) = -0.558;\ T_f = 0.0 - 0.558 = -0.558\ °C = -0.6\ °C$

(b) $\Delta T_f \text{ (nonelectrolyte)} = -1.86(0.100) = -0.186;\ T_f = 0.0 - 0.186 = -0.186\ °C = -0.2\ °C$

$T_f \text{ (measured)} = i \times T_f \text{ (nonelectrolyte)}$

From Table 13.4, i for $0.100\ m\ K_2SO_4 = 2.32$

$T_f \text{ (measured)} = 2.32(-0.186\ °C) = -0.432\ °C = -0.4\ °C$

Integrative Exercises

13.106 Because these are very dilute solutions, assume that the density of the solution ≈ the density of $H_2O \approx 1.0$ g/mL at 25 °C. Then, 100 g solution = 100 g H_2O = 0.100 kg H_2O.

(a) CF_4: $$\frac{0.0015 \text{ g } CF_4}{0.100 \text{ kg } H_2O} \times \frac{1 \text{ mol } CF_4}{88.00 \text{ g } CF_4} = 1.7 \times 10^{-4}\ m$$

$CClF_3$: $$\frac{0.009 \text{ g } CClF_3}{0.100 \text{ kg } H_2O} \times \frac{1 \text{ mol } CClF_3}{104.46 \text{ g } CClF_3} = 8.6 \times 10^{-4}\ m = 9 \times 10^{-4}\ m$$

CCl_2F_2: $$\frac{0.028 \text{ g } CCl_2F_2}{0.100 \text{ kg } H_2O} \times \frac{1 \text{ mol } CCl_2F_2}{120.9 \text{ g } CCl_2F_2} = 2.3 \times 10^{-3}\ m$$

$CHClF_2$: $$\frac{0.30 \text{ g } CHClF_2}{0.100 \text{ kg } H_2O} \times \frac{1 \text{ mol } CHClF_2}{86.47 \text{ g } CHClF_2} = 3.5 \times 10^{-2}\ m$$

(b) Dipole moment. CCl_2F_2 has the largest molar mass, but it is not the most soluble. None of the molecules are capable of hydrogen bonding with water; F atoms bound to C are not hydrogen bond acceptors and H bound to C is not a hydrogen bond donor.

(c) Air is 21% O_2 by volume. The volume of $O_2(g)$ in a baby's lungs is

0.21(15 mL) = 3.15 = 3.2 mL = 0.0032 L

Assume air pressure in the lungs is 1 atm and body temperature is 37 °C or 310 K.

$$n = \frac{PV}{RT} = 1\ \text{atm} \times \frac{\text{mol-K}}{0.08206\ \text{L-atm}} \times \frac{0.00315\ \text{L}}{310\ \text{K}} = 1.238 \times 10^{-4} = 1.2 \times 10^{-4}\ \text{mol } O_2$$

A volume of 66 mL of $O_2(g)$ dissolves in 100 mL of the fluorinated liquid. That is 66% O_2 by volume. [(66/100)100 = 66]

In a 15 mL volume of the liquid, the volume of $O_2(g)$ is

0.66(15 mL) = 9.90 = 9.9 mL = 0.0099 L

$$n = \frac{PV}{RT} = 1\ \text{atm} \times \frac{\text{mol-K}}{0.08206\ \text{L-atm}} \times \frac{0.00990\ \text{L}}{310\ \text{K}} = 3.892 \times 10^{-4} = 3.9 \times 10^{-4}\ \text{mol } O_2$$

13.109 (a) The central atom and the number of electron-pair domains about it are: (i) Cl, 4; (ii) B, 4; (iii) P, 6; (iv) Al, 4; (v) B, 4

(b) The electron-domain geometry around B in BARF is tetrahedral.

(c) The central P atom in anion (iii) has an expanded octet. As drawn, the central Cl atom in anion (i) also has an expanded octet. Note that multiple resonance structures for ClO_4^- can be drawn, including one where Cl obeys the octet rule. The structure shown in this exercise is the one that minimizes formal charge.

(d) BARF is the largest anion; it will have the strongest dispersion forces that promote solubility in nonpolar solvents.

13.113 $\chi_{CHCl_3} = \chi_{C_3H_6O} = 0.500$

(a) For an ideal solution, Raoult's law is obeyed.

$P_t = P_{CHCl_3} + P_{C_3H_6O}$; $P_{CHCl_3} = 0.5(300\ \text{torr}) = 150\ \text{torr}$

$P_{C_3H_6O} = 0.5(360\ \text{torr}) = 180\ \text{torr}$; $P_t = 150\ \text{torr} + 180\ \text{torr} = 330\ \text{torr}$

(b) The mixing of the two liquids is exothermic. According to Coulomb's law, electrostatic attractive forces lead to an overall lowering of the energy of the system. Thus, when the two liquids mix and hydrogen bonds are formed, the energy of the system is decreased and $\Delta H_{soln} < 0$.

14 Chemical Kinetics

Visualizing Concepts

14.1 *Analyze/Plan.* Consider the chemical reaction that occurs in the cylinders of an automobile engine. How are the droplets related to the reaction, and how does droplet size affect the rate of the reaction?

Solve. The reaction occurring in the cylinder is the combustion of gasoline. Gasoline is injected into the cylinders in the form of a spray, as shown in the photos. This is a heterogeneous reaction, because gasoline is a liquid and oxygen (from air) is a gas. The rate of a heterogeneous reaction depends on the surface area of the liquid or solid reactant, in this case, the surface area of the droplets in the spray. The smaller the droplets, the greater the surface area exposed to oxygen, the faster the combustion reaction.

In the case of a clogged injector, larger droplets lead to slower combustion. Uneven combustion in the various cylinders can cause the engine to run roughly and decrease fuel economy.

14.3 (a) Chemical equation (iv), $B \rightarrow 2A$, is consistent with the data. The concentration of A increases with time, and the concentration of B decreases with time, so B must be a reactant and A must be a product. The ending concentration of A is approximately twice as large as the starting concentration of B, so mole ratio of A:B is 2:1. The reaction is $B \rightarrow 2A$.

(b) Rate $= -\Delta[B]/\Delta t = ½\, \Delta[A]/\Delta t$

14.9 *Analyze/Plan.* The reaction profile has a single high point (peak), so the reaction occurs in a single step. This step is necessarily the rate-determining step. *Solve.*

(1) Total potential energy of the reactants.

(2) E_a, activation energy of the reaction. This is the difference in energy between the potential energy of the activated complex (transition state) and the potential energy of the reactants.

(3) ΔE, net energy change for the reaction. This is the difference in energy between the products and reactants. (Under appropriate conditions, this could also be ΔH.) For this reaction, the energy of products is lower than the energy of reactants, and the reaction releases energy to the surroundings.

(4) Total potential energy of the products.

14.12 (a) $NO_2 + F_2 \rightarrow NO_2F + F$

$NO_2 + F \rightarrow NO_2F$

(b) $2NO_2 + F_2 \rightarrow 2NO_2F$

(c) F is the intermediate, because it is produced and then consumed during the reaction.

(d) Rate = $k[NO_2][F_2]$

14.15 (a) $A_2 + AB + AC \rightarrow BA_2 + A + AC$

$BA_2 + A + AC \rightarrow A_2 + BA_2 + C$

net: $AB + AC \rightarrow BA_2 + C$

(b) A is the intermediate; it is produced and consumed.

(c) A_2 is the catalyst; it is consumed and reproduced.

14.16

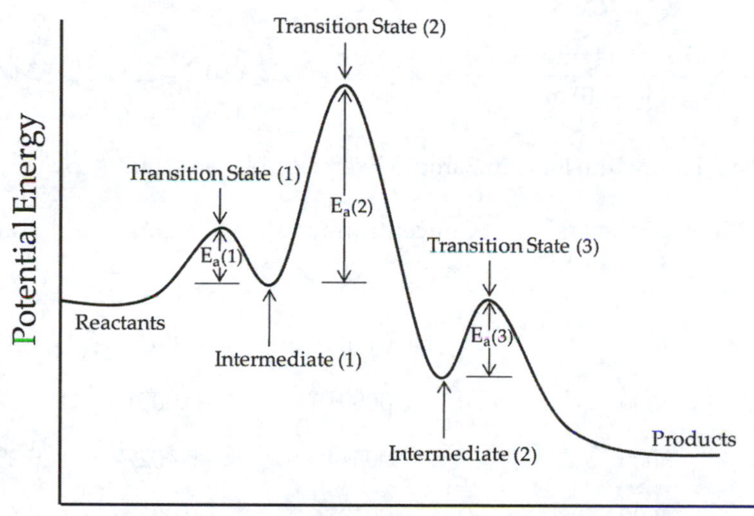

The reaction is exothermic because the energy of products is lower than the energy of reactants. The two intermediates are formed at different rates because $E_a(1) \neq E_a(2)$. To have two intermediates, the mechanism must have at least three steps.

Reaction Rates (Sections 14.1 and 14.2)

14.17 (a) *Reaction rate* is the change in the amount of products or reactants in a given amount of time; it is the speed of a chemical reaction.

(b) Rates depend on concentration of reactants, physical state (or surface area) of reactants, temperature, and reaction activation energy/presence of catalyst.

(c) No, the rate of disappearance of reactants is not necessarily the same as the rate of appearance of products. The stoichiometry of the reaction (mole ratios of reactants and products) must be known to relate rate of disappearance of reactants to rate of appearance of products.

14.19 *Analyze/Plan*. Given mol A at a series of times in minutes, calculate mol B produced, molarity of A at each time, change in M of A at each 10 min interval, and ΔM A/s. For this reaction, mol B produced equals mol A consumed. M of A or [A] = mol A/0.100 L. The average rate of disappearance of A for each 10 minute interval is

$$-\frac{\Delta[A]}{s} = -\frac{[A]_1 - [A]_0}{10 \text{ min}} \times \frac{1 \text{ min}}{60 \text{ s}}$$

Solve.

Time (min)	Mol A	(a) Mol B	[A]	Δ[A]	(b) Rate –(Δ[A]/s)
0	0.065	0.000	0.65		
10	0.051	0.014	0.51	–0.14	2.3×10^{-4}
20	0.042	0.023	0.42	–0.09	2×10^{-4}
30	0.036	0.029	0.36	–0.06	1×10^{-4}
40	0.031	0.034	0.31	–0.05	0.8×10^{-4}

(c) $$\frac{\Delta M_B}{\Delta t} = \frac{(0.029 - 0.014) \text{ mol}/0.100 \text{ L}}{(30 - 10) \text{ min}} \times \frac{1 \text{ min}}{60 \text{ s}} = 1.25 \times 10^{-4} = 1.3 \times 10^{-4} \; M/\text{s}$$

14.21 (a) *Analyze/Plan*. Follow the logic in Sample Exercises 14.1 and 14.2. *Solve.*

Time (s)	Time Interval (s)	Concentration (M)	ΔM	Rate (M/s)
0		0.0165		
2000	2000	0.0110	–0.0055	28×10^{-7}
5000	3000	0.00591	–0.0051	17×10^{-7}
8000	3000	0.00314	–0.00277	9.23×10^{-7}
12,000	4000	0.00137	–0.00177	4.43×10^{-7}
15,000	3000	0.00074	–0.00063	2.1×10^{-7}

(b) $$-\frac{\Delta M}{\Delta t} = -\frac{(0.00074 - 0.0165)\,M}{(15{,}000 - 0)\text{ s}} = 1.0507 \times 10^{-6} = 1.05 \times 10^{-6} \; M/\text{s}$$

(c) $$-\frac{\Delta M}{\Delta t} = -\frac{(0.00137 - 0.0110)\,M}{(12{,}000 - 2000)\text{ s}} = 9.63 \times 10^{-7} \; M/\text{s}$$

$$-\frac{\Delta M}{\Delta t} = -\frac{(0.00074 - 0.00314)\,M}{(15{,}000 - 8000)\text{ s}} = 3.43 \times 10^{-7} \; M/\text{s}$$

The average rate between t = 2000 and t = 12,000 s is greater. In general, the rate of a reaction decreases over time.

(d) From the slopes of the lines in the figure at right, the rates are: at 5000 s, 12×10^{-7} M/s; at 8000 s, 5.8×10^{-7} M/s.

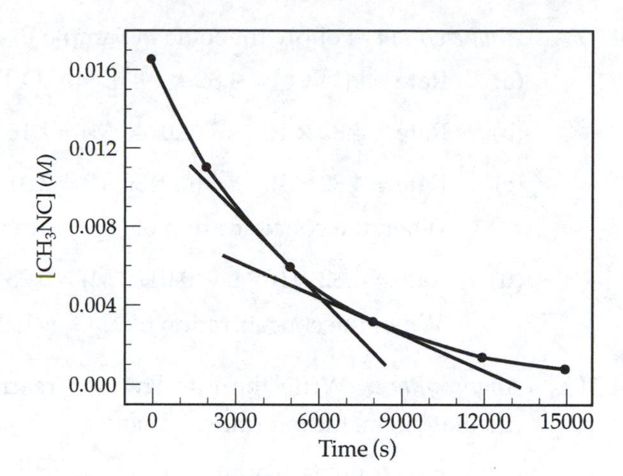

14.23 *Analyze/Plan*. Follow the logic in Sample Exercise 14.3. *Solve*.

(a) $-\Delta[H_2O_2]/\Delta t = \Delta[H_2]/\Delta t = \Delta[O_2]/\Delta t$

(b) $-\Delta[N_2O]/2\Delta t = \Delta[N_2]/2\Delta t = \Delta[O_2]/\Delta t$
$-\Delta[N_2O]/\Delta t = \Delta[N_2]/\Delta t = 2\Delta[O_2]/\Delta t$

(c) $-\Delta[N_2]/\Delta t = \Delta[NH_3]/2\Delta t$; $-\Delta[H_2]/3\Delta t = \Delta[NH_3]/2\Delta t$
$-2\Delta[N_2]/\Delta t = \Delta[NH_3]/\Delta t$; $-\Delta[H_2]/\Delta t = 3\Delta[NH_3]/2\Delta t$

(d) $-\Delta[C_2H_5NH_2]/\Delta t = \Delta[C_2H_4]/\Delta t = \Delta[NH_3]/\Delta t$

14.25 *Analyze/Plan*. Use Equation 14.4 to relate the rate of disappearance of reactants to the rate of appearance of products. Use this relationship to calculate desired quantities. *Solve*.

(a) $\Delta[H_2O]/2\Delta t = -\Delta[H_2]/2\Delta t = -\Delta[O_2]/\Delta t$
H_2 is burning, $-\Delta[H_2]/\Delta t = 0.48$ mol/s
O_2 is consumed, $-\Delta[O_2]/\Delta t = -\Delta[H_2]/2\Delta t = 0.48$ mol/s/2 = 0.24 mol/s
H_2O is produced, $+\Delta[H_2O]/\Delta t = -\Delta[H_2]/\Delta t = 0.48$ mol/s

(b) The change in total pressure is the sum of the changes of each partial pressure. NO and Cl_2 are disappearing and NOCl is appearing.
$-\Delta P_{NO}/\Delta t = -56$ torr/min
$-\Delta P_{Cl_2}/\Delta t = \Delta P_{NO}/2\Delta t = -28$ torr/min
$+\Delta P_{NOCl}/\Delta t = -\Delta P_{NO}/\Delta t = +56$ torr/min
$\Delta P_T/\Delta t = -56$ torr/min $-$ 28 torr/min + 56 torr/min = -28 torr/min

Rate Laws (Section 14.3)

14.27 *Analyze/Plan*. Follow the logic in Sample Exercise 14.5. *Solve*.

(a) If [A] is doubled, there will be no change in the rate or the rate constant. The overall rate is unchanged because [A] does not appear in the rate law; the rate constant changes only with a change in temperature.

(b) The reaction is zero order in A, second order in B, and second order overall.

(c) Units of $k = \dfrac{M/\text{s}}{M^2} = M^{-1}\,\text{s}^{-1}$

14.29 *Analyze/Plan.* Follow the logic in Sample Exercise 14.5. *Solve.*

(a) Rate = $k[N_2O_5] = 4.82 \times 10^{-3}\ s^{-1}\ [N_2O_5]$

(b) Rate = $4.82 \times 10^{-3}\ s^{-1}\ (0.0240\ M) = 1.16 \times 10^{-4}\ M/s$

(c) Rate = $4.82 \times 10^{-3}\ s^{-1}\ (0.0480\ M) = 2.31 \times 10^{-4}\ M/s$

When the concentration of N_2O_5 doubles, the rate of the reaction doubles.

(d) Rate = $4.82 \times 10^{-3}\ s^{-1}\ (0.0120\ M) = 5.78 \times 10^{-5}\ M/s$

When the concentration of N_2O_5 is halved, the rate of the reaction is halved.

14.31 *Analyze/Plan.* Write the rate law and rearrange to solve for k. Use the given data to calculate k, including units. *Solve.*

(a, b) Rate = $k[CH_3Br][OH^-]$; $k = \dfrac{\text{rate}}{[CH_3Br][OH^-]}$

$$\text{at 298 K, } k = \frac{0.0432\ M/s}{(5.0 \times 10^{-3}\ M)(0.050\ M)} = 1.7 \times 10^2\ M^{-1}s^{-1}$$

(c) Because the rate law is first order in $[OH^-]$, if $[OH^-]$ is tripled, the rate triples.

(d) If $[OH^-]$ and $[CH_3Br]$ both triple, the rate increases by a factor of (3)(3) = 9.

14.33 *Analyze/Plan.* Follow the logic in Sample Exercise 14.6. *Solve.*

(a) From the data given, when $[OCl^-]$ doubles, rate doubles. When $[I^-]$ doubles, rate doubles. The reaction is first order in both $[OCl^-]$ and $[I^-]$. Rate = $k[OCl^-][I^-]$.

(b) Using the first set of data:

$$k = \frac{\text{rate}}{[OCl^-][I^-]} = \frac{1.36 \times 10^{-4}\ M/s}{(1.5 \times 10^{-3}\ M)(1.5 \times 10^{-3}\ M)} = 60.444 = 60\ M^{-1}\ s^{-1}$$

(c) $$\text{Rate} = \frac{60.444}{M\text{-s}}(2.0 \times 10^{-3}\ M)(5.0 \times 10^{-4}\ M) = 6.0444 \times 10^{-5} = 6.0 \times 10^{-5}\ M/s$$

14.35 *Analyze/Plan.* Follow the logic in Sample Exercise 14.6 to deduce the rate law. Rearrange the rate law to solve for k and deduce units. Calculate a k value for each set of concentrations and then average the three values. *Solve.*

(a) Doubling $[NH_3]$ while holding $[BF_3]$ constant doubles the rate (experiments 1 and 2). Doubling $[BF_3]$ while holding $[NH_3]$ constant doubles the rate (experiments 4 and 5).

Thus, the reaction is first order in both BF_3 and NH_3; rate = $k[BF_3][NH_3]$.

(b) The reaction is second order overall.

(c) $$\text{From experiment 1: } k = \frac{0.2130\ M/s}{(0.250\ M)(0.250\ M)} = 3.41\ M^{-1}\ s^{-1}$$

(Any of the five sets of initial concentrations and rates could be used to calculate the rate constant k. The average of these 5 values is $k_{avg} = 3.408 = 3.41\ M^{-1}s^{-1}$.)

(d) Rate = $3.408\ M^{-1}s^{-1}(0.100\ M)(0.500\ M) = 0.1704 = 0.170\ M/s$.

14.37 *Analyze/Plan.* Follow the logic in Sample Exercise 4.6 to deduce the rate law. Rearrange the rate law to solve for k and deduce units. Calculate a k value for each set of concentrations and then average the three values. *Solve.*

(a) Increasing [NO] by a factor of 2.5 while holding $[Br_2]$ constant (experiments 1 and 2) increases the rate by a factor 6.25 or $(2.5)^2$. Increasing $[Br_2]$ by a factor of 2.5 while holding [NO] constant increases the rate by a factor of 2.5. The rate law for the appearance of NOBr is: Rate = $\Delta[NOBr]/\Delta t = k[NO]^2[Br_2]$.

(b) From experiment 1: $k_1 = \dfrac{24\ M/s}{(0.10\ M)^2\ (0.20\ M)} = 1.20 \times 10^4 = 1.2 \times 10^4\ M^{-2}\ s^{-1}$

$k_2 = 150/(0.25)^2(0.20) = 1.20 \times 10^4 = 1.2 \times 10^4\ M^{-2}\ s^{-1}$

$k_3 = 60/(0.10)^2(0.50) = 1.20 \times 10^4 = 1.2 \times 10^4\ M^{-2}\ s^{-1}$

$k_4 = 735/(0.35)^2(0.50) = 1.2 \times 10^4 = 1.2 \times 10^4\ M^{-2}\ s^{-1}$

$k_{avg} = (1.2 \times 10^4 + 1.2 \times 10^4 + 1.2 \times 10^4 + 1.2 \times 10^4)/4 = 1.2 \times 10^4\ M^{-2}\ s^{-1}$

(c) Use the reaction stoichiometry and Equation 14.4 to relate the designated rates. $\Delta[NOBr]/2\Delta t = -\Delta[Br_2]/\Delta t$; the rate of disappearance of Br_2 is half the rate of appearance of NOBr.

(d) Note that the data are given in terms of appearance of NOBr.

$$\frac{-\Delta[Br_2]}{\Delta t} = \frac{k[NO]^2\,[Br_2]}{2} = \frac{1.2 \times 10^4}{2\ M^2\ s} \times (0.075\ M)^2 \times (0.250\ M) = 8.4\ M/s$$

Change of Concentration with Time (Section 14.4)

14.39 (a) A graph of ln[A] versus time yields a straight line for a first-order reaction.

(b) On graph of ln[A] versus time, the rate constant, k, is the (–slope) of the straight line.

14.41 *Analyze/Plan.* The half-life of a first-order reaction depends only on the rate constant, $t_{1/2} = 0.693/k$. Use this relationship to calculate k for a given $t_{1/2}$, and, at a different temperature, $t_{1/2}$ given k. *Solve.*

(a) $t_{1/2} = 2.3 \times 10^5$ s; $t_{1/2} = 0.693/k$, $k = 0.693/t_{1/2}$

$k = 0.693/2.3 \times 10^5\ s = 3.0 \times 10^{-6}\ s^{-1}$

(b) $k = 2.2 \times 10^{-5}\ s^{-1}$. $t_{1/2} = 0.693/2.2 \times 10^{-5}\ s^{-1} = 3.15 \times 10^4 = 3.2 \times 10^4$ s

14.43 *Analyze/Plan.* Follow the logic in Sample Exercise 14.7. In this reaction, pressure is a measure of concentration. In (a) we are given k, $[A]_0$, t and asked to find $[A]_t$, using Equation 14.13, the integrated form of the first-order rate law. In (b), $[A_t] = 0.1[A_0]$, find t. *Solve.*

(a) $\ln P_t = -kt + \ln P_0$; $P_0 = 450$ torr; t = 60 s

$\ln P_{60} = -4.5 \times 10^{-2}\ s^{-1}(60) + \ln(450) = -2.70 + 6.109 = 3.409$

$P_{60} = 30.24 = 30$ torr

(b) $P_t = 0.10\ P_0$; $\ln(P_t/P_0) = -kt$

$\ln(0.10\ P_0/P_0) = -kt$, $\ln(0.10) = -kt$; $-\ln(0.10)/k = t$

$t = -(-2.303)/4.5 \times 10^{-2}\ s^{-1} = 51.2 = 51$ s

Check. From part (a), the pressure at 60 s is 30 torr, $P_t \sim 0.07\ P_0$. In part (b) we calculate the time where $P_t = 0.10\ P_0$ to be 51 s. This time should be smaller than 60 s, and it is. Data and results in the two parts are consistent.

14.45 *Analyze/Plan.* Given reaction order, various values for t and P_t, find the rate constant for the reaction at this temperature. For a first-order reaction, a graph of ln P versus t is linear with as slope of –k. *Solve.*

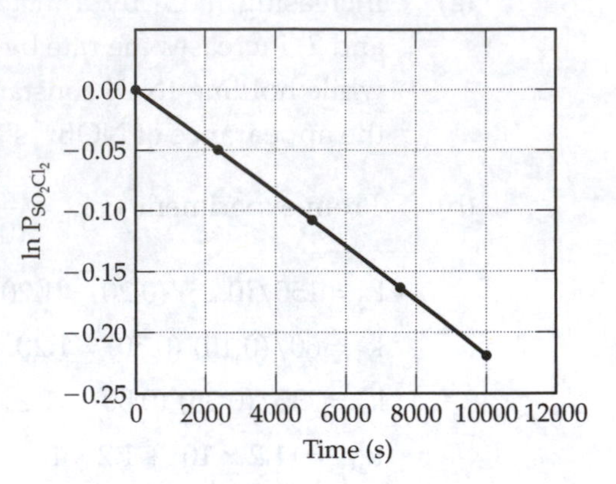

Time (s)	$P_{SO_2Cl_2}$	ln $P_{SO_2Cl_2}$
0	1.000	0
2500	0.947	–0.0545
5000	0.895	–0.111
7500	0.848	–0.165
10,000	0.803	–0.219

Graph ln $P_{SO_2Cl_2}$ versus time. (Pressure is a satisfactory unit for a gas, because the concentration in moles/liter is proportional to P.) The graph is linear with slope $-2.19 \times 10^{-5}\ s^{-1}$ as shown on the figure. The rate constant k = –slope = $2.19 \times 10^{-5}\ s^{-1}$.

14.47 *Analyze/Plan.* Given: mol A, t. Change mol to *M* at various times. Make both first- and second-order plots to see which is linear. *Solve.*

(a)

Time (min)	mol A	[A] (*M*)	ln[A]	1/mol A
0	0.065	0.65	–0.43	1.5
10	0.051	0.51	–0.67	2.0
20	0.042	0.42	–0.87	2.4
30	0.036	0.36	–1.02	2.8
40	0.031	0.31	–1.17	3.2

The plot of 1/[A] vs time is linear, so the reaction is second order in [A].

(b) For a second-order reaction, a plot of 1/[A] versus t is linear with slope k.

$k = \text{slope} = (3.2 - 2.0)\ M^{-1} / 30\ \text{min} = 0.040\ M^{-1}\ \text{min}^{-1}$

(The best fit to the line yields slope = $0.042\ M^{-1}\ \text{min}^{-1}$.)

(c) $t_{1/2} = 1/k[A]_0 = 1/(0.040\ M^{-1}\ \text{min}^{-1})(0.65\ M) = 38.46 = 38$ min

(Using the "best-fit" slope, $t_{1/2}$ = 37 min.)

14.49 *Analyze/Plan.* Make both first- and second-order plots to see which is linear. *Solve.*

(a)

Time (s)	$[NO_2](M)$	ln $[NO_2]$	$1/[NO_2]$
0.0	0.100	–2.303	10.0
5.0	0.017	–4.08	59
10.0	0.0090	–4.71	110
15.0	0.0062	–5.08	160
20.0	0.0047	–5.36	210

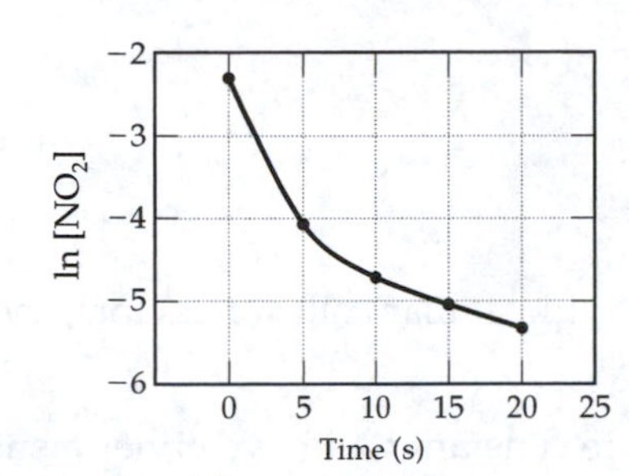

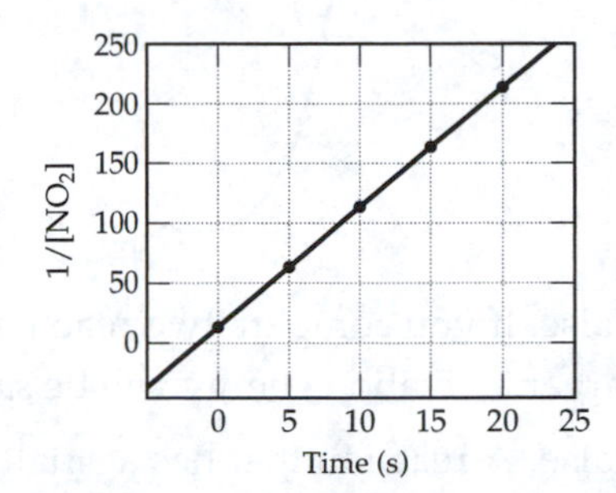

The plot of $1/[NO_2]$ versus time is linear, so the reaction is second order in NO_2.

(b) The slope of the line is $(210 - 59)\ M^{-1} / 15.0\ s = 10.07 = 10\ M^{-1}s^{-1}$ = k. (The slope of the best-fit line is $10.02 = 10\ M^{-1}s^{-1}$.)

(c) From the results above, the rate law is: rate = $k[NO_2]^2 = 10\ M^{-1}s^{-1}[NO_2]^2$
Using the rate law, calculate the rate at each of the given initial concentrations.

Rate @ 0.200 $M = 10\ M^{-1}s^{-1}[NO_2]^2 = 10\ M^{-1}s^{-1}[0.200\ M]^2 = 0.400\ M/s$

Rate @ 0.100 $M = 10\ M^{-1}s^{-1}[NO_2]^2 = 10\ M^{-1}s^{-1}[0.100\ M]^2 = 0.100\ M/s$

Rate @ 0.050 $M = 10\ M^{-1}s^{-1}[NO_2]^2 = 10\ M^{-1}s^{-1}[0.050\ M]^2 = 0.025\ M/s$

Temperature and Rate (Section 14.5)

14.51 (a) The energy of the collision and the orientation of the molecules when they collide determine whether a reaction will occur.

(b) Assuming other conditions remain the same, the rate and therefore the rate constant usually increase with an increase in reaction temperature.

(c) The fraction of molecules with energy greater than the activation energy changes most dramatically with temperature. Frequency of collision and the orientation factor are lumped into the frequency factor, A, which is considered to be constant with temperature.

14.53 *Analyze/Plan.* Given the temperature and energy, use Equation 14.20 to calculate the fraction of Ar atoms that have at least this energy. *Solve.*

$f = e^{-E_a/RT}$ $E_a = 10.0\ kJ/mol = 1.00 \times 10^4\ J/mol$; T = 400 K (127 °C)

$$-E_a/RT = -\frac{1.00 \times 10^4\ J/mol}{400\ K} \times \frac{mol\text{-}K}{8.314\ J} = -3.0070 = -3.01$$

$f = e^{-3.0070} = 4.9 \times 10^{-2}$

At 400 K, approximately 1 out of 20 molecules has this kinetic energy.

14.55 *Analyze/Plan.* Use the definitions of activation energy ($E_{max} - E_{react}$) and ΔE ($E_{prod} - E_{react}$) to sketch the graph and calculate E_a for the reverse reaction. *Solve.*

(a)

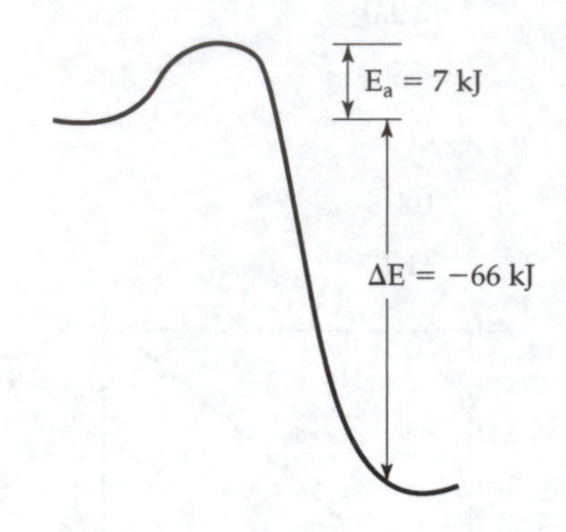

(b) E_a(reverse) = 73 kJ

14.57 (a) False. If you compare two reactions with similar collision factors, the one with the larger activation energy will be *slower*.

(b) False. A reaction that has a small rate constant will have either a small frequency factor (A), a large activation energy (E_a), or both.

(c) True.

14.59 The order of slowest reaction to fastest reaction is: rate (c) < rate (a) < rate (b). Assuming all collision factors (A) to be the same, reaction rate depends only on E_a; it is independent of ΔE.

14.61 *Analyze/Plan.* Given k_1, at T_1, calculate k_2 at T_2. Change T to Kelvins, then use the Equation 14.23 to calculate k_2. *Solve.*

$T_1 = 20\ °C + 273 = 293\ K$; $T_2 = 60\ °C + 273 = 333\ K$; $k_1 = 2.75 \times 10^{-2} s^{-1}$

(a) $$\ln\left(\frac{k_1}{k_2}\right) = \frac{E_a}{R}\left(\frac{1}{333} - \frac{1}{293}\right) = \frac{75.5 \times 10^3\ J/mol}{8.314\ J/mol}(-4.1 \times 10^{-4})$$

$$\ln(k_1/k_2) = -3.7229 = -3.7;\ k_1/k_2 = 0.0242 = 0.02;\ k_2 = \frac{0.0275\ s^{-1}}{0.0242} = 1.14 = 1\ s^{-1}$$

(b) $$\ln\left(\frac{k_1}{k_2}\right) = \frac{125 \times 10^3\ J/mol}{8.314\ J/mol}\left(\frac{1}{333} - \frac{1}{293}\right) = -6.1638 = -6.2$$

$$k_1/k_2 = 2.104 \times 10^{-3} = 2 \times 10^{-3};\ k_2 = \frac{0.0275\ s^{-1}}{2.104 \times 10^{-3}} = 13.07 = 1 \times 10\ s^{-1}$$

(c) The method in parts (a) and (b) assumes that the collision model and thus the Arrhenious equation describe the kinetics of the reactions. That is, activation energy is constant over the temperature range under consideration. There is no assumption about temperature dependence of the frequency factor, because it drops out of the difference equation by subtraction.

14.63 *Analyze/Plan*. Follow the logic in Sample Exercise 14.11. *Solve.*

k	ln k	T(K)	$1/T(\times 10^3)$
0.0521	–2.955	288	3.47
0.101	–2.293	298	3.36
0.184	–1.693	308	3.25
0.332	–1.103	318	3.14

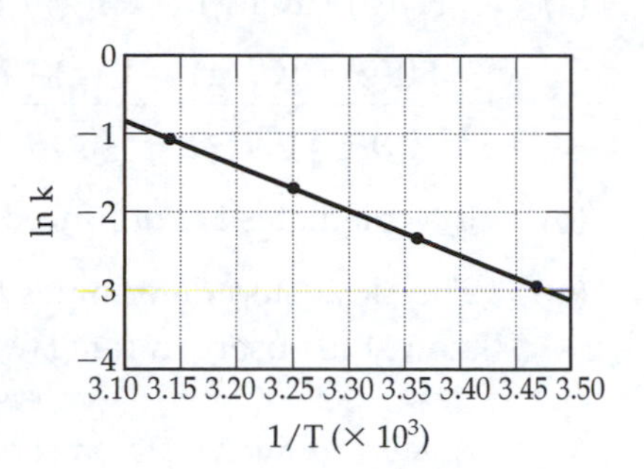

The slope, -5.64×10^3, equals $-E_a/R$. Thus,
$E_a = 5.64 \times 10^3 \times 8.314\ J/mol = 46.9\ kJ/mol$.

Reaction Mechanisms (Section 14.6)

14.65 (a) An *elementary reaction* is a process that occurs in a single event; the order is given by the coefficients in the balanced equation for the reaction.

(b) A *unimolecular* elementary reaction involves only one reactant molecule; the activated complex is derived from a single molecule. A *bimolecular* elementary reaction involves two reactant molecules in the activated complex and the overall process.

(c) A *reaction mechanism* is a series of elementary reactions that describe how an overall reaction occurs and explain the experimentally determined rate law.

(d) A *rate-determining step* is the slowest step in a reaction mechanism. It limits the overall reaction rate.

14.67 *Analyze/Plan*. Elementary reactions occur as a single step, so the molecularity is determined by the number of reactant molecules; the rate law reflects reactant stoichiometry. *Solve.*

(a) unimolecular, rate = $k[Cl_2]$

(b) bimolecular, rate = $k[OCl^-][H_2O]$

(c) bimolecular, rate = $k[NO][Cl_2]$

14.69 *Analyze/Plan*. Use the definitions of the term *intermediate*, along with the characteristics of reaction profiles, to answer the questions. *Solve.*

This is a three-step mechanism, $A \rightarrow B$, $B \rightarrow C$, and $C \rightarrow D$.

(a) There are 2 intermediates, B and C.

(b) There are 3 energy maxima in the reaction profile, so there are 3 transition states.

(c) Step $C \rightarrow D$ has the lowest activation energy, so it is fastest.

(d) The energy of D is slightly greater than the energy of A, so ΔE for the overall reaction is positive.

14.71 *Analyze/Plan.* Follow the logic in Sample Exercise 14.14. *Solve.*

(a)

$$H_2(g) + ICl(g) \rightarrow HI(g) + HCl(g)$$
$$\underline{HI(g) + ICl(g) \rightarrow I_2(g) + HCl(g)}$$
$$H_2(g) + 2\,ICl(g) \rightarrow I_2(g) + 2\,HCl(g)$$

(b) Intermediates are produced and consumed during reaction. HI is the intermediate.

(c) The slow step determines the rate law for the overall reaction. If the first step is slow, the observed rate law is: Rate = $k[H_2][ICl]$.

14.73 (a) *Analyze.* Given data on concentration of a reactant versus time, determine whether the proposed reaction mechanism is consistent with the data.

Plan. Based on the graph, decide the order of reaction with respect to [NO]. Write the two possible rate laws, depending on which step is rate-determining. Decide if one of the rate laws, and thus the mechanism, is consistent with the rate data.

Solve. The graph of 1/[NO] versus time is linear with positive slope, indicating that the reaction is second order in [NO]. The rate law will include $[NO]^2$.

If the first step is slow, the observed rate law is the rate law for this step: rate = $k[NO][Cl_2]$. Because the observed rate law is second order in [NO], the second step must be slow relative to the first step. Follow the logic in Sample Exercise 14.15 for determining the rate law of a mechanism with a fast initial step.

From the rate-determining second step, rate = $k[NOCl_2][NO]$.

Assuming the first step is a fast equilibrium, $k_1\,[NO][Cl_2] = k_{-1}\,[NOCl_2]$.

Solving for $[NOCl_2]$ in terms of $[NO][Cl_2]$, $[NOCl_2] = \frac{k_1}{k_{-1}}[NO][Cl_2]$

$$\text{Rate} = \frac{k_2 k_1}{k_{-1}}[NO][Cl_2][NO] = [NO]^2[Cl_2]$$

This rate law is second order in [NO]. It is consistent with the observed data.

(b) The linear plot guarantees that the overall rate law will include $[NO]^2$. Because the data were obtained at constant $[Cl_2]$, we have no information about reaction order with respect to $[Cl_2]$.

Catalysis (Section 14.7)

14.75 (a) A catalyst is a substance that changes (usually increases) the speed of a chemical reaction without undergoing a permanent chemical change itself.

(b) A homogeneous catalyst is in the same phase as the reactants; a heterogeneous catalyst is in a different phase and is usually a solid.

(c) A catalyst has no effect on the overall enthalpy change for a reaction. A catalyst does affect activation energy, E_a, which is one way that it changes reaction rate. It can also affect the frequency factor, A.

14.77 KBr(s) is added to $H_2O_2(aq)$ at t = 0. Assume the KBr(s) dissolves instantly. As the reaction proceeds, the Br^- catalyst is consumed and then regenerated.

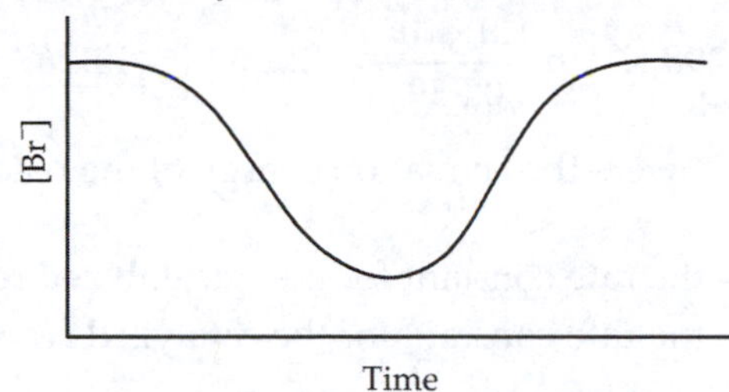

14.79 (a)

$$\begin{array}{l} 2\,[NO_2(g) + SO_2(g) \rightarrow NO(g) + SO_3(g)] \\ \quad 2\,NO(g) + O_2(g) \rightarrow 2\,NO_2(g) \\ \hline 2\,SO_2(g) + O_2(g) \rightarrow 2\,SO_3(g) \end{array}$$

(b) $NO_2(g)$ is a catalyst because it is consumed and then reproduced in the reaction sequence.

(c) NO(g) is an intermediate, because it is produced and then consumed during the reaction.

(d) Because NO_2 is in the same state as the other reactants, this is homogeneous catalysis.

14.81 (a) When using a powdered metal catalyst, only a small percentage of the metal atoms are at the surface of the bulk material and catalytically active. Use of chemically stable supports such as alumina and silica makes it possible to obtain very large surface areas per unit mass of the precious metal catalyst. This is so because the metal can be deposited in a very thin, even monomolecular, layer on the surface of the support.

(b) The greater the surface area of the catalyst, the more reaction sites, and the greater the rate of the catalyzed reaction.

14.83 As illustrated in Figure 14.23, the two C–H bonds that exist on each carbon of the ethylene molecule before adsorption are retained in the process in which a D atom is added to each C (assuming we use D_2 rather than H_2). To put two deuteriums on a single carbon, it is necessary that one of the already existing C–H bonds in ethylene be broken while the molecule is adsorbed, so the H atom moves off as an adsorbed atom, and is replaced by a D. This requires a larger activation energy than simply adsorbing C_2H_4 and adding one D atom to each carbon.

14.85 Let k and E_a equal the rate constant and activation energy without the enzyme (uncatalyzed). Let k_c and E_{ac} equal the rate constant and activation energy with the enzyme (catalyzed). A is the same for the uncatalyzed and catalyzed reactions. The difference in activation energies is $E_{ac} - E_a$, $k_c = 1.0 \times 10^6\ s^{-1}$, $k = 0.039\ s^{-1}$, T = 25 °C = 298 K.

According to Equation 14.22, $\ln k = E_a/RT + \ln A$. Subtracting ln k from ln k_c

$$\ln k_c - \ln k = \left[\frac{-E_{ac}}{RT}\right] + \ln A - \left[\frac{-E_a}{RT}\right] - \ln A$$

$$\ln(k_c/k) = \frac{E_a - E_{ac}}{RT};\ E_a - E_{ac} = RT\ln(k_c/k)$$

$$E_a - E_{ac} = \frac{8.314\ J}{mol\text{-}K} \times 298\ K \times \ln\frac{1.0\times10^6}{0.039} = 42{,}267\ J = 42.267\ kJ = 42\ kJ$$

Carbonic anyhdrase lowers the activation energy of the reaction by 42 kJ.

14.87 *Analyze/Plan.* Let k = the rate constant for the uncatalyzed reaction,
k_c = the rate constant for the catalyzed reaction

According to Equation 14.22, $\ln k = -E_a/RT + \ln A$

Subtracting ln k from ln k_c,

$$\ln k_c - \ln k = -\left[\frac{55\ kJ/mol}{RT} + \ln A\right] - \left[-\frac{95\ kJ/mol}{RT} + \ln A\right].$$ *Solve.*

(a) $RT = 8.314\ J/mol\text{-}K \times 298\ K \times 1\ kJ/1000\ J = 2.478\ kJ/mol$; ln A is the same for both reactions.

$$\ln(k_c/k) = \frac{95\ kJ/mol - 55\ kJ/mol}{2.478\ kJ/mol};\ k_c/k = 1.024\times10^7 = 1\times10^7$$

The catalyzed reaction is approximately 10,000,000 (ten million) times faster at 25 °C.

(b) $RT = 8.314\ J/mol\text{-}K \times 398\ K \times 1\ kJ/1000\ J = 3.309\ kJ/mol$

$$\ln(k_c/k) = \frac{40\ kJ/mol}{3.309\ kJ/mol};\ k_c/k = 1.778\times10^5 = 2\times10^5$$

The catalyzed reaction is 200,000 times faster at 125 °C.

Additional Exercises

14.91 (a) $$\text{Rate} = \frac{-\Delta[NO]}{2\Delta t} = \frac{-\Delta[O_2]}{\Delta t} = \frac{9.3\times10^{-5}\ M/s}{2} = 4.7\times10^{-5}\ M/s$$

(b, c) $\text{Rate} = k[NO]^2[O_2]$; $k = \text{rate}/[NO]^2[O_2]$

$$k = \frac{4.7\times10^{-5}\ M/s}{(0.040\ M)^2(0.035\ M)} = 0.8393 = 0.84\ M^{-2}\ s^{-1}$$

(d) Because the reaction is second order in NO, if the [NO] is increased by a factor of 1.8, the rate would increase by a factor of 1.8^2, or (3.24) = 3.2.

14.95 (a) Because the units of the rate constant are $M^{-1}s^{-1}$, the reaction is second order overall and second order in NO_2.

(b) If $[NO_2]_0 = 0.100\ M$ and $[NO_2]_t = 0.025\ M$, use the integrated form of the second order rate equation, $\frac{1}{[A]_t} = kt + \frac{1}{[A]_0}$, Equation 14.14, to solve for t.

$$\frac{1}{0.025\ M} = 0.63\ M^{-1}s^{-1}\ (t) + \frac{1}{0.100\ M};\ \frac{(40-10)\ M^{-1}}{0.63\ M^{-1}s^{-1}} = t = 47.62 = 48\ s.$$

14.99 *Analyze.* Given rate constants for the decay of two radioisotopes, determine half-lives, decay rates, and amount remaining after three half-lives. *Plan.* Determine reaction order. Based on reaction-order, select the appropriate relationships for (a) rate constant and half-life and (c) rate-constant, time and concentration. In this example, mass is a measure of concentration.

Solve. Decay of radioisotopes is a first-order process, because only one species is involved and the decay is not initiated by collision.

(a) For a first-order process, $t_{1/2} = 0.693/k$.

^{241}Am: $t_{1/2} = 0.693/1.6 \times 10^{-3}\ yr^{-1} = 433.1 = 4.3 \times 10^{2}\ yr$

^{125}I: $t_{1/2} = 0.693/0.011\ day^{-1} = 63.00 = 63$ days

(b) For a given sample size, half of the ^{241}Am sample decays in 433 years, whereas half of the ^{125}I sample decays in 63 days. ^{125}I decays at a much faster rate.

(c) For a first-order process, $\ln[A]_t - \ln[A]_0 = -kt$. $\ln[A]_t = -kt + \ln[A]_0$.

$[A]_0 = 1.0$ mg; $t = 3\ t_{1/2}$.

^{241}Am: $t = 3\ t_{1/2} = 3(433.1\ yr) = 1.299 \times 10^{3} = 1.3 \times 10^{3}\ yr$

$\ln[Am]_t = -1.6 \times 10^{-3}\ yr^{-1}(1.299 \times 10^{3}\ yr) - \ln(1.0) = -2.079 - 0 = -2.08$

$[Am]_t = 0.125 = 0.13$ mg

or, mass ^{241}Am remaining $= 1.0\ mg/2^{3} = 0.125 = 0.13$ mg

^{125}I: For the same size starting sample and number of elapsed half-lives, the same mass, 0.13 mg ^{125}I, will remain. (The difference is that the elapsed time of 3 half-lives (for ^{125}I is 3(63) = 189 days = 0.52 yr, versus 433 yr for ^{241}Am.)

(d) Again, for a first-order process, $\ln[A]_t - \ln[A]_0 = -kt$. $\ln[A]_t = -kt + \ln[A]_0$.

$[A]_0 = 1.0$ mg; $t = 4$ days.

$k_{Am} = 1.6 \times 10^{-3}\ yr^{-1}\ (1\ yr/365\ days) = 4.3836 \times 10^{-6} = 4.4 \times 10^{-6}\ day^{-1}$

^{241}Am: $\ln[Am]_t = -4.4 \times 10^{-6}\ day^{-1}\ (4\ days) - \ln(1.0) = -1.7 \times 10^{-5} - 0 = -1.7 \times 10^{-5}$

The amount of ^{241}Am remaining after 4 days is 0.99998 mg, to three significant figures, 1.00 mg.

^{125}I: $\ln[I]_t = -0.011\ day^{-1}\ (4\ days) - \ln(1.0) = -0.044 - 0 = -0.044$

The amount of ^{125}I remaining after 4 days is 0.956 mg.

14.103

Time (s)	$[C_5H_6]$ (*M*)	$\ln[C_5H_6]$	$1/[C_5H_6]$
0	0.0400	–3.219	25.0
50	0.0300	–3.507	33.3
100	0.0240	–3.730	41.7
150	0.0200	–3.912	50.0
200	0.0174	–4.051	57.5

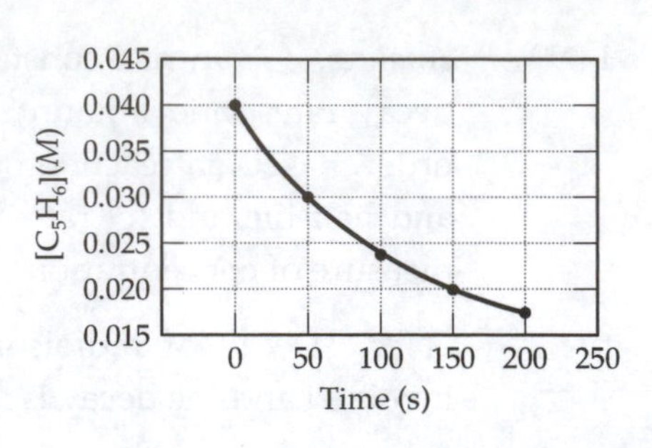

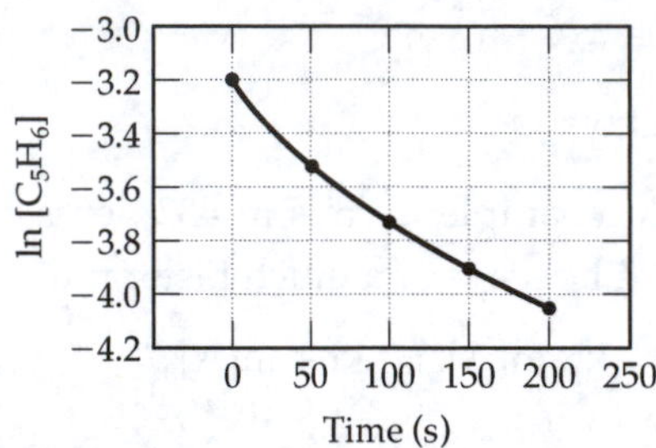

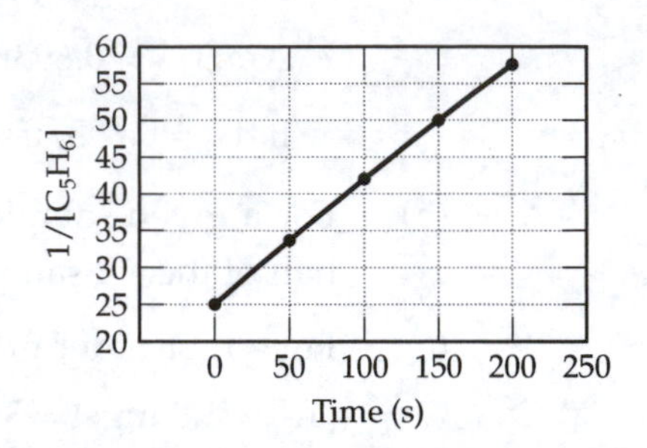

(a) The plot of $1/[C_5H_6]$ vs time is linear and the reaction is second order.

(b) The slope of the line in the plot of $1/[C_5H_6]$ vs time is the value of k.

$k = \text{slope} = (50.0 - 25.0)\ M^{-1}/(150\text{–}0)\text{s} = 0.167\ M^{-1}\ s^{-1}$

(The best-fit slope and k value is $0.163\ M^{-1}\ s^{-1}$.)

14.107 (a)

$$NO(g) + NO(g) \rightarrow N_2O_2(g)$$
$$N_2O_2(g) + H_2(g) \rightarrow N_2O(g) + H_2O(g)$$
$$\overline{2\,NO(g) + N_2O_2(g) + H_2(g) \rightarrow N_2O_2(g) + N_2O(g) + H_2O(g)}$$
$$2\,NO(g) + H_2(g) \rightarrow N_2O(g) + H_2O(g)$$

(b) First reaction: $-\Delta[NO]/\Delta t = k[NO]\,[NO] = k[NO]^2$

Second reaction: $-\Delta[H_2]/\Delta t = k[H_2][N_2O_2]$

(c) N_2O_2 is the intermediate: it is produced in the first step and consumed in the second.

(d) Because $[H_2]$ appears in the rate law, the second step must be slow relative to the first.

14.110 (a)

$$Cl_2(g) \rightleftharpoons 2Cl(g)$$
$$Cl(g) + CHCl_3(g) \rightarrow HCl(g) + CCl_3(g)$$
$$Cl(g) + CCl_3(g) \rightarrow CCl_4(g)$$
$$\overline{Cl_2(g) + 2\,Cl(g) + CHCl_3(g) + CCl_3(g) \rightarrow 2\,Cl(g) + HCl(g) + CCl_3(g) + CCl_4(g)}$$
$$Cl_2(g) + CHCl_3(g) \rightarrow HCl(g) + CCl_4(g)$$

(b) $Cl(g)$, $CCl_3(g)$

(c) Reaction 1 - unimolecular, Reaction 2 - bimolecular, Reaction 3 - bimolecular

(d) Reaction 2, the slow step, is rate determining.

(e) If Reaction 2 is rate determining, rate = $k_2[CHCl_3][Cl]$. Cl is an intermediate formed in reaction 1, an equilibrium. By definition, the rates of the forward and reverse processes are equal; $k_1[Cl_2] = k_{-1}[Cl]^2$. Solving for [Cl] in terms of $[Cl_2]$,

$$[\text{Cl}]^2 = \frac{k_1}{k_{-1}}[\text{Cl}_2];\ [\text{Cl}] = \left(\frac{k_1}{k_{-1}}[\text{Cl}_2]\right)^{1/2}$$

Substituting into the overall rate law

$$\text{rate} = k_2\left(\frac{k_1}{k_{-1}}\right)^{1/2}[\text{CHCl}_3][\text{Cl}_2]^{1/2} = k[\text{CHCl}_3][\text{Cl}_2]^{1/2}\ \text{(The overall order is 3/2.)}$$

14.115 Let k and E_a equal the rate constant and activation energy for the uncatalyzed reaction. Let k_c and E_{ac} equal the rate constant and activation energy of the enzyme-catalyzed reaction. Assume A is the same for the uncatalyzed and catalyzed reactions.

$k_c/k = 5000$, T = 37 °C = 310 K.

According to Equation 14.22, $\ln k = -E_a/RT + \ln A$. Subtracting ln k from $\ln k_c$

$$\ln k_c - \ln k = \left[\frac{-E_{ac}}{RT}\right] + \ln A - \left[\frac{-E_a}{RT}\right] - \ln A$$

$$\ln (k_c/k) = \frac{E_a - E_{ac}}{RT};\ E_a - E_{ac} = RT \ln (k_c/k)$$

$$E_a - E_{ac} = \frac{8.314\ \text{J}}{\text{mol-K}} \times 310\ \text{K} \times \ln (5000) = 2.195 \times 10^4\ \text{J/mol} = 21.95\ \text{kJ/mol} = 22\ \text{kJ/mol}$$

The enzyme must lower the activation energy by 22 kJ/mol to increase the reaction rate by a factor of 5000.

Integrative Exercises

14.120 (a) $\ln k = -E_a/RT + \ln A$, Equation 14.22. $E_a = 6.3$ kJ/mol $= 6.3 \times 10^3$ J/mol

T = 100 °C + 273 = 373 K

$$\ln k = \frac{-6.3\times10^3\ \text{J/mol}}{8.314\ \text{J/K-mol}\times 373\ \text{K}} + \ln (6.0\times10^8\, M^{-1}\text{s}^{-1})$$

$\ln k = -2.032 + 20.212 = 18.181 = 18.2$; $k = 7.87 \times 10^7 = 8 \times 10^7\ M^{-1}\text{s}^{-1}$

(b) NO, 11 valence e^-, 5.5 e^- pair (Assume the less electronegative N atom will be electron deficient.)

:Ṅ═Ö:

ONF, 18 valence e^-, 9 e^- pr

:Ö═N̈—F̤: ⟷ (:Ö̤—N̈═F̤)

The resonance form on the right is a very minor contributor to the true bonding picture, because of high formal charges and the unlikely double bond involving F.

(c) ONF has trigonal planar electron domain geometry, which leads to a "bent" structure with a bond angle of approximately 120°.

(d) [O═N···F—F]

(e) The electron deficient NO molecule is attracted to electron-rich F_2, so the driving force for formation of the transition state is greater than simple random collisions.

14.123 (a) A generic Lewis structure for a primary amine is shown below. There are four electron domains about nitrogen, so the hybridization is sp^3. The hybrid orbital picture is shown on the right.

(b) A reactant that is attracted to the lone pair of electrons on nitrogen will produce a tetrahedral intermediate. This can be a moiety with a full, partial, or even transient positive charge. Steric hindrance will not be large, because two of the atoms bound to nitrogen are small hydrogens.

15 Chemical Equilibrium

Visualizing Concepts

15.1 (a) $k_f > k_r$. According to the Arrhenius equation [14.21], $k = Ae^{-E_a/RT}$. As the magnitude of E_a increases, k decreases. On the energy profile, E_a is the difference in energy between the starting point and the energy at the top of the barrier. Clearly this difference is smaller for the forward reaction, so $k_f > k_r$.

(b) From the Equation [15.5], the equilibrium constant = k_f/k_r. Because $k_f > k_r$, the equilibrium constant for the process shown in the energy profile is greater than 1.

15.3 *Analyze*. Given box diagram and reaction type, determine whether K greater or smaller than one for the equilibrium mixture depicted in the box.

Plan. Assign species in the box to reactants and products. Write an equilibrium expression in terms of concentrations. Find the relationship between numbers of moles (molecules in the diagram) and concentration. Calculate K.

Solve. Let red = A, blue = X, red and blue pairs = AX. (The colors of A and X are arbitrary.) There are 3A, 2B, and 8AX in the box.

M = mol/L. Because each particle represents one mole, we can use numbers of particles in place of moles in the molarity formula. (The mole is a counting unit for particles, so mol ratios and particle ratios are equivalent.) V = 1 L, so in this case, [A] = number of A particles.

$$K = \frac{[AX]}{[A][X]}; \quad [AX] = 8/V = 8; [A] = 3/V = 3; [X] = 2/V = 2. \; K = \frac{8}{[3][2]} = \frac{8}{6} = 1.33$$

K is greater than one.

15.5 The reaction $A(g) + B(g) \rightleftharpoons AB(g)$ has the larger equilibrium constant. At 40s and 50s, there are more diatomic products and fewer monoatomic reactants than in the other reaction. An equilibrium constant is the ratio of concentrations of products to concentrations of reactants at equilibrium. Another way to say this is that the A(g) + B(g) reaction favors products more than the X(g) + Y(g) reaction. (The X(g) + Y(g) reaction reaches equilibrium more quickly than the A(g) + B(g) reaction, but this is a matter of kinetics, not equilibrium constants.)

15.6 *Analyze/Plan*. The reaction with the largest equilibrium constant has the largest ratio of products to reactants. Count product and reactant molecules. Calculate ratios and compare. *Solve*.

$K = \dfrac{[C_2H_4X_2]}{[C_2H_4][X_2]}$. Use numbers of molecules as an adequate measure of concentration.

(Although the volume terms don't cancel, they are the same for all parts. For the purpose of comparison, we can ignore volume.) *Solve.*

(a) 8 $C_2H_4Cl_2$, 2 Cl_2, 2 C_2H_4. $K = \frac{8}{(2)(2)} = 2$

(b) 6 $C_2H_4Br_2$, 4 Br_2, 4 C_2H_4. $K = \frac{6}{(4)(4)} = 0.375 = 0.4$

(c) 3 $C_2H_4I_2$, 7 I_2, 7 C_2H_4. $K = \frac{3}{(7)(7)} = 0.0612 = 0.06$

From the smallest to the largest equilibrium constant, (c) < (b) < (a).

Check. By inspection, there are the fewest product molecules and the most reactant molecules in (c); most product and least reactant in (a).

15.7 Statement (b) is definitely true. The reaction is a heterogeneous equilibrium for which $K_p = P_{O_2}$. The larger volume of vessel B requires that more $PbO_2(s)$ must decompose to produce the equilibrium pressure of $O_2(g)$. (After heating, both vessels will have less than the 5.0 g $PbO_2(s)$ that was present initially. If the "less" refers to the amount of solid present initially, statement (a) is also true.)

15.9 For the reaction $A_2(g) + B(g) \rightleftharpoons A(g) + AB(g)$, $\Delta n = 0$ and $K_p = K_c$. We can evaluate the equilibrium expression in terms of concentration. Also, because $\Delta n = 0$, the volume terms in the expression cancel and we can use number of particles as a measure of moles and molarity. The mixture contains 2A, 4AB, and $2A_2$.

$$K_c = \frac{[A][AB]}{[A_2][B]} = \frac{(2)(4)}{(2)(B)} = 2;\ B = 2$$

2 B atoms should be added to the diagram.

15.11 If temperature increases, K of an endothermic reaction increases and K of an exothermic reaction decreases. Calculate the value of K for the two temperatures and compare. For this reaction, $\Delta n = 0$ and $K_p = K_c$. We can ignore volume and use number of particles as a measure of moles and molarity. $K_c = [A][AB]/[A_2][B]$

(1) 300 K, 3A, 5AB, $1A_2$, 1B; $K_c = (3)(5)/(1)(1) = 15$

(2) 500 K, 1A, 3AB, $3A_2$, 3B; $K_c = (1)(3)/(3)(3) = 0.33$

K_c decreases as T increases, so the reaction is exothermic.

Equilibrium; The Equilibrium Constant (Sections 15.1–15.4)

15.13 *Analyze/Plan.* Given the forward and reverse rate constants, calculate the equilibrium constant using Equation 15.5. At equilibrium, the rates of the forward and reverse reactions are equal. Write the rate laws for the forward and reverse reactions and use their equality to answer part (b). *Solve.*

(a) $K_c = \frac{k_f}{k_r}$, Equation 15.5; $K_c = \frac{4.7 \times 10^{-3}\ s^{-1}}{5.8 \times 10^{-1}\ s^{-1}} = 8.1 \times 10^{-3}$

[For this reaction, $K_p = K_c = 8.1 \times 10^{-3}$]

(b) At equilibrium, the partial pressure of A is greater than the partial pressure of B.
$rate_f = rate_r$; $k_f[A] = k_r[B]$
At equilibrium, the two rates are equal. Because $k_f < k_r$, [A] must be greater than [B] and the partial pressure of A is greater than the partial pressure of B.

15.15 *Analyze/Plan*. Follow the logic in Sample Exercises 15.1 and 15.5. *Solve*.

(a) $K_c = \dfrac{[N_2O][NO_2]}{[NO]^3}$ (b) $K_c = \dfrac{[CS_2][H_2]^4}{[CH_4][H_2S]^2}$

(c) $K_c = \dfrac{[CO]^4}{[Ni(CO)_4]}$ (d) $K_c = \dfrac{[H^+][F^-]}{[HF]}$

(e) $K_c = \dfrac{[Ag^+]^2}{[Zn^{2+}]}$ (f) $K_c = [H^+][OH^-]$

(g) $K_c = [H^+]^2[OH^-]^2$

homogeneous: (a), (b), (d), (f), (g); heterogeneous: (c), (e)

15.17 *Analyze*. Given the value of K_c or K_p, predict the contents of the equilibrium mixture.

Plan. If K_c or $K_p >> 1$, products dominate; if K_c or $K_p << 1$, reactants dominate. *Solve*.

(a) mostly reactants ($K_c << 1$)

(b) mostly products ($K_p >> 1$)

15.19 (a) True.

(b) False. A single-headed arrow indicates that the reaction "goes to completion," that the equilibrium constant is extremely large.

(c) False. The value of the equilibrium constant gives no information about the speed of a reaction.

15.21 *Analyze/Plan*. Follow the logic in Sample Exercise 15.2. *Solve*.

$PCl_3(g) + Cl_2(g) \rightleftharpoons PCl_5(g)$, $K_c = 0.042$. $\Delta n = 1 - 2 = -1$

$K_p = K_c(RT)^{\Delta n} = 0.042(RT)^{-1} = 0.042/RT$

$$K_p = \frac{0.042}{(0.08206)(500)} = 0.001024 = 1.0 \times 10^{-3}$$

15.23 *Analyze*. Given K_c for a chemical reaction, calculate K_c for the reverse reaction.

Plan. Evaluate which species are favored by examining the magnitude of K_c. The equilibrium expressions for the reaction and its reverse are the reciprocals of each other, and the values of K_c are also reciprocal. *Solve*.

(a) For the reaction as written, $K_c < 1$, which means that reactants are favored. At this temperature, the equilibrium favors NO and Br_2.

(b) $K_c(\text{forward}) = \dfrac{[NOBr]^2}{[NO]^2[Br_2]} = 1.3 \times 10^{-2}$

$$K_c(\text{reverse}) = \frac{[NO]^2[Br_2]}{[NOBr]^2} = \frac{1}{1.3 \times 10^{-2}} = 76.92 = 77$$

(c) $K_{c2}(\text{reverse}) = \dfrac{[NO][Br_2]^{1/2}}{[NOBr]} = (K_c(\text{reverse}))^{1/2} = (76.92)^{1/2} = 8.8$

15.25 *Analyze.* Given K_p for a reaction, calculate K_p for a related reaction.

Plan. The algebraic relationship between the K_p values is the same as the algebraic relationship between equilibrium expressions.

Solve. $K_p = \frac{P_{SO_3}}{P_{SO_2} \times P_{O_2}^{1/2}} = 1.85$

(a) $K_p = \frac{P_{SO_2} \times P_{O_2}^{1/2}}{P_{SO_3}} = \frac{1}{1.85} = 0.541$

(b) $K_p = \frac{P_{SO_3}^2}{P_{SO_2}^2 \times P_{O_2}} = (1.85)^2 = 3.4225 = 3.42$

(c) $K_p = K_c(RT)^{\Delta n}$; $\Delta n = 2 - 3 = -1$; $T = 1000$ K

$K_p = K_c(RT)^{-1} = K_c/RT$; $K_c = K_p(RT)$

$K_c = 3.4225(0.08206)(1000) = 280.85 = 281$

15.27 *Analyze/Plan.* Follow the logic in Sample Exercise 15.4. *Solve.*

$$CoO(s) + H_2(g) \rightleftharpoons Co(s) + H_2O(g) \qquad K_1 = 67$$

$$Co(s) + CO_2(g) \rightleftharpoons CoO(s) + CO(g) \qquad K_2 = 1/490$$

$$CoO(s) + H_2(g) + Co(s) + CO_2(g) \rightleftharpoons Co(s) + H_2O(g) + CoO(s) + CO(g)$$

$$H_2(g) + CO_2(g) \rightleftharpoons H_2O(g) + CO(g)$$

$$K_c = K_1 \times K_2 = 67 \times \frac{1}{490} = 0.1367 = 0.14$$

15.29 *Analyze/Plan.* Follow the logic in Sample Exercise 15.5. *Solve.*

(a) $K_p = P_{O_2}$

(b) $K_c = [Hg(solv)]^4[O_2(solv)]$

Calculating Equilibrium Constants (Section 15.5)

15.31 *Analyze/Plan.* Calculate molarity of reactants and products. Follow the logic in Sample Exercise 15.7 using concentrations rather than pressures. *Solve.*

$$[CH_3OH] = \frac{0.0406 \text{ mol}}{2.00 \text{ L}} = 0.0203\ M$$

$$[CO] = \frac{0.170 \text{ mol CO}}{2.00 \text{ L}} = 0.0850\ M; \quad [H_2] = \frac{0.302 \text{ mol } H_2}{2.00 \text{ L}} = 0.151\ M$$

$$K_c = \frac{[CH_3OH]}{[CO][H_2]^2} = \frac{0.0203}{(0.0850)(0.151)^2} = 10.4743 = 10.5$$

15.33 *Analyze/Plan.* Follow the logic in Sample Exercise 15.7. *Solve.*

(a) $2NO(g) + Cl_2(g) \rightleftharpoons 2NOCl(g)$

$$K_p = \frac{P_{NOCl}^2}{P_{NO}^2 \times P_{Cl_2}} = \frac{(0.28)^2}{(0.095)^2(0.171)} = 50.80 = 51$$

(b) $K_p = K_c(RT)^{\Delta n}$; $\Delta n = 2 - 3 = -1$; $K_p = K_c(RT)^{-1} = K_c/(RT)$

$K_c = K_p(RT) = 50.80(0.08206 \times 500) = 2.1 \times 10^3$

15.35 *Analyze/Plan.* Follow the logic in Sample Exercise 15.8. Because the container volume is 1.0 L, mol = *M*. *Solve.*

(a) First calculate the change in [NO], 0.062 – 0.10 = –0.038 = –0.04 *M*. From the stoichiometry of the reaction, calculate the changes in the other pressures. Finally, calculate the equilibrium pressures.

	$2NO(g)$	+	$2H_2(g)$	$\rightleftharpoons$	$N_2(g)$ +	$2H_2O(g)$
initial	0.10 *M*		0.050 *M*		0 *M*	0.10 *M*
change	–0.038 *M*		–0.038 *M*		+0.019 *M*	+0.038 *M*
equil.	0.062 *M*		0.012 *M*		0.019 *M*	0.138 *M*

Strictly speaking, the change in [NO] has two decimal places and thus one sig fig. This limits equilibrium pressures to one sig fig for all but H_2O, and K_c to one sig fig. We compute the extra figures and then round.

(b) $$K_c = \frac{[N_2][H_2O]^2}{[NO]^2[H_2]^2} = \frac{(0.019)(0.138)^2}{(0.062)^2(0.012)^2} = \frac{(0.02)(0.14)^2}{(0.06)^2(0.01)^2} = 653.7 = 7 \times 10^2$$

15.37 *Analyze/Plan.* Follow the logic in Sample Exercise 15.8, using partial pressures, rather than concentrations. *Solve.*

(a) $$P = nRT/V; P_{CO_2} = 0.2000 \text{ mol} \times \frac{500 \text{ K}}{2.000\text{L}} \times \frac{0.08206 \text{ L-atm}}{\text{mol-K}} = 4.1030 = 4.10 \text{ atm}$$

$$P_{H_2} = 0.1000 \text{ mol} \times \frac{500 \text{ K}}{2.000\text{L}} \times \frac{0.08206 \text{ L-atm}}{\text{mol-K}} = 2.0515 = 2.05 \text{ atm}$$

$$P_{H_2O} = 0.1600 \times \frac{500 \text{ K}}{2.000\text{L}} \times \frac{0.08206 \text{ L-atm}}{\text{mol-K}} = 3.2824 = 3.28 \text{ atm}$$

(b) The change in P_{H_2O} is 3.51 – 3.28 = 0.2276 = 0.23 atm. From the reaction stoichiometry, calculate the change in the other pressures and the equilibrium pressures.

	$CO_2(g)$ +	$H_2(g)$	$\rightleftharpoons$	$CO(g)$ +	$H_2O(g)$
initial	4.10 atm	2.05 atm		0 atm	3.28 atm
change	–0.23 atm	–0.23 atm		+0.23	+0.23 atm
equil.	3.87 atm	1.82 atm		0.23 atm	3.51 atm

(c) $$K_p = \frac{P_{CO} \times P_{H_2O}}{P_{CO_2} \times P_{H_2}} = \frac{(0.23)(3.51)}{(3.87)(1.82)} = 0.1146 = 0.11$$

Without intermediate rounding, equilibrium pressures are $P_{H_2O} = 3.51$, $P_{CO} = 0.2276$, $P_{H_2} = 1.8239$, $P_{CO_2} = 3.8754$ and $K_p = 0.1130 = 0.11$, in good agreement with the value above.

(d) $K_p = K_c(RT)^{\Delta n}$; $\Delta n = 2 - 2 = 0$; $K_p = K_c(RT)^0$; $K_c = K_p = 0.11$

15.39 *Analyze/Plan*. Follow the logic in Sample Exercise 15.8. m*M* = 10^{-3} *M*

	X(aq)	+	Y(aq)	⇌	XY(aq)
initial	1.0 m*M*		1.0 m*M*		0
change	–0.80 m*M*		–0.80 m*M*		+0.80 m*M*
equil.	0.20 m*M*		0.20 m*M*		0.80 m*M*

$$K_c = \frac{[XY]}{[X][Y]} = \frac{(0.80\times10^{-3})}{(0.20\times10^{-3})(0.20\times10^{-3})} = 2.0\times10^4$$

Applications of Equilibrium Constants (Section 15.6)

15.41 (a) If $Q_c < K_c$, the reaction will proceed in the direction of more products, to the right.

(b) If $Q_c = K_c$, the system is in equilibrium; the concentrations used to calculate Q must be equilibrium concentrations.

15.43 *Analyze/Plan*. Follow the logic in Sample Exercise 15.9. We are given molarities, so we calculate Q directly and decide on the direction to equilibrium. *Solve.*

$$K_c = \frac{[CO][Cl_2]}{[COCl_2]} = 2.19\times10^{-10} \text{ at } 100\ °C$$

(a) $$Q = \frac{(3.3\times10^{-6})(6.62\times10^{-6})}{(2.00\times10^{-3})} = 1.1\times10^{-8};\ Q > K$$

The reaction will proceed left to attain equilibrium.

(b) $$Q = \frac{(1.1\times10^{-7})(2.25\times10^{-6})}{(4.50\times10^{-2})} = 5.5\times10^{-12};\ Q < K$$

The reaction will proceed right to attain equilibrium.

(c) $$Q = \frac{(1.48\times10^{-6})^2}{(0.0100)} = 2.19\times10^{-10};\ Q = K$$

The reaction is at equilibrium.

15.45 *Analyze/Plan*. We are given concentrations, so write the K_c expression and solve for $[Cl_2]$. Change molarity to partial pressure using the ideal gas equation and the definition of molarity. *Solve.*

$$K_c = \frac{[SO_2][Cl_2]}{[SO_2Cl_2]};\ [Cl_2] = \frac{K_c[SO_2Cl_2]}{[SO_2]} = \frac{(0.078)(0.108)}{0.052} = 0.16200 = 0.16\ M$$

$$PV = nRT,\ P = \frac{n}{V}RT;\ \frac{n}{V} = M;\ P = M\,RT;\ T = 100\ °C + 273 = 373\ K$$

$$P_{Cl_2} = \frac{0.16200\ \text{mol}}{L} \times \frac{0.08206\ \text{L-atm}}{\text{mol-K}} \times 373\ K = 4.959 = 5.0\ \text{atm}$$

Check. $K_c = \frac{(0.052)(0.162)}{(0.108)} = 0.078$. Our values are self-consistent.

15.47 *Analyze/Plan*. Write the equilibrium constant expression. In each case, change masses to molarities, solve for the equilibrium molarity of the desired component, and calculate mass of that substance present at equilibrium. *Solve.*

$$K_c = \frac{[Br]^2}{[Br_2]} = 1.04 \times 10^{-3}$$

$$[Br_2] = \frac{0.245 \text{ g } Br_2}{0.200 \text{ L}} \times \frac{1 \text{ mol } Br_2}{159.8 \text{ g } Br_2} = 0.007666 = 0.00767\ M$$

$$[Br] = (K_c[Br_2])^{1/2} = [(1.04 \times 10^{-3})(0.007666)]^{1/2} = 0.002824 = 0.00282\ M$$

$$\frac{0.002824 \text{ mol Br}}{\text{L}} \times 0.200 \text{ L} \times \frac{79.90 \text{ g Br}}{\text{mol}} = 0.0451 \text{ g Br(g)}$$

Check. $K_c = (0.002824)^2/(0.007666) = 1.04 \times 10^{-3}$

15.49 *Analyze/Plan*. Write the equilibrium constant expression. In each case, change masses to molarities, solve for the equilibrium molarity of the desired component, and calculate mass of that substance present at equilibrium. *Solve.*

$$K_c = \frac{[I]^2}{[I_2]} = 3.1 \times 10^{-5}$$

$$[I] = \frac{2.67 \times 10^{-2} \text{ g I}}{10.0 \text{ L}} \times \frac{1 \text{ mol I}}{126.9 \text{ g I}} = 2.1040 \times 10^{-5} = 2.10 \times 10^{-5}\ M$$

$$[I_2] = \frac{[I]^2}{K_c} = \frac{(2.104 \times 10^{-5})^2}{3.1 \times 10^{-5}} = 1.428 \times 10^{-5} = 1.43 \times 10^{-5}\ M$$

$$\frac{1.428 \times 10^{-5} \text{ mol } I_2}{\text{L}} \times 10.0 \text{ L} \times \frac{253.8 \text{ g } I_2}{\text{mol } I_2} = 0.0362 \text{ g } I_2$$

Check. $K_c = \dfrac{(2.104 \times 10^{-5})^2}{1.428 \times 10^{-5}} = 3.1 \times 10^{-5}$

15.51 *Analyze/Plan*. Follow the logic in Sample Exercise 15.11. Because molarity of NO is given directly, we can construct the equilibrium table straight away. *Solve.*

$K_c = \dfrac{[N_2][O_2]}{[NO]^2} = 2.4 \times 10^3$

	$2NO(g)$ $\rightleftharpoons$	$N_2(g)$ +	$O_2(g)$
initial	0.175 *M*	0	0
change	−2x	+x	+x
equil.	0.175 − 2x	+x	+x

$$2.4 \times 10^3 = \frac{x^2}{(0.175 - 2x)^2};\ (2.4 \times 10^3)^{1/2} = \frac{x}{0.175 - 2x}$$

$x = (2.4 \times 10^3)^{1/2}(0.175 - 2x)$; $x = 8.573 - 97.98x$; $98.98x = 8.573$, $x = 0.08662 = 0.087\ M$

$[N_2] = [O_2] = 0.087\ M$; $[NO] = 0.175 - 2(0.08662) = 0.00177 = 0.002\ M$

Check. $K_c = (0.08662)^2/(0.00177)^2 = 2.4 \times 10^3$

15.53 *Analyze/Plan*. Write the K_p expression, substitute the stated pressure relationship, and solve for P_{Br_2}. *Solve.*

$$K_p = \frac{P^2_{NO} \times P_{Br_2}}{P^2_{NOBr}}$$

When $P_{NOBr} = P_{NO}$, these terms cancel and $P_{Br_2} = K_p = 0.416$ atm. This is true for all cases where $P_{NOBr} = P_{NO}$.

15.55 (a) $CaSO_4(s) \rightleftharpoons Ca^{2+}(aq) + SO_4^{2-}(aq)$ $K_c = [Ca^{2+}][SO_4^{2-}] = 2.4 \times 10^{-5}$

At equilibrium, $[Ca^{2+}] = [SO_4^{2-}] = x$

$K_c = 2.4 \times 10^{-5} = x^2$; $x = 4.9 \times 10^{-3}$ M Ca^{2+} and SO_4^{2-}

(b) A saturated solution of $CaSO_4(aq)$ is 4.9×10^{-3} M.

1.4 L of this solution contain:

$$\frac{4.9 \times 10^{-3} \text{ mol}}{\text{L}} \times 1.4 \text{ L} \times \frac{136.14 \text{ g CaSO}_4}{\text{mol}} = 0.9337 = 0.94 \text{ g CaSO}_4$$

A bit more than 1.0 g $CaSO_4$ is needed to have some undissolved $CaSO_4(s)$ in equilibrium with 1.4 L of saturated solution.

15.57 *Analyze/Plan*. Follow the approach in Solution 15.51. Calculate [IBr] from mol IBr and construct the equilibrium table.

Solve. [IBr] = 0.500 mol/2.00 L = 0.250 M

Because no I_2 or Br_2 was present initially, the amounts present at equilibrium are produced by the reverse reaction and stoichiometrically equal. Let these amounts equal x. The amount of HBr that reacts is then 2x. Substitute the equilibrium molarities (in terms of x) into the equilibrium expression and solve for x.

$$K_c = \frac{[IBr]^2}{[I_2][Br_2]} = 280$$

	I_2 +	Br_2 $\rightleftharpoons$	2IBr
initial	0 M	0 M	0.250 M
change	+x M	+x M	–2x M
equil.	x M	x M	(0.250 – 2x) M

$K_c = 280 = \frac{(0.250 - 2x)^2}{x^2}$; taking the square root of both sides

$16.733 = \frac{0.250 - 2x}{x}$; $16.733x + 2x = 0.250$; $18.733x = 0.250$

$x = 0.013345 = 0.0133$ M; $[I_2] = [Br_2] = 0.0133$ M

$[IBr] = 0.250 - 2(0.013345) = 0.2233 = 0.223$ M

Check. $\frac{(0.2233)^2}{(0.013345)^2} = 280.$ Our values are self-consistent.

15.59 *Analyze/Plan.* Follow the logic in sample Exercise 15.11, using torr in place of *M*. We are given K_p, so we use pressure in torr directly in the equilibrium expression.

	$CH_4(g)$	+	$I_2(g)$	$\rightleftharpoons$	$CH_3I(g)$ +	$HI(g)$
initial	105.1 torr		7.96 torr		0 torr	0 torr
change	–x torr		–x torr		+x torr	+x torr
equil.	105.1–x torr		7.96–x torr		+x torr	+x torr

$$K_p = 2.26\times10^{-4} = \frac{x^2}{(105.1-x)(7.96-x)};\ x^2 = 2.26\times10^{-4}(836.6-113.1x+x^2)$$

$$0.999774\,x^2 + 0.02555\,x - 0.18907 = 0;\ x = \frac{-0.02555 \pm \sqrt{(0.02555)^2 - 4(0.999774)(-0.18907)}}{2(0.999774)}$$

x = 0.422 torr (The negative solution is not chemically meaningful.)

at equilibirum: $P_{CH_3I} = P_{HI} = 0.422$ torr; $P_{CH_4} = 104.7$ torr; $P_{I_2} = 7.54$ torr

LeChâtelier's Principle (Section 15.7)

15.61 *Analyze/Plan.* Follow the logic in Sample Exercise 15.12. *Solve.*

(a) Shift equilibrium to the right; more $SO_3(g)$ is formed, the amount of $SO_2(g)$ decreases.

(b) Heating an exothermic reaction decreases the value of K. More SO_2 and O_2 will form, the amount of SO_3 will decrease. This is fundamentally different than shifting the relative amounts of reactants and products to maintain K; here, the equilibrium position itself changes.

(c) Because $\Delta n = -1$, a change in volume will affect the equilibrium position and favor the side with more moles of gas. The amounts of SO_2 and O_2 increase and the amount of SO_3 decreases; equilibrium shifts to the left.

(d) No effect. Speeds up the forward and reverse reactions equally.

(e) No effect. The noble gas does not appear in the equilibrium expression; the partial pressures of reactants and products do not change upon addition of a noble gas.

(f) Shift equilibrium to the right; amounts of SO_2 and O_2 decrease.

15.63 *Analyze/Plan.* Given certain changes to a reaction system, determine the effect on K_p, if any. Only changes in temperature cause changes to the value of K_p. *Solve.*

(a) no effect (b) no effect (c) no effect

(d) increase equilibrium constant (e) no effect

15.65 *Analyze/Plan.* Use Hess's law, $\Delta H° = \Sigma\Delta H_f^\circ$ products $-\Sigma\Delta H_f^\circ$ reactants, to calculate $\Delta H°$. According to the sign of $\Delta H°$, describe the effect of temperature on the value of K. According to the value of Δn, describe the effect of changes to container volume. *Solve.*

(a) $\Delta H° = \Delta H_f^\circ\ NO_2(g) + \Delta H_f^\circ\ N_2O(g) - 3\Delta H_f^\circ\ NO(g)$

$\Delta H° = 33.84\ kJ + 81.6\ kJ - 3(90.37\ kJ) = -155.7\ kJ$

(b) Because the reaction is exothermic, the equilibrium constant will decrease with increasing temperature.

(c) A change in volume at constant temperature will affect the fraction of products in the equilibrium mixture because Δn does not equal zero. An increase in container volume would favor reactants, whereas a decrease in volume would favor products.

15.67 For this reaction, there are more moles of product gas than moles of reactant gas. An increase in total pressure increases the partial pressure of each gas, shifting the equilibrium toward reactants. An increase in pressure favors formation of ozone.

15.69 (a) Endothermic. Bond breaking is always an endothermic process.

(b) The equilibrium constant increases. For an endothermic reaction, heat is a "reactant." An increase in temperature and heat favors the forward reaction and the value of K_c increases.

(c) The forward rate constant increases by a larger amount than the reverse rate constant. If $K_c = k_f/k_r$ and the value of K_c increases, the value of k_f must increase by a greater amount than the value of k_r.

Additional Exercises

15.73 $CH_4(g) + H_2O(g) \rightarrow CO(g) + 3\,H_2(g)$

$$K_p = \frac{P_{CO} \times P_{H_2}^3}{P_{CH_4} \times P_{H_2O}};\ P = \frac{g\,RT}{MM\,V};\ T = 1000\ K$$

$$P_{CO} = \frac{8.62\ g}{28.01\ g/mol} \times \frac{0.08206\ L\text{-}atm}{mol\text{-}K} \times \frac{1000\ K}{5.00\ L} = 5.0507 = 5.05\ atm$$

$$P_{H_2} = \frac{2.60\ g}{2.016\ g/mol} \times \frac{0.08206\ L\text{-}atm}{mol\text{-}K} \times \frac{1000\ K}{5.00\ L} = 21.1663 = 21.2\ atm$$

$$P_{CH_4} = \frac{43.0\ g}{16.04\ g/mol} \times \frac{0.08206\ L\text{-}atm}{mol\text{-}K} \times \frac{1000\ K}{5.00\ L} = 43.9973 = 44.0\ atm$$

$$P_{H_2O} = \frac{48.4\ g}{18.02\ g/mol} \times \frac{0.08206\ L\text{-}atm}{mol\text{-}K} \times \frac{1000\ K}{5.00\ L} = 44.0811 = 44.1\ atm$$

$$K_p = \frac{(5.0507)\,(21.1663)^3}{(43.9973)\,(44.0811)} = 24.6949 = 24.7$$

$$K_p = K_c(RT)^{\Delta n},\ K_c = K_p/(RT)^{\Delta n};\ \Delta n = 4 - 2 = 2$$

$$K_c = (24.6949)/[(0.08206)(1000)]^2 = 3.6673 \times 10^{-3} = 3.67 \times 10^{-3}$$

15.76 (a) $K_p = \dfrac{P_{Br_2} \times P_{NO}^2}{P_{NOBr}^2};\ P = \dfrac{gRT}{MM \times V};\ T = 100\ °C + 273 = 373\ K$

$$P_{Br_2} = \frac{4.19\,g}{159.8\ g/mol} \times \frac{0.08206\ L\text{-}atm}{mol\text{-}K} \times \frac{373}{5.00\ L} = 0.16051 = 0.161\ atm$$

$$P_{NO} = \frac{3.08\,g}{30.01\ g/mol} \times \frac{0.08206\ L\text{-}atm}{mol\text{-}K} \times \frac{373}{5.00\ L} = 0.62828 = 0.628\ atm$$

$$P_{NOBr} = \frac{3.22\,g\,NOBr}{109.9\,g/mol} \times \frac{0.08206\text{ L-atm}}{\text{mol-K}} \times \frac{373}{5.00\,L} = 0.17936 = 0.179\text{ atm}$$

$$K_p = \frac{(0.16051)(0.62828)^2}{(0.17936)^2} = 1.9695 = 1.97 \qquad K_p = K_c(RT)^{\Delta n}, \Delta n = 3-2 = 1$$

$$K_c = K_p/RT = 1.9695/(0.08206)(373) = 0.064345 = 0.0643$$

(b) $P_t = P_{Br_2} + P_{NO} + P_{NOBr} = 0.16051 + 0.62828 + 0.17936 = 0.96815 = 0.968\text{ atm}$

(c) All NO and Br_2 present at equilibrium came from the decomposition of the original NOBr. The mass of original NOBr is the sum of the masses of all compounds at equilibrium.

Original g NOBr = 4.19 gBr_2 + 3.08 g NO + 3.22 g NOBr = 10.49 g

15.79

	$2IBr$	$\rightleftharpoons$	I_2 +	Br_2
initial	0.025 atm		0	0
change	–2x		x	x
equil.	(0.025 – 2 x) atm		x	x

$$K_p = 8.5\times10^{-3} = \frac{P_{I_2} \times P_{Br_2}}{P^2_{IBr}} = \frac{x^2}{(0.025-2x)^2};\ \text{taking the square root of both sides}$$

$$\frac{x}{0.025-2x} = (8.5\times10^{-3})^{1/2} = 0.0922;\ x = 0.0922(0.025-2x)$$

$x + 0.184\,x = 0.002305;\ 1.184\,x = 0.002305;\ x = 0.001947 = 1.9 \times 10^{-3}$

At equilibrium, $P_{I_2} = P_{Br_2} = x = 1.9\times10^{-3}$ atm

P_{IBr} at equilibrium = $0.025 - 2(1.947 \times 10^{-3}) = 0.02111 = 0.021$ atm

Check. $K_p = (0.001947)^2/(0.02111)^2 = 8.5 \times 10^{-3}$; the calculated concentrations are self-consistent.

15.82 Initial $P_{SO_3} = \dfrac{gRT}{MM\ V} = \dfrac{0.831\,g}{80.07\,g/mol} \times \dfrac{0.08206\text{ L-atm}}{\text{mol-K}} \times \dfrac{1100\,K}{1.00\,L} = 0.9368 = 0.937$ atm

	$2\,SO_3$	$\rightleftharpoons$	$2\,SO_2$	+	O_2
initial	0.9368 atm		0		0
change	–2 x		+2x		+x
equil.	0.9368–2 x		2x		x
[equil.]	0.2104 atm		0.7264 atm		0.3632 atm

$P_t = (0.9368 - 2x) + 2x + x;\ 0.9368 + x = 1.300\text{ atm};\ x = 1.300 - 0.9368 = 0.3632 = 0.363\text{ atm}$

$$K_p = \frac{P^2_{SO_2} \times P_{O_2}}{P^2_{SO_3}} = \frac{(0.7264)^2\,(0.3632)}{(0.2104)^2} = 4.3292 = 4.33$$

$K_p = K_c(RT)^{\Delta n};\ \Delta n = 3 - 2 = 1;\ K_p = K_c(RT)$

$K_c = K_p/RT = 4.3292/[(0.08206)(1100)] = 0.04796 = 0.0480$

15.85

	$CO_2(g)$	+ $H_2(g)$	$\rightleftharpoons$ $CO(g)$	+ $H_2O(g)$
initial	1.50 mol	1.50 mol	0	0
change	−x	−x	+x	+x
equil.	(1.50 – x)mol	(1.50 – x)mol	x	x

Since $\Delta n = 0$, the volume terms cancel and we can use moles in place of molarity in the K expression.

$$K_c = 0.802 = \frac{[CO][H_2O]}{[CO_2][H_2O]} = \frac{x^2}{(1.50 - x)^2}$$

Take the square root of both sides.

$(0.802)^{1/2} = x/(1.50 - x)$; $0.8955(1.50 - x) = x$

$1.3433 = 1.8955x$, $x = 0.7087 = 0.709$ mol

$[CO] = [H_2O] = 0.7087\ \text{mol}/3.00\ \text{L} = 0.236\ M$

$[CO_2] = [H_2] = (1.50 - 0.709)\text{mol}/3.00\ \text{L} = 0.264\ M$

15.93 $K_p = \frac{P_{O_2} \times P_{CO}^2}{P_{CO_2}^2} \approx 1 \times 10^{-13}$; $P_{O_2} = (0.03)(1\ \text{atm}) = 0.03\ \text{atm}$

$P_{CO} = (0.002)(1\ \text{atm}) = 0.002\ \text{atm}$; $P_{CO_2} = (0.12)(1\ \text{atm}) = 0.12\ \text{atm}$

$$Q = \frac{(0.03)(0.002)^2}{(0.12)^2} = 8.3 \times 10^{-6} = 8 \times 10^{-6}$$

Because $Q > K_p$, the system will shift to the left to attain equilibrium. Thus, a catalyst that promoted the attainment of equilibrium would result in a lower CO content in the exhaust.

Integrative Exercises

15.95 Calculate the initial $[IO_4^-]$, and then construct an equilibrium table to determine $[H_4IO_6^-]$ at equilibrium.

$M_c \times V_c = M_d \times L_d$; $\frac{0.905\ M \times 25.0\ \text{mL}}{500.0\ \text{mL}} = M_d = 0.04525 = 0.0453\ M\ IO_4^-$

	$IO_4^-(aq)$ + $2H_2O(l)$	$\rightleftharpoons$ $H_4IO_6^-(aq)$
initial	0.0453 M	0
change	−x	+x
equil.	0.0453 − x	+x

$$K_c = 3.5 \times 10^{-2} = \frac{[H_4IO_6^-]}{[IO_4^-]} = \frac{x}{(0.0453 - x)}$$

Because K_c is relatively large and $[IO_4^-]$ is relatively small, we cannot assume x is small relative to 0.0453.

$0.035(0.04525 - x) = x$; $0.001584 - 0.035\ x = x$; $0.001584 = 1.035\ x$

$x = 0.001584/1.035 = 0.001530 = 0.0015\ M\ H_4IO_6^-$ at equilibrium

16 Acid–Base Equilibria

Visualizing Concepts

16.1 *Analyze*. From the structures decide which reactant fits the description of a Brønsted–Lowry (B-L) acid, a B-L base, a Lewis acid, and a Lewis base. *Plan*. A B-L acid is an H^+ donor, and a B-L base is an H^+ acceptor. A Lewis acid is an electron pair acceptor, and a Lewis base is an electron pair donor. *Solve*.

(a) HCl is a B-L acid, because it donates H^+ during reaction. NH_3 is a B-L base, because it accepts H^+ during reaction.

(b) By virtue of its unshared electron pair, NH_3 is the electron pair donor and a Lewis base. HCl is the electron pair acceptor and a Lewis acid.

16.5 *Plan*. Strong acids are completely ionized. The acid that is least ionized is the weakest and has the smallest K_a value. At equal concentrations, the weakest acid has the smallest $[H^+]$ and highest pH. *Solve*.

(a) HY is a strong acid. There are no neutral HY molecules in the solution, only H^+ cations and Y^- anions.

(b) HX has the smallest K_a value. It has the most neutral acid molecules and fewest ions.

(c) HX has the fewest H^+ ions and, therefore, the highest pH.

16.11 *Plan*. Evaluate the structures to determine if the molecules are binary acids or oxyacids. Consider the trends in acid strength for both classes of acids. *Solve*.

(a) If X is the same atom on both molecules, the molecule (b) is more acidic. The carboxylate anion, the conjugate base of this carboxylic acid, is stabilized by resonance, whereas the conjugate base of (a) is not resonance-stabilized. Stabilization of the conjugate base causes the ionization equilibrium to favor products, and (b) is the stronger acid.

(b) Increasing the electronegativity of X increases the strength of both acids. As X becomes more electronegative and attracts more electron density, the O–H bond becomes weaker and more polar. This increases the likelihood of ionization, which increases acid strength. An electronegative X group also stabilizes the anionic conjugate bases by delocalizing the negative charge. This causes the ionization equilibrium to favor products, and the values of K_a to increase.

Arrhenius and Brønsted–Lowry Acids and Bases (Sections 16.1 and 16.2)

16.13 $HCl(g) + NH_3(g) \rightarrow NH_4^+Cl^-(s)$. HCl is the B-L (Brønsted–Lowry) acid; it donates an H^+ to NH_3 to form NH_4^+. NH_3 is the B-L base; it accepts the H^+ from HCl.

16.15 *Analyze/Plan.* Follow the logic in Sample Exercise 16.1. A conjugate base has one less H^+ than its conjugate acid. A conjugate acid has one more H^+ than its conjugate base. *Solve.*

(a) (i) IO_3^- (ii) NH_3

(b) (i) OH^- (ii) H_3PO_4

16.17 *Analyze/Plan.* Use the definitions of B-L acids and bases, and conjugate acids and bases to make the designations. Evaluate the changes going from reactant to product to inform your choices. *Solve.*

	B-L Acid	+	B-L Base	⇌	Conjugate Acid	+	Conjugate Base
(a)	$NH_4^+(aq)$		$CN^-(aq)$		$HCN(aq)$		$NH_3(aq)$
(b)	$H_2O(l)$		$(CH_3)_3N(aq)$		$(CH_3)_3NH^+(aq)$		$OH^-(aq)$
(c)	$HCOOH(aq)$		$PO_4^{3-}(aq)$		$HPO_4^{2-}(aq)$		$HCOO^-(aq)$

16.19 *Analyze/Plan.* Follow the logic in Sample Exercise 16.2. *Solve.*

(a) Acid: $HSO_3^-(aq) + H_2O(l) \rightleftharpoons SO_3^{2-}(aq) + H_3O^+(aq)$

B-L acid, B-L base, conj. base, conj. acid

Base: $HSO_3^-(aq) + H_2O(l) \rightleftharpoons H_2SO_3(aq) + OH^-(aq)$

B-L base, B-L acid, conj. acid, conj. base

(b) H_2SO_3 is the conjugate acid of HSO_3^-.

SO_3^{2-} is the conjugate base of HSO_3^-.

16.21 *Analyze/Plan.* Based on the chemical formula, decide whether the base is strong, weak, or negligible. Is it the conjugate of a strong acid (negligible base), weak acid (weak base), or negligible acid (strong base)? Also check Figure 16.4. To write the formula of the conjugate acid, add a single H and increase the particle charge by one. *Solve.*

(a) CH_3COO^-, weak base; CH_3COOH, weak acid

(b) HCO_3^-, weak base; H_2CO_3, weak acid

(c) O^{2-}, strong base; OH^-, negligible acid

(d) Cl^-, negligible base; HCl, strong acid

(e) NH_3, weak base; NH_4^+, weak acid

16.23 *Analyze/Plan.* Based on the chemical formula, determine the strength of acids and bases by checking the known strong acids (Section 16.5). Recall the paradigm "The stronger the acid, the weaker its conjugate base, and vice versa." *Solve.*

(a) HBr. It is one of the seven strong acids (Section 16.5).

(b) F^-. HCl is a stronger acid than HF, so F^- is the stronger conjugate base.

16.25 *Analyze/Plan.* Acid–base equilibria favor formation of the weaker acid and base. Compare the relative strengths of the substances acting as acids on opposite sides of the reaction arrow. (Bases can also be compared; the conclusion should be the same.) *Solve.*

	Base	+	Acid	⇌	Conjugate Acid	+	Conjugate Base
(a)	$O^{2-}(aq)$	+	$H_2O(l)$	⇌	$OH^-(aq)$	+	$OH^-(aq)$
(b)	$HS^-(aq)$	+	$CH_3COOH(aq)$	⇌	$H_2S(aq)$	+	$CH_3COO^-(aq)$
(c)	$NO_2^-(aq)$	+	$H_2O(l)$	⇌	$HNO_2(aq)$	+	$OH^-(aq)$

(a) H_2O is a stronger acid than OH^-, so the equilibrium lies to the right.

(b) CH_3COOH is a stronger acid than H_2S, so the equilibrium lies to the right.

(c) HNO_2 is a stronger acid than H_2O, so the equilibrium lies to the left.

Autoionization of Water (Section 16.3)

16.27 Statement (ii) is correct. In pure water, the only source of H^+ is the autoionization reaction, which produces equal concentrations of H^+ and OH^-. As the temperature of water changes, the value of K_w changes, and the pH at which $[H^+] = [OH^-]$ changes.

16.29 *Analyze/Plan.* Follow the logic in Sample Exercise 16.5. In pure water at 25 °C, $[H^+] = [OH^-] = 1 \times 10^{-7}$ *M*. If $[H^+] > 1 \times 10^{-7}$ *M*, the solution is acidic; if $[H^+] < 1 \times 10^{-7}$ *M*, the solution is basic. *Solve.*

(a) $[H^+] = \dfrac{K_w}{[OH^-]} = \dfrac{1.0\times10^{-14}}{4.5\times10^{-4}\ M} = 2.2\times10^{-11}\ M < 1\times10^{-7}\ M$; basic

(b) $[H^+] = \dfrac{K_w}{[OH^-]} = \dfrac{1.0\times10^{-14}}{8.8\times10^{-9}\ M} = 1.1\times10^{-6}\ M > 1\times10^{-7}\ M$; acidic

(c) $[OH^-] = 100[H^+]$; $K_w = [H^+] \times 100[H^+] = 100[H^+]^2$;

$[H^+] = (K_w/100)^{1/2} = 1.0 \times 10^{-8}\ M < 1 \times 10^{-7}\ M$; basic

16.31 *Analyze/Plan.* Follow the logic in Sample Exercise 16.4. Note that the value of the equilibrium constant (in this case, K_w) changes with temperature. *Solve.*

At 0 °C, $K_w = 1.2 \times 10^{-15} = [H^+][OH^-]$

In pure water, $[H^+] = [OH^-]$; $1.2 \times 10^{-15} = [H^+]^2$; $[H^+] = (1.2 \times 10^{-15})^{1/2}$

$[H^+] = [OH^-] = 3.5 \times 10^{-8}\ M$

The pH Scale (Section 16.4)

16.33 *Analyze/Plan.* A change of one pH unit (in either direction) is:

$\Delta pH = pH_2 - pH_1 = -(\log[H^+]_2 - \log[H^+]_1) = -\log\dfrac{[H^+]_2}{[H^+]_1} = \pm 1$. The antilog of +1 is 10; the antilog of –1 is 1×10^{-1}. Thus, a ΔpH of one unit represents an increase or decrease in $[H^+]$ by a factor of 10. *Solve.*

(a) $\Delta pH = \pm 2.00$ is a change of $10^{2.00}$; $[H^+]$ changes by a factor of 100.

(b) $\Delta pH = \pm 0.5$ is a change of $10^{0.50}$; $[H^+]$ changes by a factor of 3.2.

16.35 *Analyze/Plan*. At 25 °C, $[H^+][OH^-] = 1 \times 10^{-14}$; pH + pOH = 14. Use these relationships to complete the table. If pH < 7, the solution is acidic; if pH > 7, the solution is basic. *Solve*.

$[H^+]$	$[OH^-]$	pH	pOH	acidic or basic
7.5×10^{-3} *M*	1.3×10^{-12} *M*	2.12	11.88	acidic
2.8×10^{-5} *M*	3.6×10^{-10} *M*	4.56	9.44	acidic
5.6×10^{-9} *M*	1.8×10^{-6} *M*	8.25	5.75	basic
5.0×10^{-9} *M*	2.0×10^{-6} *M*	8.30	5.70	basic

Check. pH + pOH = 14; $[H^+][OH^-] = 1 \times 10^{-14}$

16.37 *Analyze/Plan*. Based on the pH and a new value of the equilibrium constant K_w, calculate equilibrium concentrations of $H^+(aq)$ and $OH^-(aq)$. The definition of pH remains $pH = -\log[H^+]$. *Solve*.

$pH = 7.40; [H^+] = 10^{-pH} = 10^{-7.40} = 4.0 \times 10^{-8}\ M$

$K_w = 2.4 \times 10^{-14} = [H^+][OH^-]; [OH^-] = 2.4 \times 10^{-14} / [H^+]$

$[OH^-] = 2.4 \times 10^{-14} / 4.0 \times 10^{-8} = 6.0 \times 10^{-7}\ M; pOH = -\log(6.0 \times 10^{-7}) = 6.22$

Alternately, $pH + pOH = pK_w$. At 37 °C, $pH + pOH = -\log(2.4 \times 10^{-14})$

$pH + pOH = 13.62; pOH = 13.62 - 7.40 = 6.22$

$[OH^-] = 10^{-pOH} = 10^{-6.22} = 6.0 \times 10^{-7}\ M$

16.39 *Analyze/Plan*. We are given the behavior of an unknown solution in two indicators. Use information from Figure 16.8 to answer the questions about the solution. *Solve*.

(a) Acidic. Both indicators change color at pH less than 7.

(b) The range of possible integer pH values for the solution is 4 to 6. Methyl orange changes color near pH 4 and bromthymol (or bromothymol) blue changes near pH 6.

(c) Methyl red changes color slightly above pH 5. It would help narrow the range of possible pH values.

Strong Acids and Bases (Section 16.5)

16.41 (a) True.

(b) True.

(c) False. A 1.0 *M* strong acid solution has a pH of 0.

16.43 *Analyze/Plan*. Follow the logic in Sample Exercise 16.8. Strong acids are completely ionized, so $[H^+]$ = original acid concentration and $pH = -\log[H^+]$. For the solutions obtained by dilution, use the "dilution" formula, $M_1V_1 = M_2V_2$, to calculate molarity of the acid. *Solve*.

(a) $8.5 \times 10^{-3}\ M\ HBr = 8.5 \times 10^{-3}\ M\ H^+; pH = -\log(8.5 \times 10^{-3}) = 2.07$

(b) $$\frac{1.52\ g\ HNO_3}{0.575\ L\ soln} \times \frac{1\ mol\ HNO_3}{63.02\ g\ HNO_3} = 0.041947 = 0.0419\ M\ HNO_3$$

$[H^+] = 0.0419\ M; pH = -\log(0.041947) = 1.377$

(c) $M_c \times V_c = M_d \times V_d; 0.250\ M \times 0.00500\ L = ?\ M \times 0.0500\ L$

$$M_d = \frac{0.250\ M \times 0.00500\ \text{L}}{0.0500\ \text{L}} = 0.0250\ M\ \text{HCl}$$

$[H^+] = 0.0250\ M$; pH = –log (0.0250) = 1.602

(d) $$[H^+]_{total} = \frac{\text{mol } H^+ \text{from HBr} + \text{mol } H^+ \text{ from HCl}}{\text{total L solution}}$$

$$[H^+]_{total} = \frac{(0.100\ M\ \text{HBr} \times 0.0100\ \text{L}) + (0.200\ M \times 0.0200\ \text{L})}{0.0300\ \text{L}}$$

$$[H^+]_{total} = \frac{1.00 \times 10^{-3}\ \text{mol } H^+ + 4.00 \times 10^{-3}\ \text{mol } H^+}{0.0300\ \text{L}} = 0.1667 = 0.167\ M$$

pH = –log (0.1667 M) = 0.778

16.45 *Analyze/Plan*. Follow the logic in Sample Exercise 16.9. Strong bases dissociate completely upon dissolving. pOH = $-\log[OH^-]$; pH = 14 – pOH. *Solve.*

(a) Pay attention to the formula of the base to get $[OH^-]$.

$[OH^-] = 2[Sr(OH)_2] = 2(1.5 \times 10^{-3}\ M) = 3.0 \times 10^{-3}\ M\ OH^-$

pOH = $-\log (3.0 \times 10^{-3})$ = 2.52; pH = 14 – pOH = 11.48

(b) mol/LiOH = g LiOH/molar mass LiOH. $[OH^-]$ = [LiOH].

$$\frac{2.250\ \text{g LiOH}}{0.2500\ \text{L soln}} \times \frac{1\ \text{mol LiOH}}{23.948\ \text{g LiOH}} = 0.37581 = 0.3758\ M\ \text{LiOH} = [OH^-]$$

pOH = –log (0.37581) = 0.4250; pH = 14 – pOH = 13.5750

(c) Use the dilution formula to get the [NaOH] = $[OH^-]$.

$M_c \times V_c = M_d \times V_d$; $0.175\ M \times 0.00100\ \text{L} = ?\ M \times 2.00\ \text{L}$

$$M_d = \frac{0.175\ M \times 0.00100\ \text{L}}{2.00\ \text{L}} = 8.75 \times 10^{-5}\ M\ \text{NaOH} = [OH^-]$$

pOH = $-\log (8.75 \times 10^{-5})$ = 4.058; pH = 14 – pOH = 9.942

(d) Consider total mol OH^- from KOH and $Ca(OH)_2$ as well as total solution volume.

$$[OH^-]_{total} = \frac{\text{mol } OH^- \text{ from KOH} + \text{mol } OH^- \text{ from } Ca(OH)_2}{\text{total L soln}}$$

$$[OH^-]_{total} = \frac{(0.105\ M \times 0.00500\ \text{L}) + 2(9.5 \times 10^{-2}\ M \times 0.0150\ \text{L})}{0.0200\ \text{L}}$$

$$[OH^-]_{total} = \frac{0.525 \times 10^{-3}\ \text{mol } OH^- + 2.85 \times 10^{-3}\ \text{mol } OH^-}{0.0200\ \text{L}} = 0.16875 = 0.17\ M$$

pOH = –log (0.16875) = 0.77; pH = 14 – pOH = 13.23

($9.5 \times 10^{-2}\ M$ has 2 sig figs, so the $[OH^-]$ has 2 sig figs and pH and pOH have 2 decimal places.)

16.47 *Analyze/Plan.* pH → pOH → $[OH^-] = [NaOH]$. *Solve.*

$pOH = 14 - pH = 14.00 - 11.50 = 2.50$

$pOH = 2.50 = -\log[OH^-]; [OH^-] = 10^{-2.50} = 3.2 \times 10^{-3}\ M$

$[OH^-] = [NaOH] = 3.2 \times 10^{-3}\ M$

Weak Acids (Section 16.6)

16.49 *Analyze/Plan.* Remember that K_a = [products]/[reactants]. If $H_2O(l)$ appears in the equilibrium reaction, it will **not** appear in the K_a expression, because it is a pure liquid. *Solve.*

(a) $HBrO_2(aq) \rightleftharpoons H^+(aq) + BrO_2^-(aq);\ K_a = \frac{[H^+][BrO_2^-]}{[HBrO_2]}$

$HBrO_2(aq) + H_2O(l) \rightleftharpoons H_3O^+(aq) + BrO_2^-(aq);\ K_a = \frac{[H_3O^+][BrO_2^-]}{[HBrO_2]}$

(b) $C_2H_5COOH(aq) \rightleftharpoons H^+(aq) + C_2H_5COO^-(aq);\ K_a = \frac{[H^+][C_2H_5COO^-]}{[C_2H_5COOH]}$

$C_2H_5COOH(aq) + H_2O(l) \rightleftharpoons H_3O^+(aq) + C_2H_5COO^-(aq);$

$$K_a = \frac{[H_3O^+][C_2H_5COO^-]}{[C_2H_5COOH]}$$

16.51 *Analyze/Plan.* Follow the logic in Sample Exercise 16.10. *Solve.*

$CH_3CH(OH)COH \rightleftharpoons H^+(aq) + CH_3CH(OH)COO^-(aq);\ K_a = \frac{[H^+][CH_3CH(OH)COO^-]}{[CH_3CH(OH)COOH]}$

$[H^+] = [CH_3CH(OH)COO^-] = 10^{-2.44} = 3.63 \times 10^{-3} = 3.6 \times 10^{-3}\ M$

$[CH_3CH(OH)COOH] = 0.10 - 3.63 \times 10^{-3} = 0.0964 = 0.096\ M$

$$K_a = \frac{(3.63 \times 10^{-3})^2}{(0.0964)} = 1.4 \times 10^{-4}$$

16.53 *Analyze/Plan.* Write the equilibrium reaction and the K_a expression. Use percent ionization to get equilibrium concentration of $[H^+]$, and by stoichiometry, $[X^-]$ and [HX]. Calculate K_a *Solve.*

$[H^+] = 0.110 \times [CH_2ClCOOH]_{initial} = 0.0110\ M$

	$CH_2ClCOOH(aq)$	$\rightleftharpoons$	$H^+(aq)$	+	$CH_2ClCOO^-(aq)$
initial	0.100 *M*		0		0
equil.	0.089 *M*		0.0110 *M*		0.0110 *M*

$$K_a = \frac{[H^+][CH_2ClCOO^-]}{[CH_2ClCOOH]} = \frac{(0.0110)^2}{0.089} = 1.4 \times 10^{-3}$$

16.55 *Analyze/Plan*. Write the equilibrium reaction and the K_a expression.

$[H^+] = 10^{-pH} = [CH_3COO^-]$; $[CH_3COOH] = x - [H^+]$.

Substitute into the K_a expression and solve for x. *Solve*.

$[H^+] = 10^{-pH} = 10^{-2.90} = 1.26 \times 10^{-3} = 1.3 \times 10^{-3}\ M$

$$K_a = 1.8 \times 10^{-5} = \frac{[H^+][CH_3COO^-]}{[CH_3COOH]} = \frac{(1.26 \times 10^{-3})^2}{(x - 1.26 \times 10^{-3})}$$

$1.8 \times 10^{-5}\ (x - 1.26 \times 10^{-3}) = (1.26 \times 10^{-3})^2$;

$1.8 \times 10^{-5}\ x = 1.585 \times 10^{-6} + 2.266 \times 10^{-8} = 1.608 \times 10^{-6}$;

$x = 0.08931 = 0.089\ M\ CH_3COOH$

16.57 *Analyze/Plan*. Follow the logic in Sample Exercise 16.12. Write K_a, construct the equilibrium table, solve for $x = [H^+]$, and then get equilibrium $[C_6H_5COO^-]$ and $[C_6H_5COOH]$ by substituting $[H^+]$ for x. *Solve*.

	$C_6H_5COOH(aq)$	$\rightleftharpoons$	$H^+(aq)$	+	$C_6H_5COO^-(aq)$
initial	$0.050\ M$		0		0
equil.	$(0.050 - x)\ M$		$x\ M$		$x\ M$

$$K_a = \frac{[H^+][C_6H_5COO^-]}{[C_6H_5COOH]} = \frac{x^2}{(0.050 - x)} \approx \frac{x^2}{0.050} = 6.3 \times 10^{-5}$$

$x^2 = 0.050\ (6.3 \times 10^{-5})$; $x = 1.8 \times 10^{-3}\ M = [H^+] = [H_3O^+] = [C_6H_5COO^-]$

$[C_6H_5COOH] = 0.050 - 0.0018 = 0.048\ M$

Check. $\dfrac{1.8 \times 10^{-3}\ M\ H^+}{0.050\ M\ C_6H_5COOH} \times 100 = 3.6\%$ ionization; the approximation is valid

16.59 *Analyze/Plan*. Follow the logic in Sample Exercise 16.12. *Solve*.

(a)

	$C_2H_5COOH(aq)$	$\rightleftharpoons$	$H^+(aq)$	+	$C_2H_5COO^-(aq)$
initial	$0.095\ M$		0		0
equil.	$(0.095 - x)\ M$		$x\ M$		$x\ M$

$$K_a = \frac{[H^+][C_2H_5COO^-]}{[C_2H_5COOH]} = \frac{x^2}{(0.095 - x)} \approx \frac{x^2}{0.095} = 1.3 \times 10^{-5}$$

$x^2 = 0.095(1.3 \times 10^{-5})$; $x = 1.111 \times 10^{-3} = 1.1 \times 10^{-3}\ M\ H^+$; pH = 2.95

Check. $\dfrac{1.1 \times 10^{-3}\ M\ H^+}{0.095\ M\ C_2H_5COOH} \times 100 = 1.2\%$ ionization; the approximation is valid

(b) $$K_a = \frac{[H^+][CrO_4^{2-}]}{[HCrO_4^-]} = \frac{x^2}{(0.100 - x)} \approx \frac{x^2}{0.100} = 3.0 \times 10^{-7}$$

$x^2 = 0.100(3.0 \times 10^{-7})$; $x = 1.732 \times 10^{-4} = 1.7 \times 10^{-4}\ M\ H^+$

$pH = -\log(1.732 \times 10^{-4}) = 3.7614 = 3.76$

Check. $\dfrac{1.7 \times 10^{-4}\ M\ H^+}{0.100\ M\ HCrO_4^-} \times 100 = 0.17\%$ ionization; the approximation is valid

(c) Follow the logic in Sample Exercise 16.15. $pOH = -\log[OH^-]$; $pH = 14 - pOH$

	$C_5H_5N(aq) + H_2O(l)$	$\rightleftharpoons$	$C_5H_5NH^+(aq)$	+	$OH^-(aq)$
initial	0.120 *M*		0		0
equil.	(0.120 – x) *M*		x *M*		x *M*

$$K_b = \frac{[C_5H_5NH^+][OH^-]}{[C_5H_5N]} = \frac{x^2}{(0.120-x)} \approx \frac{x^2}{0.120} = 1.7\times10^{-9}$$

$x^2 = 0.120(1.7 \times 10^{-9})$; $x = 1.428 \times 10^{-5} = 1.4 \times 10^{-5}$ *M* OH^-; pH = 9.15

Check. $\frac{1.4\times10^{-5}\ M\ OH^-}{0.120\ M\ C_5H_5N}\times100 = 0.012\%$ ionization; the approximation is valid

16.61 *Analyze/Plan.* $K_a = 10^{-pKa}$. Follow the logic in Sample Exercise 16.13. *Solve.*

Let $[H^+] = [NC_7H_4SO_3^-] = z$. $K_a = \text{antilog}(-2.32) = 4.79 \times 10^{-3} = 4.8 \times 10^{-3}$.

$\frac{z^2}{0.10-z} = 4.79\times10^{-3}$. As K_a is relatively large, solve the quadratic.

$z^2 + 4.79 \times 10^{-3}\ z - 4.79 \times 10^{-4} = 0$

$$z = \frac{-4.79\times10^{-3} \pm \sqrt{(4.79\times10^{-3})^2 - 4(1)(-4.79\times10^{-4})}}{2(1)} = \frac{-4.79\times10^{-3} \pm \sqrt{1.937\times10^{-3}}}{2}$$

$z = 1.96 \times 10^{-2} = 2.0 \times 10^{-2}$ *M* H^+; $pH = -\log(1.96 \times 10^{-2}) = 1.71$

16.63 *Analyze/Plan.* Follow the logic in Sample Exercise 16.12 and 16.13. *Solve.*

(a)

	$HN_3(aq)$	$\rightleftharpoons$	$H^+(aq)$	+	$N_3^-(aq)$
initial	0.400 *M*		0		0
equil.	(0.400 – x) *M*		x *M*		x *M*

$$K_a = \frac{[H^+][N_3^-]}{[HN_3]} = 1.9\times10^{-5};\ \frac{x^2}{(0.400-x)} \approx \frac{x^2}{0.400} = 1.9\times10^{-5}$$

$$x = 0.00276 = 2.8\times10^{-3}\ M = [H^+];\ \%\ \text{ionization} = \frac{2.76\times10^{-3}}{0.400}\times100 = 0.69\%$$

(b) $1.9\times10^{-5} \approx \frac{x^2}{0.100}$; $x = 0.00138 = 1.4\times10^{-3}\ M\ H^+$

$$\%\ \text{ionization} = \frac{1.38\times10^{-3}\ M\ H^+}{0.100\ M\ HN_3}\times100 = 1.4\%$$

(c) $1.9\times10^{-5} \approx \frac{x^2}{0.0400}$; $x = 8.72\times10^{-4} = 8.7\times10^{-4}\ M\ H^+$

$$\%\ \text{ionization} = \frac{8.72\times10^{-4}\ M\ H^+}{0.0400\ M\ HN_3}\times100 = 2.2\%$$

Check. Notice that a tenfold dilution [part (a) versus part (c)] leads to a slightly more than threefold increase in percent ionization.

16.65 *Analyze/Plan.* Follow the logic in Sample Exercise 16.14. Citric acid is a triprotic acid with three K_a values that do not differ by more than 10^3. We must consider all three steps. Also, $C_6H_5O_7^{3-}$ is only produced in step 3. *Solve.*

$$H_3C_6H_5O_7(aq) \rightleftharpoons H^+(aq) + H_2C_6H_5O_7^-(aq) \qquad K_{a1} = 7.4 \times 10^{-4}$$

$$H_2C_6H_5O_7^-(aq) \rightleftharpoons H^+(aq) + HC_6H_5O_7^{2-}(aq) \qquad K_{a2} = 1.7 \times 10^{-5}$$

$$HC_6H_5O_7^{2-}(aq) \rightleftharpoons H^+(aq) + C_6H_5O_7^{3-}(aq) \qquad K_{a3} = 4.0 \times 10^{-7}$$

(a) To calculate the pH of a 0.040 *M* solution, assume initially that only the first ionization is important:

	$H_3C_6H_5O_7(aq)$	$\rightleftharpoons$	$H^+(aq)$	+	$H_2C_6H_5O_7^-(aq)$
initial	0.040 *M*		0		0
equil.	(0.040 – x) *M*		x *M*		x *M*

$$K_{a1} = \frac{[H^+][H_2C_6H_5O_7^-]}{[H_3C_6H_5O_7]} = \frac{x^2}{(0.040 - x)} = 7.4 \times 10^{-4}$$

$x^2 = (0.040 - x)(7.4 \times 10^{-4})$; $x^2 \approx (0.040)(7.4 \times 10^{-4})$; $x = 0.00544 = 5.4 \times 10^{-3}$ *M*

As this value for x is rather large in relation to 0.040, a better approximation for x can be obtained by substituting this first estimate into the expression for x^2, and then solving again for x:

$x^2 = (0.040 - x)(7.4 \times 10^{-4}) = (0.040 - 5.44 \times 10^{-3})(7.4 \times 10^{-4})$

$x^2 = 2.557 \times 10^{-5}$; $x = 5.057 \times 10^{-3} = 5.1 \times 10^{-3}$ *M*

(This is the same result obtained from the quadratic formula.)

The correction to the value of x, though not large, is significant. Does the second ionization produce a significant additional concentration of H^+?

	$H_2C_6H_5O_7^-(aq)$	$\rightleftharpoons$	$H^+(aq)$	+	$HC_6H_5O_7^{2-}(aq)$
initial	5.1×10^{-3} *M*		5.1×10^{-3} *M*		0
equil.	$(5.1 \times 10^{-3} - y)$		$(5.1 \times 10^{-3} + y)$		y

$$K_{a2} = \frac{[H^+][HC_6H_5O_7^{2-}]}{[H_2C_6H_5O_7^-]} = 1.7 \times 10^{-5};\ \frac{(5.1 \times 10^{-3} + y)(y)}{(5.1 \times 10^{-3} - y)} = 1.7 \times 10^{-5}$$

Assume that y is small relative to 5.1×10^{-3}; that is, that additional ionization of $H_2C_6H_5O_7^-$ is small, then

$$\frac{(5.1 \times 10^{-3})y}{(5.1 \times 10^{-3})} = 1.7 \times 10^{-5}\ M;\ y = 1.7 \times 10^{-5}\ M$$

This value is indeed small compared to 5.1×10^{-3} *M*; $[H^+]$ and pH are determined by the first ionization step. pH = $-\log(5.057 \times 10^{-3}) = 2.30$.

(b) Yes. We started the calculation by assuming that only the first step made a significant contribution to $[H^+]$ and pH. Calculation proved this assumption to be true. Next, we assumed $[H^+]$ from the first ionization was small relative to 0.040 *M* citric acid; this assumption was not valid. Finally, we assumed that additional ionization of $H_2C_6H_5O_7^-$ was small, which was true.

(c) The concentration of citrate ion, $[C_6H_5O_7^{3-}]$, is much less than $[H^+]$. Because the second ionization does not contribute significantly to $[H^+]$, we know that $[HC_6H_5O_7^{2-}]$ is less than $[H^+]$. The third ionization is even less extensive, so $[C_6H_5O_7^{3-}]$ is much less than $[H^+]$.

Weak Bases (Section 16.7)

16.67 (a) $HONH_3^+$

(b) When hydroxylamine acts as a base, the nitrogen atom accepts a proton.

(c) $14\ e^-$, $7\ e^-$ pairs

$$H{-}\ddot{\underset{\cdot\cdot}{O}}{-}\ddot{\underset{|}{N}}{-}H \quad \left[H{-}\ddot{\underset{\cdot\cdot}{O}}{-}\overset{H}{\underset{H}{N}}{-}H\right]^+ \quad \left[H{-}\overset{H}{\underset{\cdot\cdot}{O}}{-}\ddot{\underset{H}{N}}{-}H\right]^+$$

FC on N is +1 FC on O is +1

In neutral hydroxylamine, both O and N have zero formal charges. Nitrogen is less electronegative than oxygen and more likely to share a lone pair of electrons with an incoming (and electron-deficient) H^+. The resulting cation with the +1 formal charge on N is more stable than the one with the +1 formal charge on O.

16.69 *Analyze/Plan*. Remember that K_b = [products]/[reactants]. If $H_2O(l)$ appears in the equilibrium reaction, it will not appear in the K_b expression, because it is a pure liquid. *Solve.*

(a) $(CH_3)_2NH(aq) + H_2O(l) \rightleftharpoons (CH_3)_2NH_2^+(aq) + OH^-(aq)$; $K_b = \dfrac{[(CH_3)_2NH_2^+][OH^-]}{[(CH_3)_2NH]}$

(b) $CO_3^{2-}(aq) + H_2O(l) \rightleftharpoons HCO_3^-(aq) + OH^-(aq)$; $K_b = \dfrac{[HCO_3^-][OH^-]}{[CO_3^{2-}]}$

(c) $HCOO^-(aq) + H_2O(l) \rightleftharpoons HCOOH(aq) + OH^-(aq)$; $K_b = \dfrac{[HCOOH][OH^-]}{[HCOO^-]}$

16.71 *Analyze/Plan*. Follow the logic in Sample Exercise 16.15. *Solve.*

	$C_2H_5NH_2(aq)$ + $H_2O(l)$	$\rightleftharpoons$	$C_2H_5NH_3^+(aq)$	+	$OH^-(aq)$
initial	0.075 *M*		0		0
equil.	(0.075 – x) *M*		x *M*		x *M*

$$K_b = \frac{[C_2H_5NH_3^+][OH^-]}{[C_2H_5NH_2]} = \frac{(x)(x)}{(0.075 - x)} = \frac{x^2}{0.075} = 6.4 \times 10^{-4}$$

$x^2 = 0.075\ (6.4 \times 10^{-4})$; $x = [OH^-] = 6.9 \times 10^{-3}\ M$; pH = 11.84

Check. $\dfrac{6.9 \times 10^{-3}\ M\ OH^-}{0.075\ M\ C_2H_5NH_2} \times 100 = 9.2\%$ ionization; the assumption is not valid

To obtain a more precise result, the K_b expression is rewritten in standard quadratic form and solved via the quadratic formula.

$$\frac{x^2}{0.075-x} = 6.4\times10^{-4};\ x^2 + 6.4\times10^{-4}\ x - 4.8\times10^{-5} = 0$$

$$x = \frac{-b\pm\sqrt{b^2-4ac}}{2a} = \frac{-6.4\times10^{-4}\pm\sqrt{(6.4\times10^{-4})^2 - 4(1)(-4.8\times10^{-5})}}{2}$$

$x = 6.62\times10^{-3} = 6.6\times10^{-3}$ *M* OH^-; pOH = 2.18; pH = 14.00 – pOH = 11.82

Note that the pH values obtained using the two algebraic techniques are very similar.

16.73 *Analyze/Plan*. Based on the pH and initial concentration of base, calculate all equilibrium concentrations. pH → pOH → $[OH^-]$ at equilibrium. Construct the equilibrium table and calculate other equilibrium concentrations. Substitute into the K_b expression and calculate K_b. *Solve*.

(a) $[OH^-] = 10^{-pOH}$; pOH = 14 – pH = 14.00 – 11.33 = 2.67

$[OH^-] = 10^{-2.67} = 2.138\times10^{-3} = 2.1\times10^{-3}$ *M*

	$C_{10}H_{15}ON(aq)$ + $H_2O(l)$	⇌	$C_{10}H_{15}ONH^+(aq)$	+	$OH^-(aq)$
initial	0.035 *M*		0		0
equil.	0.033 *M*		2.1×10^{-3} *M*		2.1×10^{-3} *M*

(b) $$K_b = \frac{[C_{10}H_{15}ONH^+][OH^-]}{[C_{10}H_{15}ON]} = \frac{(2.138\times10^{-3})^2}{(0.03286)} = 1.4\times10^{-4}$$

The K_a – K_b Relationship; Acid–Base Properties of Salts (Sections 16.8 and 16.9)

16.75 *Analyze/Plan*. Refer to Equation 16.6 and Sample Exercise 16.17. *Solve*.

(a) $C_6H_5OH(aq) + H_2O(l) \rightleftharpoons H_3O^+(aq) + C_6H_5O^-(aq)$

(b) $K_b = K_w/K_a = 1.0\times10^{-14} / 1.3\times10^{-10} = 7.7\times10^{-5}$

(c) Phenol is a stronger acid than water. The benchmark for acid strength in water is 1.0×10^{-14}. All acids listed in Table D.1 of Appendix D are stronger acids than water. (The one notable exception is K_{a2} for H_2S. HS^-, the product of the first ionization of H_2S, has a K_a value of 1×10^{-19}.)

16.77 *Analyze/Plan*. Based on K_a, determine relative strengths of the acids and their conjugate bases. The greater the magnitude of K_a, the stronger the acid and the weaker the conjugate base. K_b (conjugate base) = K_w/K_a. *Solve*.

(a) Acetic acid is stronger, because it has the larger K_a value.

(b) Hypochlorite ion is the stronger base because the weaker acid, hypochlorous acid, has the stronger conjugate base.

(c) K_b for CH_3COO^- = K_w/K_a for CH_3COOH = $1.0\times10^{-14}/1.8\times10^{-5} = 5.6\times10^{-10}$

K_b for ClO^- = K_w/K_a for HClO = $1\times10^{-14}/3.0\times10^{-8} = 3.3\times10^{-7}$

Note that K_b for ClO^- is greater than K_b for CH_3COO^-.

16.79 *Analyze.* When the solute in an aqueous solution is a salt, evaluate the acid/base properties of the component ions.

(a) *Plan.* NaBrO is a soluble salt and, thus, a strong electrolyte. When it is dissolved in H_2O, it dissociates completely into Na^+ and BrO^-. [NaBrO] = $[Na^+]$ = $[BrO^-]$ = 0.10 *M*. Na^+ is the conjugate acid of the strong base NaOH and, thus, does not influence the pH of the solution. BrO^-, on the other hand, is the conjugate base of the weak acid HBrO and *does* influence the pH of the solution. Like any other weak base, it hydrolyzes water to produce $OH^-(aq)$. Solve the equilibrium problem to determine $[OH^-]$. *Solve.*

	$BrO^-(aq) + H_2O(l)$	$\rightleftharpoons$	$HBrO(aq)$	+	$OH^-(aq)$
initial	0.10 *M*		0		0
equil.	(0.10 – x) *M*		x *M*		x *M*

$$K_b \text{ for } BrO^- = \frac{[HBrO][OH^-]}{[BrO^-]} = \frac{K_w}{K_a \text{ for } HBrO} = \frac{1\times10^{-14}}{2.5\times10^{-9}} = 4.00\times10^{-6} = 4.0\times10^{-6}$$

$4.00\times10^{-6} = \dfrac{(x)(x)}{(0.10-x)}$; assume the percent of BrO^- that hydrolyzes is small

$x^2 = 0.10\,(4.00\times10^{-6})$; $x = [OH^-] = 6.32\times10^{-4} = 6.3\times10^{-4}$ *M*

pOH = 3.20; pH = 14 – 3.20 = 10.80

(b) *Plan.* $NaHS(aq) \rightarrow Na^+(aq) + HS^-(aq)$

HS^- is the conjugate base of H_2S and its hydrolysis reaction will determine the $[OH^-]$ and pH of the solution [see similar explanation for NaBrO in part (a)]. We will assume the process $HS^-(aq) \rightleftharpoons H^+(aq) + S^-(aq)$ will not significantly affect the $[OH^-]$ in solution because K_{a2} for H_2S is so small. Solve the equilibrium problem for $[OH^-]$. *Solve.*

	$HS^-(aq) + H_2O(l)$	$\rightleftharpoons$	$H_2S(aq)$	+	$OH^-(aq)$
initial	0.080 *M*		0		0
equil.	(0.080 – x) *M*		x		x

$$K_b = \frac{[H_2S][OH^-]}{[HS^-]} = \frac{K_w}{K_a \text{ for } H_2S} = \frac{1.0\times10^{-14}}{9.5\times10^{-8}} = 1.053\times10^{-7} = 1.1\times10^{-7}$$

$1.053\times10^{-7} = \dfrac{x^2}{(0.080-x)}$; $x^2 = 0.080\,(1.053\times10^{-7})$; $x = 9.177\times10^{-5} = 9.2\times10^{-5}\ M\ OH^-$

(Assume x is small compared to 0.080); pOH = 4.04; pH = 14 – 4.04 = 9.96

Check. $\dfrac{9.2\times10^{-5}\ M\ OH^-}{0.080\ M\ HS^-}\times100 = 0.12\%$ hydrolysis; the approximation is valid

(c) *Plan.* For the two salts present, Na^+ and Ca^{2+} are negligible acids. NO_2^- is the conjugate base of HNO_2 and will determine the pH of the solution. *Solve.*

Calculate total $[NO_2^-]$ present initially.

$[NO_2^-]_{total} = [NO_2^-]$ from $NaNO_2$ + $[NO_2^-]$ from $Ca(NO_2)_2$

$[NO_2^-]_{total} = 0.10\ M + 2(0.20\ M) = 0.50\ M$

The hydrolysis equilibrium is:

	$NO_2^-(aq)$ + $H_2O(l)$	⇌	$HNO_2(aq)$	+	$OH^-(aq)$
initial	0.50 M		0		0
equil.	(0.50 – x) M		x M		x M

$$K_b = \frac{[HNO_2][OH^-]}{[NO_2^-]} = \frac{K_w}{K_a \text{ for } HNO_2} = \frac{1.0\times10^{-14}}{4.5\times10^{-4}} = 2.22\times10^{-11} = 2.2\times10^{-11}$$

$$2.2 \times 10^{-11} = \frac{x^2}{(0.50-x)} \approx \frac{x^2}{0.50};\ x^2 = 0.50\,(2.22 \times 10^{-11})$$

$x = 3.33 \times 10^{-6} = 3.3 \times 10^{-6}\ M\ OH^-$; pOH = 5.48; pH = 14 – 5.48 = 8.52

16.81 *Analyze/Plan*. The salt dissociates to form Na^+ and CH_3COO^-. Na^+ is a negligible base; the hydrolysis equilibrium of CH_3COO^- determines the pH of the solution.

Solve. The hydrolysis equilibrium is:

$$CH_3COO^-(aq) + H_2O(l) \rightleftharpoons CH_3COOH(aq) + OH^-(aq)$$

$$K_b = \frac{[CH_3COOH][OH^-]}{[CH_3COO^-]} = \frac{K_w}{K_a \text{ for } CH_3COOH} = \frac{1.0\times10^{-14}}{1.8\times10^{-5}} = 5.56\times10^{-10} = 5.6\times10^{-10}$$

$[CH_3COO^-] = x$; $[OH^-] = [CH_3COOH] = 10^{-pOH}$

pOH = 14.00 – pH = 14.00 – 9.70 = 4.30; $[OH^-] = 10^{-4.30} = 5.012 \times 10^{-5} = 5.0 \times 10^{-5}$

$$K_b = 5.56\times10^{-10} = \frac{(5.012\times10^{-5})^2}{x};\quad x = 4.518 = 4.5\ M\ NaCH_3COO$$

Note that no assumption was required in this calculation.

16.83 *Analyze/Plan*. Based on the formula of a salt, predict whether an aqueous solution will be acidic, basic, or neutral. Evaluate the acid–base properties of both ions and determine the overall effect on solution pH. *Solve*.

(a) acidic; NH_4^+ is a weak acid, Br^- is negligible.

(b) acidic; Fe^{3+} is a highly charged metal cation and a Lewis acid; Cl^- is negligible.

(c) basic; CO_3^{2-} is the conjugate base of HCO_3^-; Na^+ is negligible.

(d) neutral; both K^+ and ClO_4^- are negligible.

(e) acidic; $HC_2O_4^-$ is amphoteric, but K_a for the acid dissociation (6.4×10^{-5}) is much greater than K_b for the base hydrolysis ($1.0 \times 10^{-14} / 5.9 \times 10^{-2} = 1.7 \times 10^{-13}$).

16.85 *Plan*. Estimate pH using relative base strength and then calculate to confirm prediction. NaCl is a neutral salt, so it is not the unknown. The unknown is a relatively weak base, because a pH of 8.08 is not very basic. Because F^- is a weaker base than OCl^-, the unknown is probably NaF. Calculate K_b for the unknown from the data provided. *Solve*.

$[OH^-] = 10^{-pOH}$; pOH = 14.00 – pH = 14.00 – 8.08 = 5.92

$[OH^-] = 10^{-5.92} = 1.202 \times 10^{-6} = 1.2 \times 10^{-6}\ M = [HX]$

$[NaX] = [X^-] = 0.050 \text{ mol salt}/0.500 \text{ L} = 0.10\ M$

$$K_b = \frac{[OH^-][HX]}{[X^-]} = \frac{(1.202\times10^{-6})^2}{(0.10-1.2\times10^{-6})} = \frac{(1.202\times10^{-6})^2}{0.10} = 1.4\times10^{-11}$$

K_b for $F^- = K_w/K_a$ for $HF = 1.0\times10^{-14}/6.8\times10^{-4} = 1.5\times10^{-11}$

The unknown is NaF.

Acid–Base Character and Chemical Structure (Section 16.10)

16.87 (a) HNO_3 is a stronger acid than HNO_2 because it has one more nonprotonated oxygen atom and, thus, a higher oxidation number on N.

(b) H_2S is a stronger acid than H_2O. For binary hydrides, acid strength increases going down a family.

(c) H_2SO_4 is a stronger acid than H_2SeO_4. For oxyacids, the greater the electronegativity of the central atom, the stronger the acid.

(d) CCl_3COOH is stronger than CH_3COOH because the electronegative Cl atoms withdraw electron density from other parts of the molecule, which weakens the O–H bond and makes H^+ easier to remove. Also, the electronegative Cl delocalizes negative charge on the carboxylate anion. This stabilizes the conjugate base, favoring products in the ionization equilibrium and increasing K_a.

16.89 (a) BrO^- (HClO is the stronger acid because Cl is more electronegative than Br, so BrO^- is the stronger base.)

(b) BrO^- ($HBrO_2$ has more nonprotonated O atoms and is the stronger acid, so BrO^- is the stronger base.)

(c) HPO_4^{2-} (larger negative charge, greater attraction for H^+)

16.91 (a) True.

(b) False. In a series of acids that have the same central atom, acid strength increases with the number of nonprotonated oxygen atoms bonded to the central atom.

(c) False. H_2Te is a stronger acid than H_2S because the H–Te bond is longer, weaker, and more easily ionized than the H–S bond. Binary hydride acid strength increase going down a family.

Lewis Acids and Bases (Section 16.11)

16.93 $NH_3(aq) + H_2O(l) \rightleftharpoons NH_4^+(aq) + OH^-(aq)$

Ammonia, NH_3, acts as an Arrhenius base because it increases the concentration of hydroxide ion, OH^-, in aqueous solution. It acts like a Brønsted-Lowry base because it is a proton, H^+, acceptor. It acts like a Lewis base because it is an electron pair donor. If a substance is an Arrhenius base, it must also be a Brønsted–Lowry base and a Lewis base.

16.95 *Analyze/Plan.* Identify each reactant as an electron pair donor (Lewis base) or electron pair acceptor (Lewis acid). Remember that a Brønsted–Lowry acid is necessarily a Lewis acid, and a Brønsted–Lowry base is necessarily a Lewis base (Solution 16.93). *Solve.*

	Lewis Acid	Lewis Base
(a)	$Fe(ClO_4)_3$ or Fe^{3+}	H_2O
(b)	H_2O	CN^-
(c)	BF_3	$(CH_3)_3N$
(d)	HIO	NH_2^-

16.97 (a) Cu^{2+}, higher cation charge

(b) Fe^{3+}, higher cation charge

(c) Al^{3+}, smaller cation radius, same charge

Additional Exercises

16.101 $H_3C_6H_5O_7 + CH_3NH_2 \rightarrow CH_3NH_3^+ + H_2C_6H_5O_7^-$

citric acid methylamine odorless salt

$$H_3C_6H_5O_7 \rightleftharpoons H^+ + H_2C_6H_5O_7^- \qquad K_{a1} = 7.4 \times 10^{-4}$$

$$CH_3NH_2 + H_2O \rightleftharpoons CH_3NH_3^+ + OH^- \qquad K_b = 4.4 \times 10^{-4}$$

$$H^+ + OH^- \rightleftharpoons H_2O \qquad 1/K_w = 1/1.0 \times 10^{-14}$$

$$H_3C_6H_5O_7 + CH_3NH_2 + H_2O + H^+ + OH^- \rightleftharpoons H_2C_6H_5O_7^- + CH_3NH_3^+ + H^+ + OH^- + H_2O$$

$$H_3C_6H_5O_7 + CH_3NH_2 \rightleftharpoons H_2C_6H_5O_7^- + CH_3NH_3^+$$

$$K = \frac{K_{a_1} \times K_b}{K_w} = \frac{(7.4\times10^{-4})(4.4\times10^{-4})}{1.0\times10^{-14}} = 3.256\times10^7 = 3.3\times10^7$$

16.106 Assume T = 25 °C. If $[OH^-] = 2.5 \times 10^{-9}$ *M*, pOH = 8.60 and pH = 5.40. This does not make sense (!) because NaOH is a strong base. Usually, we assume that $[H^+]$ and $[OH^-]$ from the autoionization of water do not contribute to the overall $[H^+]$ and $[OH^-]$. However, for acid or base solute concentrations less than 1×10^{-6} *M*, the autoionization of water produces significant $[H^+]$ and $[OH^-]$ and we must consider it when calculating pH.

	$H_2O(l)$	$\rightleftharpoons$	$[H^+]$	+	$[OH^-]$
initial	C		0		2.5×10^{-9} *M*
equil.	C		x		$(x + 2.5 \times 10^{-9})$ *M*

$$K_w = 1.0 \times 10^{-14} = [H^+][OH^-] = (x)(x + 2.5 \times 10^{-9}); \; x^2 + 2.5 \times 10^{-9}\,x - 1.0 \times 10^{-14} = 0$$

From the quadratic formula, $$x = \frac{-2.5\times10^{-9} \pm \sqrt{(2.5\times10^{-9})^2 - 4(1)(-1\times10^{-14})}}{2(1)}$$

$$= 9.876 \times 10^{-8} = 9.9 \times 10^{-8}\ M\ H^+$$

$[H^+] = 9.9 \times 10^{-8}$ *M*; $[OH^-] = (9.876 \times 10^{-8} + 2.5 \times 10^{-9}) = 1.013 \times 10^{-7} = 1.0 \times 10^{-7}$ *M*

pH = 7.0054 = 7.01

Check: $[9.876 \times 10^{-8}][1.013 \times 10^{-7}] = 1.0 \times 10^{-14}$. Now our answer makes sense. The very small concentration of OH^- from the solute raises the solution pH to slightly more than 7.

16.109 (a) $K_b = K_w/K_a$; $pK_b = 14 - pK_a$; $pK_b = 14 - 4.84 = 9.16$

(b) K_a for butyric acid (buCOOH) is $10^{-4.84} = 1.4454 \times 10^{-5} = 1.4 \times 10^{-5}$

$$K_a = \frac{[H^+][buCOO^-]}{[buCOOH]}; \ [H^+] = [buCOO^-] = x; \ [buCOOH] = 0.050 - x$$

$$1.4454 \times 10^{-5} = \frac{x^2}{0.050 - x}; \text{ assume x is small relative to 0.050}$$

$x^2 = 7.227 \times 10^{-7}$; $x = [H^+] = 8.501 \times 10^{-4} = 8.5 \times 10^{-4}$ *M* H^+; pH = 3.07

(This represents 1.7% ionization, so the approximation is valid.)

(c) K_b for butyrate anion ($buCOO^-$) is $10^{-9.16} = 6.918 = 6.918 \times 10^{-10} = 6.9 \times 10^{-10}$

$$K_b = \frac{[OH^-][buCOOH]}{[buCOO^-]}; \ [OH^-] = [buCOOH] = x; \ [buCOO^-] = 0.050 - x$$

$$6.918 \times 10^{-10} = \frac{x^2}{0.050 - x}; \text{ assume x is small relative to 0.050}$$

$x^2 = 3.459 \times 10^{-11}$; $x = [OH^-] = 5.881 \times 10^{-6} = 5.9 \times 10^{-6}$ *M* OH^-

pOH = 5.23; pH = 8.77

16.113 Call each compound in the neutral form Q.

Then, $Q(aq) + H_2O(l) \rightleftharpoons QH^+ (aq) + OH^-$. $K_b = [QH^+][OH^-]/[Q]$

The ratio in question is $[QH^+]/[Q]$, which equals $K_b/[OH^-]$ for each compound.
At pH = 2.5, pOH = 11.5, $[OH^-]$ = antilog (–11.5) = $3.16 \times 10^{-12} = 3 \times 10^{-12}$ *M*. Now calculate $K_b/[OH^-]$ for each compound:

Nicotine $\dfrac{[QH^+]}{[Q]} = 7 \times 10^{-7}/3.16 \times 10^{-12} = 2 \times 10^5$

Caffeine $\dfrac{[QH^+]}{[Q]} = 4 \times 10^{-14}/3.16 \times 10^{-12} = 1 \times 10^{-2}$

Strychnine $\dfrac{[QH^+]}{[Q]} = 1 \times 10^{-6}/3.16 \times 10^{-12} = 3 \times 10^5$

Quinine $\dfrac{[QH^+]}{[Q]} = 1.1 \times 10^{-6}/3.16 \times 10^{-12} = 3.5 \times 10^5$

For all the compounds except caffeine, the protonated form has a much higher concentration than the neutral form. However, for caffeine, a very weak base, the neutral form dominates.

Integrative Exercises

16.116 At 25 °C, $[H^+] = [OH^-] = 1.0 \times 10^{-7}$ *M*

$$\frac{1.0\times10^{-7}\text{ mol H}^+}{1\text{L H}_2\text{O}}\times0.0010\text{ L}\times\frac{6.022\times10^{23}\text{ H}^+\text{ ions}}{\text{mol H}^+}=6.0\times10^{13}\text{ H}^+\text{ ions}$$

16.119 *Analyze.* If pH were directly related to CO_2 concentration, this exercise would be simple. Unfortunately, we must solve the equilibrium problem for the diprotic acid H_2CO_3 to calculate $[H^+]$ and pH. We are given ppm CO_2 in the atmosphere at two different times and the pH that corresponds to one of these CO_2 levels. We are asked to find pH at the other atmospheric CO_2 level.

Plan. Assume all dissolved CO_2 is present as H_2CO_3 (aq) (Sample Exercise 16.14). $pH \rightarrow [H^+] \rightarrow [H_2CO_3]$. Although H_2CO_3 is a diprotic acid, the two K_a values differ by more than 10^3, so we can ignore the second ionization when calculating $[H_2CO_3]$. Change 380 ppm CO_2 to pressure and calculate the Henry's law constant for CO_2. Calculate the dissolved $[CO_2] = [H_2CO_3]$ at 315 ppm and then solve the K_{a1} expression for $[H^+]$ and pH. *Solve.*

(a) $H_2CO_3(aq) \rightleftharpoons H^+(aq) + HCO_3^-(aq)$

$$K_{a1}=4.3\times10^{-7}=\frac{[H^+][HCO_3^-]}{[H_2CO_3]};\ [H^+]=10^{-5.4}=3.98\times10^{-6}=4\times10^{-6}\ M$$

$[H^+] = [HCO_3^-]$; $[H_2CO_3] = x - 4 \times 10^{-6}$

$$4.3\times10^{-7}=\frac{(3.98\times10^{-6})^2}{(x-3.98\times10^{-6})};\ 4.3\times10^{-7}\ x=1.585\times10^{-11}+1.711\times10^{-12}$$

$x = 1.756 \times 10^{-11}/4.3 \times 10^{-7} = 4.084 \times 10^{-5} = 4 \times 10^{-5}$ *M* H_2CO_3

380 ppm = 380 mol $CO_2/1 \times 10^6$ mol air = 0.000380 mol % CO_2

Because of the properties of gases, mol % = pressure %. $P_{CO_2} = 0.000380$ atm. According to Equation 13.4, $S_{CO_2} = kP_{CO_2}$;

4.084×10^{-5} mol/L = k(3.80×10^{-4} atm); k = 0.1075 = 0.1 mol/L-atm.

$$\text{Forty years ago, } S_{CO_2}=\frac{0.1075\text{ mol}}{\text{L-atm}}\times3.15\times10^{-4}\text{ atm}=3.385\times10^{-5}$$

$$=3\times10^{-5}\ M$$

Now solve K_{a1} for $[H^+]$ at this $[H_2CO_3]$. $[H^+] = x$.

We cannot assume x is small, because $[H_2CO_3]$ is so low.

$4.3 \times 10^{-7} = x^2/(3.385 \times 10^{-5} - x)$; $x^2 + 4.3 \times 10^{-7}\ x - 1.456 \times 10^{-11} = 0$

$$x=\frac{-4.3\times10^{-7}\pm\sqrt{(4.3\times10^{-7})^2-4(-1.456\times10^{-11})}}{2}=\frac{-4.3\times10^{-7}+7.644\times10^{-6}}{2}$$

$= 3.607 \times 10^{-6} = 4 \times 10^{-6}$ *M* H^+; $[H^+] = 4 \times 10^{-6}$ *M*; pH = 5.443 = 5.4

(Note that, to the precision that the pH data is reported, the change in atmospheric CO_2 leads to no change in pH.)

(b) From part (a), $[H_2CO_3]$ today = 4.084×10^{-5} *M*

$$\frac{4.084\times10^{-5}\ \text{mol}\ H_2CO_3}{1\ \text{L}}\times 20.0\ \text{L} = 8.168\times10^{-4} = 8\times10^{-4}\ \text{mol}\ CO_2$$

$$V = \frac{nRT}{P} = 8.168\times10^{-4}\ \text{mol}\times\frac{298\ \text{K}}{1.0\ \text{atm}}\times\frac{0.08206\ \text{L-atm}}{\text{mol-K}} = 0.01997 = 0.02\ \text{L} = 20\ \text{mL}$$

16.123 Rx 1: $\Delta H = D(H–F) + 2D(H–O) – 3D(H–O) = D(H–F) – D(H–O)$

$\Delta H = 567\ \text{kJ} – 463\ \text{kJ} = 104\ \text{kJ}$

Rx 2: $\Delta H = D(H–Cl) + 2D(H–O) – 3D(H–O) = D(H–Cl) – D(H–O)$

$\Delta H = 431\ \text{kJ} – 463\ \text{kJ} = –32\ \text{kJ}$

The reaction involving HCl is exothermic, whereas the reaction involving HF is endothermic, owing to the smaller bond dissociation enthalpy of H–Cl. HCl is a stronger acid than HF, and the enthalpy of ionization for HCl is exothermic, whereas that of HF is endothermic. This is consistent with the trend in acid strength for binary acids with heavy atoms (X) in the same family. That is, the longer and weaker the H–X bond, the stronger the acid (and the more exothermic the ionization reaction).

16.126 (a) (i) $HCO_3^-(aq) \rightleftharpoons H^+(aq) + CO_3^{2-}(aq)$ $\quad K_1 = K_{a2}$ for $H_2CO_3 = 5.6\times10^{-11}$

$H^+(aq) + OH^-(aq) \rightleftharpoons H_2O(l)$ $\quad K_2 = 1/K_w = 1\times10^{14}$

$HCO_3^-(aq) + OH^-(aq) \rightleftharpoons CO_3^{2-}(aq) + H_2O(l)$ $\quad K = K_1 \times K_2 = 5.6\times10^{3}$

(ii) $NH_4^+(aq) \rightleftharpoons H^+(aq) + NH_3(aq)$ $\quad K_1 = K_a$ for $NH_4^+ = 5.6\times10^{-10}$

$CO_3^{2-}(aq) + H^+(aq) \rightleftharpoons HCO_3^-(aq)$ $\quad K_2 = 1/K_{a2}$ for $H_2CO_3 = 1.8\times10^{10}$

$NH_4^+(aq) + CO_3^{2-}(aq) \rightleftharpoons HCO_3^-(aq) + NH_3(aq)$ $\quad K = K_1 \times K_2 = 10$

(b) Both (i) and (ii) have $K > 1$, although $K = 10$ is not *much* greater than 1. Both could be written with a single arrow. (This is true in general when a strong acid or strong base, $H^+(aq)$ or $OH^-(aq)$, is a reactant.)

17 Additional Aspects of Aqueous Equilibria

Visualizing Concepts

17.1 *Analyze.* Given diagrams showing equilibrium mixtures of HX and X^- with different compositions, decide which has the highest pH. HX is a weak acid and X^- is its conjugate base. *Plan.* Evaluate the contents of the boxes. Use acid–base equilibrium principles to relate $[H^+]$ to box composition. *Solve.*

Use the following acid ionization equilibrium to describe the mixtures:
$HX(aq) \rightleftharpoons H^+(aq) + X^-(aq)$. Each box has 4 HX molecules, but differing amounts of X^- ions. The greater the amount of X^- (conjugate base), for the same amount of HX (weak acid), the lower the amount of H^+ and the higher the pH. The middle box, with most X^-, has least H^+ and highest pH.

17.7 *Analyze.* Given two titration curves where 0.10 *M* NaOH is the titrant, decide which represents the more concentrated acid, and which the stronger acid.

Plan. For equal volumes of acid, concentration is related to volume of titrant (0.10 *M* NaOH) at the equivalence points. To determine K_a, pH = pK_a half-way to the equivalence point.

Solve.

(a) Both acids have one ionizable hydrogen, because there is one "jump" in each titration curve. For equal volumes of acid, and the same titrant, the more concentrated acid requires a greater volume of titrant to reach equivalence. The equivalence point of the blue curve is at 25 mL NaOH and of the red curve is at 35 mL NaOH. The red acid is more concentrated.

(b) According to the Henderson–Hasselbach equation, $pH = pK_a + \log\frac{[\text{conj. base}]}{[\text{conj. acid}]}$.

At half-way to the equivalence point, [conj. acid] = [conj. base] and pH = K_a of the conjugate acid. For the blue curve, half-way is 12.5 mL NaOH. The pH at this volume is approximately 7.0. For the red curve, half-way is 17.5 mL NaOH. The pH at this volume is approximately 4.2. A pK_a of 7 corresponds to K_a of 1×10^{-7}, whereas pK_a of 4.2 corresponds to K_a of 6×10^{-5}. The red acid has the larger K_a value.

Note that the stronger acid, the one with the larger K_a value, has a larger change in pH (jump) at the equivalence point. Also note that initial acid pH was not a definitive measure of acid strength, because the acids have different starting concentrations. Both K_a values and concentration contribute to solution pH.

17 Additional Aspects of Aqueous Equilibria

Solutions to Red Exercises

17.9 *Analyze/Plan*. Common anions or cations decrease the solubility of salts. Ions that participate in acid–base or complex ion equilibria increase solubility. *Solve*.

(a) CO_2^{3-} from $BaCO_3$ reacts with H^+ from HNO_3, causing solubility of $BaCO_3$ to increase with increasing HNO_3 concentration. This behavior matches the right diagram.

(b) Extra CO_2^{3-} from Na_2CO_3 decreases the solubility of $BaCO_3$. Solubility of $BaCO_3$ decreases as $[Na_2CO_3]$ increases. This behavior matches the left diagram.

(c) $NaNO_3$ has no common ions, nor does it enter into acid–base or complex ion equilibria with Ba^{2+} or CO_3^{2-}; it does not affect the solubility of $BaCO_3$. This behavior is shown in the center diagram.

The Common-Ion Effect (Section 17.1)

17.13 Statement (a) is most correct. The common ion can be either the cation or anion of a salt. The common-ion effect applies to all ions if they are "common" to the salt in question, but not to noncommon ions. Common ions do no affect the equilibrium constant.

17.15 *Analyze/Plan*. Follow the logic in Sample Exercise 17.1. *Solve*.

(a)

	$C_2H_5COOH(aq)$	$\rightleftharpoons$	$H^+(aq)$	+	$C_2H_5COO^-(aq)$
i	0.085 *M*				0.060 *M*
c	−x		+x		+x
e	(0.085 − x) *M*		+x *M*		(0.060 + x) *M*

$$K_a = 1.3\times10^{-5} = \frac{[H^+][C_2H_5COO^-]}{[C_2H_5COOH]} = \frac{(x)(0.060+x)}{(0.085-x)}$$

Assume x is small compared to 0.060 and 0.085.

$$1.3\times10^{-5} = \frac{0.060\,x}{0.085};\ x = 1.8\times10^{-5} = [H^+],\ pH = 4.73$$

Check. Because the extent of ionization of a weak acid or base is suppressed by the presence of a conjugate salt, the 5% rule usually holds true in buffer solutions.

(b)

	$(CH_3)_3N(aq) + H_2O(l)$	$\rightleftharpoons$	$(CH_3)_3NH^+(aq)$	+	$OH^-(aq)$
i	0.075 *M*		0.10 *M*		
c	−x		+x		+x
e	(0.075 − x) *M*		(0.10 + x) *M*		+x *M*

$$K_b = 6.4\times10^{-5} = \frac{[OH^-][(CH_3)_3NH^+]}{[(CH_3)_3N]} = \frac{(x)(0.10+x)}{(0.075-x)} \approx \frac{0.10\,x}{0.075}$$

$x = 4.8\times10^{-5} = [OH^-]$, pOH = 4.32, pH = 14.00 − 4.32 = 9.68

Check. In a buffer, if [conj. acid] > [conj. base], pH < pK_a of the conj. acid. If [conj. acid] < [conj. base], pH > pK_a of the conj. acid. In this buffer, pK_a of $(CH_3)_3NH^+$ is 9.81. $[(CH_3)_3NH^+] > [(CH_3)_3N]$ and pH = 9.68, less than 9.81.

(c) mol = $M \times$ L; mol CH_3COOH = 0.15 $M \times$ 0.0500 L = 7.5×10^{-3} mol

mol CH_3COO^- = 0.20 $M \times$ 0.0500 L = 0.010 mol

	$CH_3COOH(aq)$	$\rightleftharpoons$	$H^+(aq)$	+	$CH_3COO^-(aq)$
i	7.5×10^{-3} mol		0		0.010 mol
c	−x		+x		+x
e	$(7.5 \times 10^{-3} - x)$ mol		+x mol		(0.010 + x) mol

$[CH_3COOH(aq)] = (7.5 \times 10^{-3} - x)$ mol/0.1000 L;

$[CH_3COO^-(aq)] = (0.010 + x)$ mol/0.1000 L

$$K_a = 1.8 \times 10^{-5} = \frac{[H^+][CH_3COO^-]}{[CH_3COOH]} = \frac{(x)(0.010 + x)/0.1000\text{ L}}{(0.0075 - x)/0.1000\text{ L}} \approx \frac{x(0.010)}{0.0075}$$

$x = 1.35 \times 10^{-5}$ M = 1.4×10^{-5} M H^+; pH = 4.87

Check. pK_a for CH_3COOH = 4.74. $[CH_3COO^-] > [CH_3COOH]$, pH of buffer = 4.87, greater than 4.74.

17.17 *Analyze/Plan.* We are asked to calculate % ionization of (a) a weak acid and (b) a weak acid in a solution containing a common ion, its conjugate base. Calculate % ionization as in Sample Exercise 16.13. In part (b), the concentration of the common ion is 0.085 M, not x, as in part (a). *Solve.*

$$buCOOH(aq) \rightleftharpoons H^+(aq) + buCOO^-(aq) \quad K_a = \frac{[H^+][buCOO^-]}{[buCOOH]} = 1.5 \times 10^{-5}$$

equil (a)	0.0075 − x M	x M	x M
equil (b)	0.0075 − x M	x M	0.085 + x M

(a) $$K_a = 1.5 \times 10^{-5} = \frac{x^2}{0.0075 - x} \approx \frac{x^2}{0.0075}; \; x = [H^+] = 3.354 \times 10^{-4} = 3.4 \times 10^{-4}\ M\ H^+$$

$$\%\text{ ionization} = \frac{3.4 \times 10^{-4}\ M\ H^+}{0.0075\ M\text{ buCOOH}} \times 100 = 4.5\%\text{ ionization}$$

(b) $$K_a = 1.5 \times 10^{-5} = \frac{(x)(0.085 + x)}{0.0075 - x} \approx \frac{0.085x}{0.0075}; \; x = 1.3 \times 10^{-6}\ M\ H^+$$

$$\%\text{ ionization} = \frac{1.3 \times 10^{-6}\ M\ H^+}{0.0075\ M\text{ buCOOH}} \times 100 = 0.018\%\text{ ionization}$$

Check. Percent ionization is much smaller when the "common ion" is present.

Buffered Solutions (Section 17.2)

17.19 Only solution (a) is a buffer. CH_3COOH and CH_3COONa are a weak conjugate acid/conjugate base pair that acts as a buffer; CH_3COOH reacts with added base and CH_3COO^- reacts with added acid, leaving $[H^+]$ relatively unchanged. Solution (b) contains only a weak acid, which has no capacity to react with added acid. Although solution (c) contains a conjugate acid/conjugate base pair, Cl^- is a negligible base. In general, the conjugate bases of strong acids are negligible and mixtures of strong acids and their conjugate salts do not act as buffers.

17 Additional Aspects of Aqueous Equilibria

Solutions to Red Exercises

17.21 *Analyze/Plan.* Follow the logic in Sample Exercise 17.3. Assume that % ionization is small in these buffers (Solutions 17.17 and 17.18). *Solve.*

(a) $$K_a = \frac{[H^+][CH_3CH(OH)COO^-]}{[CH_3CH(OH)COOH]};\ [H^+] = \frac{[K_a][CH_3CH(OH)COOH]}{[CH_3CH(OH)COO^-]}$$

$$[H^+] = \frac{1.4 \times 10^{-4}\ (0.12)}{(0.11)};\ [H^+] = 1.53 \times 10^{-4} = 1.5 \times 10^{-4}\ M;\ pH = 3.82$$

(b) mol = $M \times$ L; total volume = 85 mL + 95 mL = 180 mL

$$[H^+] = \frac{K_a[CH_3CH(OH)COOH]}{[CH_3CH(OH)COO^-]} = \frac{1.4 \times 10^{-4}(0.13\ M \times 0.085\ L)/0.180\ L}{(0.15\ M \times 0.095\ L)/0.180\ L}$$

$$[H^+] = \frac{1.4 \times 10^{-4}\ (0.13 \times 0.085)}{(0.15 \times 0.095)};\ [H^+] = 1.086 \times 10^{-4} = 1.1 \times 10^{-4}\ M;\ pH = 3.96$$

17.23 (a) *Analyze/Plan.* Follow the logic in Sample Exercises 17.1 and 17.3. As in Sample Exercise 17.1, start by calculating concentrations of the components. *Solve.*

$$CH_3COOH(aq) \rightleftharpoons H^+(aq) + CH_3COO^-(aq);\ K_a = 1.8 \times 10^{-5} = \frac{[H^+][CH_3COO^-]}{[CH_3COOH]}$$

$[CH_3COOH] = 0.150\ M$

$$[CH_3COO^-] = \frac{20.0\ g\ CH_3COONa}{0.500\ L\ soln} \times \frac{1\ mol\ CH_3COONa}{82.04\ g\ CH_3COONa} = 0.488\ M$$

$$[H^+] = \frac{K_a[CH_3COOH]}{[CH_3COO^-]} = \frac{1.8 \times 10^{-5}(0.150 - x)}{(0.488 + x)} \approx \frac{1.8 \times 10^{-5}(0.150)}{(0.488)}$$

$[H^+] = 5.533 \times 10^{-6} = 5.5 \times 10^{-6}\ M$, pH = 5.26

(b) *Plan.* On the left side of the equation, write all ions present in solution after HCl or NaOH is added to the buffer. Using acid–base properties and relative strengths, decide which ions will combine to form new products. *Solve.*

$$Na^+(aq) + CH_3COO^-(aq) + H^+(aq) + Cl^-(aq) \rightarrow CH_3COOH(aq) + Na^+(aq) + Cl^-(aq)$$

(c) $$CH_3COOH(aq) + Na^+(aq) + OH^-(aq) \rightarrow CH_3COO^-(aq) + H_2O(l) + Na^+(aq)$$

17.25 *Analyze/Plan.* Follow the logic in Sample Exercises 16.12 and 17.5. *Solve.*

(a) $$K_a = 6.8 \times 10^{-4} = \frac{x^2}{1.00 - x} = \frac{x^2}{1.00};\ x = [H^+] = 0.02608 = 0.026\ M;\ pH = 1.58$$

There is 2.6% ionization, so the approximation is valid.

(b) In this problem, $[F^-]$ is the unknown.

pH = 3.00, $[H^+] = 10^{-3.00} = 1.0 \times 10^{-3}$; [HF] = 1.00 – 0.0010 = 0.999 M

$$K_a = 6.8 \times 10^{-4} = \frac{1.0 \times 10^{-3}\ [F^-]}{0.999};\ [F^-] = 0.6793 = 0.68\ M$$

$$\frac{0.6793\ mol\ NaF}{1\ L} \times \frac{41.990\ g\ NaF}{1\ mol\ NaF} \times 1.25\ L = 35.654 = 36\ g\ NaF$$

17 Additional Aspects of Aqueous Equilibria

Solutions to Red Exercises

17.27 *Analyze/Plan.* Follow the logic in Sample Exercises 17.3 and 17.6. *Solve.*

(a) $K_a = \dfrac{[H^+][CH_3COO^-]}{[CH_3COOH]}; [H^+] = \dfrac{K_a[CH_3COOH]}{[CH_3COO^-]}$

$$[H^+] \approx \frac{1.8 \times 10^{-5}\,(0.10)}{(0.13)} = 1.385 \times 10^{-5} = 1.4 \times 10^{-5}\ M;\ pH = 4.86$$

(b)

$CH_3COOH(aq)$	+	$KOH(aq)$	$\rightarrow$	$CH_3COO^-(aq) + H_2O(l) + K^+(aq)$
0.10 mol		0.02 mol		0.13 mol
–0.02 mol		–0.02 mol		+0.02 mol
0.08 mol		0 mol		0.15 mol

$$[H^+] = \frac{1.8 \times 10^{-5}\,(0.08\ mol/1.00\ L)}{(0.15\ mol/1.00\ L)} = 9.60 \times 10^{-6} = 1 \times 10^{-5}\ M;\ pH = 5.02 = 5.0$$

(c)

$CH_3COO^-(aq)$	+	$HNO_3(aq)$	$\rightarrow$	$CH_3COOH(aq) + NO_3^-(aq)$
0.13 mol		0.02 mol		0.10 mol
–0.02 mol		–0.02 mol		+0.02 mol
0.11 mol		0 mol		0.12 mol

$$[H^+] = \frac{1.8 \times 10^{-5}\,(0.12\ mol/1.00\ L)}{(0.11\ mol/1.00\ L)} = 1.96 \times 10^{-5} = 2.0 \times 10^{-5}\ M;\ pH = 4.71$$

17.29 *Analyze/Plan.* Calculate the [conj. base]/[conj. acid] ratio in the H_2CO_3/HCO_3^- blood buffer. Write the acid dissociation equilibrium and K expression. Find K_a for H_2CO_3 in Appendix D.1. Calculate $[H^+]$ from the pH and solve for the ratio. *Solve.*

$$H_2CO_3(aq) \rightleftharpoons H^+(aq) + HCO_3^-(aq) \quad K_a = \frac{[H^+][HCO_3^-]}{[H_2CO_3]};\ \frac{[HCO_3^-]}{[H_2CO_3]} = \frac{K_a}{[H^+]}$$

(a) at $pH = 7.4, [H^+] = 10^{-7.4} = 4.0 \times 10^{-8}\ M;\ \dfrac{[HCO_3^-]}{[H_2CO_3]} = \dfrac{4.3 \times 10^{-7}}{4.0 \times 10^{-8}} = 11$

(b) at $pH = 7.1, [H^+] = 7.9 \times 10^{-8}\ M;\ \dfrac{[HCO_3^-]}{[H_2CO_3]} = 5.4$

17.31 *Analyze.* Given six solutions, decide which two should be used to prepare a pH 3.50 buffer. Calculate the volumes of the two 0.10 *M* solutions needed to make approximately 1 L of buffer.

Plan. A buffer must contain a conjugate acid/conjugate base (CA/CB) pair. By examining the chemical formulas, decide which pairs of solutions could be used to make a buffer. If there is more than one possible pair, calculate pK_a for the acids. A buffer is most effective when its pH is within 1 pH unit of pK_a for the conjugate acid component. Select the pair with pK_a nearest to 3.50. Use Equation 17.9 to calculate the [CB]/[CA] ratio and the volumes of 0.10 *M* solutions needed to prepare 1 L of buffer. *Solve.*

There are three CA/CB pairs:

HCOOH/HCOONa, $pK_a = 3.74$

CH_3COOH/CH_3COONa, $pK_a = 4.74$

H_3PO_4/NaH_2PO_4, $pK_a = 2.12$

The most appropriate solutions are HCOOH/HCOONa, because pK_a for HCOOH is nearest to 3.50.

$$pH = pK_a + \log\frac{[CB]}{[CA]}; \quad 3.50 = 3.7447 + \log\frac{[HCOONa]}{[HCOOH]}$$

$$\log\frac{[HCOONa]}{[HCOOH]} = -0.2447; \quad \frac{[HCOONa]}{[HCOOH]} = 0.5692 = 0.57$$

Because we are making a total of 1 L of buffer,

let y = vol HCOONa and (1 – y) = vol HCOOH.

$$0.5692 = \frac{[HCOONa]}{[HCOOH]} = \frac{(0.10\ M \times y)/1L}{[0.10\ M \times (1-y)]/1\ L}; \quad 0.5692[0.10(1-y)] = 0.10\ y;$$

$0.05692 = 0.15692\ y;\ y = 0.3627 = 0.36\ L$

360 mL of 0.10 *M* HCOONa, 640 mL of 0.10 *M* HCOOH

Check. The pH of the buffer is less than pK_a for the conjugate acid, indicating that the amount of CA in the buffer is greater than the amount of CB. This agrees with our result.

Acid–Base Titrations (Section 17.3)

17.33 (a) Curve B. The initial pH is lower and the equivalence point region is steeper.

(b) pH at the approximate equivalence point of curve A = 8.0

pH at the approximate equivalence point of curve B = 7.0

(c) Volume of base required to reach the equivalence point depends only on moles of acid present; it is independent of acid strength. Because acid B requires 40 mL and acid A requires only 30 mL, more moles of acid B are being titrated. For equal volumes of A and B, the concentration of acid B is greater.

(d) pK_a of the weak acid is approximately 4.5. In the titration of a weak acid, pH equals pK_a of the weak acid at the volume half-way to the equivalence point. On curve A, the equivalence point is at 30 mL, half-way is 15 mL, and the pH there is 4.5.

17.35 (a) False. The same volume of NaOH(aq) is required to reach the equivalence point of both titrations, because moles of acid to be titrated are the same in both flasks.

(b) True. CH_3COONa, the salt formed in the titration of CH_3COOH, produces a basic solution, whereas $NaNO_3$, formed in the titration of HNO_3, produces a neutral solution.

(c) True. Even though the pH values at the equivalence points of the two titrations are different, phenolphthalein changes color over a wide range of pH values and is appropriate for both titrations.

17 Additional Aspects of Aqueous Equilibria

17.37 *Analyze.* Given reactants, predict whether pH at the equivalence point of a titration is less than, equal to, or greater than 7.

Plan. At the equivalence point of a titration, only product is present in solution; there is no excess of either reactant. Determine the product of each reaction and whether a solution of it is acidic, basic, or neutral. *Solve.*

(a) $NaHCO_3(aq) + NaOH(aq) \rightarrow Na_2CO_3(aq) + H_2O(l)$

At the equivalence point, the major species in solution are Na^+ and CO_3^{2-}. Na^+ is negligible and CO_3^{2-} is the CB of HCO_3^-. The solution is basic, above pH 7.

(b) $NH_3(aq) + HCl(aq) \rightarrow NH_4Cl(aq)$

At the equivalence point, the major species are NH_4^+ and Cl^-. Cl^- is negligible and NH_4^+ is the CA of NH_3. The solution is acidic, below pH 7.

(c) $KOH(aq) + HBr(aq) \rightarrow KBr(aq) + H_2O(l)$

At the equivalence point, the major species are K^+ and Br^-; both are negligible. The solution is at pH 7.

17.39 The second color change, from yellow to blue near pH = 8.5, is more suitable for the titration of a weak acid with a strong base. The salt present at the equivalence point of this type of titration produces a slightly basic solution. The second color change of thymol blue is in the correct pH range to show (indicate) the equivalence point.

17.41 *Analyze/Plan.* We are asked to calculate the volume of 0.0850 *M* NaOH required to titrate various acid solutions to their equivalence point. At the equivalence point, moles base added equals moles acid initially present. Solve the stoichiometry problem, recalling that mol = $M \times$ L. In part (c), calculate molarity of HCl from g/L and proceed as outlined earlier. *Solve.*

(a) $$40.0 \text{ mL } HNO_3 \times \frac{0.0900 \text{ mol } HNO_3}{1000 \text{ mL soln}} \times \frac{1 \text{ mol NaOH}}{1 \text{ mol } HNO_3} \times \frac{1000 \text{ mL soln}}{0.0850 \text{ mol NaOH}}$$

$$= 42.353 = 42.4 \text{ mL NaOH soln}$$

(b) $$35.0 \text{ mL } CH_3COOH \times \frac{0.0850\ M\ CH_3COOH}{1000 \text{ mL soln}} \times \frac{1 \text{ mol NaOH}}{1 \text{ mol } CH_3COOH} \times \frac{1000 \text{ mL soln}}{0.0850 \text{ mol NaOH}}$$

$$= 35.0 \text{ mL NaOH soln}$$

(c) $$\frac{1.85 \text{ g HCl}}{1 \text{ L soln}} \times \frac{1 \text{ mol HCl}}{36.46 \text{ g HCl}} = 0.05074 = 0.0507\ M \text{ HCl}$$

$$50.0 \text{ mL HCl} \times \frac{0.05074 \text{ mol HCl}}{1000 \text{ mL}} \times \frac{1 \text{ mol NaOH}}{1 \text{ mol HCl}} \times \frac{1000 \text{ mL soln}}{0.0850 \text{ mol NaOH}}$$

$$= 29.847 = 29.8 \text{ mL NaOH soln}$$

17.43 *Analyze/Plan*. Follow the logic in Sample Exercise 17.7 for the titration of a strong acid with a strong base. *Solve*.

moles $H^+ = M_{HBr} \times L_{HBr} = 0.200\ M \times 0.0200\ L = 4.00 \times 10^{-3}$ mol

moles $OH^- = M_{NaOH} \times L_{NaOH} = 0.200\ M \times L_{NaOH}$

	mL_{HBr}	mL_{NaOH}	Total Volume	Moles H^+	Moles OH^-	Molarity Excess Ion	pH
(a)	20.0	15.0	35.0	4.00×10^{-3}	3.00×10^{-3}	$0.0286(H^+)$	1.544
(b)	20.0	19.9	39.9	4.00×10^{-3}	3.98×10^{-3}	$5 \times 10^{-4}(H^+)$	3.3
(c)	20.0	20.0	40.0	4.00×10^{-3}	4.00×10^{-3}	$1 \times 10^{-7}(H^+)$	7.0
(d)	20.0	20.1	40.1	4.00×10^{-3}	4.02×10^{-3}	$5 \times 10^{-4}(OH^-)$	10.7
(e)	20.0	35.0	55.0	4.00×10^{-3}	7.00×10^{-3}	$0.0545(OH^-)$	12.737

molarity of excess ion = moles ion/total vol in L

(a) $$\frac{4.00 \times 10^{-3} \text{mol H}^+ - 3.00 \times 10^{-3} \text{ mol OH}^-}{0.0350 \text{ L}} = 0.0286\ M\ H^+$$

(b) $$\frac{4.00 \times 10^{-3} \text{ mol H}^+ - 3.98 \times 10^{-3} \text{ mol OH}^-}{0.0339 \text{ L}} = 5.01 \times 10^{-4} = 5 \times 10^{-4}\ M\ H^+$$

(c) equivalence point, mol H^+ = mol OH^-

NaBr does not hydrolyze, so $[H^+] = [OH^-] = 1 \times 10^{-7}\ M$

(d) $$\frac{4.02 \times 10^{-3} \text{mol OH}^- - 4.00 \times 10^{-3} \text{mol H}^+}{0.0401 \text{ L}} = 4.99 \times 10^{-4} = 5 \times 10^{-4}\ M\ OH^-$$

(e) $$\frac{7.00 \times 10^{-3} \text{mol OH}^- - 4.00 \times 10^{-3} \text{mol H}^+}{0.0550 \text{ L}} = 0.054545 = 0.0545\ M\ OH^-$$

17.45 *Analyze/Plan*. Follow the logic in Sample Exercise 17.8 for the titration of a weak acid with a strong base. *Solve*.

(a) At 0 mL, only weak acid, CH_3COOH, is present in solution. Using the acid ionization equilibrium

	$CH_3COOH(aq)$	$\rightleftharpoons$	$H^+(aq)$	+	$CH_3COO^-(aq)$
initial	$0.150\ M$		0		0
equil.	$0.150 - x\ M$		$x\ M$		$x\ M$

$$K_a = \frac{[H^+][CH_3COO^-]}{[CH_3COOH]} = 1.8 \times 10^{-5} \text{ (Appendix D)}$$

$$1.8 \times 10^{-5} = \frac{x^2}{(0.150 - x)} \approx \frac{x^2}{0.150};\ x^2 = 2.7 \times 10^{-6};\ x = [H^+] = 0.001643$$

$$= 1.6 \times 10^{-3}\ M;\ \text{pH} = 2.78$$

(b–f) Calculate the moles of each component after the acid–base reaction takes place.
Moles CH_3COOH originally present = $M \times L = 0.150\ M \times 0.0350\ L = 5.25 \times 10^{-3}$ mol.
Moles NaOH added = $M \times L = 0.150\ M \times$ y mL.

		NaOH(aq) +	CH_3COOH (aq) →	CH_3COONa(aq) + H_2O(l)
		($0.150\ M \times 0.0175$ L) =		
(b)	before rx	2.625×10^{-3} mol	5.25×10^{-3} mol	
	after rx	**0**	**2.625×10^{-3} mol**	**2.63×10^{-3} mol**
		($0.150\ M \times 0.0345$ L) =		
(c)	before rx	5.175×10^{-3} mol	5.25×10^{-3} mol	
	after rx	**0**	**0.075×10^{-3} mol**	**5.18×10^{-3} mol**
		($0.150\ M \times 0.0350$ L) =		
(d)	before rx	5.25×10^{-3} mol	5.25×10^{-3} mol	
	after rx	**0**	**0**	**5.25×10^{-3} mol**
		($0.150\ M \times 0.0355$ L) =		
(e)	before rx	5.325×10^{-3} mol	5.25×10^{-3} mol	
	after rx	**0.075×10^{-3} mol**	**0**	**5.25×10^{-3} mol**
		($0.150\ M \times 0.0500$ L) =		
(f)	before rx	7.50×10^{-3} mol	5.25×10^{-3} mol	
	after rx	**2.25×10^{-3} mol**	**0**	**5.25×10^{-3} mol**

Calculate the molarity of each species (M = mol/L) and solve the appropriate equilibrium problem in each part.

(b) $V_T = 35.0$ mL CH_3COOH + 17.5 mL NaOH = 52.5 mL = 0.0525 L

$$[CH_3COOH] = \frac{2.625 \times 10^{-3}\ \text{mol}}{0.0525} = 0.0500\ M$$

$$[CH_3COO^-] = \frac{2.625 \times 10^{-3}\ \text{mol}}{0.0525} = 0.0500\ M$$

$$CH_3COOH(aq) \rightleftharpoons H^+(aq) + CH_3COO^-(aq)$$

equil. $0.0500 - x\ M$ $x\ M$ $0.0500 + x\ M$

$$K_a = \frac{[H^+][CH_3COO^-]}{[CH_3COOH]};\ [H^+] = \frac{K_a[CH_3COOH]}{[CH_3COO^-]}$$

$$[H^+] = \frac{1.8 \times 10^{-5}(0.0500 - x)}{(0.0500 + x)} = 1.8 \times 10^{-5}\ M\ H^+;\ pH = 4.74$$

(c)
$$[CH_3COOH] = \frac{7.5 \times 10^{-5}\ \text{mol}}{0.0695\ L} = 0.001079 = 1.1 \times 10^{-3}\ M$$

$$[CH_3COO^-] = \frac{5.175 \times 10^{-3}\ \text{mol}}{0.0695\ L} = 0.07446 = 0.074\ M$$

$$[H^+] = \frac{1.8 \times 10^{-5}\ (1.079 \times 10^{-3} - x)}{(0.07446 + x)} \approx 2.6 \times 10^{-7}\ M\ H^+;\ pH = 6.58$$

(d) At the equivalence point, only CH_3COO^- is present.

$$[CH_3COO^-] = \frac{5.25 \times 10^{-3} \text{ mol}}{0.0700 \text{ L}} = 0.0750 \, M$$

The pertinent equilibrium is the base hydrolysis of CH_3COO^-.

	$CH_3COO^-(aq) + H_2O(l)$	$\rightleftharpoons$	$CH_3COOH(aq)$ +	$OH^-(aq)$
initial	0.0750 M		0	0
equil.	0.0750 – x M		x	x

$$K_b = \frac{K_w}{K_a \text{ for } CH_3COOH} = \frac{1.0 \times 10^{-14}}{1.8 \times 10^{-5}} = 5.56 \times 10^{-10} = 5.6 \times 10^{-10} = \frac{[CH_3COOH][OH^-]}{[CH_3COO^-]}$$

$$5.56 \times 10^{-10} = \frac{x^2}{0.0750 - x}; \; x^2 \approx 5.56 \times 10^{-10}(0.0750); \; x = 6.458 \times 10^{-6}$$

$$= 6.5 \times 10^{-6} \, M \, OH^-$$

$pOH = -\log(6.458 \times 10^{-6}) = 5.19$; $pH = 14.00 - pOH = 8.81$

(e) After the equivalence point, the excess strong base determines the pOH and pH. The $[OH^-]$ from the hydrolysis of CH_3COO^- is small and can be ignored.

$$[OH^-] = \frac{0.075 \times 10^{-3} \text{ mol}}{0.0705 \text{ L}} = 1.064 \times 10^{-3} = 1.1 \times 10^{-3} \, M; \; pOH = 2.97$$

$$pH = 14.00 - 2.97 = 11.03$$

(f) $$[OH^-] = \frac{2.25 \times 10^{-3} \text{ mol}}{0.0850 \text{ L}} = 0.0265 \, M \, OH^-; \; pOH = 1.577; \; pH = 14.00 - 1.577 = 12.423$$

17.47 *Analyze/Plan.* Calculate the pH at the equivalence point for the titration of several bases with 0.200 M HBr. The volume of 0.200 M HBr required in all cases equals the volume of base and the final volume = $2V_{base}$. The concentration of the salt produced at the

equivalence point is $\dfrac{0.200 \, M \times V_{base}}{2 \, V_{base}} = 0.100 \, M$.

In each case, identify the salt present at the equivalence point, determine its acid–base properties (Section 16.9), and solve the pH problem. *Solve.*

(a) NaOH is a strong base; the salt present at the equivalence point, NaBr, does not affect the pH of the solution. 0.100 M NaBr, pH = 7.00.

(b) $HONH_2$ is a weak base, so the salt present at the equivalence point is $HONH_3^+Br^-$. This is the salt of a strong acid and a weak base, so it produces an acidic solution.

0.100 M $HONH_3^+Br^-$; $HONH_3^+(aq) \rightleftharpoons H^+(aq) + HONH_2$

	$HONH_3^+(aq)$	$H^+(aq)$	$HONH_2$
[equil.]	0.100 – x	x	x

$$K_a = \frac{[H^+][HONH_2]}{[HONH_3^+]} = \frac{K_w}{K_b} = \frac{1.0 \times 10^{-14}}{1.1 \times 10^{-8}} = 9.09 \times 10^{-7} = 9.1 \times 10^{-7}$$

Assume x is small with respect to [salt].

$K_a = x^2/0.100$; $x = [H^+] = 3.02 \times 10^{-4} = 3.0 \times 10^{-4}$ *M*, pH = 3.52

(c) $C_6H_5NH_2$ is a weak base and $C_6H_5NH_3^+Br^-$ is an acidic salt.

0.100 *M* $C_6H_5NH_3^+Br^-$. Proceeding as in (b):

$$K_a = \frac{[H^+][C_6H_5NH_2]}{[C_6H_5NH_3^+]} = \frac{K_w}{K_b} = 2.33 \times 10^{-5} = 2.3 \times 10^{-5}$$

$[H^+]^2 = 0.100(2.33 \times 10^{-5})$; $[H^+] = 1.52 \times 10^{-3} = 1.5 \times 10^{-3}$ *M*, pH = 2.82

Solubility Equilibria and Factors Affecting Solubility (Sections 17.4 and 17.5)

17.49 (a) True.

(b) False. The solubility product of a slightly soluble salt is the square of the solubility if the salt contains one cation and one anion.

(c) False. The common-ion effect is in play for solubility equilibria as well as acid–base equilibria.

(d) True. The common-ion effect does not change the equilibrium constant.

17.51 *Analyze/Plan.* Follow the example in Sample Exercise 17.10. *Solve.*

$K_{sp} = [Ag^+][I^-]$; $K_{sp} = [Sr^{2+}][SO_4^{2-}]$; $K_{sp} = [Fe^{2+}][OH^-]^2$; $K_{sp} = [Hg_2^{2+}][Br^-]^2$

17.53 *Analyze/Plan.* Follow the logic in Sample Exercise 17.11. *Solve.*

(a) $CaF_2(s) \rightleftharpoons Ca^{2+}(aq) + 2\,F^-(aq)$; $K_{sp} = [Ca^{2+}][F^-]^2$

The molar solubility is the moles of CaF_2 that dissolve per liter of solution. Each mole of CaF_2 produces **1** mol $Ca^{2+}(aq)$ and **2** mol $F^-(aq)$.

$[Ca^{2+}] = 1.24 \times 10^{-3}$ *M*; $[F^-] = 2 \times 1.24 \times 10^{-3}$ *M* $= 2.48 \times 10^{-3}$ *M*

$K_{sp} = (1.24 \times 10^{-3})(2.48 \times 10^{-3})^2 = 7.63 \times 10^{-9}$

(b) $SrF_2(s) \rightleftharpoons Sr^{2+}(aq) + 2\,F^-(aq)$; $K_{sp} = [Sr^{2+}][F^-]^2$

Transform the gram solubility to molar solubility.

$$\frac{1.1 \times 10^{-2}\text{ g SrF}_2}{0.100\text{ L}} \times \frac{1\text{ mol SrF}_2}{125.6\text{ g SrF}_2} = 8.76 \times 10^{-4} = 8.8 \times 10^{-4}\text{ mol SrF}_2/\text{L}$$

$[Sr^{2+}] = 8.76 \times 10^{-4}$ *M*; $[F^-] = 2(8.76 \times 10^{-4}$ *M*)

$K_{sp} = (8.76 \times 10^{-4})(2(8.76 \times 10^{-4}))^2 = 2.7 \times 10^{-9}$

(c) $Ba(IO_3)_2(s) \rightleftharpoons Ba^{2+}(aq) + 2\,IO_3^-(aq)$; $K_{sp} = [Ba^{2+}][IO_3^-]^2$

Because 1 mole of dissolved $Ba(IO_3)_2$ produces 1 mole of Ba^{2+}, the molar solubility of $Ba(IO_3)_2 = [Ba^{2+}]$. Let $x = [Ba^{2+}]$; $[IO_3^-] = 2x$.

$K_{sp} = 6.0 \times 10^{-10} = (x)(2x)^2$; $4x^3 = 6.0 \times 10^{-10}$; $x^3 = 1.5 \times 10^{-10}$; $x = 5.3 \times 10^{-4}$ *M*

The molar solubility of $Ba(IO_3)_2$ is 5.3×10^{-4} mol/L.

17.55 *Analyze/Plan*. Given gram solubility of a compound, calculate K_{sp}. Write the dissociation equilibrium and K_{sp} expression. Change gram solubility to molarity of the individual ions, taking the stoichiometry of the compound into account. Calculate K_{sp}. *Solve*.

$CaC_2O_4(s) \rightleftharpoons Ca^{2+}(aq) + C_2O_4{}^{2-}(aq)$; $K_{sp} = [Ca^{2+}][C_2O_4{}^{2-}]$

$$[Ca^{2+}] = [C_2O_4{}^{2-}] = \frac{0.0061 \text{ g } CaC_2O_4}{1.00 \text{ L soln}} \times \frac{1 \text{ mol } CaC_2O_4}{128.1 \text{ g } CaC_2O_4} = 4.76 \times 10^{-5} = 4.8 \times 10^{-5}\ M$$

$K_{sp} = (4.76 \times 10^{-5}\ M)(4.76 \times 10^{-5}\ M) = 2.3 \times 10^{-9}$

17.57 *Analyze/Plan*. Follow the logic in Sample Exercises 17.12 and 17.13. *Solve*.

(a) $AgBr(s) \rightleftharpoons Ag^+(aq) + Br^-(aq)$; $K_{sp} = [Ag^+][Br^-] = 5.0 \times 10^{-13}$

molar solubility = x = $[Ag^+] = [Br^-]$; $K_{sp} = x^2$

$x = (5.0 \times 10^{-13})^{1/2}$; $x = 7.1 \times 10^{-7}$ mol AgBr/L

(b) Molar solubility = x = $[Br^-]$; $[Ag^+] = 0.030\ M + x$

$K_{sp} = (0.030 + x)(x) \approx 0.030(x)$

$5.0 \times 10^{-13} = 0.030(x)$; $x = 1.7 \times 10^{-11}$ mol AgBr/L

(c) Molar solubility = x = $[Ag^+]$

There are two sources of Br^-: NaBr(0.10 *M*) and AgBr(x *M*)

$K_{sp} = (x)(0.10 + x)$; assume x is small compared to 0.10 *M*.

$5.0 \times 10^{-13} = 0.10\ (x)$; $x \approx 5.0 \times 10^{-12}$ mol AgBr/L

17.59 *Analyze/Plan*. Given a saturated solution of CaF_2 in contact with undissolved $CaF_2(s)$, consider the effect of adding $CaCl_2(s)$. The two salts have the Ca^{2+} ion in common.

Solve. As $CaCl_2$ is added, $[Ca^{2+}]$ increases, K_{sp} is exceeded, and additional CaF_2 precipitates until equilibrium is reestablished. At the new equilibrium position:

(a) Increase. The additional Ca^{2+} from $CaCl_2$ decreases the solubility of CaF_2.

(b) Increase. We have added $CaCl_2$, which contains Ca^{2+}.

(c) Decrease. After $CaCl_2$ is added, $CaF_2(s)$ precipitates, which decreases $[F^-]$.

17.61 *Analyze/Plan*. We are asked to calculate the solubility of a slightly soluble hydroxide salt at various pH values. This is a common ion problem; pH tells us not only $[H^+]$ but also $[OH^-]$, which is an ion common to the salt. Use pH to calculate $[OH^-]$ and then proceed as in Sample Exercise 17.13. *Solve*.

$Mn(OH)_2(s) \rightleftharpoons Mn^{2+}(aq) + 2\ OH^-(aq)$; $K_{sp} = 1.6 \times 10^{-13}$

Because $[OH^-]$ is set by the pH of the solution, the solubility of $Mn(OH)_2$ is just $[Mn^{2+}]$.

(a) pH = 7.0, pOH = 14 – pH = 7.0, $[OH^-] = 10^{-pOH} = 1.0 \times 10^{-7}\ M$

$$K_{sp} = 1.6 \times 10^{-13} = [Mn^{2+}](1.0 \times 10^{-7})^2;\ [Mn^{2+}] = \frac{1.6 \times 10^{-13}}{1.0 \times 10^{-14}} = 16\ M$$

$$\frac{16 \text{ mol } Mn(OH)_2}{1 \text{ L}} \times \frac{88.95 \text{ g } Mn(OH)_2}{1 \text{ mol } Mn(OH)_2} = 1423 = 1.4 \times 10^3 \text{ g } Mn(OH)_2/\text{L}$$

Check. Note that the solubility of $Mn(OH)_2$ in pure water is 3.6×10^{-5} *M*, and the pH of the resulting solution is 9.0. The relatively low pH of a solution buffered to pH 7.0 actually increases the solubility of $Mn(OH)_2$.

(b) pH = 9.5, pOH = 4.5, $[OH^-] = 3.16 \times 10^{-5} = 3.2 \times 10^{-5}$ *M*

$$K_{sp} = 1.6\times10^{-13} = [Mn^{2+}](3.16\times10^{-5})^2;\ [Mn^{2+}] = \frac{1.6\times10^{-13}}{1.0\times10^{-9}} = 1.6\times10^{-4}\ M$$

1.6×10^{-4} *M* $Mn(OH)_2 \times 88.95$ g/mol = 0.0142 = 0.014 g/L

(c) pH = 11.8, pOH = 2.2, $[OH^-] = 6.31 \times 10^{-3} = 6.3 \times 10^{-3}$ *M*

$$K_{sp} = 1.6\times10^{-13} = [Mn^{2+}](6.31\times10^{-3})^2;\ [Mn^{2+}] = \frac{1.6\times10^{-13}}{3.98\times10^{-5}} = 4.0\times10^{-9}\ M$$

4.02×10^{-9} *M* $Mn(OH)_2 \times 88.95$ g/mol $= 3.575 \times 10^{-7} = 3.6 \times 10^{-7}$ g/L

17.63 *Analyze/Plan.* Follow the logic in Sample Exercise 17.14. *Solve.*

If the anion of the salt is the conjugate base of a weak acid, it will combine with H^+, reducing the concentration of the free anion in solution, thereby causing more salt to dissolve. More soluble in acid: (a) $ZnCO_3$, (b) ZnS, (d) AgCN, (e) $Ba_3(PO_4)_2$.

17.65 *Analyze/Plan.* Follow the logic in Sample Exercise 17.15. *Solve.*

The formation equilibrium is

$$Ni^{2+}(aq) + 6NH_3(aq) \rightleftharpoons Ni(NH_3)_6{}^{2+}(aq)\quad K_f = \frac{[Ni(NH_3)_6^{2+}]}{[Ni^{2+}][NH_3]^6} = 1.2\times10^9$$

$$1.25\text{ g NiCl}_2 \times \frac{1\text{ mol NiCl}_2}{129.62\text{ g NiCl}_2} \times \frac{1}{0.1000\text{ L}} = 0.096436 = 0.0964\ M\ \text{NiCl}_2$$

Assume that nearly all the Ni^{2+} is in the form $Ni(NH_3)_6{}^{2+}$.

$[Ni(NH_3)_6{}^{2+}] = 0.0964$ *M*; $[Ni^{2+}] = x$; $[NH_3] = 0.20$ *M*

$$1.2\times10^9 = \frac{(0.0964)}{x(0.20)^6};\ x = 1.26\times10^{-6} = 1.3\times10^{-6}\ M = [Ni^{2+}]$$

$[Ni(NH_3)_6{}^{2+}] = 0.0964\ M - 1.26 \times 10^{-6}\ M = 0.0964\ M$

Note that our assumption that most of the Ni^{2+} is present as $Ni(NH_3)_6{}^{2+}$ is true.

17.67 *Analyze/Plan.* Calculate the solubility of AgI in pure water according to the method in Sample Exercise 17.12. Obtain K_{eq} for the complexation reaction, making use of pertinent K_{sp} and K_f values from Appendix D.3 and Table 17.1. Write the dissociation equilibrium for AgI and the formation reaction for $Ag(CN)_2{}^-$. Use algebra to manipulate these equations and their associated equilibrium constants to obtain the desired reaction and its equilibrium constant. Finally, use this K_{eq} value to calculate the solubility of AgI in 0.100 *M* NaCN solution. *Solve.*

(a) $AgI(s) \rightleftharpoons Ag^+(aq) + I^-(aq)$; $K_{sp} = [Ag^+][I^-] = 8.3 \times 10^{-17}$

molar solubility = $x = [Ag^+] = [I^-]$; $K_{sp} = x^2$

$x = (8.3 \times 10^{-17})^{1/2}$; $x = 9.1 \times 10^{-9}$ mol AgI/L

(b)

$$AgI(s) \rightleftharpoons Ag^+(aq) + I^-(aq)$$
$$Ag^+(aq) + 2\,CN^-(aq) \rightleftharpoons Ag(CN)_2^-(aq)$$

$$AgI(s) + 2\,CN^-(aq) \rightleftharpoons Ag(CN)_2^-(aq) + I^-(aq)$$

$$K = K_{sp} \times K_f = [Ag^+][I^-] \times \frac{[Ag(CN)_2^-]}{[Ag^+][CN^-]^2} = (8.3 \times 10^{-17})(1 \times 10^{21}) = 8 \times 10^4$$

(c) K is much greater than one for the reaction of AgI(s) with CN^-. This means that the reaction goes to completion. For a AgI(s) in 0.100 *M* NaCN solution, CN^- is the limiting reactant. Two moles of CN^- react with one mole of AgI, so the solublility of AgI in 0.100 *M* NaCN is (0.100/2) = 0.0500 *M*.

Precipitation and Separation of Ions (Section 17.6)

17.69 *Analyze/Plan*. Follow the logic in Sample Exercise 17.16. Precipitation conditions: will Q (see Chapter 15) exceed K_{sp} for the compound? *Solve.*

(a) In base, Ca^{2+} can form $Ca(OH)_2(s)$.

$Ca(OH)_2(s) \rightleftharpoons Ca^{2+}(aq) + 2\,OH^-(aq)$; $K_{sp} = [Ca^{2+}][OH^-]^2$

$Q = [Ca^{2+}][OH^-]^2$; $[Ca^{2+}] = 0.050\,M$; pOH = 14 – 8.0 = 6.0; $[OH^-] = 1.0 \times 10^{-6}\,M$

$Q = (0.050)(1.0 \times 10^{-6})^2 = 5.0 \times 10^{-14}$; $K_{sp} = 6.5 \times 10^{-6}$ (Appendix D.3)

$Q < K_{sp}$, no $Ca(OH)_2$ precipitates.

(b) $Ag_2SO_4(s) \rightleftharpoons 2\,Ag^+(aq) + SO_4^{2-}(aq)$; $K_{sp} = [Ag+]^2[SO_4^{2-}]$

$$[Ag^+] = \frac{0.050\,M \times 100\text{ mL}}{110\text{ mL}} = 4.545 \times 10^{-2} = 4.5 \times 10^{-2}\,M$$

$$[SO_4^{2-}] = \frac{0.050\,M \times 10\text{ mL}}{110\text{ mL}} = 4.545 \times 10^{-3} = 4.5 \times 10^{-3}\,M$$

$Q = (4.545 \times 10^{-2})^2 (4.545 \times 10^{-3}) = 9.4 \times 10^{-6}$; $K_{sp} = 1.5 \times 10^{-5}$

$Q < K_{sp}$, no Ag_2SO_4 precipitates.

17.71 *Analyze/Plan*. We are asked to calculate pH necessary to precipitate $Mn(OH)_2(s)$ if the resulting Mn^{2+} concentration is ≤ 1 μg/L.

$Mn(OH)_2(s) \rightleftharpoons Mn^{2+}(aq) + 2\,OH^-(aq)$; $K_{sp} = [Mn^{2+}][OH^-]^2 = 1.6 \times 10^{-13}$

At equilibrium, $[Mn^{2+}][OH^-]^2 = 1.6 \times 10^{-13}$. Change concentration $Mn^{2+}(aq)$ to mol/L and solve for $[OH^-]$. *Solve.*

$$\frac{1\,\mu g\,Mn^{2+}}{1\text{ L}} \times \frac{1 \times 10^{-6}\text{ g}}{1\,\mu g} \times \frac{1\text{ mol }Mn^{2+}}{54.94\text{ g }Mn^{2+}} = 1.82 \times 10^{-8} = 2 \times 10^{-8}\,M\,Mn^{2+}$$

$1.6 \times 10^{-13} = (1.82 \times 10^{-8})[OH^-]^2$; $[OH^-]^2 = 8.79 \times 10^{-6}$; $[OH^-] = 2.96 \times 10^{-3} = 3 \times 10^{-3}\,M$

pOH = 2.53; pH = 14 – 2.53 = 11.47 = 11.5

17.73 *Analyze/Plan.* We are asked which ion will precipitate first from a solution containing $Pb^{2+}(aq)$ and $Ag^+(aq)$ when $I^-(aq)$ is added. Follow the logic in Sample Exercise 17.17. Calculate $[I^-]$ needed to initiate precipitation of each ion. The cation that requires lower $[I^-]$ will precipitate first. *Solve.*

$$Ag^+: K_{sp} = [Ag^+][I^-]; 8.3\times10^{-17} = (2.0\times10^{-4})[I^-]; [I^-] = \frac{8.3\times10^{-17}}{2.0\times10^{-4}} = 4.2\times10^{-13}\ M\ I^-$$

$$Pb^{2+}: K_{sp} = [Pb^{2+}][I^-]^2; 7.9\times10^{-9} = (1.5\times10^{-3})[I^-]^2;\ [I^-] = \left(\frac{7.9\times10^{-9}}{1.5\times10^{-3}}\right)^{1/2} = 2.3\times10^{-3}\ M\ I^-$$

AgI will precipitate first, at $[I^-] = 4.2\times10^{-13}$ *M*.

17.75 *Analyze/Plan.* We are asked which ion will precipitate first when dilute $Ag^+(aq)$ is added to a solution containing 0.20 *M* CrO_4^{2-}, 0.10 *M* CO_3^{2-}, and 0.10 *M* Cl^-. The anions are present at different concentrations and their silver compounds have different stoichiometry, so we cannot directly compare K_{sp} values. Follow the logic in Sample Exercise 17.17. Calculate $[Ag^+]$ needed to initiate precipitation of each ion. The anion that requires lowest $[Ag^+]$ will precipitate first, and so on. *Solve.*

Ag_2CrO_4: $K_{sp} = [Ag^+]^2[CrO_4^{2-}] = 1.2\times10^{-12}$

$1.2\times10^{-12} = [Ag^+]^2(0.20);\ [Ag^+]^2 = 6.0\times10^{-12};\ [Ag^+] = 2.4\times10^{-6}\ M$

Ag_2CO_3: $K_{sp} = [Ag^+]^2[CO_3^{2-}] = 8.1\times10^{-12}$

$8.1\times10^{-12} = [Ag^+]^2(0.10);\ [Ag^+]^2 = 8.1\times10^{-11};\ [Ag^+] = 9.0\times10^{-6}\ M$

AgCl: $K_{sp} = [Ag^+][Cl^-] = 1.8\times10^{-10}$

$1.8\times10^{-10} = [Ag^+](0.010);\ [Ag^+] = 1.8\times10^{-8}$

AgCl requires the smallest $[Ag^+]$ for precipitation and it will precipitate first. The other two will precipitate almost simultaneously.

Qualitative Analysis for Metallic Elements (Section 17.7)

17.77 *Analyze/Plan.* Use Figure 17.23 and the description of the five qualitative analysis "groups" in Section 17.7 to analyze the given data. Ag^+ is in Group 1, Al^{3+} is in Group 3, Mg^{2+} is in Group 4, and Na^+ is in Group 5. *Solve.*

The first two experiments eliminate Group 1 and 2 ions (Figure 17.23). The presence of a precipitate after the third experiment means that a Group 3 cation is present, in this case Al^{3+}. The fact that no insoluble phosphates form in the filtrate from the third experiment rules out Group 4 ions. Ag^+ (Group 1) and Mg^{2+} (Group 4) are definitely absent. Al^{3+} (Group 3) is definitely present and Na^+ (Group 5) is possibly present.

17.79 *Analyze/Plan.* We are asked to devise a procedure to separate various pairs of ions in aqueous solutions. In each case, refer to Figure 17.23 to find a set of conditions where the solubility of the two ions differs. Construct a procedure to generate these conditions. *Solve.*

(a) Cd^{2+} is in Gp. 2, but Zn^{2+} is not. Make the solution acidic using 0.2 *M* HCl; saturate with H_2S. CdS will precipitate, but ZnS will not.

(b) $Cr(OH)_3$ is amphoteric, but $Fe(OH)_3$ is not. Add excess base; $Fe(OH)_3(s)$ precipitates, but Cr^{3+} forms the soluble complex $Cr(OH)_4^-$.

(c) Mg^{2+} is a member of Gp. 4, but K^+ is not. Add $(NH_4)_2HPO_4$ to a basic solution; Mg^{2+} precipitates as $MgNH_4PO_4$, but K^+ remains in solution.

(d) Ag^+ is a member of Gp. 1, but Mn^{2+} is not. Add 6 *M* HCl; precipitate Ag^+ as AgCl(s); Mn^{2+} remains soluble.

17.81 (a) Because phosphoric acid is a weak acid, the concentration of free PO_4^{3-}(aq) in an aqueous phosphate solution is low except in strongly basic media. In less basic media, the solubility product of the phosphates of interest is not exceeded.

(b) K_{sp} for those cations in Group 3 is much larger. Thus, to exceed K_{sp}, a higher $[S^{2-}]$ is required. This is achieved by making the solution more basic.

(c) They should all redissolve in strongly acidic solution; for example, in 12 *M* HCl (the chlorides of all Group 3 metals are soluble).

Additional Exercises

17.83 *Analyze/Plan.* Follow the approach for deriving the Henderson–Hasselbach (H–H) equation from the K_a expression shown in Section 17.2. Begin with a general K_b expression. *Solve.*

$$B(aq) + H_2O(l) \rightleftharpoons BH^+(aq) + OH^-(aq); \quad K_b = \frac{[BH^+][OH^-]}{[B]}$$

pOH = –log[OH^-]; rearrange K_a to solve for [OH^-]

$$[OH^-] = \frac{K_b[B]}{[BH^+]}; \text{ take the –log of both sides}$$

$$-\log[OH^-] = -\log K_b + (-\log[B] - (-\log[BH^+]))$$

$$pOH = pK_b + \log[BH^+] - \log[B]$$

$$pOH = pK_b + \log\frac{[BH^+]}{[B]}$$

17.89

$$\frac{0.15 \text{ mol } CH_3COOH}{1 \text{ L soln}} \times 0.750 \text{ L} = 0.1125 = 0.11 \text{ mol } CH_3COOH$$

$$0.1125 \text{ mol } CH_3COOH \times \frac{60.05 \text{ g } CH_3COOH}{1 \text{ mol } CH_3COOH} \times \frac{1 \text{ g gl acetic acid}}{0.99 \text{ g } CH_3COOH} \times \frac{1.00 \text{ mL gl acetic acid}}{1.05 \text{ g gl acetic acid}} = 6.5 \text{ mL glacial acetic acid}$$

At pH 4.50, $[H^+] = 10^{-4.50} = 3.16 \times 10^{-5} = 3.2 \times 10^{-5}$ *M*; this is small compared to 0.15 *M* CH_3COOH.

$$K_a = \frac{(3.16 \times 10^{-5})[CH_3COO^-]}{0.15} = 1.8 \times 10^{-5}; [CH_3COO^-] = 0.0854 = 0.085\ M$$

$$\frac{0.0854 \text{ mol } CH_3COONa}{1 \text{ L soln}} \times 0.750 \text{ L} \times \frac{82.03 \text{ g } CH_3COONa}{1 \text{ mol } CH_3COONa} = 5.253 = 5.25 \text{ g } CH_3COONa$$

17.91 (a) For a monoprotic acid (one H^+ per mole of acid), at the equivalence point moles OH^- added = moles H^+ originally present

$M_B \times V_B$ = g acid/molar mass

$$MM = \frac{\text{g acid}}{M_B \times V_B} = \frac{0.1687 \text{ g}}{0.1150\, M \times 0.0155 \text{ L}} = 94.642 = 94.6 \text{ g/mol}$$

(b) $$\text{initial mol HA} = \frac{0.1687 \text{ g}}{94.642 \text{ g/mol}} = 1.783 \times 10^{-3} = 1.78 \times 10^{-3} \text{ mol HA}$$

mol OH^- added to pH 2.85 = $0.1150\, M \times 0.00725 \text{ L} = 8.338 \times 10^{-4}$
$= 8.34 \times 10^{-4}$ mol OH^-

	HA(aq)	+	NaOH(aq)	→	NaA(aq) + H_2O
before rx	1.783×10^{-3} mol		0.834×10^{-3} mol		0
change	-0.834×10^{-3} mol		-0.834×10^{-3} mol		0.834×10^{-3} mol
after rx	0.949×10^{-3} mol		0		0.834×10^{-3} mol

$$[HA] = \frac{9.49 \times 10^{-4} \text{ mol}}{0.0325 \text{ L}} = 0.02919 = 0.0292\, M$$

$$[A^-] = \frac{8.34 \times 10^{-4} \text{ mol}}{0.0325 \text{ L}} = 0.02565 = 0.0257\, M$$

$$[H^+] = 10^{-2.85} = 1.413 \times 10^{-3} = 1.4 \times 10^{-3}$$

The mixture after reaction (a buffer) can be described by the acid dissociation equilibrium.

	HA(aq)	⇌	H^+(aq)	+	A^-(aq)
initial	0.0292 *M*		0		0.0257 *M*
equil.	$(0.0292 - 1.4 \times 10^{-3}\, M)$		$1.4 \times 10^{-3}\, M$		$(0.0257 + 1.4 \times 10^{-3})\, M$

$$K_a = \frac{[H^+][A^-]}{[HA]} \approx \frac{(1.413 \times 10^{-3})(0.02707)}{(0.02778)} = 1.4 \times 10^{-3}$$

(Although we have carried extra figures through the calculation to avoid rounding errors, the data dictate an answer with 2 sig figs.)

17.97 (a) $PbCO_3(s) \rightleftharpoons Pb^{2+}(aq) + CO_3^{2-}(aq)$

$K_{sp} = [Pb^{2+}][CO_3^{2-}] = 7.4 \times 10^{-14}$. molar solubility = s = $[Pb^{2+}] = [CO_3^{2-}]$

$K_{sp} = s^2 = 7.4 \times 10^{-14}$. s = $[Pb^{2+}] = 2.7203 \times 10^{-7} = 2.7 \times 10^{-7}\, M$

(b) For very dilute aqueous solutions, assume the solution density is 1.0 g/mL.

$$\text{ppb} = \frac{\text{g solute}}{10^9 \text{ g solution}} = \frac{1 \times 10^{-6} \text{ g solute}}{1 \times 10^3 \text{ g solution}} = \frac{\mu\text{g solute}}{\text{L solution}}$$

$$\frac{2.7203 \times 10^{-7} \text{ mol } Pb^{2+}}{\text{L}} \times \frac{207.2 \text{ g } Pb^{2+}}{1 \text{ mol } Pb^{2+}} \times \frac{1\, \mu\text{g}}{1 \times 10^{-6} \text{ g}} = \frac{56.365\, \mu\text{g } Pb^{2+}}{\text{L}} = 56 \text{ ppb}$$

(c) The solubility of $PbCO_3$ increases as pH is lowered. When pH is lowered, $[H^+]$ increases. The $H^+(aq)$ reacts with $CO_3^{2-}(aq)$ to form HCO_3^- and $H_2CO_3(aq)$. This shifts the solubility equilibrium to the right and increases the solubility of $PbCO_3$.

(d) A saturated solution of lead carbonate, with a lead concentration of 56 ppb, exceeds the EPA acceptable lead level of 15 ppb.

17.102 *Analyze/Plan.* Calculate the solubility of $Mg(OH)_2$ in 0.50 *M* NH_4Cl. Find K_{sp} for $Mg(OH)_2$ in Appendix D.3. NH_4^+ is a weak acid, which will increase the solubility of $Mg(OH)_2$. Combine the various interacting equilibria to obtain an overall reaction. Calculate K for this reaction and use it to calculate solubility (s) for $Mg(OH)_2$ in 0.50 *M* NH_4Cl. *Solve.*

$$Mg(OH)_2(s) \rightleftharpoons Mg^{2+}(aq) + 2\,OH^-(aq) \qquad K_{sp}$$

$$2\,NH_4^+(aq) \rightleftharpoons 2\,NH_3(aq) + 2\,H^+(aq) \qquad K_a$$

$$2\,H^+(aq) + 2\,OH^-(aq) \rightleftharpoons 2\,H_2O(l) \qquad 1/K_w$$

$$Mg(OH)_2(s) + 2\,NH_4^+(aq) + 2\,H^+(aq) + 2\,OH^-(aq) \rightleftharpoons Mg^{2+}(aq) + 2\,NH_3(aq) + 2\,OH^-(aq) + 2\,H^+(aq) + 2\,H_2O(l)$$

$$Mg(OH)_2(s) + 2\,NH_4^+(aq) \rightleftharpoons Mg^{2+}(aq) + 2\,NH_3(aq) + 2\,H_2O(l)$$

$$K = \frac{[Mg^{2+}][NH_3]^2}{[NH_4^+]^2} = \frac{K_{sp} \times K_a^{\ 2}}{K_w^{\ 2}}; \ K_a \text{ for } NH_4^+ = \frac{K_w}{K_b \text{ for } NH_3}; \ \frac{K_a}{K_w} = \frac{1}{K_b}$$

$$K = \frac{K_{sp} \times K_a^2}{K_w^2} = \frac{K_{sp}}{K_b^2} = \frac{1.8 \times 10^{-11}}{(1.8 \times 10^{-5})^2} = 5.556 \times 10^{-2} = 5.6 \times 10^{-2}$$

Let $[Mg^{2+}] = s$, $[NH_3] = 2s$, $[NH_4^+] = 0.50 - 2s$

$$K = 5.6 \times 10^{-2} = \frac{[Mg^{2+}][NH_3]^2}{[NH_4^+]^2} = \frac{s\,(2s)^2}{(0.5 - 2s)^2} = \frac{4s^3}{0.25 - 2s + 4s^2}$$

$5.6 \times 10^{-2}(0.25 - 2s + 4s^2) = 4s^3$; $4s^3 - 0.222s^2 + 0.111s - 1.39 \times 10^{-2} = 0$

Clearly, 2s is not small relative to 0.50. Solving the third-order equation, s = 0.1054 = 0.11 *M*. The solubility of $Mg(OH)_2$ in 0.50 *M* NH_4Cl is 0.11 mol/L.

Check. Substitute s = 0.1054 into the K expression.

$$K = \frac{4(0.1054)^3}{[0.50 - 2(0.1054)]^2} = 5.6 \times 10^{-2}.$$

The solubility and K value are consistent, to the precision of the K_{sp} and K_b values.

17.108

$$Zn(OH)_2(s) \rightleftharpoons Zn^{2+}(aq) + 2\,OH^-(aq) \qquad K_{sp} = 3.0 \times 10^{-16}$$

$$Zn^{2+}(aq) + 4\,OH^-(aq) \rightleftharpoons Zn(OH)_4^{2-}(aq) \qquad K_f = 4.6 \times 10^{17}$$

$$Zn(OH)_2(s) + 2\,OH^-(aq) \rightleftharpoons Zn(OH)_4^{2-}(aq) \qquad K = K_{sp} \times K_f = 138 = 1.4 \times 10^2$$

$$K = 138 = 1.4 \times 10^2 = \frac{[Zn(OH)_4^{2-}]}{[OH^-]^2}$$

If 0.015 mol $Zn(OH)_2$ dissolves, 0.015 mol $Zn(OH)_4^{2-}$ should be present at equilibrium.

$$[OH^-]^2 = \frac{(0.015)}{138};\,[OH^-] = 1.043 \times 10^{-2}\ M;\ [OH^-] \geq 1.0 \times 10^{-2}\ M \text{ or pH} \geq 12.02$$

Integrative Exercises

17.111 (a) For a monoprotic acid (one H^+ per mole of acid), at the equivalence point moles OH^- added = moles H^+ originally present

$M_B \times V_B$ = g acid/molar mass

$$MM = \frac{\text{g acid}}{M_B \times V_B} = \frac{0.1044\,\text{g}}{0.0500\ M \times 0.02210\ \text{L}} = 94.48 = 94.5\ \text{g/mol}$$

(b) 11.05 mL is exactly half-way to the equivalence point (22.10 mL). When half of the unknown acid is neutralized, $[HA] = [A^-]$, $[H^+] = K_a$ and $pH = pK_a$.

$K_a = 10^{-4.89} = 1.3 \times 10^{-5}$

(c) From Appendix D, Table D.1, acids with K_a values close to 1.3×10^{-5} are

Name	**K_a**	**Formula**	**Molar Mass**
propionic	1.3×10^{-5}	C_2H_5COOH	74.1
butanoic	1.5×10^{-5}	C_3H_7COOH	88.1
acetic	1.8×10^{-5}	CH_3COOH	60.1
hydroazoic	1.9×10^{-5}	HN_3	43.0

Of these, butanoic has the closest match for K_a and molar mass, but the agreement is not good.

17.113 Calculate the initial M of aspirin in the stomach and solve the equilibrium problem to find equilibrium concentrations of $C_8H_7O_2COOH$ and $C_8H_7O_2COO^-$. At pH = 2, $[H^+] = 1 \times 10^{-2}$.

$$\frac{325\text{ mg}}{\text{tablet}} \times 2\text{ tablets} \times \frac{1\text{ g}}{1000\text{ mg}} \times \frac{1\text{ mol } C_8H_7O_2COOH}{180.2\text{ g } C_8H_7O_2COOH} \times \frac{1}{1\text{ L}} = 3.61 \times 10^{-3} = 4 \times 10^{-3}\ M$$

	$C_8H_7O_2COOH(aq)$	$\rightleftharpoons$	$C_8H_7O_2COO^-(aq)$	+	$H^+(aq)$
initial	$3.61 \times 10^{-3}\ M$		0		$1 \times 10^{-2}\ M$
equil.	$(3.61 \times 10^{-3} - x)\ M$		$x\ M$		$(1 \times 10^{-2} + x)\ M$

$$K_a = 3 \times 10^{-5} = \frac{[H^+][C_8H_7O_2COO^-]}{[C_8H_7O_2COOH]} = \frac{(0.01 + x)(x)}{(3.61 \times 10^{-3} - x)} \approx \frac{0.01\,x}{3.61 \times 10^{-3}}$$

$x = [C_8H_7O_2COO^-] = 1.08 \times 10^{-5} = 1 \times 10^{-5}\ M$

$$\%\ \text{ionization} = \frac{1.08\times10^{-5}\ M\ C_8H_7O_2COO^-}{3.61\times10^{-3}\ M\ C_8H_7O_2COOH}\times100 = 0.3\%$$

(% ionization is small, so the approximation was valid.)

% aspirin molecules = 100.0% – 0.3% = 99.7% molecules

17.117 For very dilute aqueous solutions, assume the solution density is 1 g/mL.

$$\text{ppb} = \frac{\text{g solute}}{10^9\ \text{g solution}} = \frac{1\times10^{-6}\ \text{g solute}}{1\times10^3\ \text{g solution}} = \frac{\mu\text{g solute}}{\text{L solution}}$$

(a) $K_{sp} = [Ag^+][Cl^-] = 1.8\times10^{-10}$; $[Ag^+] = (1.8\times10^{-10})^{1/2} = 1.34\times10^{-5} = 1.3\times10^{-5}\ M$

$$\frac{1.34\times10^{-5}\ \text{mol}\ Ag^+}{\text{L}}\times\frac{107.9\ \text{g}\ Ag^+}{1\ \text{mol}\ Ag^+}\times\frac{1\ \mu\text{g}}{1\times10^{-6}\ \text{g}} = \frac{1.4\times10^3\ \mu\text{g}\ Ag^+}{\text{L}}$$

$$= 1.4\times10^3\ \text{ppb} = 1.4\ \text{ppm}$$

(b) $K_{sp} = [Ag^+][Br^-] = 5.0\times10^{-13}$; $[Ag^+] = (5.0\times10^{-13})^{1/2} = 7.07\times10^{-7} = 7.1\times10^{-7}\ M$

$$\frac{7.07\times10^{-7}\ \text{mol}\ Ag^+}{\text{L}}\times\frac{107.9\ \text{g}\ Ag^+}{1\ \text{mol}\ Ag^+}\times\frac{1\ \mu\text{g}}{1\times10^{-6}\ \text{g}} = 76\ \text{ppb}$$

(c) $K_{sp} = [Ag^+][I^-] = 8.3\times10^{-17}$; $[Ag^+] = (8.3\times10^{-17})^{1/2} = 9.11\times10^{-9} = 9.1\times10^{-9}\ M$

$$\frac{9.11\times10^{-9}\ \text{mol}\ Ag^+}{\text{L}}\times\frac{107.9\ \text{g}\ Ag^+}{1\ \text{mol}\ Ag^+}\times\frac{1\ \mu\text{g}}{1\times10^{-6}\ \text{g}} = 0.98\ \text{ppb}$$

AgBr(s) would maintain $[Ag^+]$ in the correct range.

18 Chemistry of the Environment

Visualizing Concepts

18.1 *Analyze.* Given that one mole of an ideal gas at 1 atm and 298 K occupies 22.4 L, is the volume of one mole of ideal gas in the middle of the stratosphere greater or less than 22.4 L?

Plan. Consider the relationship between pressure, temperature, and volume of an ideal gas. Use Figure 18.1 to estimate the pressure and temperature in the middle of the stratosphere, and compare the two sets of temperature and pressure.

Solve. According to the ideal-gas law, $PV = nRT$, so $V = nRT/P$. Because n and R are constant for this exercise, V is proportional to T/P.

(a) Greater. The stratosphere ranges from 10 to 50 km, so the middle is at approximately 30 km. At this altitude, $T \approx 230$ K, $P \approx 40$ torr (from Figure 18.1). Because we are comparing T/P ratios, either atm or torr can be used as pressure units; we will use torr.

At sea level: $T/P = 298\ \text{K}/760\ \text{torr} = 0.39$

At 30 km: $T/P = 230\ \text{K}/40\ \text{torr} = 5.75$

The proportionality constant (T/P) is much greater at 30 km than sea level, so the volume of 1 mol of an ideal gas is greater at this altitude. The decrease in temperature at 30 km is more than offset by the substantial decrease in pressure.

(b) No. Volume is proportional to T/P, not simply T. The relative volumes of one mole of an ideal gas at 50 km and 85 km depend on the temperature and pressure at the two altitudes. From Figure 18.1,

50 km: $T \approx 270$ K, $P \approx 20$ torr, $T/P = 270\ \text{K}/20\ \text{torr} = 13.5$

85 km: $T \approx 190$ K, $P < 0.01$ torr, $T/P = 190\ \text{K}/0.01\ \text{torr} = 19{,}000$

Again, the slightly lower temperature at 85 km is more than offset by a much lower pressure. One mole of an ideal gas will occupy a much larger volume at 85 km than 50 km.

(c) The thermosphere, stratopause, and low-altitude troposphere. Gases behave most ideally at high temperature and low pressure. Pressure is minimum and temperature is high in the thermosphere. The stratopause (the boundary between the stratosphere and mesosphere) and the troposphere at low altitude are other regions with temperature maxima and relatively low pressures.

18.3 (a) A = troposphere, 0–10 km; B = stratosphere, 12–50 km; C = mesosphere, 50–85 km.

(b) Ozone is a pollutant in the troposphere and filters UV radiation in the stratosphere.

(c) Infrared radiation from Earth is most strongly reflected back in the troposphere.

(d) Assuming the "boundary" between the stratosphere and mesosphere is at 50 km, only region C in the diagram is involved in an aurora borealis.

(e) The concentration of water vapor is greatest near Earth's surface in region A and decreases with altitude. Water's single bonds are susceptible to photodissociation in regions B and C, so its concentration is likely to be very low in these regions. The relative concentration of CO_2, with strong double bonds, increases in regions B and C, because it is less susceptible to photodissociation.

18.5 The Sun.

18.7 $CO_2(g)$ dissolves in seawater to form $H_2CO_3(aq)$. The basic pH of the ocean encourages ionization of $H_2CO_3(aq)$ to form $HCO_3^-(aq)$ and $CO_3^{2-}(aq)$. Under the correct conditions, carbon is removed from the ocean as $CaCO_3(s)$ (sea shells, coral, chalk cliffs). As carbon is removed, more $CO_2(g)$ dissolves to maintain the balance of complex and interacting acid–base and precipitation equilibria.

18.9 *Plan.* Follow the yellow arrows on the diagram to find potential routes for environmental contamination at a fracking well site. *Solve.*

Above ground, the two main avenues for escape of contaminants are leakage and evaporation. Methane, hydrogen sulfide, and other petroleum compounds that are gases at atmospheric conditions can leak from the well head. Heavier petroleum compounds (those containing from three to six carbon atoms) are also released by fracking, mix with the resulting aqueous solution, and end up in the wastewater ponds. Volatile organic compounds produced by fracking and from the fracking liquid can evaporate from the ponds. Wastewwater ponds, if unlined, can seep into nearby water sources or over flow due to rainfall.

Below ground, petroleum gases and fracking liquid can migrate into groundwater, both deep and shallow aquifers. (Fracking mobilizes petroleum and gases, which enables these underground "leaks.")

The extent of underground leaking depends on the integrity of the well plumbing and casings, as well as the real permeability of the "impermeable" layers. Although there are several avenues for contamination, they can be minimized by careful attention to geology, engineering, and oversight.

Earth's Atmosphere (Section 18.1)

18.11 (a) The temperature profile of the atmosphere (Figure 18.1) is the basis of its division into regions. The center of each peak or trough in the temperature profile corresponds to a new region.

(b) Troposphere, 0–12 km; stratosphere, 12–50 km; mesosphere, 50–85 km; thermosphere, 85–110 km.

18.13 *Analyze/Plan.* Given O_3 concentration in ppm, calculate partial pressure. Use the definition of ppm to get mol fraction O_3. For gases, mole fraction = pressure fraction. Use the ideal-gas law to find mol O_3/L air and Avogadro's number to get molecules.

$$P_{O_3} = X_{O_3} \times P_{atm};\ 0.441 \text{ ppm } O_3 = \frac{0.441 \text{ mol } O_3}{1\times10^6 \text{ mol air}} = 4.41\times10^{-7} = X_{O_3}$$ *Solve.*

(a) $P_{O_3} = X_{O_3} \times P_{atm} = 4.41\times10^{-7}(0.67 \text{ atm}) = 2.955\times10^{-7} = 3.0\times10^{-7}$ atm

(b) $$n = \frac{PV}{RT} = \frac{2.955\times10^{-7} \text{ atm} \times 1.0 \text{ L}}{298 \text{ K}} \times \frac{\text{mol-K}}{0.08206 \text{ L-atm}} = 1.208\times10^{-8} = 1.2\times10^{-8} \text{ mol } O_3$$

$$1.208\times10^{-8} \text{ mol } O_3 \times \frac{6.022\times10^{23} \text{ molecules}}{\text{mol}} = 7.277\times10^{15} = 7.3\times10^{15}\ O_3 \text{ molecules}$$

18.15 *Analyze/Plan.* Given CO concentration in ppm, calculate number of CO molecules in 1.0 L air at given conditions. ppm CO $\rightarrow X_{O_3} \rightarrow$ atm CO $\rightarrow$ mol CO $\rightarrow$ molecules CO. Use the ideal-gas law to change atm CO to mol CO and Avogadro's number to get molecules. *Solve.*

$$3.5 \text{ ppm CO} = \frac{3.5 \text{ mol CO}}{1\times10^6 \text{ mol air}} = 3.5\times10^{-6} = X_{CO}$$

$$P_{CO} = X_{CO} \times P_{atm} = 3.5\times10^{-6} \times 759 \text{ torr} \times \frac{1 \text{ atm}}{760 \text{ torr}} = 3.495\times10^{-6} = 3.5\times10^{-6} \text{ atm}$$

$$n_{CO} = \frac{P_{CO}V}{RT} = \frac{3.495\times10^{-6} \text{ atm} \times 1.0 \text{ L}}{295 \text{ K}} \times \frac{\text{mol-K}}{0.08206 \text{ L-atm}} = 1.444\times10^{-7} = 1.4\times10^{-7} \text{ mol CO}$$

$$1.444\times10^{-7} \text{ mol CO} \times \frac{6.022\times10^{23} \text{ molecules}}{\text{mol}} = 8.695\times10^{16} = 8.7\times10^{16} \text{ CO molecules}$$

18.17 *Analyze/Plan.* Given bond dissociation energy in kJ/mol, calculate the wavelength of a single photon that will rupture a C–Br bond. kJ/mol $\rightarrow$ J/molecule. $\lambda = hc/E$. ($\lambda = hc/E$ describes the energy/wavelength relationship of a single photon.) *Solve.*

(a) $$\frac{276 \times 10^3 \text{ J}}{1 \text{ mol}} \times \frac{1 \text{ mol}}{6.022\times10^{23} \text{ molecules}} = 4.583\times10^{-19} = 4.58\times10^{-19} \text{ J/molecule}$$

$$\lambda = \frac{hc}{E} = \frac{(6.626\times10^{-34} \text{ J-sec})(3.00\times10^8 \text{ m/sec})}{4.583\times10^{-19} \text{ J}} = 4.337\times10^{-7} \text{ m} = 434 \text{ nm}$$

(b) This 434 nm wavelength is visible electromagnetic radiation.

18.19 (a) *Photodissociation* is cleavage of the O=O bond such that two neutral O atoms are produced: $O_2(g) \rightarrow 2O(g)$.

Photoionization is absorption of a photon with sufficient energy to eject an electron from an O_2 molecule: $O_2(g) + h\nu \rightarrow O_2^+ + e^-$.

(b) Photoionization of O_2 requires 1205 kJ/mol. Photodissociation requires only 495 kJ/mol. At lower elevations, solar radiation with wavelengths corresponding to 1205 kJ/mol or shorter has already been absorbed, whereas the longer wavelength radiation has passed through relatively well. Below 90 km, the increased concentration of O_2 and the availability of longer wavelength radiation cause the photodissociation process to dominate.

18.21 (a) A wavelength of 145 nm is in the ultraviolet portion of the electromagnetic spectrum. (See Figure 6.4.)

(b) *Analyze/Plan*. $E = hc/\lambda$. 145 nm = 1.45×10^{-7} m. Change J/photon to kJ/mol. Compare to the bond energy of O_2, 495 kJ/mol. *Solve.*

$$E = \frac{hc}{\lambda} = \frac{(6.626 \times 10^{-34} \text{ J-sec})(3.00 \times 10^{8} \text{ m/sec})}{1.45 \times 10^{-7} \text{ m}} = 1.37 \times 10^{-18} \text{ J/photon}$$

$$\frac{1.37 \times 10^{-18} \text{ J}}{\text{photon}} \times \frac{1 \text{ kJ}}{1000 \text{ J}} \times \frac{6.022 \times 10^{23} \text{ photons}}{1 \text{ mol}} = 825.55 = 826 \text{ kJ/mol}$$

The 145-nm photon has more than enough energy to photodissociate O_2.

According to Table 18.3, the photoionization energy of O_2 is 1205 kJ/mol. The 145-nm photon does not have enough energy to photoionize O_2.

Human Activities and Earth's Atmosphere (Section 18.2)

18.23 The oxidation state of oxygen in O_3, O_2, and O is zero (0). Reactions in which oxygen changes only from one of these species to another do not involve changes in oxidation state. Examples: $O(g) + O(g) \rightarrow O_2(g)$; $2\,O_3(g) \rightarrow 3\,O_2(g)$.

Ozone depletion reactions that involve a halogen oxide such as ClO do involve a change in oxidation state for oxygen. In ClO, the oxidation state of oxygen is either +1 or +2, but it is not zero. A reaction involving ClO and one of the oxygen species with a zero oxidation state does involve a change in the oxidation state of oxygen atoms.

18.25 (a) A *chlorofluorocarbon* is a compound that contains chlorine, fluorine, and carbon. A *hydrofluorocarbon* contains hydrogen, fluorine, and carbon; it contains hydrogen in place of chlorine.

(b) CFCs are harmful because they undergo photodissociation to produce Cl atoms that catalyze the destructions of ozone. HFCs are potentially less harmful to the ozone layer because they contain no C–Cl bonds. Their relatively stronger C–F bonds require more energy to undergo photodissociation, energy that is unlikely to be available in the stratosphere. (HFCs are still a powerful greenhouse gas. Montreal Protocal members have recently agreed to limit HFCs as well as CFCs.)

18.27 (a) *Analyze/Plan*. Given bond enthalpies in kJ/mol, calculate the maximum wavelength of a single photon that will rupture a C–F and a C–Cl bond, respectively. kJ/mol → J/molecule. $\lambda = hc/E$. ($\lambda = hc/E$ describes the energy/wavelength relationship of a single photon.) *Solve.*

$$\frac{485 \times 10^{3} \text{ J}}{1 \text{ mol}} \times \frac{1 \text{ mol}}{6.022 \times 10^{23} \text{ C-F bonds}} = 8.054 \times 10^{-19} = 8.05 \times 10^{-19} \text{ J/C-F bond}$$

$$\lambda = \frac{hc}{E} = \frac{(6.626 \times 10^{-34} \text{ J-sec})(3.00 \times 10^{8} \text{ m/sec})}{8.054 \times 10^{-19} \text{ J}} = 2.47 \times 10^{-7} \text{ m} = 247 \text{ nm}$$

$$\frac{328 \times 10^{3} \text{ J}}{1 \text{ mol}} \times \frac{1 \text{ mol}}{6.022 \times 10^{23} \text{ C-Cl bonds}} = 5.447 \times 10^{-19} = 5.45 \times 10^{-19} \text{ J/C-Cl bond}$$

$$\lambda = \frac{hc}{E} = \frac{(6.626 \times 10^{-34} \text{ J-sec})(3.00 \times 10^{8} \text{ m/sec})}{5.447 \times 10^{-19} \text{ J}} = 3.65 \times 10^{-7} \text{ m} = 365 \text{ nm}$$

(b) The maximum wavelength that can dissociate a C–F bond is 247 nm; shorter wavelengths are also effective. Since most wavelengths shorter than 240 nm are absorbed in the upper atmosphere, few effective photons will reach the lower atmosphere. We don't expect the photodissociation of C–F bonds to be significant in the lower atmosphere. (The dissociation of C–Cl bonds will be significant.)

18.29 *Analyze/Plan.* Write and balance equations for the reaction of NO(g) and NO_2(g) with water. First assume a simple acid–base reaction with a single product, HNO_3(aq); these equations can't be balanced. Reaction of NO(g) and NO_2(g) to produce HNO_3(aq) [or HNO_2(aq)] are redox reactions. An Internet search reveals that the reactions are as shown below. *Solve.*

$2\,NO_2(g) + H_2O(l) \rightleftharpoons HNO_2(aq) + HNO_3(aq)$

$2\,NO(g) + O_2(aq) + H_2O(l) \rightleftharpoons HNO_2(aq) + HNO_3(aq)$ or

$4\,NO_2(g) + O_2(aq) + 2\,H_2O(l) \rightleftharpoons 4\,HNO_3(aq)$

$4\,NO(g) + 3\,O_2(aq) + 2\,H_2O(l) \rightleftharpoons 4\,HNO_3(aq)$

18.31 (a) Acid rain is primarily H_2SO_4(aq).

$H_2SO_4(aq) + CaCO_3(s) \rightarrow CaSO_4(s) + H_2O(l) + CO_2(g)$

(b) The $CaSO_4$(s) would be much less reactive with acidic solution, because it would require a strongly acidic solution to shift the relevant equilibrium to the right.

$CaSO_4(s) + 2\,H^+(aq) \rightleftharpoons Ca^{2+}(aq) + 2\,HSO_4^-(aq)$

Note, however, that $CaSO_4$(s) is brittle and easily dislodged; it provides none of the structural strength of limestone.

18.33 *Analyze/Plan.* Given wavelength of a photon, place it in the electromagnetic spectrum, calculate its energy in kJ/mol, and compare it to an average bond dissociation energy. Use Figure 6.4; E(J/photon) = hc/λ. J/photon → kJ/mol. *Solve.*

(a) Ultraviolet (Figure 6.4).

(b) $$E_{photon} = hc/\lambda = \frac{6.626\times10^{-34}\ \text{J-s}\times3.00\times10^{8}\ \text{m/s}}{335\times10^{-9}\ \text{m}} = 5.934\times10^{-19}$$

$$= 5.93\times10^{-19}\ \text{J/photon}$$

$$\frac{5.934\times10^{-19}\ \text{J}}{1\ \text{photon}}\times\frac{6.022\times10^{23}\ \text{photons}}{1\ \text{mol}}\times\frac{1\ \text{kJ}}{1000\ \text{J}} = 357\ \text{kJ/mol}$$

(c) The average C–H bond energy from Table 8.3 is 413 kJ/mol. The energy calculated in part (b), 357 kJ/mol, is the energy required to break 1 mol of C–H bonds in formaldehyde, CH_2O. The C–H bond energy in CH_2O must be less than the "average" C–H bond energy.

(d)
```
    :O:                    :O:
    ||                     ||
H—C—H + hν  ——→  H—C· + H·
```

18.35 (a) Four sources transfer energy to the atmosphere, surface radiation, evapotranspiration, incoming solar radiation, and convective heating. Surface radiation makes the largest contribution. The total energy absorbed by the atmosphere is [350 + 78 + 67 + 24] = 519 W/m^2.

(b) Of the 519 W/m^2 transferred to the atmosphere, 324 W/m^2 are radiated back to the surface. The percentage is (324/519) × 100 = 62.4%.

Earth's Water (Section 18.3)

18.37 *Analyze/Plan*. Given salinity and density, calculate molarity. A salinity of 5.6 denotes that there are 5.6 g of dry salt per kg of water. 1.03 g/mL = 1.03 kg/L *Solve.*

$$\frac{5.6 \text{ g NaCl}}{1 \text{ kg soln}} \times \frac{1.03 \text{ kg soln}}{1 \text{ L soln}} \times \frac{1 \text{ mol NaCl}}{58.44 \text{ g NaCl}} \times \frac{1 \text{ mol Na}^+}{1 \text{ mol NaCl}} = 0.0987 = 0.099\ M \text{ Na}^+$$

18.39 *Analyze/Plan*. Given the power of sunlight per square meter striking Earth's surface, the enthalpy of evaporation of water, and specific heat capacity of water, calculate the amount of energy delivered by the Sun over a 12-hour day. Use this amount of energy to calculate: (a) how many grams of water can be evaporated and (b) the temperature of a 10.0 cm by 1 square meter volume of water after 12 hours in the sunlight, assuming no evaporation. Calculate the mass of this volume of water using density at 25 °C. *Solve.*

(a) $$\frac{168 \text{ W}}{\text{m}^2} \times \frac{1 \text{ J/s}}{1 \text{ W}} = \frac{168 \text{ J}}{\text{m}^2\text{-s}}$$

$$\frac{168 \text{ J}}{\text{m}^2\text{-s}} \times 1.00 \text{ m}^2 \times 12 \text{ h} \times \frac{60 \text{ min}}{1 \text{ h}} \times \frac{60 \text{ s}}{1 \text{ min}} \times \frac{1 \text{ kJ}}{1000 \text{ J}} = 7257.6 = 7.26 \times 10^3 \text{ kJ}$$

$$7257.6 \text{ kJ} \times \frac{1 \text{ mol H}_2\text{O}}{40.67 \text{ kJ}} \times \frac{18.02 \text{ g H}_2\text{O}}{1 \text{ mol H}_2\text{O}} = 3215.7 = 3.22 \times 10^3 \text{ g H}_2\text{O}$$

(b) $$1.00 \text{ m}^2 \times 10.0 \text{ cm} \times \frac{(100)^2 \text{ cm}^2}{1 \text{ m}^2} \times \frac{0.99707 \text{ g}}{1 \text{ cm}^3} = 99{,}707 = 9.97 \times 10^4 \text{ gH}_2\text{O}$$

$$7257.6 \text{ kJ} \times \frac{1000 \text{ J}}{1 \text{ kJ}} \times \frac{1 \text{ g-°C}}{4.184 \text{ J}} \times \frac{1}{99{,}707 \text{ g H}_2\text{O}} = 17.397 = 17.4 \text{ °C}$$

The final temperature is 26 °C + 17.4 °C = 43.4 °C.

18.41 *Analyze/Plan*. g $Mg(OH)_2$ → mol $Mg(OH)_2$ → mol ratio → mol CaO → g CaO. *Solve.*

$$1000 \text{ lb Mg(OH)}_2 \times \frac{453.6 \text{ g}}{\text{lb}} \times \frac{1 \text{ mol Mg(OH)}_2}{58.33 \text{ g Mg(OH)}_2} \times \frac{1 \text{ mol CaO}}{1 \text{ mol Mg(OH)}_2} \times \frac{56.08 \text{ g CaO}}{1 \text{ mol CaO}} = 4.361 \times 10^5 \text{ g CaO}$$

18.43 *Analyze/Plan*. Use molar concentrations of the six major ions in seawater from Table 18.5. Calculate charge in coulombs by multiplying molarity × integer charge × Faraday's constant (coulombs/mol). Sum the coulombic charges of the anions, the cations, and compare. *Solve.*

$$\frac{0.55 \text{ mol Cl}^-}{\text{L seawater}} \times \frac{1 \text{ mol charge}}{\text{mol Cl}^-} \times \frac{9.64853365 \text{ C}}{\text{mol}} = 5.30669351 = 5.3 \text{ C}$$

$$\frac{0.028 \text{ mol } SO_4^{2-}}{\text{L seawater}} \times \frac{2 \text{ mol charge}}{\text{mol } SO_4^{2-}} \times \frac{9.64853365 \text{ C}}{\text{mol}} = 0.54031788 = 0.54 \text{ C}$$

$$\frac{0.47 \text{ mol } Na^+}{\text{L seawater}} \times \frac{1 \text{ mol charge}}{\text{mol } K^+} \times \frac{9.64853365 \text{ C}}{\text{mol}} = 4.53481082 = 4.5 \text{ C}$$

$$\frac{0.054 \text{ mol } Mg^{2+}}{\text{L seawater}} \times \frac{2 \text{ mol charge}}{\text{mol } Mg^{2+}} \times \frac{9.64853365 \text{ C}}{\text{mol}} = 1.04204163 = 1.0 \text{ C}$$

$$\frac{0.010 \text{ mol } Ca^{2+}}{\text{L seawater}} \times \frac{2 \text{ mol charge}}{\text{mol } Mg^{2+}} \times \frac{9.64853365 \text{ C}}{\text{mol}} = 0.19297067 = 0.19 \text{ C}$$

$$\frac{0.010 \text{ mol } K^+}{\text{L seawater}} \times \frac{1 \text{ mol charge}}{\text{mol } K^+} \times \frac{9.64853365 \text{ C}}{\text{mol}} = 0.0964853365 = 0.096 \text{ C}$$

anion charge: [5.30669351 + 0.54031788] = 5.84701139 = 5.8 C

cation charge: [4.53481082 + 1.04204164 + 0.19297067 + 0.0964853365] = 5.86630846 = 5.9 C

The two numbers vary in the third significant figure. This is not surprising, because the molarities of the various ions are given to two significant figures.

Human Activities and Water Quality (Section 18.4)

18.45 *Analyze/Plan*. Given temperature and the concentration difference between the two solutions, (ΔM = 0.22 − 0.01 = 0.21 M), calculate the minimum pressure for reverse osmosis. Use the relationship $\Pi = MRT$ from Section 13.5. This is the pressure required to halt osmosis from the more dilute (0.01 M) to the more concentrated (0.22 M) solution. Slightly more pressure will initiate reverse osmosis. *Solve*.

$$\Pi = \Delta MRT = \frac{0.21 \text{ mol}}{\text{L}} \times \frac{0.08206 \text{ L-atm}}{\text{mol-K}} \times 298 \text{ K} = 5.135 = 5.1 \text{ atm}$$

The minimum pressure required to initiate reverse osmosis is greater than 5.1 atm.

18.47 *Analyze/Plan*. Under aerobic conditions, excess oxygen is present and decomposition leads to oxidized products, the element in its maximum oxidation state combined with oxygen. Under anaerobic conditions, little or no oxygen is present so decomposition leads to reduced products, the element in its minimum oxidation state combined with hydrogen. *Solve*.

(a) CO_2, HCO_3^-, H_2O, SO_4^{2-}, NO_3^-.

(b) $CH_4(g)$, $H_2S(g)$, $NH_3(g)$.

18.49 *Analyze/Plan*. Given the balanced equation, calculate the amount of one reactant required to react exactly with a certain amount of the other reactants. Solve the stoichiometry problem. g $C_{18}H_{29}SO_3^-$ → mol → mol ratio → mol O_2 → g O_2. *Solve*.

$$10.0 \text{ g } C_{18}H_{29}SO_3^- \times \frac{1 \text{ mol } C_{18}H_{29}SO_3^-}{325 \text{ g } C_{18}H_{29}SO_3^-} \times \frac{51 \text{ mol } O_2}{2 \text{ mol } C_{18}H_{29}SO_3^-} \times \frac{32.0 \text{ g } O_2}{1 \text{ mol } O_2} = 25.1 \text{ g } O_2$$

Notice that the mass of O_2 required is 2.5 times greater than the mass of biodegradable material.

18.51 *Analyze/Plan*. The reaction is metathesis. *Solve*.

$Mg^{2+}(aq) + Ca(OH)_2(s) \rightarrow Mg(OH)_2(s) + Ca^{2+}(aq)$

[The excess $Ca^{2+}(aq)$ is removed as $CaCO_3$ by naturally occurring bicarbonate or added Na_2CO_3.]

18.53 (a) *Trihalomethanes* are a class of molecules with one central carbon atom bound to one hydrogen and three halogen atoms. They are produced by the reaction of dissolved chlorine with organic matter naturally present in water, and are by-products of water disinfection via chlorination.

(b)

$CHCl_3$ (H–C bonded to three Cl) and $CHCl_2Br$ (H–C bonded to two Cl and one Br)

Green Chemistry (Section 18.5)

18.55 The fewer steps in a process, the less waste (solvents as well as unusable by-products) is generated. It is probably true that a process with fewer steps requires less energy at the site of the process, and it is certainly true that the less waste the process generates, the less energy is required to clean or dispose of the waste.

18.57 (a) cyclohexanone + H–O–O–H ⟶ caprolactone (seven-membered ring: C(=O)–O–CH_2–CH_2–CH_2–CH_2–CH_2) + H_2O

$$C_6H_{10}O + H{-}O{-}O{-}H \longrightarrow C_6H_{10}O_2 + H_2O$$

(b)

- **Prevention (1).** The alternative process eliminates production of 3-chlorobenzoic acid by-product, chlorine-containing waste that must be treated.
- **Atom economy (2)**. Most of the starting atoms are in the final product.
- **Less hazardous chemical synthesis (3)** and **Inherently safer for accident prevention (12)**. The starting material of the alternative process, if it is less concentrated than 30% by mass, is not shock sensitive, and the by-product is nontoxic water. The low molar mass of water means that a small amount of "waste" is generated.
- **Catalysis (9)** and **Design for energy efficiency (6)**. The alternative process is catalyzed, which could mean that the process will be more energy efficient than the Baeyer–Villiger reaction (see Solution 18.58).
- **Raw materials should be renewable (7)**. The catalyst can be recovered from the reaction mixture and reused. We don't have information about solvents or other auxiliary substances.

18.59 (a) Water as a solvent is much "greener" than benzene, which is a known carcinogen. Water fits criteria: (5) safer solvent, (7) renewable feedstock, and (12) inherently safer for accident prevention.

(b) Reaction temperature of 500 K rather than 1000 K is "greener," according to criteria (6) design for energy efficiency and (12) inherently safer chemistry for accident prevention. Also, low temperature is less likely to produce undesirable by-products that have to be separated and treated as waste, which fits criterion (1).

(c) Sodium chloride as a by-product rather than chloroform ($CHCl_3$) is "greener," according to criteria: (1) prevention, (3) less hazardous chemical systems, and (12) inherently safer.

Additional Exercises

18.64

$$2[Cl(g) + O_3(g) \rightarrow ClO(g) + O_2(g)] \quad [18.7]$$
$$2[ClO(g) + h\nu \rightarrow O(g) + Cl(g)] \quad [18.9]$$
$$O(g) + O(g) \rightarrow O_2(g)$$
$$2\,Cl(g) + 2\,O_3(g) + 2\,ClO(g) + 2\,O(g) \rightarrow 2\,ClO(g) + 3\,O_2(g) + 2\,Cl(g)$$
$$2\,O_3(g) \xrightarrow{Cl} 3\,O_2(g) \quad [18.10]$$

Note that Cl(g) fits the definition of a catalyst in this reaction.

18.67 (a) A CFC has C–Cl bonds and C–F bonds. In an HFC, the C–Cl bonds are replaced by C–H bonds.

(b) The longer a halogen-containing molecule exists in the stratosphere, the greater the likelihood that it will encounter light with energy sufficient to dissociate a carbon–halogen bond. Free halogen atoms catalyze the destruction of ozone.

(c) The bond dissociation enthalpy of a C–F bond is 485 kJ/mol, much more than for a C–Cl bond, 328 kJ/mol (Table 8.3). Although HFCs have long lifetimes in the stratosphere, it is infrequent that light with energy sufficient to dissociate a C–F bond will reach an HFC molecule. F atoms are much less likely than Cl atoms to be produced by photodissociation in the stratosphere.

(d) The main disadvantage of HFCs as replacements for CFCs is that they are potent greenhouse gases. Although HFCs are far less threatening to stratospheric ozone, they may contribute to global climate change.

18.69 (a) $CH_4(g) + 2\,O_2(g) \rightarrow CO_2(g) + 2\,H_2O(g)$

(b) $2\,CH_4(g) + 3\,O_2(g) \rightarrow 2\,CO(g) + 4\,H_2O(g)$

(c) vol $CH_4 \rightarrow$ vol $O_2 \rightarrow$ volume air ($X_{O_2} = 0.20948$)

Equal volumes of gases at the same temperature and pressure contain equal numbers of moles (Avogadro's law). If 2 moles of O_2 are required for 1 mole of CH_4, 2.0 L of pure O_2 are needed to burn 1.0 L of CH_4.

$$\text{vol } O_2 = X_{O_2} \times \text{vol}_{air} = \frac{\text{vol } O_2}{X_{O_2}} = \frac{2.0\text{ L}}{0.20948} = 9.5\text{ L air}$$

18.73 Given 168 watts/m^2 at 10% efficiency, find the land area needed to produce 12,000 megawatts. 13,200 megawatts = 13,200 × 10^6 = 1.32 × 10^{10} watts.

168 watts/m^2 (0.10) = 16.8 watts/m^2 solar energy possible with current technology.

$$1.32\times10^{10}\ \text{watts}\times\frac{1\ \text{m}^2}{16.8\ \text{watts}}=7.857\times10^8=7.9\times10^8\ \text{m}^2$$

The land area of New York City is 830 km^2, which is 830 × 10^6 m^2. The area needed for solar energy harvesting to provide peak power would then be $\frac{7.857\times10^8\ \text{m}^2}{830\times10^6\ \text{m}^2}=0.95$ times the land area of New York City.

18.75 (a) CO_3^{2-} is a relatively strong Brønsted–Lowry base and produces OH^- in aqueous solution according to the hydrolysis reaction:

$$CO_3^{2-}(aq) + H_2O(l) \rightleftharpoons HCO_3^-(aq) + OH^-(aq),\quad K_b = 1.8\times10^{-4}$$

If $[OH^-(aq)]$ is sufficient for the reaction quotient, Q, to exceed K_{sp} for $Mg(OH)_2$, the solid will precipitate.

(b)

$$\frac{125\ \text{mg Mg}^{2+}}{1\ \text{kg soln}}\times\frac{1\ \text{g Mg}^{2+}}{1000\ \text{mg Mg}^{2+}}\times\frac{1.00\ \text{kg soln}}{1.00\ \text{L soln}}\times\frac{1\ \text{mol Mg}^{2+}}{24.305\ \text{g Mg}^{2+}}=5.143\times10^{-3}$$

$$=5.14\times10^{-3}\ M\ \text{Mg}^{2+}$$

$$\frac{4.0\ \text{g Na}_2\text{CO}_3}{1.0\ \text{L soln}}\times\frac{1\ \text{mol CO}_3^{2-}}{106.0\ \text{g Na}_2\text{CO}_3}=0.03774=0.038\ M\ \text{CO}_3^{2-}$$

$$K_b=1.8\times10^{-4}=\frac{[HCO_3^-][OH^-]}{[CO_3^{2-}]}\approx\frac{x^2}{0.03774};\ x=[OH^-]=2.606\times10^{-3}$$

$$=2.6\times10^{-3}\ M$$

(This represents 6.9% hydrolysis, but the result will not be significantly different using the quadratic formula.)

$Q = [Mg^{2+}][OH^-]^2 = (5.143\times10^{-3})(2.606\times10^{-3})^2 = 3.5\times10^{-8}$

K_{sp} for $Mg(OH)_2 = 1.6\times10^{-12}$; $Q > K_{sp}$, so $Mg(OH)_2$ will precipitate.

Integrative Exercises

18.79 (a)

$$8{,}376{,}726\ \text{tons coal}\times\frac{83\ \text{ton C}}{100\ \text{ton coal}}\times\frac{44.01\ \text{ton CO}_2}{12.01\ \text{ton C}}=2.5\times10^7\ \text{ton CO}_2$$

$$8{,}376{,}726\ \text{tons coal}\times\frac{2.5\ \text{ton S}}{100\ \text{ton coal}}\times\frac{64.07\ \text{ton SO}_2}{32.07\ \text{ton S}}=4.2\times10^5\ \text{ton SO}_2$$

(b) $CaO(s) + SO_2(g) \rightarrow CaSO_3(s)$

$$4.18\times10^5\ \text{ton SO}_2\times\frac{55\ \text{ton SO}_2\ \text{removed}}{100\ \text{ton SO}_2\ \text{produced}}\times\frac{120.15\ \text{ton CaSO}_3}{64.07\ \text{ton SO}_2}$$

$$=4.3\times10^5\ \text{ton CaSO}_3$$

18.82 (a) $H—\ddot{\underset{\cdot\cdot}{O}}—H \longrightarrow H\cdot + \cdot\ddot{\underset{\cdot\cdot}{O}}—H$

(b) $\Delta H = 2D(O—H) - D(O—H) = D(O—H) = 463 \text{ kJ/mol}$

$$\frac{463 \text{ kJ}}{\text{mol } H_2O} \times \frac{1 \text{ mol } H_2O}{6.022 \times 10^{23} \text{ molecules}} \times \frac{1000 \text{ J}}{\text{kJ}} = 7.688 \times 10^{-19}$$

$$= 7.69 \times 10^{-19} \text{ J}/H_2O\text{molecule}$$

$$\lambda = \frac{hc}{\Delta E} = \frac{6.626 \times 10^{-34} \text{ J-sec} \times 2.998 \times 10^{8} \text{ m/s}}{7.688 \times 10^{-19} \text{ J}} = 2.58 \times 10^{-7} \text{ m} = 258 \text{ nm}$$

This wavelength is in the UV region of the spectrum, close to the visible.

(c)

$$OH(g) + O_3(g) \rightarrow HO_2(g) + O_2(g)$$
$$HO_2(g) + O(g) \rightarrow OH(g) + O_2(g)$$
$$\overline{OH(g) + O_3(g) + HO_2(g) + O(g) \rightarrow HO_2(g) + 2\,O_2(g) + OH(g)}$$
$$O_3(g) + O(g) \rightarrow 2\,O_2(g)$$

OH(g) is the catalyst in this overall reaction, another pathway for the destruction of ozone.

18.84 (i) $ClO(g) + O_3(g) \rightarrow ClO_2(g) + O_2(g)$

$\Delta H_i = \Delta H_f^\circ\ ClO_2(g) + \Delta H_f^\circ\ O_2(g) - \Delta H_f^\circ\ ClO(g) - \Delta H_f^\circ\ O_3(g)$

$\Delta H_i = 102 + 0 - 101 - (142.3) = -141 \text{ kJ}$

(ii) $ClO_2(g) + O(g) \rightarrow ClO(g) + O_2(g)$

$\Delta H_{ii} = \Delta H_f^\circ\ ClO(g) + \Delta H_f^\circ\ O_2(g) - \Delta H_f^\circ\ ClO_2(g) - \Delta H_f^\circ\ O(g)$

$\Delta H_{ii} = 101 + 0 - 102 - (247.5) = -249 \text{ kJ}$

(overall) $ClO(g) + O_3(g) + ClO_2(g) + O(g) \rightarrow ClO_2(g) + O_2(g) + ClO(g) + O_2(g)$

$$O_3(g) + O(g) \rightarrow 2O_2(g)$$

$\Delta H = \Delta H_i + \Delta H_{ii} = -141 \text{ kJ} + (-249) \text{ kJ} = -390 \text{ kJ}$

Because the enthalpies of both (i) and (ii) are distinctly exothermic, it is possible that the $ClO - ClO_2$ pair could be a catalyst for the destruction of ozone.

18.87 (a) Holding one reactant concentration constant and changing the other, evaluate the effect this has on the initial rate. Use these observations to write the rate law.

Compare Experiments 1 and 3. $[O_3]$ is constant, [H] doubles, initial rate doubles. The reaction is first order in [H].

Compare Experiments 2 and 1. [H] is constant, $[O_3]$ doubles, initial rate doubles. The reaction is first order in $[O_3]$.

rate = $k[O_3][H]$

(b) Calculate a value for the rate constant for each experiment and average them to obtain a single representative value.

rate = k[O_3][H]; k = rate/[O_3][H]

$$k_1 = \frac{1.88\times10^{-14}\ M/s}{(5.17\times10^{-33}\ M)(3.22\times10^{-26}\ M)} = 1.1293\times10^{44} = 1.13\times10^{44}$$

$$k_2 = \frac{9.44\times10^{-15}\ M/s}{(2.59\times10^{-33}\ M)(3.25\times10^{-26}\ M)} = 1.1215\times10^{44} = 1.12\times10^{44}$$

$$k_3 = \frac{3.77\times10^{-14}\ M/s}{(5.19\times10^{-33}\ M)(6.46\times10^{-26}\ M)} = 1.1245\times10^{44} = 1.12\times10^{44}$$

$$k_{avg} = (1.1293\times10^{44} + 1.1215\times10^{44} + 1.1245\times10^{44})/3 = 1.1251\times10^{44} = 1.13\times10^{44}\ M^{-1}\ s^{-1}$$

18.91 (a) Process (i) is greener, because it does not involve the toxic reactant phosgene ($COCl_2$) and the by-product is water, not HCl.

(b) Reaction (i): C in CO_2 is linear with *sp* hybridization; C in R–N=C=O is linear with *sp* hybridization; C in the urethane monomer is trigonal planar with sp^2 hybridization. Reaction (ii): C in $COCl_2$ is trigonal planar with sp^2 hybridization; C in R–N=C=O is linear with *sp* hybridization; C in the urethane monomer is trigonal planar with sp^2 hybridization.

(c) Traditionally, industrial processes are conducted at higher temperatures to speed up reactions and encourage formation of product. However, this is not a green solution, because it requires additional energy. Using Le Châtelier's principle, we could either "push" or "pull" the reaction toward products. The "push" requires that we increase the amount of reactants, again not a green approach. The greenest way to promote formation of the isocyanate is to "pull" the reaction forward by removing by-product from the reaction mixture. In reaction (i), remove water; in reaction (ii), remove HCl.

19 Chemical Thermodynamics

Visualizing Concepts

19.1 (a)

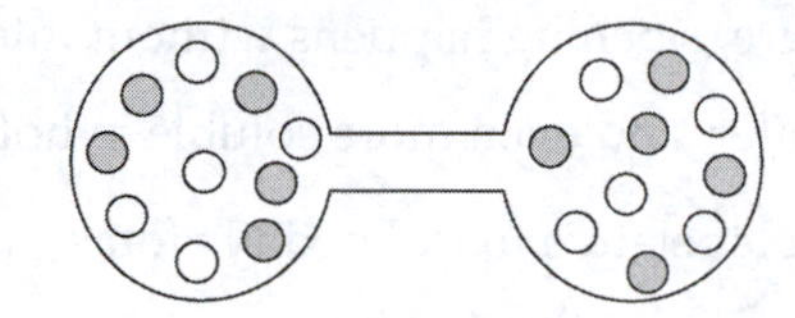

(b) ΔS is positive, because the disorder of the system increases. Each gas has greater motional freedom as it expands into the second bulb, and there are many more possible arrangements for the mixed gases.

By definition, ideal gases experience no attractive or repulsive intermolecular interactions, so ΔH for the mixing of ideal gases is zero, assuming heat exchange only between the two bulbs.

(c) The process is spontaneous and, therefore, irreversible. It is inconceivable that the gases would reseparate.

(d) The entropy change of the surroundings is related to ΔH for the system. Because we are mixing ideal gases and $\Delta H = 0$, ΔH_{surr} is also zero, assuming heat exchange only between the two bulbs.

19.4 Both ΔH and ΔS for this reaction are positive.

The reaction involves breaking five blue–blue and twenty blue–red bonds and then forming twenty blue–red bonds. The net change is breaking five blue–blue bonds. Enthalpies for bond breaking (Section 8.8) are always positive.

In the depicted reaction, both reactants and products are in the gas phase (they are far apart and randomly placed). There are twice as many molecules (or moles) of gas in the products, so ΔS is positive for this reaction.

19.7 (a) At 300 K, $\Delta H = T\Delta S$. Because $\Delta G = \Delta H - T\Delta S$, $\Delta G = 0$ at this point. When $\Delta G = 0$, the system is at equilibrium.

(b) The reaction is spontaneous when ΔG is negative. This condition is met when $T\Delta S > \Delta H$. From the diagram, $T\Delta S > \Delta H$ when $T > 300$ K. The reaction is spontaneous at temperatures above 300 K.

19.10 (a) True. When $\Delta G = 0$, the reaction is at equilibrium.

(b) False. At equilibrium, there is a mixture of reactants and products.

(c) False. There are fewer moles of gas in the products than the reactants, so ΔS is negative.

(d) False. At the left and right extremes of the graph, reactants and products are gases at 1 atm pressure; they are in their standard states. The quantity "x" is the difference in free energy between reactants and products in their standard states, ΔG°.

(e) True. ΔG is a measure of the driving force for a reaction to reach equilibrium.

Spontaneous Processes (Section 19.1)

19.11 *Analyze/Plan.* Follow the logic in Sample Exercise 19.1. *Solve.*

(a) Spontaneous; at ambient temperature, ripening happens without intervention.

(b) Spontaneous; sugar is soluble in water, and even more soluble in hot coffee.

(c) Spontaneous; N_2 molecules are stable relative to isolated N atoms.

(d) Spontaneous; under certain atmospheric conditions, lightning occurs.

(e) Nonspontaneous; CO_2 and H_2O are in contact continuously at atmospheric conditions in nature and do not form CH_4 and O_2.

19.13 (a) True, assuming the conditions are the same for the forward and reverse reactions.

(b) False. A spontaneous process occurs without outside intervention. This definition says nothing about how quickly the process occurs. Spontaneity is a thermodynamic property, while rate is a kinetic property.

(c) False. All spontaneous processes are real processes and real processes are irreversible.

(d) True.

(e) False. The maximum amount of work can be accomplished by a reversible process.

19.15 *Analyze/Plan.* Define the system and surroundings. Use the appropriate definition to answer the specific questions. *Solve.*

(a) Water is the system. Heat must be added to the system to evaporate the water. The process is endothermic.

(b) At 1 atm, the reaction is spontaneous at temperatures above 100 °C.

(c) At 1 atm, the reaction is nonspontaneous at temperatures below 100 °C.

(d) The two phases are in equilibrium at 100 °C.

19.17 *Analyze/Plan.* Consider the definitions of the terms *reversible*, *isothermal*, and *state function* to answer the questions. *Solve.*

(a) No. Temperature is a state function, so a change in temperature does not depend on pathway. (As we are talking about an ideal gas, a reversible pathway may be possible for this change in state. An irreversible pathway is always possible.)

(b) No. An isothermal process occurs at constant temperature.

(c) No. ΔE is a state function. ΔE = q + w; q and w are not state functions. Their values do depend on path, but their sum, ΔE, does not.

19.19 *Analyze/Plan.* Define the system and surroundings. Use the appropriate definition to answer the specific questions. *Solve.*

(a) An ice cube can melt reversibly at the conditions of temperature and pressure where the solid and liquid are in equilibrium. At 1 atm external pressure, the normal melting point of water is 0 °C.

(b) No. We know that melting is endothermic, so ΔH for melting ice is a positive, nonzero value. Also, since ΔH is a state function, the nonzero value is independent of path. Whether the ice cube melts reversibly or irreversibly, ΔH for the process is not zero.

Entropy and the Second Law of Thermodynamics (Section 19.2)

19.21 (a) True.

(b) False. For a reversible process, the entropy change of the universe is zero.

(c) True.

(d) False. For a reversible process, the entropy change to the system need not be zero, but it must be matched by the entropy change to the surroundings.

19.23 (a) $Br_2(l) \rightarrow Br_2(g)$, entropy increases, more mol gas in products, greater motional freedom.

(b) $$\Delta S = \frac{\Delta H}{T} = \frac{29.6 \text{ kJ}}{\text{mol } Br_2(l)} \times 1.00 \text{ mol } Br_2(l) \times \frac{1}{(273.15 + 58.8)\text{K}} \times \frac{1000 \text{ J}}{1 \text{ kJ}} = 89.2 \text{ J/K}$$

19.25 (a) False. The second law states that entropy is conserved for a reversible process.

(b) True. In a reversible process, $\Delta S_{sys} + \Delta S_{surr} = 0$. If ΔS_{sys} is positive, ΔS_{surr} must be negative.

(c) False. Because ΔS_{univ} must be positive for a spontaneous process, ΔS_{surr} must be greater than –4.2 J/K. Entropy is not conserved for a spontaneous process.

19.27 *Analyze.* Consider ΔS for the isothermal expansion of 0.200 mol of an ideal gas at 27 °C and an initial volume of 10.0 L.

(a) Whenever an ideal gas expands isothermally, we expect an increase in entropy, or positive ΔS, owing to the greater volume available for motion of the particles.

(b) *Plan.* Use the relationship $\Delta S_{sys} = nR \ln(V_2/V_1)$, Equation 19.3.

Solve. $\Delta S_{sys} = 0.200 \text{ (8.314 J/mol-K)}(\ln [18.5 \text{ L}/10.0 \text{ L}]) = 1.02 \text{ J/K}$.

Check. We expect ΔS to be positive when the motional freedom of a gas increases, and our calculation agrees with this prediction.

(c) No. The temperature at which the expansion occurs is not needed to calculate the entropy change, as long as the process is isothermal.

The Molecular Interpretation of Entropy and the Third Law of Thermodynamics (Section 19.3)

19.29 (a) Yes, the expansion is spontaneous.

(b) The ideal gas is the system, and everything else, including the vessel containing the vacuum, is the surroundings. There is literally nothing inside the vessel containing the vacuum, no gas molecules and no physical barriers. As the ideal gas expands into the vacuum, there is nothing for it to "push back," so no work is done. Mathematically, $w = -P_{ext}\Delta V$. Because the gas expands into a vacuum, $P_{ext} = 0$ and $w = 0$.

(c) Entropy. The "driving force" for the expansion of the gas is the increase in entropy associated with greater volume, more motional freedom, and more possible positions for the gas particles.

19.31 (a) The higher the temperature, the broader the distribution of molecular speeds and kinetic energies available to the particles. At higher temperature, the wider range of accessible kinetic energies leads to more microstates for the system.

(b) A decrease in volume reduces the number of possible positions for the particles and leads to fewer microstates for the system.

(c) Going from liquid to gas, particles have greater translational motion, which increases the number of positions available to the particles and the number of microstates for the system.

19.33 *Analyze/Plan*. Consider the conditions that lead to an increase in entropy: more mol gas in products than reactants, increase in volume of sample and, therefore, number of possible arrangements, more motional freedom of molecules, and so on. *Solve*.

(a) More gaseous particles means more possible arrangements and greater disorder; ΔS is positive.

(b) S_{sys} increases when a banana ripens. Starches are polysaccharides, large molecules that break down into sugars, smaller monosaccharides and disaccharides, when ripening occurs. Formation of the sugars increases the number of molecules and the entropy.

S_{sys} clearly increases in 19.11 (b), where there is an increase in volume and possible arrangements for the sample.

In 19.11 (c), the system goes from two moles of gaseous reactants to one mole of gaseous products, and S_{sys} decreases.

In 19.11 (d), the entropy of the universe clearly increases, but the definition of the system in a lightning strike is more problematic.

In 19.11 (e), the specified state is room temperature and 1 atm pressure. This means that H_2O is present as a liquid; there is then 1 mol of gaseous reactants (CO_2) and 3 mol of gaseous products (CH_4 and 2 O_2), so S_{sys} increases. (The reaction is not spontaneous because of the very large positive ΔH_{sys} for the reaction as written.)

19.35 *Analyze/Plan.* Consider the conditions that lead to an increase in entropy: more mol gas in products than reactants, increase in volume of sample and, therefore, number of possible arrangements, more motional freedom of molecules, and so on. *Solve.*

(a) S increases; translational motion is greater in the liquid than the solid.

(b) S decreases; volume and translational motion decrease going from the gas to the liquid.

(c) S increases; volume and translational motion are greater in the gas than the solid.

19.37 (a) False. The entropy of a pure crystalline substance at absolute zero is zero.

(b) True.

(c) False. Monoatomic gases have no rotational or vibrational states.

(d) True.

19.39 *Analyze/Plan.* Consider the factors that lead to higher entropy: more mol gas in products than reactants, increase in volume of sample and, therefore, number of possible arrangements, more motional freedom of molecules, and so on. *Solve.*

(a) Ar(g) (gases have higher entropy due primarily to much larger volume)

(b) He(g) at 1.5 atm (larger volume and more motional freedom)

(c) 1 mol of Ne(g) in 15.0 L (larger volume provides more motional freedom)

(d) $CO_2(g)$ (more motional freedom)

19.41 *Analyze/Plan.* Consider the markers of an increase in entropy for a chemical reaction: liquids or solutions formed from solids, gases formed from either solids or liquids, increase in mol gas during reaction. *Solve.*

(a) ΔS negative (moles of gas decrease)

(b) ΔS positive (gas produced, increased disorder)

(c) ΔS negative (moles of gas decrease)

(d) ΔS is small and probably positive [moles of gas same in reactants and products, $H_2O(g)$ is more structurally complex than $H_2(g)$]

Entropy Changes in Chemical Reactions (Section 19.4)

19.43 (a)

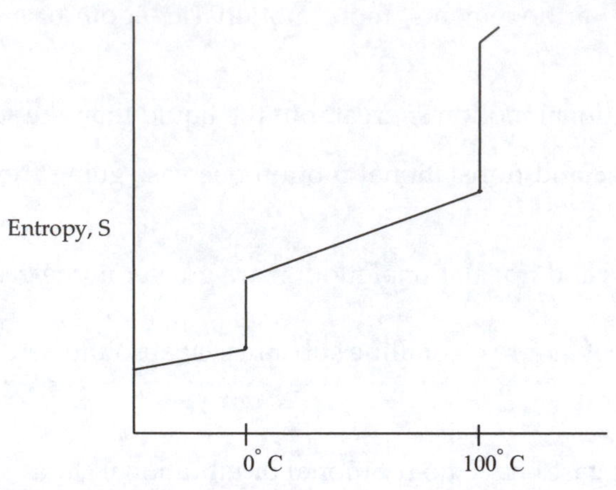

(b) Boiling water, at 100 °C, has a much larger entropy change than melting ice at 0 °C. Before and after melting, H_2O molecules are touching. And there is actually a small decrease in volume going from solid to liquid water. Boiling drastically increases the distance between molecules and the volume of the sample. The increase in available molecular positions is much greater for boiling than melting, so the entropy change is also greater.

19.45 *Analyze/Plan*. Given two molecules in the same state, predict which will have the higher molar entropy. In general, for molecules in the same state, the more atoms in the molecule, the more degrees of freedom, the greater the number of microstates, and the higher the standard entropy, S°.

(a) $C_2H_6(g)$ has more degrees of freedom and larger S°.

(b) $CO_2(g)$ has more degrees of freedom and larger S°.

19.47 *Analyze/Plan*. Consider the conditions that lead to an increase in entropy: more mol gas in products than reactants, increase in volume of sample and, therefore, number of possible arrangements, more motional freedom of molecules, and so on. *Solve*.

(a) Sc(g) will have the higher standard entropy at 25 °C. In general, the gas phase of a substance has a larger S° than the solid phase because of the greater volume and motional freedom of the molecules. Sc(s), 34.6 J/mol-K; Sc(g), 174.7 J/mol-K.

(b) $NH_3(g)$ will have the higher standard entropy at 25 °C. Molecules in the gas phase have more motional freedom than molecules in solution. $NH_3(g)$, 192.5 J/mol-K; $NH_3(aq)$, 111.3 J/mol-K.

(c) $O_3(g)$ will have the higher standard entropy at 25 °C. The triatomic molecule will have more vibrational degrees of freedom than the diatomic molecule. $O_2(g)$, 205.0 J/K; $O_3(g)$, 237.6 J/K.

(d) C(graphite) will have the higher standard entropy at 25 °C. Diamond is a network covalent solid with each C atom tetrahedrally bound to four other C atoms. Graphite consists of sheets of fused planar 6-membered rings with each C atom bound in a trigonal planar arrangement to three other C atoms. The internal entropy in graphite is greater because there is translational freedom among the planar sheets of C atoms while there is very little vibrational freedom within the network covalent diamond lattice. C(diamond), 2.43 J/mol-K; C(graphite) 5.69 J/mol-K.

19.49 For elements with similar structures, the heavier the atoms, the lower the vibrational frequencies at a given temperature. This means that more vibrations can be accessed at a particular temperature resulting in a greater absolute entropy for the heavier elements.

19.51 *Analyze/Plan.* Follow the logic in Sample Exercise 19.5. *Solve.*

(a) $\Delta S° = S°\ C_2H_6(g) - S°\ C_2H_4(g) - S°\ H_2(g)$
$= 229.5 - 219.4 - 130.58 = -120.5\ J/K$

$\Delta S°$ is negative because there are fewer moles of gas in the products.

(b) $\Delta S° = 2S°\ NO_2(g) - \Delta S°\ N_2O_4(g) = 2(240.45) - 304.3 = +176.6\ J/K$

$\Delta S°$ is positive because there are more moles of gas in the products.

(c) $\Delta S° = \Delta S°\ BeO(s) + \Delta S°\ H_2O(g) - \Delta S°\ Be(OH)_2(s)$
$= 13.77 + 188.83 - 50.21 = +152.39\ J/K$

$\Delta S°$ is positive because the product contains more total particles and more moles of gas.

(d) $\Delta S° = 2S°\ CO_2(g) + 4S°\ H_2O(g) - 2S°\ CH_3OH(g) - 3S°\ O_2(g)$
$= 2(213.6) + 4(188.83) - 2(237.6) - 3(205.0) = +92.3\ J/K$

$\Delta S°$ is positive because the product contains more total particles and more moles of gas.

Gibbs Free Energy (Sections 19.5 and 19.6)

19.53 (a) Yes. $\Delta G = \Delta H - T\Delta S$

(b) No. If ΔG is positive, the process is nonspontaneous.

(c) No. There is no relationship between ΔG and rate of reaction. A spontaneous reaction, one with a $-\Delta G$, may occur at a very slow rate. For example, $2\ H_2(g) + O_2(g) \rightarrow 2\ H_2O(g)$, $\Delta G = -457$ kJ is very slow if not initiated by a spark.

19.55 *Analyze/Plan.* Consider the definitions of $\Delta H°$, $\Delta S°$, and $\Delta G°$, along with sign conventions. $\Delta G° = \Delta H° - T\Delta S°$. *Solve.*

(a) $\Delta H°$ is negative; the reaction is exothermic.

(b) $\Delta S°$ is negative; the reaction leads to decrease in disorder (increase in order) of the system.

(c) $\Delta G° = \Delta H° - T\Delta S° = -35.4\ kJ - 298\ K\ (-0.0855\ kJ/K) = -9.921 = -9.9\ kJ$

(d) At 298 K, $\Delta G°$ is negative. If all reactants and products are present in their standard states, the reaction is spontaneous (in the forward direction) at this temperature.

19.57 *Analyze/Plan.* Follow the logic in Sample Exercises 19.6 and 19.7. Calculate $\Delta H°$ according to Equation 5.31, $\Delta S°$ by Equation 19.8, and $\Delta G°$ by Equation 19.14. Then, use $\Delta H°$ and $\Delta S°$ to calculate $\Delta G°$ using Equation 19.12, $\Delta G° = \Delta H° - T\Delta S°$. *Solve.*

(a) $\Delta H° = 2(-268.61) - [0 + 0] = -537.22\ kJ$

$\Delta S° = 2(173.51) - [130.58 + 202.7] = 13.74 = 13.7\ J/K$

$\Delta G° = 2(-270.70) - [0 + 0] = -541.40\ kJ$

$\Delta G° = -537.22\ kJ - 298(0.01374)\ kJ = -541.31\ kJ$

(b) $\Delta H° = -106.7 - [0 + 2(0)] = -106.7$ kJ

$\Delta S° = 309.4 - [5.69 + 2(222.96)] = -142.21 = -142.2$ J/K

$\Delta G° = -64.0 - [0 + 2(0)] = -64.0$ kJ

$\Delta G° = -106.7$ kJ $- 298(-0.14221)$ kJ $= -64.3$ kJ

(c) $\Delta H° = 2(-542.2) - [2(-288.07) + 0] = -508.26 = -508.3$ kJ

$\Delta S° = 2(325) - [2(311.7) + 205.0] = -178.4 = -178$ J/K

$\Delta G° = 2(-502.5) - [2(-269.6) + 0] = -465.8$ kJ

$\Delta G° = -508.26$ kJ $- 298(-0.1784)$ kJ $= -455.097 = -455.1$ kJ

(The discrepancy in $\Delta G°$ values is because of experimental uncertainties in the tabulated thermodynamic data.)

(d) $\Delta H° = -84.68 + 2(-241.82) - [2(-201.2) + 0] = -165.92 = -165.9$ kJ

$\Delta S° = 229.5 + 2(188.83) - [2(237.6) + 130.58] = 1.38 = 1.4$ J/K

$\Delta G° = -32.89 + 2(-228.57) - [2(-161.9) + 0] = -166.23 = -166.2$ kJ

$\Delta G° = -165.92$ kJ $- 298(0.00138)$ kJ $= -166.33 = -166.3$ kJ

19.59 *Analyze/Plan*. Follow the logic in Sample Exercise 19.7. *Solve*.

(a) $\Delta G° = 2\Delta G°\ SO_3(g) - [2\Delta G°\ SO_2(g) + \Delta G°\ O_2(g)]$
$= 2(-370.4) - [2(-300.4) + 0] = -140.0$ kJ, spontaneous

(b) $\Delta G° = 3\Delta G°\ NO(g) - [\Delta G°\ NO_2(g) + \Delta G°\ N_2O(g)]$
$= 3(86.71) - [51.84 + 103.59] = +104.70$ kJ, nonspontaneous

(c) $\Delta G° = 4\Delta G°\ FeCl_3(s) + 3\Delta G°\ O_2(g) - [6\Delta G°\ Cl_2(g) + 2\Delta G°\ Fe_2O_3(s)]$
$= 4(-334) + 3(0) - [6(0) + 2(-740.98)] = +146$ kJ, nonspontaneous

(d) $\Delta G° = \Delta G°\ S(s) + 2\Delta G°\ H_2O(g) - [\Delta G°\ SO_2(g) + 2\Delta G°\ H_2(g)]$
$= 0 + 2(-228.57) - [(-300.4) + 2(0)] = -156.7$ kJ, spontaneous

19.61 *Analyze/Plan*. Follow the logic in Sample Exercise 19.8(a). *Solve*.

(a) $2\ C_8H_{18}(l) + 25\ O_2(g) \rightarrow 16\ CO_2(g) + 18\ H_2O(l)$

(b) Because there are more moles of gas in the reactants, $\Delta S°$ is negative, which makes $-T\Delta S$ positive. $\Delta G°$ is less negative than $\Delta H°$. (This argument is true for the reaction as written. If the products are all in the gas phase, there are more moles of gas in the products and $\Delta G°$ is more negative than $\Delta H°$.)

19.63 *Analyze/Plan*. Based on the signs of ΔH and ΔS for a particular reaction, assign a category from Table 19.3 to each reaction. *Solve*.

(a) (iii) ΔG is negative at low temperatures, positive at high temperatures. That is, the forward reaction is spontaneous at lower temperatures but not spontaneous at higher temperatures.

(b) (ii) ΔG is positive at all temperatures. The forward reaction is nonspontaneous at all temperatures.

(c) (iv) ΔG is positive at low temperatures, negative at high temperatures. That is, the forward reaction will proceed spontaneously at high temperature, but not at low temperature.

19.65 *Analyze/Plan*. We are told that the reaction is barely spontaneous and endothermic, and asked to estimate the sign and magnitude of ΔS. If a reaction is spontaneous, $\Delta G < 0$. Use this information with Equation 19.11 to solve the problem. *Solve*.

At 390 K, $\Delta G < 0$; $\Delta G = \Delta H - T\Delta S < 0$

$23.7\ \text{kJ} - 390\ \text{K}\ (\Delta S) < 0$; $23.7\ \text{kJ} < 390\ \text{K}\ (\Delta S)$; $\Delta S > 23.7\ \text{kJ}/390\ \text{K}$

$\Delta S > 0.06077\ \text{kJ/K}$ or $\Delta S > 60.8\ \text{J/K}$

The reaction is spontaneous and endothermic, so the sign of ΔS must be positive. Because the reaction is "barely" spontaneous, the magnitude will not be much greater than 61 J/K.

19.67 *Analyze/Plan*. Use Equation 19.11 to calculate T when $\Delta G = 0$. This is similar to calculating the temperature of a phase transition in Sample Exercise 19.10. Use Table 19.3 to determine whether the reaction is spontaneous or nonspontaneous above this temperature. *Solve*.

(a) $\Delta G = \Delta H - T\Delta S$; $0 = -32\ \text{kJ} - T(-98\ \text{J/K})$; $32 \times 10^3\ \text{J} = T(98\ \text{J/K})$

$T = 32 \times 10^3\ \text{J}/(98\ \text{J/K}) = 326.5 = 330\ \text{K}$

(b) Nonspontaneous. The sign of ΔS is negative, so as T increases, ΔG becomes more positive.

19.69 *Analyze/Plan*. Given a chemical equation and thermodynamic data (values of ΔH_f°, ΔG_f°, and S°) for reactants and products, predict the variation of ΔG° with temperature and calculate ΔG° at 800 K and 1000 K. Use Equations 5.31 and 19.8 to calculate ΔH° and ΔS°, respectively; use these values to calculate ΔG° at various temperatures, using Equation 19.12. The signs of ΔH° and ΔS° determine the variation of ΔG° with temperature. *Solve*.

(a) Calculate ΔH° and ΔS° to determine the sign of $T\Delta S^\circ$.

$\Delta H^\circ = 3\Delta H^\circ\ NO(g) - \Delta H^\circ\ NO_2(g) - \Delta H^\circ\ N_2O(g)$

$= 3(90.37) - 33.84 - 81.6 = 155.7\ \text{kJ}$

$\Delta S^\circ = 3S^\circ\ NO(g) - S^\circ\ NO_2(g) - S^\circ\ N_2O(g)$

$= 3(210.62) - 240.45 - 220.0 = 171.4\ \text{J/K}$

$\Delta G^\circ = \Delta H^\circ - T\Delta S^\circ$. Because ΔS° is positive, $-T\Delta S^\circ$ becomes more negative as T increases and ΔG° becomes more negative.

(b) $\Delta G^\circ = \Delta H^\circ - T\Delta S^\circ = 155.7\ kJ - (800\ K)(0.1714\ kJ/K)$

$\Delta G^\circ = 155.7\ kJ - 137\ kJ = 19\ kJ$

Because ΔG° is positive at 800 K, the reaction is not spontaneous at this temperature.

(c) $\Delta G^\circ = 155.7\ kJ - (1000\ K)(0.1714\ kJ/K) = 155.7\ kJ - 171.4\ kJ = -15.7\ kJ$

ΔG° is negative at 1000 K and the reaction is spontaneous at this temperature.

19.71 *Analyze/Plan*. Follow the logic in Sample Exercise 19.10. *Solve*.

(a) $\Delta S^\circ_{vap} = \Delta H^\circ_{vap}/T_b;\ T_b = \Delta H^\circ_{vap}/\Delta S^\circ_{vap}$

$\Delta H^\circ_{vap} = \Delta H^\circ\ C_6H_6(g) - \Delta H^\circ\ C_6H_6(l) = 82.9 - 49.0 = 33.9\ kJ$

$\Delta S^\circ_{vap} = S^\circ\ C_6H_6(g) - S^\circ\ C_6H_6(l) = 269.2 - 172.8 = 96.4\ J/K$

$T_b = 33.9 \times 10^3\ J/96.4\ J/K = 351.66 = 352\ K = 79\ ^\circ C$

(b) From the *Handbook of Chemistry and Physics*, 74th edition, $T_b = 80.1\ ^\circ C$. The values are remarkably close; the small difference is due to deviation from ideal behavior by $C_6H_6(g)$ and experimental uncertainty in both the boiling point measurement and the thermodynamic data.

19.73 *Analyze/Plan*. We are asked to write a balanced equation for the combustion of acetylene, calculate ΔH° for this reaction, and calculate maximum useful work possible by the system. Combustion is combination with O_2 to produce CO_2 and H_2O. Calculate ΔH° using data from Appendix C and Equation 5.31. The maximum obtainable work is ΔG (Equation 19.18), which can be calculated from data in Appendix C and Equation 19.14. *Solve*.

(a) $C_2H_2(g) + 5/2\ O_2(g) \rightarrow 2\ CO_2(g) + H_2O(l)$

(b) $\Delta H^\circ = 2\Delta H^\circ\ CO_2(g) + \Delta H^\circ\ H_2O(l) - \Delta H^\circ\ C_2H_2(g) - 5/2\Delta H^\circ\ O_2(g)$

$= 2(-393.5) - 285.83 - 226.77 - 5/2(0)$

$= -1299.6$ kJ produced/mol C_2H_2 burned

(c) $w_{max} = \Delta G^\circ = 2\Delta G^\circ\ CO_2(g) + \Delta G^\circ\ H_2O(l) - \Delta G^\circ\ C_2H_2(g) - 5/2\ \Delta G^\circ\ O_2(g)$

$= 2(-394.4) - 237.13 - 209.2 - 5/2(0) = -1235.1\ kJ$

The negative sign indicates that the system does work on the surroundings; the system can accomplish a maximum of 1235.1 kJ of work on its surroundings.

Free Energy and Equilibrium (Section 19.6)

19.75 *Analyze/Plan*. We are given a chemical reaction and asked to predict the effect of the partial pressure of $O_2(g)$ on the value of ΔG for the system. Consider the relationship $\Delta G = \Delta G^\circ + RT \ln Q$, where Q is the reaction quotient. *Solve*.

(a) ΔG decreases. $O_2(g)$ appears in the denominator of Q for this reaction. An increase in pressure of O_2 decreases Q and ΔG gets smaller or becomes more negative. Increasing the concentration or partial pressure of a reactant increases the tendency for a reaction to occur.

(b) ΔG increases. $O_2(g)$ appears in the numerator of Q for this reaction. Increasing the pressure of O_2 increases Q and ΔG becomes more positive. Increasing the concentration or partial pressure of a product decreases the tendency for the reaction to occur.

(c) ΔG increases. $O_2(g)$ appears in the numerator of Q for this reaction. An increase in pressure of O_2 increases Q and ΔG becomes more positive. Because pressure of O_2 is raised to the third power in Q, an increase in pressure of O_2 will have the largest effect on ΔG for this reaction. Increasing the concentration or partial pressure of a product decreases the tendency for the reaction to occur.

19.77 *Analyze/Plan*. Given a chemical reaction, we are asked to calculate $\Delta G°$ from Appendix C data and ΔG for a given set of initial conditions. Use Equation 19.14 to calculate $\Delta G°$ and Equation 19.19 to calculate ΔG. Follow the logic in Sample Exercise 19.11 when calculating ΔG. *Solve.*

(a) $\Delta G° = \Delta G°\ N_2O_4(g) - 2\Delta G°\ NO_2(g) = 98.28 - 2(51.84) = -5.40\ kJ$

(b) $\Delta G = \Delta G° + RT \ln P_{N_2O_4} / P^2_{NO_2}$

$$= -5.40\ kJ + \frac{8.314 \times 10^{-3}\ kJ}{mol\text{-}K} \times 298\ K \times \ln[1.60/(0.40)^2] = 0.3048 = 0.30\ kJ$$

19.79 *Analyze/Plan*. Given a chemical reaction, we are asked to calculate K using ΔG_f° data from Appendix C. Calculate $\Delta G°$ using Equation 19.14. Then, $\Delta G° = -RT \ln K$, Equation 19.20; $\ln K = -\Delta G°/RT$ *Solve.*

(a) $\Delta G° = 2\Delta G°\ HI(g) - \Delta G°\ H_2(g) - \Delta G°\ I_2(g)$

$= 2(1.30) - 0 - 19.37 = -16.77\ kJ$

$$\ln K = \frac{-(-16.77\ kJ) \times 10^3\ J/kJ}{8.314\ J/K \times 298\ K} = 6.76876 = 6.769; \quad K = 870$$

(b) $\Delta G° = \Delta G°\ C_2H_4(g) + \Delta G°\ H_2O(g) - \Delta G°\ C_2H_5OH(g)$

$= 68.11 - 228.57 - (-168.5) = 8.04 = 8.0\ kJ$

$$\ln K = \frac{-(8.04\ kJ) \times 10^3\ J/kJ}{8.314\ J/K \times 298\ K} = -3.24511 = -3.25; \ K = 0.039$$

(c) $\Delta G° = \Delta G°\ C_6H_6(g) - 3\Delta G°\ C_2H_2(g) = 129.7 - 3(209.2) = -497.9\ kJ$

$$\ln K = \frac{-\Delta G°}{RT} = \frac{-(-497.9\ kJ) \times 10^3\ J/KJ}{8.314\ J/K \times 298\ K} = 200.963 = 201.0; \ K = 2 \times 10^{87}$$

19.81 *Analyze/Plan*. Given a chemical reaction and thermodynamic data in Appendix C, calculate the equilibrium pressure of $CO_2(g)$ at two temperatures. $K = P_{CO_2}$. Calculate $\Delta G°$ at the two temperatures using $\Delta G° = \Delta H° - T\Delta S°$ and then calculate K and P_{CO_2}. *Solve.*

$\Delta H° = \Delta H°\ BaO(s) + \Delta H°\ CO_2(g) - \Delta H°\ BaCO_3(s)$
$= -553.5 + -393.5 - (-1216.3) = +269.3\ kJ$

$\Delta S° = S°\ BaO(s) + S°\ CO_2(g) - S°\ BaCO_3(s)$
$= 70.42 + 213.6 - 112.1 = 171.92\ J/K = 0.1719\ kJ/K$

(a) ΔG at 298 K = 269.3 kJ – 298 K (0.17192 kJ/K) = 218.07 = 218.1 kJ

$$\ln K = \frac{-\Delta G^o}{RT} = \frac{-218.07 \times 10^3 \text{ J}}{8.314 \text{ J/K} \times 298 \text{ K}} = -88.017 = -88.02$$

$K = 6.0 \times 10^{-39}$; $\quad P_{CO_2} = 6.0 \times 10^{-39}$ atm

(b) ΔG at 1100 K = 269.3 kJ – 1100 K (0.17192 kJ) = 80.19 = +80.2 kJ

$$\ln K = \frac{-\Delta G^o}{RT} = \frac{-80.19 \times 10^3 \text{J}}{8.314 \text{ J/K} \times 1100 \text{ K}} = -8.768 = -8.77$$

$K = 1.6 \times 10^{-4}$; $\quad P_{CO_2} = 1.6 \times 10^{-4}$ atm

19.83 *Analyze/Plan.* Given an acid dissociation equilibrium and the corresponding K_a value, calculate $\Delta G°$ and ΔG for a given set of concentrations. Use Equation 19.20 to calculate $\Delta G°$ and Equation 19.19 to calculate ΔG. *Solve.*

(a) $HNO_2(aq) \rightleftharpoons H^+(aq) + NO_2^-(aq)$

(b) $\Delta G° = -RT \ln K_a = -(8.314 \times 10^{-3})(298) \ln (4.5 \times 10^{-4}) = 19.0928 = 19.1$ kJ

(c) $\Delta G = 0$ at equilibrium

(d) $\Delta G = \Delta G° + RT \ln Q$

$$= 19.09 \text{ kJ} + (8.314 \times 10^{-3})(298) \ln \frac{(5.0 \times 10^{-2})(6.0 \times 10^{-4})}{0.20} = -2.725 = -2.7 \text{ kJ}$$

Additional Exercises

19.85 (a) The thermodynamic quantities T, E, and S are state functions. T is directly related to the distribution of molecular speeds, which does not depend on the path from one state to another.

(b) The quantities q and w do depend on the path taken from one state to another.

(c) There is only one *reversible* path between states.

(d) Isothermal processes occur at constant T. As the process is reversible, q is q_{rev} and w is w_{max}.

$$\Delta E = q_{rev} + w_{max}; \quad \Delta S = \frac{q_{rev}}{T}$$

19.94 (a) $1/2\, N_2(g) + 3/2\, H_2(g) \rightarrow NH_3(g)$

$C(s) + 1/2\, O_2(g) \rightarrow CO(g)$

(b) In the first reaction, there are fewer moles of gas in the products than reactants, so ΔS is negative. If $\Delta G_f^o = \Delta H_f^o - T\Delta S_f^o$ and ΔS_f^o is negative, $-T\Delta S_f^o$ is positive and ΔG_f^o is more positive than ΔH_f^o.

(c) Condition (iii), when ΔS_f^o is negative.

19.99 (a) First, calculate $\Delta G°$ for each reaction:

For $C_6H_{12}O_6(s) + 6\,O_2(g) \rightleftharpoons 6\,CO_2(g) + 6\,H_2O(l)$ (A)

$\Delta G° = 6(-237.13) + 6(-394.4) - (-910.4) + 6(0) = -2878.8$ kJ

For $C_6H_{12}O_6(s) \rightleftharpoons 2\,C_2H_5OH(l) + 2\,CO_2(g)$ (B)

$\Delta G° = 2(-394.4) + 2(-174.8) - (-910.4) = -228.0$ kJ

For (A), $\ln K = 2879 \times 10^3/(8.314)(298) = 1162$; $K = 5 \times 10^{504}$

For (B), $\ln K = 228 \times 10^3/(8.314)(298) = 92.026 = 92.0$; $K = 9 \times 10^{39}$

(b) Both these values for K are unimaginably large. However, K for reaction (A) is larger, because $\Delta G°$ is more negative. The magnitude of the work that can be accomplished by coupling a reaction to its surroundings is measured by ΔG. According to these calculations, considerably more work can in principle be obtained from reaction (A), because $\Delta G°$ is more negative.

Integrative Exercises

19.103 (a) At the boiling point, vaporization is a reversible process, so $\Delta S^o_{vap} = \Delta H^o_{vap}/T$.

acetone: $\Delta S^o_{vap} = \Delta H^o_{vap}/T = (29.1\text{ kJ/mol})/329.25\text{ K} = 88.4\text{ J/mol-K}$

dimethyl ether: $\Delta S^o_{vap} = (21.5\text{ kJ/mol})/248.35\text{ K} = 86.6\text{ J/mol-K}$

ethanol: $\Delta S^o_{vap} = (38.6\text{ kJ/mol})/351.6\text{ K} = 110\text{ J/mol-K}$

octane: $\Delta S^o_{vap} = (34.4\text{ kJ/mol})/398.75\text{ K} = 86.3\text{ J/mol-K}$

pyridine: $\Delta S^o_{vap} = (35.1\text{ kJ/mol})/388.45\text{ K} = 90.4\text{ J/mol-K}$

(b) Ethanol is the only liquid listed that does not follow *Trouton's rule* and it is also the only substance that exhibits hydrogen bonding in the pure liquid. Hydrogen bonding leads to more ordering in the liquid state and a greater than usual increase in entropy upon vaporization. The rule appears to hold for liquids with London dispersion forces (octane) and ordinary dipole-dipole forces (acetone, dimethyl ether, pyridine), but not for those with hydrogen bonding.

(c) Owing to strong hydrogen bonding interactions, water probably does not obey Trouton's rule.

From Appendix B, ΔH^o_{vap} at 100 °C = 40.67 kJ/mol.

$\Delta S^o_{vap} = (40.67\text{ kJ/mol})/373.15\text{ K} = 109.0\text{ J/mol-K}$

(d) Use $\Delta S^o_{vap} = 88\text{ J/mol-K}$, the middle of the range for Trouton's rule, to estimate ΔH^o_{vap} for chlorobenzene.

$\Delta H^o_{vap} = \Delta S^o_{vap} \times T = 88\text{ J/mol-K} \times 404.95\text{ K} = 36\text{ kJ/mol}$

19.110 (a) $K = P^2_{NO_2} / P_{N_2O_4}$

Assume equal amounts means equal number of moles. For gases, P = n(RT/V). In an equilibrium mixture, RT/V is a constant, so moles of gas are directly proportional to partial pressure. Gases with equal partial pressures will have equal moles of gas present. The condition $P_{NO_2} = P_{N_2O_4}$ leads to the expression $K = P_{NO_2}$. The value of K then depends on P_t for the mixture. For any particular value of P_t, the condition of equal moles of the two gases can be achieved at some temperature. For example, $P_{NO_2} = P_{N_2O_4} = 1.0$ atm, $P_t = 2.0$ atm.

$$K = \frac{(1.0)^2}{1.0} = 1.0; \;\; \ln K = 0; \;\; \Delta G^\circ = 0 = \Delta H^\circ - T\Delta S^\circ; \;\; T = \Delta H^\circ / \Delta S^\circ$$

$$\Delta H^\circ = 2\Delta H^\circ \; NO_2(g) - \Delta H^\circ \; N_2O_4(g) = 2(33.84) - 9.66 = +58.02 \text{ kJ}$$

$$\Delta S^\circ = 2S^\circ \; NO_2(g) - S^\circ \; N_2O_4(g) = 2(240.45) - 304.3 = 0.1766 \text{ kJ/K}$$

$$T = \frac{58.02 \text{ kJ}}{0.1766 \text{ kJ/K}} = 328.5 \text{ K or } 55.5 \text{ °C}$$

(b) $P_t = 1.00$ atm; $P_{N_2O_4} = x$, $P_{NO_2} = 2x$; $x + 2x = 1.00$ atm

$x = P_{N_2O_4} = 0.3333 = 0.333$ atm; $P_{NO_2} = 0.6667 = 0.667$ atm

$$K = \frac{(0.6667)^2}{0.3333} = 1.334 = 1.33; \qquad \Delta G^\circ = -RT \ln K = \Delta H^\circ - T\Delta S^\circ$$

$-(8.314 \times 10^{-3} \text{ kJ/K})(\ln 1.334)\,T = 58.02 \text{ kJ} - (0.1766 \text{ kJ/K})\,T$

$(-0.00239 \text{ kJ/K})\,T + (0.1766 \text{ kJ/K})\,T = 58.02 \text{ kJ}$

$(0.1742 \text{ kJ/K})\,T = 58.02 \text{ kJ}$; $T = 333.0$ K

(c) $P_t = 10.00$ atm; $x + 2x = 10.00$ atm

$x = P_{N_2O_4} = 3.3333 = 3.333$ atm; $P_{NO_2} = 6.6667 = 6.667$ atm

$$K = \frac{(6.6667)^2}{3.3333} = 13.334 = 13.33; \;\; -RT \ln K = \Delta H^\circ - T\Delta S^\circ$$

$-(8.314 \times 10^{-3} \text{ kJ/K})(\ln 13.334)\,T = 58.02 \text{ kJ} - (0.1766 \text{ kJ/K})\,T$

$(-0.02154 \text{ kJ/K})\,T + (0.1766 \text{ kJ/K})\,T = 58.02 \text{ kJ}$

$(0.15506 \text{ kJ/K})\,T = 58.02 \text{ kJ}$; $T = 374.2$ K

20 Electrochemistry

Visualizing Concepts

20.1 In this analogy, the electron is analogous to the proton (H^+). Just as acid–base reactions can be viewed as proton-transfer reactions, redox reactions can be viewed as electron-transfer reactions.

Oxidizing agents are reduced; they gain electrons. A strong oxidizing agent would be analogous to a strong base.

20.3 *Analyze/Plan.* Apply the definitions of oxidation, reduction, anode and cathode to the diagram. Recall relationship between atomic and ionic size from Chapter 7. *Solve.*

(a) Oxidation. The gray spheres are uniformly sized and closely aligned; they represent an elemental solid. The diagram shows atoms from the surface of the solid going into solution. In a voltaic cell, this happens when metal atoms on an electrode surface are oxidized. They lose electrons, form cations, and move into solution.

(b) Anode. Oxidation occurs at the anode.

(c) When a neutral atom loses a valence electron, Z_{eff} for the remaining electrons increases, and the radius of the resulting cation is smaller than the radius of the neutral atom. The neutral atoms in the electrode are represented by larger spheres than the cations moving into solution.

20.7 *Analyze.* Given a redox reaction with a negative E°, answer questions regarding ΔG°, the equilibrium constant (K), and work (w). *Plan.* $\Delta G° = -nFE°$; $\Delta G° = -RT \ln K$; $w_{max} = -nFE$. *Solve.*

(a) The signs of ΔG° and E° are opposite. If E° is negative, ΔG° is positive. (The reaction is not spontaneous in the forward direction.)

(b) If ΔG° is positive, ln K is negative and $K < 1$. Also, K is less than one for a nonspontaneous reaction.

(c) No. If E° is negative, the sign of w is positive. A positive value for w means that work is done on the system by the surroundings. An electrochemical cell based on this reaction cannot accomplish work on its surroundings.

20.10 (a) Zinc is the anode. Metallic zinc cannot be reduced; it must be oxidized. Oxidation occurs at the anode.

(b) The energy density of the silver oxide battery is most similar to the nickel-cadmium battery. Energy density (see Figure 20.23) is related to the molar masses of the electrode materials and voltage per cell of the battery. The molar mass of (Ag_2O + Zn) is most like that of (Ni + Cd). The cell potential for a Ni-Cd battery is 1.3 V; for the reaction in the silver oxide battery, the cell potential is about 1.1 V. The Li-ion (3.7 V) and lead-acid (2.0 V) batteries have larger cell potentials. (The standard reduction potential of Ag_2O is 0.34 V. From *Handbook of Chemistry and Physics*, 74th edition)

Oxidation–Reduction Reactions (Section 20.1)

20.13 (a) *Oxidation* is the loss of electrons.

(b) The electrons appear on the products side (right side) of an oxidation half-reaction.

(c) The *oxidant* is the reactant that is reduced; it gains the electrons that are lost by the substance being oxidized.

(d) An *oxidizing agent* is the substance that promotes oxidation. That is, it gains electrons that are lost by the substance being oxidized. It is the same as the oxidant.

20.15 (a) True.

(b) False. Fe^{3+} is reduced to Fe^{2+}, so it is the oxidizing agent, and Co^{2+} is the reducing agent.

(c) True.

20.17 *Analyze/Plan*. Given a chemical equation, we are asked to determine the oxidation number of each reactant and product, and the total number of electrons transferred in the overall reaction. Assign oxidation numbers according to the rules given in Section 4.4. To find electrons transferred, count the electrons gained by an atom that is reduced and multiply by the number of atoms reduced. Check by counting electrons for the atom oxidized. *Solve.*

(a) (i) Reactants: I, +5; O, –2; C +2; O, –2. Products: I, 0; C, +4; O, –2

(ii) The total number of electrons transferred is 10. (Each I atom gains 5 electrons, for a total of 10 electrons gained; each C atom loses 2 electrons for a total of 10 electrons lost; electrons balance.)

(b) (i) Reactants: Hg, +2; N, –2; H, +1. Products: Hg, 0; N, 0; H, +1.

(ii) The total number of electrons transferred is 4. (Each Hg gains 2 electrons for a total of 4 electrons gained; each N loses 2 electrons for a total of 4 electrons lost; electrons balance.)

(c) (i) Reactants: H, +1; S, –2; N, +5; O, –2. Products: S, 0; N, +2; O, –2; H, +1; O, –2.

(ii) The total electrons transferred is 6. (Each N atom gains 3 electrons for a total of 6 electrons gained; each S atom loses 2 electrons for a total of 6 electrons lost; electrons balance.)

20.19 (a) No oxidation–reduction.

(b) I is oxidized from –1 to +5; Cl is reduced from +1 to –1.

(c) S is oxidized from +4 to +6; N is reduced from +5 to +2.

Balancing Oxidation–Reduction Reactions (Section 20.2)

20.21 *Analyze/Plan.* Write the balanced chemical equation and assign oxidation numbers. The substance oxidized is the reductant and the substance reduced is the oxidant. *Solve.*

(a) $TiCl_4(g) + 2\ Mg(l) \rightarrow Ti(s) + 2\ MgCl_2(l)$

(b) Mg(l) is oxidized; $TiCl_4(g)$ is reduced.

(c) Mg(l) is the reductant; $TiCl_4(g)$ is the oxidant.

20.23 *Analyze/Plan.* Follow the logic in Sample Exercises 20.2 and 20.3. If the half-reaction occurs in basic solution, balance as in acid, then add OH^- to each side. *Solve.*

(a) $Sn^{2+}(aq) \rightarrow Sn^{4+}(aq) + 2\ e^-$, oxidation

(b) $TiO_2(s) + 4\ H^+(aq) + 2\ e^- \rightarrow Ti^{2+}(aq) + 2\ H_2O(l)$, reduction

(c) $ClO_3^-(aq) + 6\ H^+(aq) + 6\ e^- \rightarrow Cl^-(aq) + 3\ H_2O(l)$, reduction

(d) $N_2(g) + 8\ H^+(aq) + 6\ e^- \rightarrow 2\ NH_4^+(aq)$, reduction

(e) $4\ OH^-(aq) \rightarrow O_2(g) + 2\ H_2O(l) + 4\ e^-$, oxidation

(f) $SO_3^{2-}(aq) + 2\ OH^-(aq) \rightarrow SO_4^{2-}(aq) + H_2O(l) + 2\ e^-$, oxidation

(g) $N_2(g) + 6\ H_2O + 6\ e^- \rightarrow 2\ NH_3(g) + 6\ OH^-(aq)$, reduction

20.25 *Analyze/Plan.* Follow the logic in Sample Exercises 20.2 and 20.3 to balance the given equations. Use the method in Sample Exercise 20.1 to identify oxidizing and reducing agents. *Solve.*

(a) $Cr_2O_7^{2-}(aq) + I^-(aq) + 8\ H^+ \rightarrow 2\ Cr^{3+}(aq) + IO_3^-(aq) + 4\ H_2O(l)$
oxidizing agent, $Cr_2O_7^{2-}$; reducing agent, I^-

(b) The half-reactions are:

$$4\ [MnO_4^-(aq) + 8\ H^+(aq) + 5\ e^- \rightarrow Mn^{2+}(aq) + 4\ H_2O(l)]$$
$$5\ [CH_3OH(aq) + H_2O(l) \rightarrow HCOOH(aq) + 4\ H^+(aq) + 4\ e^-]$$
$$4\ MnO_4^-(aq) + 5\ CH_3OH(aq) + 12\ H^+(aq) \rightarrow 4\ Mn^{2+}(aq) + 5\ HCOOH(aq) + 11\ H_2O(l)$$

oxidizing agent, MnO_4^-; reducing agent, CH_3OH

(c)
$$I_2(s) + 6\ H_2O(l) \rightarrow 2\ IO_3^-(aq) + 12\ H^+(aq) + 10\ e^-$$
$$5\ [OCl^-(aq) + 2\ H^+(aq) + 2\ e^- \rightarrow Cl^-(aq) + H_2O(l)$$
$$I_2(s) + 5\ OCl^-(aq) + H_2O(l) \rightarrow 2\ IO_3^-(aq) + 5\ Cl^-(aq) + 2\ H^+(aq)]$$

oxidizing agent, OCl^-; reducing agent, I_2

(d)

$$As_2O_3(s) + 5\ H_2O(l) \rightarrow 2\ H_3AsO_4(aq) + 4\ H^+(aq) + 4\ e^-$$
$$2\ NO_3^-(aq) + 6\ H^+(aq) + 4\ e^- \rightarrow N_2O_3(aq) + 3\ H_2O(l)$$
$$As_2O_3(s) + 2\ NO_3^-(aq) + 2\ H_2O(l) + 2\ H^+(aq) \rightarrow 2\ H_3AsO_4(aq) + N_2O_3(aq)$$

oxidizing agent, NO_3^-; reducing agent, As_2O_3

(e)

$$2\ [MnO_4^-(aq) + 2\ H_2O(l) + 3\ e^- \rightarrow MnO_2(s) + 4\ OH^-]$$
$$Br^-(aq) + 6\ OH^-(aq) \rightarrow BrO_3^-(aq) + 3\ H_2O(l) + 6\ e^-$$
$$2\ MnO_4^-(aq) + Br^-(aq) + H_2O(l) \rightarrow 2\ MnO_2(s) + BrO_3^-(aq) + 2\ OH^-(aq)$$

oxidizing agent, MnO_4^-; reducing agent, Br^-

(f) $Pb(OH)_4^{2-}(aq) + ClO^-(aq) \rightarrow PbO_2(s) + Cl^-(aq) + 2\ OH^-(aq) + H_2O(l)$

oxidizing agent, ClO^-; reducing agent, $Pb(OH)_4^{2-}$

Voltaic Cells (Section 20.3)

20.27 (a) False. Oxidation occurs at the anode; reduction occurs at the cathode.

(b) True. The terms voltaic and galvanic are interchangeable.

(c) True. The oxidation half-reaction at the anode generates free electrons which flow from the anode to the cathode.

20.29 *Analyze/Plan.* Follow the logic in Sample Exercise 20.4. *Solve.*

(a) Fe(s) is oxidized, $Ag^+(aq)$ is reduced.

(b) $Ag^+(aq) + 1\ e^- \rightarrow Ag(s)$; $Fe(s) \rightarrow Fe^{2+}(aq) + 2\ e^-$

(c) Fe(s) is the anode, Ag(s) is the cathode.

(d) Fe(s) is negative; Ag(s) is positive.

(e) Electrons flow from the Fe(–) electrode toward the Ag(+) electrode.

(f) Cations migrate toward the Ag(s) cathode; anions migrate toward the Fe(s) anode.

Cell Potentials under Standard Conditions (Section 20.4)

20.31 (a) One *volt* is the potential energy difference required to impart 1 J of energy to a charge of 1 coulomb. 1 V = 1 J/C.

(b) Yes. All voltaic cells involve a spontaneous redox reaction that produces a positive cell potential or emf.

20.33 (a) $2\ H^+(aq) + 2\ e^- \rightarrow H_2(g)$

(b) $H_2(g) \rightarrow 2\ H^+(aq) + 2\ e^-$

(c) A *standard* hydrogen electrode is a hydrogen electrode where the components are at standard conditions, 1 *M* $H^+(aq)$ and $H_2(g)$ at 1 atm, such that $E^\circ = 0$ V.

20.35 *Analyze/Plan.* Follow the logic in Sample Exercise 20.5. *Solve.*

(a) The two half-reactions are:

$Tl^{3+}(aq) + 2\ e^- \rightarrow Tl^+(aq)$ cathode $E^\circ_{red} = ?$

$2\ [Cr^{2+}(aq) \rightarrow Cr^{3+}(aq) + e^-]$ anode $E^\circ_{red} = -0.41\ V$

(b)

$$E^\circ_{cell} = E^\circ_{red}\ (\text{cathode}) - E^\circ_{red}\ (\text{anode})$$

$$1.19\ V = E^\circ_{red} - (-0.41\ V)$$

$$E^\circ_{red} = 1.19\ V - 0.41\ V = 0.78\ V$$

(c)

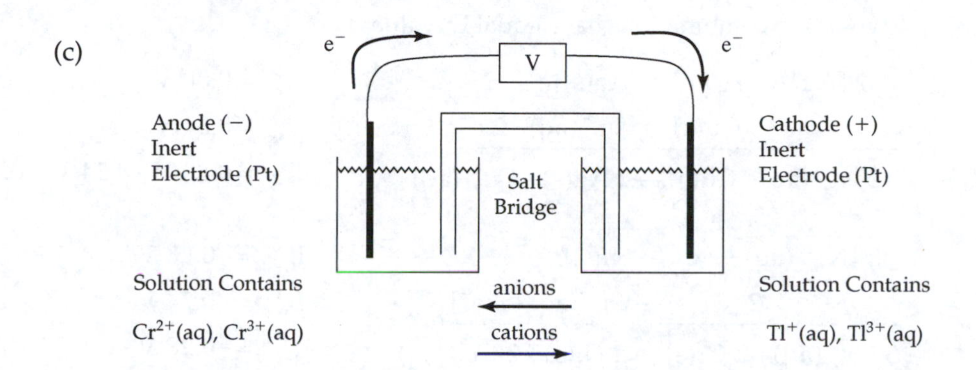

Note that because $Cr^{2+}(aq)$ is readily oxidized, it would be necessary to keep oxygen out of the left-hand cell compartment.

20.37 *Analyze/Plan.* Follow the logic in Sample Exercise 20.6. *Solve.*

(a) $Cl_2(g) \rightarrow 2\ Cl^-(aq) + 2\ e^-$ $E^\circ_{red} = 1.36\ V$

$I_2(s) + 2\ e^- \rightarrow 2\ I^-(aq)$ $E^\circ_{red} = 0.54\ V$

$E° = 1.36\ V - 0.54\ V = 0.82\ V$

(b) $Ni(s) \rightarrow Ni^{2+}(aq) + 2\ e^-$ $E^\circ_{red} = -0.28\ V$

$2\ [Ce^{4+}(aq) + 1\ e^- \rightarrow Ce^{3+}(aq)]$ $E^\circ_{red} = 1.61\ V$

$E° = 1.61\ V - (-0.28\ V) = 1.89\ V$

(c) $Fe(s) \rightarrow Fe^{2+}(aq) + 2\ e^-$ $E^\circ_{red} = -0.44\ V$

$2\ [Fe^{3+}(aq) + 1\ e^- \rightarrow Fe^{2+}(aq)]$ $E^\circ_{red} = 0.77\ V$

$E° = 0.77\ V - (-0.44\ V) = 1.21\ V$

(d) $2\ [NO_3^-(aq) + 4\ H^+ + 3\ e^- \rightarrow NO(g) + 2\ H_2O(l)]$ $E^\circ_{red} = 0.96\ V$

$3\ [Cu(s) \rightarrow Cu^{2+}(aq) + 2\ e^-]$ $E^\circ_{red} = 0.34\ V$

$E° = 0.96\ V - (0.34\ V) = 0.62\ V$

20.39 *Analyze/Plan*. Given four half-reactions, find E^{o}_{red} from Appendix E and combine them to obtain a desired E_{cell}. (a) The largest E_{cell} will combine the half-reaction with the most positive E^{o}_{red} as the cathode reaction and the one with the most negative E^{o}_{red} as the anode reaction. (b) The smallest positive E^{o}_{cell} will combine two half-reactions whose E^{o}_{red} values are closest in magnitude **and** sign. *Solve.*

(a)

$$3\,[Ag^{+}(aq) + 1\,e^{-} \rightarrow Ag(s)] \qquad E^{o}_{red} = 0.80$$
$$Cr(s) \rightarrow Cr^{3+}(aq) + 3\,e^{-} \qquad E^{o}_{red} = -0.74$$
$$3\,Ag^{+}(aq) + Cr(s) \rightarrow 3\,Ag(s) + Cr^{3+}(aq) \qquad E^{o} = 0.80 - (-0.74) = 1.54\ V$$

(b) Two of the combinations have equal E° values.

$$2\,[Ag^{+}(aq) + 1\,e^{-} \rightarrow Ag(s)] \qquad E^{o}_{red} = 0.80\ V$$
$$Cu(s) \rightarrow Cu^{2+}(aq) + 2\,e^{-} \qquad E^{o}_{red} = 0.34\ V$$
$$2\,Ag^{+}(aq) + Cu(s) \rightarrow 2\,Ag(s) + Cu^{2+}(aq) \qquad E^{o} = 0.80\ V - 0.34\ V = 0.46\ V$$

$$3\,[Ni^{2+}(aq) + 2\,e^{-} \rightarrow Ni(s)] \qquad E^{o}_{red} = -0.28\ V$$
$$2\,[Cr(s) \rightarrow Cr^{3+}(aq) + 3\,e^{-}] \qquad E^{o}_{red} = -0.74\ V$$
$$3\,Ni^{2+}(aq) + 2\,Cr(s) \rightarrow 3\,Ni(s) + 2\,Cr^{3+}(aq) \qquad E^{o} = -0.28\ V - (-0.74\ V) = 0.46\ V$$

20.41 *Analyze/Plan*. Given the description of a voltaic cell, answer questions about this cell. Combine ideas in Sample Exercises 20.4 and 20.7. The reduction half-reactions are:

$$Cu^{2+}(aq) + 2\,e^{-} \rightarrow Cu(s) \qquad E° = 0.34\ V$$
$$Sn^{2+}(aq) + 2\,e^{-} \rightarrow Sn(s) \qquad E° = -0.14\ V$$

Solve.

(a) It is evident that Cu^{2+} is more readily reduced. Therefore, Cu serves as the cathode, Sn as the anode.

(b) The copper electrode gains mass as Cu is plated out, the Sn electrode loses mass as Sn is oxidized.

(c) The overall cell reaction is $Cu^{2+}(aq) + Sn(s) \rightarrow Cu(s) + Sn^{2+}(aq)$

(d) $E° = 0.34\ V - (-0.14\ V) = 0.48\ V$

Strengths of Oxidizing and Reducing Agents (Section 20.4)

20.43 *Analyze/Plan*. The more readily a substance is oxidized, the stronger it is as a reducing agent. In each case choose the half-reaction with the more negative reduction potential and the given substance on the right. *Solve.*

(a) Mg(s) (–2.37 V vs. –0.44 V)

(b) Ca(s) (–2.87 V vs. –1.66 V)

(c) H_2(g, acidic) (0.00 V vs. 0.14 V)

(d) IO_3^- (aq) [Both IO_3^- (aq) and BrO_3^-(aq) are good oxidizing agents, but IO_3^-(aq) has the smaller positive reduction potential. (1.20 V vs. 1.52 V)]

20.45 *Analyze/Plan.* If the substance is on the left of a reduction half-reaction, it will be an oxidant; if it is on the right, it will be a reductant. *Solve.*

(a) $Cl_2(aq)$: oxidant (on the left, $E^\circ_{red} = 1.36$ V)

(b) MnO_4^-(aq, acidic): oxidant (on the left, $E^\circ_{red} = 1.51$ V)

(c) Ba(s): reductant (on the right, $E^\circ_{red} = -2.90$ V)

(d) Zn(s): reductant (on the right, $E^\circ_{red} = -0.76$ V)

20.47 *Analyze/Plan.* Follow the logic in Sample Exercise 20.8. *Solve.*

(a) Arranged in order of increasing strength as oxidizing agents (and increasing reduction potential):

$Cu^{2+}(aq) < O_2(g) < Cr_2O_7^{2-}(aq) < Cl_2(g) < H_2O_2(aq)$

(b) Arranged in order of increasing strength as reducing agents (and decreasing reduction potential):

$H_2O_2(aq) < I^-(aq) < Sn^{2+}(aq) < Zn(s) < Al(s)$

20.49 *Analyze/Plan.* To reduce Eu^{3+} to Eu^{2+}, we need an oxidizing agent, one of the reduced species from Appendix E. It must have a greater tendency to be oxidized than Eu^{3+} has to be reduced. That is, E°_{red} must be more negative than –0.43 V. *Solve.*

Any of the **reduced** species in Appendix E from a half-reaction with a reduction potential more negative than –0.43 V will reduce Eu^{3+} to Eu^{2+}. From the list of possible reductants in the exercise, Al and $H_2C_2O_4$ will reduce Eu^{3+} to Eu^{2+}.

Free Energy and Redox Reactions (Section 20.5)

20.51 *Analyze/Plan.* In each reaction, $Fe^{2+} \rightarrow Fe^{3+}$ will be the oxidation half-reaction and one of the other given half-reactions will be the reduction half-reaction. Follow the logic in Sample Exercise 20.10 to calculate E°, ΔG° and K for each reaction. *Solve.*

(a) $2\,Fe^{2+}(aq) + S_2O_6^{2-}(aq) + 4\,H^+(aq) \rightarrow 2\,Fe^{3+}(aq) + 2\,H_2SO_3(aq)$

$E^\circ = 0.60\text{ V} - 0.77\text{ V} = -0.17\text{ V}$

$2\,Fe^{2+}(aq) + N_2O(aq) + 2\,H^+(aq) \rightarrow 2\,Fe^{3+}(aq) + N_2(g) + H_2O(l)$

$E^\circ = -1.77\text{ V} - 0.77\text{ V} = -2.54\text{ V}$

$Fe^{2+}(aq) + VO_2^+(aq) + 2\,H^+(aq) \rightarrow Fe^{3+}(aq) + VO^{2+}(aq) + H_2O(l)$

$E^\circ = 1.00\text{ V} - 0.77\text{ V} = +0.23\text{ V}$

(b) $\Delta G^\circ = -nFE^\circ$ For the first reaction,

$$\Delta G^\circ = -2\text{ mol} \times \frac{96{,}485\text{ J}}{1\text{ V-mol}} \times (-0.17\text{ V}) = 3.280 \times 10^4 = 3.3 \times 10^4\text{ J or 33 kJ}$$

For the second reaction, $\Delta G^\circ = -2(96{,}485)(-2.54) = 4.901 \times 10^5 = 4.90 \times 10^2$ kJ

For the third reaction, $\Delta G^\circ = -1(96{,}485)(0.23) = -2.22 \times 10^4\text{ J} = -22$ kJ

(c) $\Delta G° = -RT \ln K; \ln K = -\Delta G°/RT$

For the first reaction,

$$\ln K = \frac{-3.281 \times 10^4 \text{ J}}{(8.314 \text{ J/mol-K})(298 \text{ K})} = -13.243 = -13; \ K = e^{-13.2428} = 1.78 \times 10^{-6} = 2 \times 10^{-6}$$

[Convert ln to log; the number of decimal places in the log is the number of sig figs in the result.]

For the second reaction,

$$\ln K = \frac{-4.902 \times 10^5 \text{ J}}{8.314 \text{ J/mol-K} \times 298 \text{ K}} = -197.86 = -198; \ K = e^{-198} = 1.23 \times 10^{-86} = 10^{-86}$$

For the third reaction,

$$\ln K = \frac{-(-2.22 \times 10^4 \text{ J})}{8.314 \text{ J/mol-K} \times 298 \text{ K}} = 8.958 = 9.0; \ K = e^{9.0} = 7.77 \times 10^3 = 8 \times 10^3$$

Check. The equilibrium constants calculated here are indicators of equilibrium position, but are not particularly precise numerical values.

20.53 *Analyze/Plan.* Given K, calculate $\Delta G°$ and E°. Reverse the logic in Sample Exercise 20.10. According to Equation 19.20, $\Delta G° = -RT \ln K$. According to Equation 20.12, $\Delta G° = -nFE°$, $E° = -\Delta G°/nF$. *Solve.*

$K = 1.5 \times 10^{-4}$

$\Delta G° = -RT \ln K = -(8.314 \text{ J/mol-K})(298) \ln (1.5 \times 10^{-4}) = 2.181 \times 10^4 \text{ J} = 21.8 \text{ kJ}$

$E° = -\Delta G°/nF; n = 2; F = 96.5 \text{ kJ/mol } e^-$

$$E° = \frac{-21.81 \text{ kJ}}{2 \text{ mol } e^- \times 96.5 \text{ kJ/V-mol } e^-} = -0.113 \text{ V}$$

Check. The unit of $\Delta G°$ is actually kJ/mol, which means kJ per "mole of reaction," or for the reaction as written. Because we do not have a specific reaction, we interpret the unit as referring to the overall reaction.

20.55 *Analyze/Plan.* Given E^{o}_{red} values for half reactions, calculate the value of K for a given redox reaction. Follow the logic in Sample Exercise 20.10. For each reaction, calculate E° from E^{o}_{red}, then use Equation 20.13 to calculate K.

Solve. $E° = \frac{RT}{nF} \ln K; \ \ln K = \frac{nF}{RT} E°; \ \frac{F}{RT} = \frac{96{,}485 \text{ J/V-mol}}{8.314 \text{ J/K-mol} \times 298.15 \text{ K}} = 38.924/\text{V}$

(a) $E° = -0.28 - (-0.44) = 0.16 \text{ V}, n = 2 \ (Ni^{2+} + 2 e^- \rightarrow Ni)$
$\ln K = 2(38.924/\text{V})(0.16 \text{ V}) = 12.456 = 12; \ K = 2.57 \times 10^5 = 3 \times 10^5$

(b) $E° = 0 - (-0.28) = 0.28 \text{ V}; n = 2 \ (2H^+ + 2 e^- \rightarrow H_2)$
$\ln K = 2(38.924/\text{V})(0.28 \text{ V}) = 21.797 = 22; \ K = 2.93 \times 10^9 = 3 \times 10^9$

(c) $E° = 1.51 - 1.07 = 0.44 \text{ V}; n = 10 \ (2 MnO_4^- + 10 e^- \rightarrow 2 Mn^{+2})$
$\ln K = 10(38.924/\text{V})(0.44 \text{ V}) = 171.265 = 170; \ K = 2.40 \times 10^{74} = 10^{74}$

Check. Note that small differences in E° values lead to large changes in the magnitude of K. Sig fig rules limit the precision of K values.

20.57 *Analyze/Plan*. At 298 K, $\ln K = n(38.924/V)E°$. See Solution 20.55 for a more complete development. *Solve*.

(a) $\ln K = 1(38.924/V)0.177\ V = 6.890 = 6.89;\ K = 982 = 9.8\times10^2$

(b) $\ln K = 2(38.924/V)0.177\ V = 13.779 = 13.8;\ K = 9.64\times10^5 = 1\times10^6$

(c) $\ln K = 3(38.924/V)0.177\ V = 20.669 = 20.7;\ K = 9.47\times10^8 = 1\times10^9$

20.59 *Analyze/Plan*. Given a spontaneous chemical reaction, calculate the maximum possible work for a given amount of reactant at standard conditions. Separate the equation into half-reactions and calculate cell emf. Use Equation 20.14, $w_{max} = -nFE$, to calculate maximum work. At standard conditions, $E = E°$. *Solve*.

$$I_2(s) + 2e^- \rightarrow 2\ I^-(aq) \qquad E^o_{red} = 0.54\ V$$

$$Sn(s) \rightarrow Sn^{2+}(aq) + 2\ e^- \qquad E^o_{red} = -0.14\ V$$

$$I_2(s) + Sn(s) \rightarrow 2\ I^-(aq) + Sn^{2+}(aq) \qquad E° = 0.54\ V - (-0.14\ V) = 0.68\ V$$

$$w_{max} = -2(96.5)(0.68) = -131.24 = 1.3 \times 10^2\ kJ/mol\ Sn$$

$$\frac{-131.24\ kJ}{mol\ Sn(s)} \times \frac{1\ mol\ Sn}{118.71\ g\ Sn} \times 75.0\ g\ Sn \times \frac{1000\ J}{kJ} = -8.3 \times 10^4\ J$$

Check. The (–) sign indicates that work is done by the cell.

Cell EMF under Nonstandard Conditions (Section 20.6)

20.61 (a) In the Nernst equation, $Q = 1$ if all reactants and products are at standard conditions.

(b) Yes. The Nernst equation is applicable to cell emf at nonstandard conditions, so it must be applicable at temperatures other than 298 K. There are two terms in the Nernst equation. First, values of E° at temperatures other than 298 K are required. Then, in the form of Equation 20.16, there is a variable for T in the second term. In the short-hand form of Equation 20.18, the value 0.0592 assumes 298 K. A different coefficient would apply to cells at temperatures other than 298 K.

20.63 *Analyze/Plan*. Given a circumstance, determine its effect on cell emf. Each circumstance changes the value of Q. An increase in Q reduces emf; a decrease in Q increases emf. *Solve*.

$$Zn(s) + 2H^+(aq) \rightarrow Zn^{2+}(aq) + H_2(g);\ E = E° - \frac{0.0592}{n}\log Q;\ \ Q = \frac{[Zn^{2+}]\ P_{H_2}}{[H^+]^2}$$

(a) P_{H_2} increases, Q increases, E decreases

(b) $[Zn^{2+}]$ increases, Q increases, E decreases

(c) $[H^+]$ decreases, Q increases, E decreases

(d) No effect; does not appear in the Nernst equation

20.65 *Analyze/Plan*. Follow the logic in Sample Exercise 20.11. *Solve*.

(a)

$$Ni^{2+}(aq) + 2\ e^- \rightarrow Ni(s) \qquad E^o_{red} = -0.28\ V$$

$$Zn(s) \rightarrow Zn^{2+}(aq) + 2\ e^- \qquad E^o_{red} = -0.76\ V$$

$$Ni^{2+}(aq) + Zn(s) \rightarrow Ni(s) + Zn^{2+}(aq) \qquad E° = -0.28 - (-0.76) = 0.48\ V$$

(b) $E = E^\circ - \frac{0.0592}{n} \log \frac{[Zn^{2+}]}{[Ni^{2+}]}$; $n = 2$

$$E = 0.48 - \frac{0.0592}{2} \log \frac{(0.100)}{(3.00)} = 0.48 - \frac{0.0592}{2} \log (0.0333)$$

$$E = 0.48 - \frac{0.0592(-1.477)}{2} = 0.48 + 0.0437 = 0.5237 = 0.52 \text{ V}$$

(c) $E = 0.48 - \frac{0.0592}{2} \log \frac{(0.900)}{(0.200)} = 0.48 - 0.0193 = 0.4607 = 0.46 \text{ V}$

20.67 *Analyze/Plan.* Follow the logic in Sample Exercise 20.11. *Solve.*

(a)

$$4\,[Fe^{2+}(aq) \rightarrow Fe^{3+}(aq) + 1\,e^-] \qquad E^\circ_{red} = 0.77 \text{ V}$$

$$O_2(g) + 4\,H^+(aq) + 4\,e^- \rightarrow 2\,H_2O(l) \qquad E^\circ_{red} = 1.23 \text{ V}$$

$$4\,Fe^{2+}(aq) + O_2(g) + 4\,H^+(aq) \rightarrow 4\,Fe^{3+}(aq) + 2\,H_2O(l) \qquad E^\circ = 1.23 - 0.77 = 0.46 \text{ V}$$

(b) $E = E^\circ - \frac{0.0592}{n} \log \frac{[Fe^{3+}]^4}{[Fe^{2+}]^4[H^+]^4 P_{O_2}}$; $n = 4$, $[H^+] = 10^{-3.50} = 3.2 \times 10^{-4}\ M$

$$E = 0.46 \text{ V} - \frac{0.0592}{4} \log \frac{(0.010)^4}{(1.3)^4(3.2 \times 10^{-4})^4(0.50)} = 0.46 - \frac{0.0592}{4} \log (7.0 \times 10^5)$$

$$E = 0.46 - \frac{0.0592}{4}(5.845) = 0.46 - 0.0865 = 0.3735 = 0.37 \text{ V}$$

20.69 *Analyze/Plan.* We are given a concentration cell with Zn electrodes. Use the definition of a concentration cell in Section 20.6 to answer the stated questions. Use Equation 20.18 to calculate the cell emf. For a concentration cell, Q = [dilute]/[concentrated]. *Solve.*

(a) The compartment with the more dilute solution will be the anode. That is, the compartment with $[Zn^{2+}] = 1.00 \times 10^{-2}\ M$ is the anode.

(b) Because the oxidation half-reaction is the opposite of the reduction half-reaction, E° is zero.

(c) $E = E^\circ - \frac{0.0592}{n} \log Q$; $Q = [Zn^{2+}, \text{dilute}]/[Zn^{2+}, \text{conc.}]$

$$E = 0 - \frac{0.0592}{2} \log \frac{(1.00 \times 10^{-2})}{(1.8)} = 0.0668 \text{ V}$$

(d) In the anode compartment, $Zn(s) \rightarrow Zn^{2+}(aq)$, so $[Zn^{2+}]$ increases from $1.00 \times 10^{-2}\ M$. In the cathode compartment, $Zn^{2+}(aq) \rightarrow Zn(s)$, so $[Zn^{2+}]$ decreases from 1.8 *M*.

20.71 *Analyze/Plan.* Follow the logic in Sample Exercise 20.12. *Solve.*

$$E = E^\circ - \frac{0.0592}{2} \log \frac{[P_{H_2}][Zn^{2+}]}{[H^+]^2}; \; E^\circ = 0.0 \text{ V} - (-0.76 \text{ V}) = 0.76 \text{ V}$$

$$0.684 = 0.76 - \frac{0.0592}{2} \times (\log [P_{H_2}][Zn^{2+}] - 2 \log [H^+])$$

$$= 0.76 - \frac{0.0592}{2} \times (-0.5686 - 2 \log [H^+])$$

$$0.684 = 0.76 + 0.0168 + 0.0592 \log [H^+]; \; \log [H^+] = \frac{0.684 - 0.0168 - 0.76}{0.0592}$$

$\log [H^+] = -1.5676 = -1.6$; $[H^+] = 0.0271 = 0.03\ M$; pH = 1.6

Batteries and Fuel Cells (Section 20.7)

20.73 *Analyze/Plan.* Given mass of a reactant (Pb), calculate mass of product (PbO_2), and coulombs of charge transferred. This is a stoichiometry problem; we need the balanced equation for the chemical reaction that occurs in the lead-acid battery.

The overall cell reaction is:

$Pb(s) + PbO_2(s) + 2\,H^+(aq) + 2\,HSO_4^-(aq) \rightarrow 2\,PbSO_4(s) + 2\,H_2O(l)$ *Solve.*

(a) $g\ Pb \rightarrow mol\ Pb \rightarrow mol\ PbO_2 \rightarrow g\ PbO_2$

$$402\ g\ Pb \times \frac{1\ mol\ Pb}{207.2\ g\ Pb} \times \frac{1\ mol\ PbO_2}{1\ mol\ Pb} \times \frac{239.2\ g\ PbO_2}{1\ mol\ PbO_2} = 464\ g\ PbO_2$$

(b) From the half-reactions for the lead-acid battery, 2 mol electrons are transferred for each mol of Pb reacted. From Section 20.5, 96,485 C/mol e^-.

$$402\ g\ Pb \times \frac{1\ mol\ Pb}{207.2\ g\ Pb} \times \frac{2\ mol\ e^-}{1\ mol\ Pb} \times \frac{96,485\ C}{1\ mol\ e^-} = 374,392 = 3.74 \times 10^5\ C$$

20.75 *Analyze/Plan.* We are given a redox reaction and asked to write half-reactions, calculate E°, and indicate whether Li(s) is the anode or cathode. Determine which reactant is oxidized and which is reduced. Separate into half-reactions, find E°_{red} for the half-reactions from Appendix E and calculate E°. *Solve.*

(a) Li(s) is oxidized at the anode.

(b)

$$Ag_2CrO_4(s) + 2\,e^- \rightarrow 2\,Ag(s) + CrO_4^{2-}(aq) \qquad E^\circ_{red} = 0.45\ V$$
$$2\,[Li(s) \rightarrow Li^+(aq) + 1\,e^-] \qquad E^\circ_{red} = -3.05\ V$$
$$Ag_2CrO_4(s) + 2\,Li(s) \rightarrow 2\,Ag(s) + CrO_4^-(aq) + 2\,Li^+(aq)$$

$E^\circ = 0.45\ V - (-3.05\ V) = 3.50\ V$

(c) The emf of the battery, 3.5 V, is exactly the standard cell potential calculated in part (b).

(d) For this battery at ambient conditions, $E \approx E^\circ$, so $\log Q \approx 0$. This makes sense because all reactants and products in the battery are solids and thus present in their standard states. Assuming that E° is relatively constant with temperature, the value of the second term in the Nernst equation is ≈ 0 at 37 °C, and E ≈ 3.5 V.

20.77 *Analyze/Plan.* (a) Consider the function of Zn in an alkaline battery. What effect would it have on the redox reaction and cell emf if Cd replaces Zn? (b) Both batteries contain Ni. What is the difference in environmental impact between Cd and the metal hydride? *Solve.*

(a) E°_{red} for Cd (–0.40 V) is less negative than E°_{red} for Zn (–0.76 V), so E_{cell} will have a smaller (less positive) value.

(b) NiMH batteries use an alloy such as $ZrNi_2$ as the anode material. This eliminates the use and concomitant disposal problems associated with Cd, a toxic heavy metal.

20.79 (a) The oxidation states of the elements in $LiCoO_2$ are +1, +3 and –2, respectively. At any point in the operation of the battery, the positive and negative charges must balance. On charging, the oxidation state of some of the cobalt ions increases to +4, which releases some of the Li^+ ions to migrate back to the anode. If approximately half of the Li^+ ions migrate, the same number of Co^{3+} ions are oxidized to the +4 state. When the battery is fully charged, the mole ratios of the elements in the cathode are: 0.5 mol Li^+ to 0.5 mol Co^{3+} to 0.5 mol Co^{4+} to 2 mol O^{2-}. The material in the cathode is $LiCo_2O_4$.

(b) The molar mass of $LiCoO_2$ is 97.87 g/mol.

$$10 \text{ g LiCoO}_2 \times \frac{1 \text{ mol LiCoO}_2}{97.87 \text{ g LiCoO}_2} \times \frac{1 \text{ mol Li}}{1 \text{ mol LiCoO}_2} = 0.1022 = 0.10 \text{ mol Li}^+$$

Half of the Li^+, 0.051 mol, migrates during a full charge.

$$0.051 \text{ mol Li}^+ \times \frac{1 \text{ mol e}^-}{1 \text{ mol Li}^+} \times \frac{96,485 \text{ C}}{1 \text{ mol e}^-} = 4,921 = 4.9 \times 10^3 \text{ C}$$

20.81 *Analyze/Plan.* Write the balanced equation for the spontaneous reaction in a hydrogen fuel cell. Separate the overall reaction into half-reactions in acid solution and basic solution, find the matching standard reduction potentials in Appendix E and calculate standard voltages. *Solve.*

(a) The spontaneous reaction in the hydrogen fuel cell is hydrogen gas plus oxygen gas makes water.

(b)
$$O_2(g) + 4\,H^+(aq) + 4\,e^- \rightarrow 2\,H_2O(l) \qquad E^\circ_{red} = 1.23 \text{ V}$$
$$2\,H_2(g) \rightarrow 4\,H^+(aq) + 4\,e^- \qquad E^\circ_{red} = 0.00 \text{ V}$$
$$O_2(g) + 2\,H_2(g) \rightarrow 2\,H_2O(l) \qquad E^\circ = 1.23 - 0.00 = 1.23 \text{ V}$$

Corrosion (Section 20.8)

20.83 *Analyze/Plan.* (a) Decide which reactant is oxidized and which is reduced. Write the balanced half-reactions and assign the appropriate one as anode and cathode. (b) Write the balanced half-reaction for $Fe^{2+}(aq) \rightarrow Fe_2O_3 \cdot 3H_2O$. Use the reduction half-reaction from part (a) to obtain the overall reaction. *Solve.*

(a) anode: $Fe(s) \rightarrow Fe^{2+}(aq) + 2\,e^-$

cathode: $O_2(g) + 4\,H^+(aq) + 4\,e^- \rightarrow 2\,H_2O(l)$

(b) $2\,Fe^{2+}(aq) + 6\,H_2O(l) \rightarrow Fe_2O_3 \cdot 3H_2O(s) + 6\,H^+(aq) + 2\,e^-$

$O_2(g) + 4\,H^+(aq) + 4\,e^- \rightarrow 2\,H_2O(l)$

(Multiply the oxidation half-reaction by two to balance electrons and obtain the overall balanced reaction.)

20.85 (a) A "sacrificial anode" is a metal that is oxidized in preference to another when the two metals are coupled in an electrochemical cell; the sacrificial anode has a more negative E°_{red} than the other metal. In this case, Mg acts as a sacrificial anode because it is oxidized in preference to the pipe metal; it is sacrificed to preserve the pipe.

(b) E°_{red} for Mg^{2+} is –2.37 V, more negative than most metals present in pipes, including Fe ($E^\circ_{red} = -0.44$ V) and Zn ($E^\circ_{red} = -0.76$ V).

20.87 *Analyze/Plan.* Positive and negative ion charges must balance in neutral compounds. Use that fact along with oxidation numbers to answer the questions. *Solve.*

(a) +3

(b) +8/3 or +2.67

(c) 2 Fe(III) and 1 Fe(II)

Electrolysis; Electrical Work (Section 20.9)

20.89 (a) *Electrolysis* is an electrochemical process driven by an outside energy source.

(b) Electrolysis reactions are, by definition, nonspontaneous.

(c) $2\,Cl^-(l) \rightarrow Cl_2(g) + 2\,e^-$

(d) When an aqueous solution of NaCl undergoes electrolysis, sodium metal is not formed because H_2O is preferentially reduced to form $H_2(g)$.

20.91 *Analyze/Plan.* Follow the logic in Sample Exercise 20.14, paying close attention to units. Coulombs = amps-s; because this is a 3 e^- reduction, each mole of Cr(s) requires 3 Faradays. *Solve.*

(a) $$7.60\ A \times 2.00\ d \times \frac{24\ h}{1\ d} \times \frac{60\ min}{1\ h} \times \frac{60\ s}{1\ min} \times \frac{1\ C}{1\ amp\text{-}s} \times \frac{1\ F}{96{,}485\ C} \times \frac{1\ mol\ Cr}{3\ F} \times \frac{52.00\ g\ Cr}{1\ mol\ Cr} = 236\ g\ Cr(s)$$

(b) $$0.250\ mol\ Cr \times \frac{3\ F}{1\ mol\ Cr} \times \frac{96{,}485\ C}{F} \times \frac{1\ amp\text{-}s}{1\ C} \times \frac{1}{8.00\ h} \times \frac{1\ h}{60\ min} \times \frac{1\ min}{60\ s} = 2.51\ A$$

20.93 *Analyze/Plan.* Follow the logic in Sample Exercise 20.14, paying close attention to units. Take the 85% efficiency into account. Li^+ is reduced at the anode; Cl^- is oxidized at the anode. *Solve.*

(a) If the cell is 85% efficient, $$\frac{96{,}485\ C}{F} \times \frac{1\ F}{0.85\ mol} = 1.13512 \times 10^5$$
$$= 1.1 \times 10^5\ C/mol\ Li \text{ is required.}$$

$$7.5 \times 10^4\ A \times 24\ h \times \frac{3600\ s}{1\ h} \times \frac{1 C}{1\ amp\text{-}s} \times \frac{1 mol\ Li}{1.13512 \times 10^5\ C} \times \frac{6.94\ g\ Li}{1\ mol\ Li}$$
$$= 3.962 \times 10^5 = 4.0 \times 10^5\ g\ Li$$

(b) $E^o_{cell} = E^o_{red}(\text{cathode}) - E^o_{red}(\text{anode}) = -3.05\ V - (1.36\ V) = -4.41\ V$

The minimum voltage required to drive the reaction is the magnitude of E^o_{cell}, 4.41 V.

20.95 Refer to Table 4.5, "The Activity Series of the Metals." Gold is the least active metal on this table, less active than copper. This means that gold is more difficult to oxidize than copper (and that E^o_{red} for Au^{3+} is more positive than E^o_{red} for Cu^{2+}). When a mixture of copper and gold is refined by electrolysis, Cu is oxidized from the anode, but any metallic gold present in the mixture is not oxidized, so it accumulates near the anode.

Additional Exercises

20.97 (a)

$$\begin{array}{ll} Ni^+(aq) + 1\ e^- & \rightarrow Ni(s) \\ Ni^+(aq) & \rightarrow Ni^{2+}(aq) + 1\ e^- \\ \hline 2\ Ni^+(aq) & \rightarrow Ni(s) + Ni^{2+}(aq) \end{array}$$

(b)

$$\begin{array}{l} MnO_4^{2-}(aq) + 4\ H^+(aq) + 2\ e^- \rightarrow MnO_2(s) + 2\ H_2O(l) \\ 2\ [MnO_4^{2-}(aq) \rightarrow MnO_4^-(aq) + 1\ e^-] \\ \hline 3\ MnO_4^{2-}(aq) + 4\ H^+(aq) \rightarrow 2\ MnO_4^-(aq) + MnO_2(s) + 2\ H_2O(l) \end{array}$$

(c)

$$\begin{array}{l} H_2SO_3(aq) + 4\ H^+(aq) + 4\ e^- \rightarrow S(s) + 3\ H_2O(l) \\ 2\ [H_2SO_3(aq) + H_2O(l) \rightarrow HSO_4^-(aq) + 3\ H^+(aq) + 2\ e^-] \\ \hline 3\ H_2SO_3(aq) \rightarrow S(s) + 2\ HSO_4^-(aq) + 2\ H^+(aq) + H_2O(l) \end{array}$$

(d)

$$\begin{array}{l} Cl_2(aq) + 2\ H_2O(l) \rightarrow 2\ ClO^-(aq) + 4\ H^+(aq) + 2\ e^- \\ 4\ OH^-(aq) \qquad\qquad + 4\ OH^-(aq) \\ \hline Cl_2(aq) + 4\ OH^-(aq) \rightarrow 2\ ClO^-(aq) + 2\ H_2O(l) + 2\ e^- \\ Cl_2(aq) + 2\ e^- \rightarrow 2\ Cl^-(aq) \\ \hline 1/2\ [2Cl_2(aq) + 4\ OH^-(aq) \rightarrow 2\ Cl^-(aq) + 2\ ClO^-(aq) + 2\ H_2O(l)] \\ Cl_2(aq) + 2\ OH^-(aq) \rightarrow Cl^-(aq) + ClO^-(aq) + H_2O(l) \end{array}$$

20.99 We need in each case to determine whether E° is positive (spontaneous) or negative (nonspontaneous).

(a)

$$\begin{array}{ll} I_2(s) + 2\ e^- \rightarrow 2\ I^-(aq) & E^\circ_{red} = 0.54\ V \\ Sn(s) \rightarrow Sn^{2+}(aq) + 2\ e^- & E^\circ_{red} = -0.14\ V \\ \hline Sn(s) + I_2(s) \rightarrow Sn^{2+}(aq) + 2\ I^-(aq) & E^\circ = 0.54 - (-0.14) = 0.68\ V, \text{ spontaneous} \end{array}$$

(b)

$$\begin{array}{ll} Ni^{2+}(aq) + 2\ e^- \rightarrow Ni(s) & E^\circ_{red} = -0.28\ V \\ 2\ I^-(aq) \rightarrow I_2(s) + 2\ e^- & E^\circ_{red} = 0.54\ V \\ \hline Ni^{2+}(aq) + 2\ I^-(aq) \rightarrow Ni(s) + I_2(s) & E^\circ = -0.28 - 0.54 = -0.82\ V, \text{ nonspontaneous} \end{array}$$

(c)

$$\begin{array}{ll} 2\ [Ce^{4+}(aq) + 1\ e^- \rightarrow Ce^{3+}(aq) & E^\circ_{red} = 1.61\ V \\ H_2O_2(aq) \rightarrow O_2(g) + 2\ H^+(aq) + 2\ e^- & E^\circ_{red} = 0.68\ V \\ \hline 2\ Ce^{4+}(aq) + H_2O_2(aq) \rightarrow 2\ Ce^{3+}(aq) + O_2(g) + 2\ H^+(aq) & E^\circ = 1.61 - 0.68 = 0.93\ V, \text{ spontaneous} \end{array}$$

(d)

$$\begin{array}{ll} Cu^{2+}(aq) + 2\ e^- \rightarrow Cu(s) & E^\circ_{red} = 0.34\ V \\ Sn^{2+}(aq) \rightarrow Sn^{4+}(aq) + 2\ e^- & E^\circ_{red} = 0.15\ V \\ \hline Cu^{2+}(aq) + S\ n^{2+}(aq) \rightarrow Cu(s) + Sn^{4+}(aq) & E^\circ = 0.34 - 0.15 = 0.19\ V, \text{ spontaneous} \end{array}$$

20.103 Assume that room temperature is 298 K. Use the relationship developed in Solution 20.55 to calculate K from E° at 298 K. Use data from Appendix E to calculate E° for the disproportionation.

$$\begin{array}{ll} Cu^{+}(aq) + 1\ e^{-} \rightarrow Cu(s) & E^{\circ}_{red} = 0.52\ V \\ Cu^{+}(aq) \rightarrow Cu^{2+}(aq) + 1\ e^{-} & E^{\circ}_{red} = 0.15\ V \\ \hline 2\ Cu^{+}(aq) \rightarrow Cu(s) + Cu^{2+}(aq) & E^{\circ} = 0.52\ V - 0.15\ V = 0.37\ V \end{array}$$

$$E^{\circ} = \frac{0.0592}{n} \log K, \log K = \frac{nE^{\circ}}{0.0592} = \frac{1 \times 0.37}{0.0592} = 6.25 = 6.3$$

$$K = 10^{6.25} = 1.778 \times 10^{6} = 2 \times 10^{6}$$

20.106 (a) By analogy to O in hydroxide (OH^-) and in alcohol (R–OH), the oxidation state of S in a thiol (R–SH) is –2.

(b) By analogy to O in peroxide (–O–O–), the oxidation state of S in a disulfide (–S–S–) is –1.

(c) When two thiols react to form a disulfide, the oxidation number of S changes from –2 to –1. The oxidation number of S becomes more positive and the thiols are oxidized.

(d) Converting a disulfide to two thiols is the reverse of the process described in part (c). In order to reduce the disulfide, a reducing agent must be added to the solution.

(e) When two thiols (R–SH) react to form a disulfide (R–S–S–R), the H atoms are probably removed by base before the oxidizing agent does its work.

20.109 *Analyze.* Given mass of aluminum desired, applied voltage, and electrolysis efficiency, calculate kWh of electricity required. *Plan.* Beginning with mass Al and paying attention to units, calculate coulombs required if the process is 100% efficient. Then, take efficiency into account and use V, C, and the relationship between J and kWh to calculate kWh required. *Solve.*

$$1.0 \times 10^{3}\ kg\ Al \times \frac{1000\ g}{1\ kg} \times \frac{1\ mol\ Al}{26.98\ g\ Al} \times \frac{3\ F}{1\ mol\ Al} \times \frac{96,485\ C}{F} = 1.073 \times 10^{10} = 1.1 \times 10^{10}\ C$$

If the cell is 45% efficient, $(1.073 \times 10^{10}/0.45) = 2.384 \times 10^{10} = 2.4 \times 10^{10}$ C is required to plate the bumper.

$$4.50\ V \times 2.384 \times 10^{10} C \times \frac{1\ J}{1\ C\text{-}V} \times \frac{1\ kWh}{3.6 \times 10^{6}\ J} = 29,801 = 3.0 \times 10^{4}\ kWh$$

Integrative Exercises

20.114 (a)

$$\begin{array}{ll} Ag^+(aq) + e^- \rightarrow Ag(s) & E^\circ_{red} = 0.80\ V \\ Fe^{2+}(aq) \rightarrow Fe^{3+}(aq) + 1\ e^- & E^\circ_{red} = 0.77\ V \\ \hline Ag^+(aq) + Fe^{2+}(aq) \rightarrow Ag(s) + Fe^{3+}(aq) & E^\circ = 0.80\ V - 0.77\ V = 0.03\ V \end{array}$$

(b) $Ag^+(aq)$ is reduced at the cathode and $Fe^{2+}(aq)$ is oxidized at the anode.

(c) $\Delta G^\circ = -nFE^\circ = -(1)(96.5)(0.03) = -2.895 = 3\ kJ$

$\Delta S^\circ = S^\circ\ Ag(s) + S^\circ\ Fe^{3+}(aq) - S^\circ Ag^+(aq) - S^\circ\ Fe^{2+}(aq)$

$= 42.55\ J + 293.3\ J - 73.93\ J - 113.4\ J = 148.5\ J$

$\Delta G^\circ = \Delta H^\circ - T\Delta S^\circ$. Because ΔS° is positive, ΔG° will become more negative and E° will become more positive as temperature is increased.

20.117

$$\begin{array}{ll} AgSCN(s) + e^- \rightarrow Ag(s) + SCN^-(aq) & E^\circ_{red} = 0.09\ V \\ Ag(s) \rightarrow Ag^+(aq) + e^- & E^\circ_{red} = 0.80\ V \\ \hline AgSCN(s) \rightarrow Ag^+(aq) + SCN^-(aq) & E^\circ = 0.09 - 0.80 = -0.71\ V \end{array}$$

$$E^\circ = \frac{0.0592}{n} \log K_{sp};\ \log K_{sp} = \frac{(-0.71)(1)}{0.0592} = -11.993 = -12$$

$$K_{sp} = 10^{-11.993} = 1.02 \times 10^{-12} = 10^{-12}$$

21 Nuclear Chemistry

Visualizing Concepts

21.1 *Analyze.* Given the name and mass number of a nuclide, decide if it lies within the belt of stability. If not, suggest a process that moves it toward the belt.

Plan. Calculate the number of protons and neutrons in each nuclide. Locate this point on Figure 21.1. If the point is above the belt, beta decay increases protons and decreases neutrons, decreasing the neutron-to-proton ratio. If the point is below the belt, either positron emission or neutron capture decreases protons and increases neutrons, increasing the neutron-to-proton ratio. *Solve.*

(a) ^{24}Ne: 10 p, 14 n, just above the belt of stability. Reduce the neutron-to-proton ratio via beta decay.

(b) ^{32}Cl: 17 p, 15 n, just below the belt of stability. Increase the neutron-to-proton ratio via positron emission or orbital electron capture.

(c) ^{108}Sn: 50 p, 58 n, just below the belt of stability. Increase the neutron-to-proton ratio via positron emission or orbital electron capture.

(d) ^{216}Po: 84 p, 132 n, just beyond the belt of stability. Nuclei with atomic numbers ≥ 84 tend to decay via alpha emission, which decreases both protons and neutrons.

21.6 *Analyze/Plan.* Write the balanced equation for the decay. Nuclear decay is a first-order process; use appropriate relationships for first-order processes to determine $t_{1/2}$, k and remaining ^{88}Mo after 12 minutes. *Solve.*

(a) $t_{1/2}$ is the time required for half of the original nuclide to decay. Relative to the graph, this is the time when the amount of ^{88}Mo is reduced from 1.0 to 0.5. This time is 7 minutes.

(b) For a first-order process, $t_{1/2} = 0.693/k$ or $k = 0.693/t_{1/2}$.
$k = 0.693/7 \text{ min} = 0.0990 = 0.1 \text{ min}^{-1}$

(c) From the graph, the fraction of ^{88}Mo remaining after 12 min is 0.3/1.0 = 0.3
Check. $\ln(N_t/N_o) = -kt = -(0.099)(12) = -1.188$; $N_t/N_o = e^{-1.188} = 0.30$.

(d) $^{88}_{42}Mo \rightarrow {}^{88}_{41}Nb + {}^{0}_{+1}e$

21.7 *Analyze/Plan.* Atomic number is number of protons, mass number is (protons + neutrons). Chemical symbol is determined by atomic number. Beta decay increases the number of protons, whereas mass number stays constant. Positron emission decreases the number of protons, whereas mass number stays constant. Half-life, $t_{1/2}$, is the time required to reduce the amount of radioactive material by half. *Solve.*

(a) ${}^{10}_{5}B$, ${}^{11}_{5}B$; ${}^{12}_{6}C$, ${}^{13}_{6}C$; ${}^{14}_{7}N$, ${}^{15}_{7}N$; ${}^{16}_{8}O$, ${}^{17}_{8}O$, ${}^{18}_{8}O$; ${}^{19}_{9}F$

(b) On the diagram, ${}^{14}_{6}C$ is the only radioactive nuclide above and left of the band of stable red nuclides. ${}^{14}_{6}C$ will reduce its neutron-to-proton ratio by beta decay.

${}^{14}_{6}C \rightarrow {}^{14}_{7}N + {}^{0}_{-1}e$

(c) On the diagram, radioactive nuclides right and below the band of stable red nuclides are likely to increase their neutron-to-proton ratio via positron emission. To be useful for positron emission tomography, the nuclides must have a half-life on the order of minutes. (Nuclides with very fast decay disappear before they can be imaged. Those with longer half-lives linger in the patient.) Four radioactive nuclides fit these criteria: ${}^{11}_{6}C$, ${}^{13}_{7}N$, ${}^{15}_{8}O$, ${}^{18}_{9}F$

(d) If 12.5% of the isotope remains, this amounts to three half-lives of decay. If the total decay time is 1 hour, 1/3 of this time is 20 min. The radionuclide with a half-life of 20 min is ${}^{11}_{6}C$.

Radioactivity and Nuclear Equations (Section 21.1)

21.9 *Analyze/Plan.* Given various nuclide descriptions, determine the number of protons and neutrons in each nuclide. The left superscript is the mass number, protons plus neutrons. If there is a left subscript, it is the atomic number, the number of protons. Protons can always be determined from the chemical symbol; all isotopes of the same element have the same number of protons. A number following the element name, as in part (c) is the mass number. *Solve.*

p = protons, n = neutrons, e = electrons; number of protons = atomic number;
number of neutrons = mass number – atomic number

(a) ${}^{56}_{24}Cr$: 24p, 32n (b) ${}^{193}Tl$: 81p, 112n (c) ${}^{38}Ar$: 18p, 20n

21.11 *Analyze/Plan.* See definitions in Section 21.1. In each case, the left superscript is mass number, the left subscript is related to atomic number. *Solve.*

(a) ${}^{1}_{0}n$ (b) ${}^{4}_{2}He$ or α (c) ${}^{0}_{0}\gamma$ or γ

21.13 *Analyze/Plan.* Follow the logic in Sample Exercises 21.1 and 21.2. Pay attention to definitions of decay particles in Table 21.2 as well as conservation of mass and charge. *Solve.*

(a) ${}^{90}_{37}Rb \rightarrow {}^{90}_{38}Sr + {}^{0}_{-1}e$ (b) ${}^{72}_{34}Se + {}^{0}_{-1}e$ (orbital electron) $\rightarrow {}^{72}_{33}As$

(c) ${}^{76}_{36}Kr \rightarrow {}^{76}_{35}Br + {}^{0}_{+1}e$ (d) ${}^{226}_{88}Ra \rightarrow {}^{222}_{86}Rn + {}^{4}_{2}He$

21.15 *Analyze/Plan.* Using definitions of the decay processes and conservation of mass number and atomic number, work backwards to the reactants in the nuclear reactions. *Solve.*

(a) ${}^{211}_{82}Pb \rightarrow {}^{211}_{83}Bi + {}^{0}_{-1}e$ (b) ${}^{50}_{25}Mn \rightarrow {}^{50}_{24}Cr + {}^{0}_{+1}e$

(c) ${}^{179}_{74}W + {}^{0}_{-1}e \rightarrow {}^{179}_{73}Ta$ (d) ${}^{230}_{90}Th \rightarrow {}^{226}_{88}Ra + {}^{4}_{2}He$

21.17 *Analyze/Plan.* Given the starting and ending nuclides in a nuclear decay sequence, we are asked to determine the number of alpha and beta emissions. Use the total change in A and Z, along with definitions of alpha and beta decay, to answer the question. *Solve.*

The total mass number change is (235–207) = 28. As each alpha particle emission decreases the mass number by four, whereas emission of a beta particle does not correspond to a mass change, there are 7 alpha particle emissions. The change in atomic number in the series is 10. Each alpha particle results in an atomic number lower by two. The 7 alpha particle emissions alone would cause a decrease of 14 in atomic number. Each beta particle emission raises the atomic number by one. To obtain the observed lowering of 10 in the series, there must be 4 beta emissions.

Patterns of Nuclear Stability (Section 21.2)

21.19 *Analyze/Plan.* Follow the logic in Sample Exercise 21.3, paying attention to the guidelines for neutron-to-proton ratio. *Solve.*

(a) ${}^{8}_{5}B$ - low neutron/proton ratio, positron emission (for low atomic numbers, positron emission is more common than orbital electron capture)

(b) ${}^{68}_{29}Cu$ - high neutron/proton ratio, beta emission

(c) ${}^{32}_{15}P$ - slightly high neutron/proton ratio, beta emission

(d) ${}^{39}_{17}Cl$ - high neutron/proton ratio, beta emission

21.21 *Analyze/Plan.* Use the criteria listed in Table 21.4. *Solve.*

(a) ${}^{39}_{19}K$; stable, odd proton, even neutron; 20 neutrons is a magic number.

${}^{40}_{19}K$; radioactive, odd proton, odd neutron

(b) ${}^{209}_{83}Bi$; stable, odd proton, even neutron; 126 neutrons is a magic number.

${}^{208}_{83}Bi$; radioactive, odd proton, odd neutron

(c) ${}^{58}_{28}Ni$; stable, even proton, even neutron

${}^{65}_{28}Ni$; radioactive, high neutron/proton ratio

21.23 *Analyze/Plan.* For each nuclide, determine the number of protons and neutrons and decide if they are magic numbers. *Solve.*

(a) ${}^{4}_{2}He$, both (b) ${}^{18}_{8}O$ has a magic number of protons, but not neutrons

(c) ${}^{40}_{20}Ca$, both (d) ${}^{66}_{30}Zn$ has neither (e) ${}^{208}_{82}Pb$, both

21.25 Statement (a). Because of its magic numbers, the alpha particle is a very stable emitted particle, which makes alpha emission a favorable process. The proton has no magic numbers, and is an odd proton–even neutron particle. Its instability does not encourage proton emission as a process.

Nuclear Transmutations (Section 21.3)

21.27 Statement (b) is the best explanation. Because they are electrically neutral, neutrons are not repelled by the positively charged nucleus being bombarded.

Statement (a) does not apply and statements (c) and (d) are false. Neutrons are not attracted to the nucleus at long distances.

21.29 *Analyze/Plan*. Determine A and Z for the missing particle by conservation principles. Find the appropriate symbol for the particle. *Solve.*

(a) $^{252}_{98}Cf + ^{10}_{5}B \rightarrow 3\ ^{1}_{0}n + ^{259}_{103}Lr$ (b) $^{2}_{1}H + ^{3}_{2}He \rightarrow ^{4}_{2}He + ^{1}_{1}H$

(c) $^{1}_{1}H + ^{11}_{5}B \rightarrow 3\ ^{4}_{2}He$ (d) $^{122}_{53}I \rightarrow ^{122}_{54}Xe + ^{0}_{-1}e$

(e) $^{59}_{26}Fe \rightarrow ^{0}_{-1}e + ^{59}_{27}Co$

21.31 *Analyze/Plan*. Follow the logic in Sample Exercise 21.4, paying attention to conservation of A and Z. *Solve.*

(a) $^{238}_{92}U + ^{4}_{2}He \rightarrow ^{241}_{94}Pu + ^{1}_{0}n$ (b) $^{14}_{7}N + ^{4}_{2}He \rightarrow ^{17}_{8}O + ^{1}_{1}H$

(c) $^{56}_{26}Fe + ^{4}_{2}He \rightarrow ^{60}_{29}Cu + ^{0}_{-1}e$

Rates of Radioactive Decay (Section 21.4)

21.33 (a) True. $k = 0.693/t_{1/2}$. The decay rate constant, k, and half-life, $t_{1/2}$ are inversely related.

(b) False. If X is not radioactive, it does not spontaneously decay and its half-life is essentially infinity.

(c) True. Changes in the amount of A would be measurable over the 40-year time frame, whereas changes in the amount of X would be very small and difficult to detect.

21.35 *Analyze/Plan*. The half-life is 12.3 yr. Use $t_{1/2}$ to calculate k and the mass fraction (N_t/N_0) remaining after 50 yr.

Solve. $k = 0.693/t_{1/2} = 0.693/12.3\ \text{yr} = 0.056341 = 0.0563\ \text{yr}^{-1}$

$$\ln\frac{N_t}{N_0} = -kt; \quad \ln\frac{N_t}{N_0} = -0.056341\ \text{yr}^{-1}(50\ \text{yr}) = -2.81707 = -2.8; \quad \frac{N_t}{N_0} = 0.05978 = 0.06$$

When the watch is 50 years old, only 6% (or 6.0%) of the tritium remains. The dial will be dimmed by 94%.

21.37 *Analyze/Plan.* We are given half-life of cobalt-60, and replacement time when the activity of the sample is 75% of the initial value. Consider the rate law for (first-order) nuclear decay: $\ln(N_t/N_o) = -kt$. *Solve.*

$k = 0.693 / t_{1/2} = 0.693/5.26 \text{ yr} = 0.1317 = 0.132 \text{ yr}^{-1}$; $N_t/N_o = 0.75$

$$t = \frac{-1}{k} \ln\frac{N_t}{N_o} = -(1/0.1317 \text{ yr}^{-1}) \ln (0.75) = 2.18 \text{ yr} = 26.2 \text{ mo} = 797 \text{ d.}$$

The source will have to be replaced sometime in the summer of 2018, probably in August.

21.39 (a) *Analyze/Plan.* ${}^{226}_{88}\text{Ra} \rightarrow {}^{222}_{86}\text{Rn} + {}^{4}_{2}\text{He}$

1 alpha particle is produced for each ^{226}Ra that decays. Rate = kN. Calculate the number of ^{226}Ra particles in the 10.0 mg sample. Calculate $t_{1/2}$ and k in min, then rate in dis/min, then the number of disintegrations in 5.0 min. *Solve.*

$$10.0 \text{ mg Ra} \times \frac{1 \text{ g}}{1000 \text{ mg}} \times \frac{1 \text{ mol Ra}}{226 \text{ g Ra}} \times \frac{6.022\times10^{23} \text{ Ra atoms}}{1 \text{ mol Ra}} = 2.6646\times10^{19} = 2.66\times10^{19} \text{ atoms}$$

Calculate k in min^{-1}. $1600 \text{ yr} \times \frac{365 \text{ d}}{1 \text{ yr}} \times \frac{24 \text{ h}}{1 \text{ d}} \times \frac{60 \text{ min}}{1 \text{ h}} = 8.410\times10^{8} \text{ min}^{-1}$

$$k = \frac{0.693}{t_{1/2}} = \frac{0.693}{8.410\times10^{8} \text{ min}} = 8.241\times10^{-10} \text{ min}^{-1}$$

Rate = kN = $(8.241 \times 10^{-10} \text{ min}^{-1})(2.6646 \times 10^{19} \text{ atoms}) = 2.20 \times 10^{10}$ atoms/min

$(2.20 \times 10^{10} \text{ atoms/min})(5.0 \text{ min}) = 1.1 \times 10^{11}$ ^{226}Ra atoms decay in 5.0 min

1.1×10^{11} alpha particles emitted in 5.0 min

(b) *Plan.* From part (a), the rate is 2.20×10^{10} disintegrations/min. Change this to dis/s and apply the definition $1 \text{ Ci} = 3.7 \times 10^{10}$ dis/s.

$$\frac{2.20\times10^{10} \text{ dis}}{1 \text{ min}} \times \frac{1 \text{ min}}{60 \text{ s}} \times \frac{1 \text{ Ci}}{3.7\times10^{10} \text{ dis/s}} \times \frac{1000 \text{ mCi}}{\text{Ci}} = 9.891 = 9.9 \text{ mCi}$$

21.41 *Analyze/Plan.* Calculate k in yr^{-1} and solve Equation 21.20 for t. $N_o = 16.3$/min/g, $N_t = 9.7$/min/g. *Solve.*

$k = 0.693/t_{1/2} = 0.693/5715 \text{ yr} = 1.213 \times 10^{-4} = 1.21 \times 10^{-4} \text{ yr}^{-1}$

$$t = \frac{-1}{k} \ln\frac{N_t}{N_o} = \frac{-1}{1.213\times10^{-4} \text{ yr}^{-1}} \ln\frac{9.7}{16.3} = 4.280\times10^{3} = 4.3\times10^{3} \text{ yr}$$

21.43 *Analyze/Plan.* Follow the procedure outlined in Sample Exercise 21.6. If the mass of ^{40}Ar is 4.2 times that of ^{40}K, the original mass of ^{40}K must have been 4.2 + 1 = 5.2 times the amount of ^{40}K present now. *Solve.*

$k = 0.693/1.27 \times 10^{9} \text{ yr} = 5.457 \times 10^{-10} = 5.46 \times 10^{-10} \text{ yr}^{-1}$

$$t = \frac{-1}{5.457\times10^{-10} \text{ yr}^{-1}} \times \ln\frac{1}{(5.2)} = 3.0\times10^{9} \text{ yr}$$

Energy Changes in Nuclear Reactions (Section 21.6)

21.45 *Analyze/Plan*. Use Equation 21.22 to change 0.1 mg to energy. *Solve*.

$$\Delta E = c^2\Delta m = (2.9979246 \times 10^8 \text{ m/s})^2 \times 0.1 \text{ mg} \times \frac{1 \text{ g}}{1000 \text{ mg}} \times \frac{1 \text{ kg}}{1000 \text{ g}} \times \frac{1 \text{ kJ}}{1000 \text{ J}} = 9 \times 10^6 \text{ kJ}$$

21.47 *Analyze/Plan*. Given the mass of an ^{27}Al atom, subtract the mass of 13 electrons to get the mass of an ^{27}Al nucleus. Calculate the mass difference between the ^{27}Al nucleus and the separate nucleons, convert this to energy using Equation 21.22. Use the molar mass of ^{27}Al to calculate the energy required for 100 g of ^{27}Al. *Solve*.

The mass of an electron is 5.485799×10^{-4} amu (inside back cover of the text). The mass of a ^{27}Al nucleus is then 26.9815386 amu $-13(5.485799 \times 10^{-4}$ amu) = 26.9744071 amu. $\Delta m = 13(1.0072765 \text{ amu}) + 14(1.0086649 \text{ amu}) - 26.9744071 \text{ amu} = 0.2414960 \text{ amu}$.

$$\Delta E = (2.9979246 \times 10^8 \text{ m/s})^2 \times 0.2414960 \text{ amu} \times \frac{1 \text{ g}}{6.0221421 \times 10^{23} \text{ amu}} \times \frac{1 \text{ kg}}{1 \times 10^3 \text{ g}}$$

$$= 3.604129 \times 10^{-11} = 3.604129 \times 10^{-11} \text{ J}/^{27}\text{Al nucleus required}$$

If the mass change for a single ^{27}Al nucleus is 0.2414960 amu, the mass change for 1 mole of ^{27}Al is 0.2414960 g.

$$\Delta E = 100 \text{ g } ^{27}\text{Al} \times \frac{1 \text{ mol } ^{27}\text{Al}}{26.9815386 \text{ g } ^{27}\text{Al}} \times \frac{0.241960}{\text{mol } ^{27}\text{Al}} \times \frac{1 \text{ kg}}{1000 \text{ g}} \times (2.9979246 \times 10^8 \text{ m/s})^2$$

$$= 8.044234 \times 10^{13} \text{ J} = 8.044234 \times 10^{10} \text{ kJ}/100 \text{ g } ^{27}\text{Al}$$

21.49 *Analyze/Plan*. Given atomic mass, subtract mass of the electrons to get nuclear mass. Calculate the nuclear binding energy by finding the mass difference between the nucleus and the separate nucleons and converting this to energy using Equation 21.22. Divide by the total number of nucleons to find binding energy per nucleon. *Solve*.

(a) Nuclear mass

^{2}H: 2.014102 amu $- 1(5.485799 \times 10^{-4}$ amu) = 2.013553 amu

^{4}He: 4.002602 amu $- 2(5.485799 \times 10^{-4}$ amu) = 4.001505 amu

^{6}Li: 6.0151228 amu $- 3(5.4857991 \times 10^{-4}$ amu) = 6.0134771 amu

(b) Nuclear binding energy

^{2}H: $\Delta m = 1(1.0072765) + 1(1.0086649) - 2.013553 = 0.002388$ amu

$$\Delta E = 0.002388 \text{ amu} \times \frac{1 \text{ g}}{6.022142 \times 10^{23} \text{ amu}} \times \frac{1 \text{ kg}}{1000 \text{ g}} \times \frac{8.987551 \times 10^{16} \text{ m}^2}{\text{s}^2}$$

$$= 3.564490 \times 10^{-13} = 3.564 \times 10^{-13} \text{ J}$$

^{4}He: $\Delta m = 2(1.0072765) + 2(1.0086649) - 4.001505 = 0.030378$ amu

$$\Delta E = 0.030378 \text{ amu} \times \frac{1 \text{ g}}{6.022142 \times 10^{23} \text{ amu}} \times \frac{1 \text{ kg}}{1000 \text{ g}} \times \frac{8.987551 \times 10^{16} \text{ m}^2}{\text{s}^2}$$

$$= 4.533636 \times 10^{-12} = 4.5336 \times 10^{-12} \text{ J}$$

^{6}Li: $\Delta m = 3(1.0072765) + 3(1.0086649) - 6.0134771 = 0.0343471$ amu

$$\Delta E = 0.0343471 \text{ amu} \times \frac{1 \text{ g}}{6.022142 \times 10^{23} \text{ amu}} \times \frac{1 \text{ kg}}{1000 \text{ g}} \times \frac{8.987551 \times 10^{16} \text{ m}^2}{\text{s}^2}$$

$$= 5.126021 \times 10^{-12} = 5.12602 \times 10^{-12} \text{ J}$$

(c) Binding energy per nucleon

^{2}H: 3.564490×10^{-13} J/2 nucleons = 1.782×10^{-13} J/nucleon

^{4}He: 4.533636×10^{-12} J/4 nucleons = 1.1334×10^{-12} J/nucleon

^{6}Li: 5.126021×10^{-12} J/6 nucleons = 8.54337×10^{-13} J/nucleon

(d) This trend in binding energy/nucleon agrees with the curve in Figure 21.12, which shows an irregular increase in binding energy/nucleon up to atomic number 56. The anomalously high value for ^{4}He calculated above is also apparent on the figure.

21.51 *Analyze/Plan*. Use Equation 21.22 to calculate the mass equivalence of the solar radiation. *Solve*.

(a) $$\frac{1.07 \times 10^{16} \text{ kJ}}{1 \text{ min}} \times \frac{60 \text{ min}}{1 \text{ h}} \times \frac{24 \text{ h}}{1 \text{ d}} = 1.541 \times 10^{19} \frac{\text{kJ}}{\text{d}} = 1.54 \times 10^{22} \text{ J/d}$$

$$\Delta m = \frac{1.541 \times 10^{22} \text{ kg-m}^2/\text{s}^2/\text{d}}{(2.998 \times 10^8 \text{ m/s})^2} = 1.714 \times 10^5 = 1.71 \times 10^5 \text{ kg/d}$$

(b) *Analyze/Plan*. Calculate the mass change in the given nuclear reaction, then a conversion factor for g ^{235}U to mass equivalent. *Solve*.

$\Delta m = 140.8833 + 91.9021 + 2(1.0086649) - 234.9935 = -0.19077 = -0.1908$ amu

Converting from atoms to moles and amu to grams, it requires 1.000 mol or 235.0 g ^{235}U to produce energy equivalent to a change in mass of 0.1908 g.

0.10% of 1.714×10^5 kg is 1.714×10^2 kg = 1.714×10^5 g

$$1.714 \times 10^5 \text{ g} \times \frac{235.0 \text{ g } ^{235}\text{U}}{0.1908 \text{ g}} = 2.111 \times 10^8 = 2.1 \times 10^8 \text{ g } ^{235}\text{U}$$

(This is about 230 tons of ^{235}U *per day*.)

21.53 Nucleus (b), ^{51}V, should possess the greatest mass defect per nucleon. Figure 21.12 shows that the binding energy per nucleon (which gives rise to the mass defect) is greatest for nuclei with mass numbers around 50.

Nuclear Power and Radioisotopes (Sections 21.7, 21.8, and 21.9)

21.55 (a) NaI is a good source of iodine, because it is a strong electrolyte and completely dissociated into ions in aqueous solution. The $I^-(aq)$ are mobile and immediately available for bio-uptake. They do not need to be digested or processed in the body before uptake can occur. Also, iodine is a large percentage of the total mass of NaI.

(b) After ingestion, $I^-(aq)$ must enter the bloodstream, travel to the thyroid and then be absorbed. This requires a finite amount of time. A Geiger counter placed near the thyroid immediately after ingestion will register background, then gradually increase in signal until the concentration of $I^-(aq)$ in the thyroid reaches a maximum. Then, over time, iodine-131 decays, and the signal decreases.

(c) *Analyze/Plan.* The half-life of iodine-131 is 8.02 days. Use $t_{1/2}$ to calculate the decay rate constant, k. Then solve Equation 21.20 for t. $N_o = 0.12$ (12% of ingested iodine absorbed); $N_t = 1.2 \times 10^{-5}$ (0.01% of the original amount taken up by the thyroid).

Solve. $k = 0.693\ t_{1/2} = 0.693/8.02\ d = 0.086409 = 0.0864\ d^{-1}$

$\ln(N_t/N_o) = -kt;\ t = -\ln(N_t/N_o)/k$

$$t = \frac{-\ln(1.2 \times 10^{-5}/0.12)}{0.086409\ d^{-1}} = 106.59 = 107\ d$$

Check. N_t is given to 1 sig fig, so 1×10^2 days may be a more correct representation of the time frame for decay.

21.57 (a) Characteristics (ii) and (iv) are required for a fuel in a nuclear power plant.

Two or more neutrons (ii) are required so that a nuclear chain reaction occurs. Fission after absorption of a slow neutron (iv) is the nuclear process that produces energy in all current nuclear power plants. Gamma radiation (i) is produced by most nuclear decay processes and is a non-specific characteristic. Short half-life (iii) would require that fuel be replaced too frequently.

(b) ^{235}U

21.59 The *control rods* in a nuclear reactor regulate the flux of neutrons to keep the reaction chain self-sustaining and also prevent the reactor core from overheating. They are composed of materials such as boron or cadmium that absorb neutrons.

21.61 (a) ${}^{2}_{1}H + {}^{2}_{1}H \rightarrow {}^{3}_{2}He + {}^{1}_{0}n$

(b) ${}^{239}_{92}U + {}^{1}_{0}n \rightarrow {}^{133}_{51}Sb + {}^{98}_{41}Nb + 9\ {}^{1}_{0}n$

21.63 *Analyze/Plan.* At these temperatures, assume the reaction occurs between nuclei rather than atoms. From Table 21.7, the nuclear mass of ${}^{4}_{2}He$ is 4.00150. The nuclear mass of ${}^{1}_{1}H$ is simply the mass of a proton, 1.007276467 amu. Note that ${}^{0}_{+1}e$ is a positron, which has the same mass as an electron, 5.4857991×10^{-4} amu. Calculate the difference in mass between product and reactant nuclei, and the energy released by this mass change. Do the calculation in terms of moles and grams, rather than nuclei and amu. 1 mol amu = 1 g. *Solve.*

$\Delta m = 4.00150 + 2(5.4857991 \times 10^{-4}) - 4(1.007276467) = -0.0265087 = -0.02651\ g$

If the reaction is run with 1 mol of ${}_1^1H$, the mass change is (0.0265087/4) = 0.0066272 = 0.006627 g

$$\Delta E = c^2\Delta m = 0.0066272\ g \times \frac{1\ kg}{1000\ g} \times \frac{8.987551\times10^{16}\ m^2}{s^2} = 5.95621\times10^{11}$$

$$= 5.956\times10^{11}\ J = 5.956\times10^{8}\ kJ$$

21.65 (a) A *boiling water reactor* does not use a secondary coolant.

(b) A *fast breeder reactor* creates more fissionable material than it consumes.

(c) A *gas-cooled reactor* uses a gas as a primary coolant.

21.67 *Analyze/Plan.* Hydroxyl radical is electrically neutral but has an unpaired electron, $\cdot OH$. Hydroxide is an anion, OH^-. *Solve.*

Hydrogen abstraction: $RCOOH + \cdot OH \rightarrow RCOO\cdot + H_2O$

Deprotonation: $RCOOH + OH^- \rightarrow RCOO^- + H_2O$

Hydroxyl radical is more toxic to living systems, because it produces other radicals when it reacts with molecules in the organism. This often starts a disruptive chain of reactions, each producing a different free radical.

Hydroxide ion, OH^-, on the other hand, will be readily neutralized in the buffered cell environment. Its most common reaction is ubiquitous and innocuous:

$H^+ + OH^- \rightarrow H_2O$. The acid–base reactions of OH^- are usually much less disruptive to the organism than the chain of redox reactions initiated by $\cdot OH$ radical.

21.69 *Analyze/Plan.* Use definitions of the various radiation units and conversion factors to calculate the specified quantities. Pay particular attention to units. *Solve.*

(a) 1 Ci = 3.7×10^{10} disintegrations(dis)/s; 1 Bq = 1 dis/s

$$14.3\ mCi \times \frac{1\ Ci}{1000\ m\ Ci} \times \frac{3.7\times10^{10}\ dis/s}{Ci} = 5.29\times10^{8} = 5.3\times10^{8}\ dis/s = 5.3\times10^{8}\ Bq$$

(b) 1 rad = 1×10^{-2} J/kg; 1 Gy = 1 J/kg = 100 rad. From part (a), the activity of the source is 5.3×10^8 dis/s.

$$5.29\times10^{8}\ dis/s \times 14.0\ s \times 0.35 \times \frac{9.12\times10^{-13}\ J}{dis} \times \frac{1}{0.385\ kg} = 6.14\times10^{-3} = 6.1\times10^{-3}\ J/kg$$

$$6.1\times10^{-3}\ J/kg \times \frac{1\ rad}{1\times10^{-2}\ J/kg} \times \frac{1000\ mrad}{rad} = 6.1\times10^{2}\ mrad$$

$$6.1\times10^{-3}\ J/kg \times \frac{1\ Gy}{1\ J/kg} = 6.1\times10^{-3}\ Gy$$

(c) rem = rad (RBE); Sv = Gy (RBE), where 1 Sv = 100 rem

mrem = 6.14×10^2 mrad (9.5) = 5.83×10^3 = 5.8×10^3 mrem (or 5.8 rem)

Sv = 6.14×10^{-3} Gy (9.5) = 5.83×10^{-2} = 5.8×10^{-2} Sv

Additional Exercises

21.72 $^{222}_{86}\text{Rn} \rightarrow \text{X} + 3\,^{4}_{2}\text{He} + 2\,^{0}_{-1}\text{e}$

This corresponds to a reduction in mass number of (3 × 4 =) 12 and a reduction in atomic number of (3 × 2 − 2) = 4. The stable nucleus is $^{210}_{82}\text{Pb}$. (This is part of the sequence in Figure 21.1.)

21.74 (a) $^{36}_{17}\text{Cl} \rightarrow ^{36}_{18}\text{Ar} + ^{0}_{-1}\text{e}$

(b) According to Table 21.4, nuclei with even numbers of both protons and neutrons, or an even number of one kind of nucleon, are more stable. ^{35}Cl and ^{37}Cl both have an odd number of protons *but* an even number of neutrons. ^{36}Cl has an odd number of protons and neutrons (17 p, 19 n), so it is less stable than the other two isotopes. Also, ^{37}Cl has 20 neutrons, a nuclear closed shell.

21.76 (a) $^{6}_{3}\text{Li} + ^{56}_{28}\text{Ni} \rightarrow ^{62}_{31}\text{Ga}$

(b) $^{40}_{20}\text{Ca} + ^{248}_{96}\text{Cm} \rightarrow ^{147}_{62}\text{Sm} + ^{141}_{54}\text{Xe}$

(c) $^{88}_{38}\text{Sr} + ^{84}_{36}\text{Kr} \rightarrow ^{116}_{46}\text{Pd} + ^{56}_{28}\text{Ni}$

(d) $^{40}_{20}\text{Ca} + ^{238}_{92}\text{U} \rightarrow ^{70}_{30}\text{Zn} + 4\,^{1}_{0}\text{n} + 2\,^{102}_{41}\text{Nb}$

21.82 Because of the relationship $\Delta E = \Delta mc^2$, the mass defect (Δm) is directly related to the binding energy (ΔE) of the nucleus.

^{7}Be: 4p, 3n; 4(1.0072765) + 3(1.0086649) = 7.05510 amu

Total mass defect = 7.0551 − 7.0147 = 0.0404 amu

0.0404 amu/7 nucleons = 5.77×10^{-3} amu/nucleon

$$\Delta E = \Delta m \times c^2 = \frac{5.77 \times 10^{-3}\ \text{amu}}{\text{nucleon}} \times \frac{1\,\text{g}}{6.022 \times 10^{23}\ \text{amu}} \times \frac{1\,\text{kg}}{1 \times 10^{3}\ \text{g}} \times \frac{8.988 \times 10^{16}\ \text{m}^2}{\text{s}^2}$$

$$= \frac{5.77 \times 10^{-3}\ \text{amu}}{\text{nucleon}} \times \frac{1.4925 \times 10^{-10}\ \text{J}}{1\,\text{amu}} = 8.612 \times 10^{-13} = 8.61 \times 10^{-13}\ \text{J/nucleon}$$

^{9}Be: 4p, 5n; 4(1.0072765) + 5(1.0086649) = 9.07243 amu

Total mass defect = 9.0724 − 9.0100 = 0.06243 = 0.0624 amu

0.0624 amu/9 nucleons = $6.937 \times 10^{-3} = 6.94 \times 10^{-3}$ amu/nucleon

6.937×10^{-3} amu/nucleon × 1.4925×10^{-10} J/amu = $1.035 \times 10^{-12} = 1.04 \times 10^{-12}$ J/nucleon

^{10}Be: 4p, 6n; 4(1.0072765) + 6(1.0086649) = 10.0811 amu

Total mass defect = 10.0811 − 10.0113 = 0.0698 amu

0.0698 amu/10 nucleons = 6.98×10^{-3} amu/nucleon

6.98×10^{-3} amu/nucleon × 1.4925×10^{-10} J/amu = $1.042 \times 10^{-12} = 1.04 \times 10^{-12}$ J/nucleon

The binding energies/nucleon for ^{9}Be and ^{10}Be are very similar; that for ^{10}Be is slightly higher.

Integrative Exercises

21.88 Calculate the amount of energy produced by the nuclear fusion reaction, the enthalpy of combustion, $\Delta H°$, of C_8H_{18}, and then the mass of C_8H_{18} required.

Δm for the reaction $4\ {}^1_1H \rightarrow {}^4_2He + 2\ {}^{\ 0}_{+1}e$ is:

$4(1.00782) - 4.00260\text{ amu} - 2(5.4858 \times 10^{-4}\text{ amu}) = 0.027583 = 0.02758\text{ amu}$

$$\Delta E = \Delta mc^2 = 0.027583\text{ amu} \times \frac{1\text{ g}}{6.02214 \times 10^{23}\text{ amu}} \times \frac{1\text{ kg}}{1000\text{ g}} \times (2.9979246 \times 10^8\text{ m/s})^2$$
$$= 4.11654 \times 10^{-12} = 4.117 \times 10^{-12}\text{ J/4 }{}^1H\text{ nuclei}$$

$$1.0\text{ g }{}^1H \times \frac{1\ {}^1H\text{ nucleus}}{1.00782\text{ amu}} \times \frac{6.02214 \times 10^{23}\text{ amu}}{\text{g}} \times \frac{4.11654 \times 10^{-12}\text{ J}}{4\ {}^1H\text{ nuclei}}$$
$$= 6.1495 \times 10^{11}\text{ J} = 6.1 \times 10^8\text{ kJ produced by the fusion of 1.0 g }{}^1H.$$

$C_8H_{18}(l) + 25/2\ O_2(g) \rightarrow 8\ CO_2(g) + 9\ H_2O(g)$

$\Delta H° = 8(-393.5\text{ kJ}) + 9(-241.82\text{ kJ}) - (-250.1\text{ kJ}) = -5074.3\text{ kJ}$

$$6.1495 \times 10^8\text{ kJ} \times \frac{1\text{ mol }C_8H_{18}(l)}{5074.3\text{ kJ}} \times \frac{114.231\text{ g }C_8H_{18}}{\text{mol }C_8H_{18}} = 1.384 \times 10^7\text{ g} = 1.4 \times 10^4\text{ kg }C_8H_{18}$$

14,000 kg $C_8H_{18}(l)$ would have to be burned to produce the same amount of energy as fusion of 1.0 g ${}^1H$.

22 Chemistry of the Nonmetals

Visualizing Concepts

22.1 Statement (c) is correct. C_2H_4, the structure on the left, is the stable compound. Carbon, with a relatively small covalent radius owing to its location in the second row of the periodic chart, is able to closely approach other atoms. This close approach enables significant π overlap, so carbon can form strong multiple bonds to satisfy the octet rule. Silicon, in the third row of the periodic table, has a covalent radius too large for significant π overlap. Si does not form stable multiple bonds and Si_2H_4 is unstable.

22.3 *Analyze.* The structure is a trigonal bipyramid where one of the five positions about the central atom is occupied by a lone pair, often called a see-saw.

Plan A: Count the valence electrons in each molecule, draw a correct Lewis structure, and count the electron domains about the central atom.

Plan B: Molecules (a)–(d) each contain four F atoms bound to a central atom through a single bond (F is unlikely to form multiple bonds because of its high electronegativity). This represents 16 electron pairs; the fifth position is occupied by a lone pair, for a total of 17 e^- pairs. A valence e^- count for (a)–(d) will tell us which molecules are likely to have the designated structure. Molecule (e), $HClO_4$, is not exactly of the type AX_4, so a Lewis structure will be required. *Solve.*

(a) XeF_4 36 e^-, 16 e^- pairs. Plan B predicts that this molecule *will not* adopt the see-saw structure.

6 e^- domains about the Xe
octahedral domain geometry
square planar structure

(b) BrF_4^+ 34 e^-, 17 e^- pairs; structure *will* be see-saw

5 e^- domains about Br trigonal bipyramidal domain geometry
see-saw structure

(c) SiF_4 32 e^-, 16 e^- pairs; structure *will not* be see-saw

```
       :F:
        |
  :F — Si — F:
        |
       :F:
```

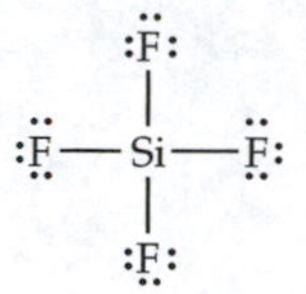

4 e^- domains about Si
tetrahedral domain geometry and structure

(d) $TeCl_4$ 34 e^-, 17 e^- pairs; structure *will* be see-saw

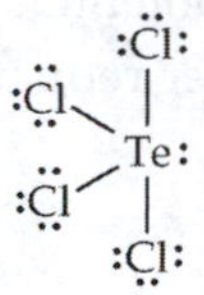

```
       :Cl:
  :Cl\  |
      Te:
  :Cl/  |
       :Cl:
```

5 e^- domains about Te
trigonal bipyramidal domain geometry
see-saw structure

(e) $HClO_4$ 32 e^-, 16 e^- pairs; structure *will not* be see-saw

```
       :O:
        |
  :O — Cl — O — H
        |
       :O:
```

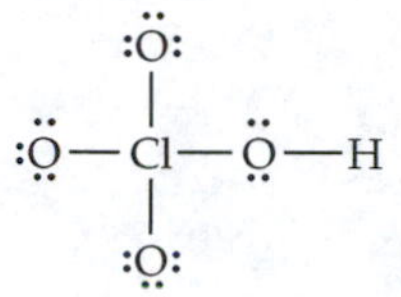

($HClO_4$ is an oxyacid, so H is bound to O, not Cl. Other Lewis structures that optimize formal charges are possible; structure predictions are the same.)

22.6 The graph is applicable only to (c) density. Density depends on both atomic mass and volume (radius). Both increase going down a family, but atomic mass increases to a greater extent. Density, the ratio of mass to volume, increases going down the family; this trend is consistent with the data in the figure.

According to periodic trends, (a) electronegativity and (b) first ionization energy both decrease rather than increase going down the family. According to Table 22.5, both (d) X–X single bond enthalpy and (e) electron affinity are somewhat erratic, with the trends decreasing from S to Po, and anomalous values for the properties of O, probably owing to its small covalent radius.

22.9 The compound on the left, with the strained three-membered ring, will be the most generally reactive. For central atoms with four electron domains*, idealized bond angles are 109°. From left to right, the bond angles in the three molecules pictured are 60°, 90°, and 108°. The larger the deviation from ideal bond angles, the more strain in the molecule and the more generally reactive it is.

*For the stick structures shown in the exercise, each line represents a C–C single bond and the intersection of two lines is a C atom. To determine the number of electron domains about each atom, visualize or draw the hydrogen atoms and nonbonded electron pairs in each molecule. Alternatively, note that both C and O atoms form only single bonds, so hybridization must be sp^3 and idealized bond angles are 109°.

Periodic Trends and Chemical Reactions (Section 22.1)

22.11 *Analyze/Plan.* Use the color-coded periodic chart on the front-inside cover of the text to classify the given elements. *Solve.*

Metals: (b) Sr, (c) Mn, (e) Na; nonmetals: (a) P, (d) Se, (f) Kr; metalloids: none

22.13 *Analyze/Plan.* Follow the logic in Sample Exercise 22.1. *Solve.*

(a) O (b) Br (c) Ba

(d) O (e) Co (f) Br

22.15 Statements (b) and (d) are true.

Statement (a) is false because a nitrogen atom is too small to accommodate five fluorine atoms around it. Statement (c) is false because the reduction potential of Cl_2 is larger than the reduction potential of I_2. The substance with the smaller reduction potential is easier to oxidize.

22.17 *Analyze/Plan.* Follow the logic in Sample Exercise 22.2. *Solve.*

(a) $NaOCH_3(s) + H_2O(l) \rightarrow NaOH(aq) + CH_3OH(aq)$

(b) $CuO(s) + 2\,HNO_3(aq) \rightarrow Cu(NO_3)_2(aq) + H_2O(l)$

(c) $WO_3(s) + 3\,H_2(g) \xrightarrow{\Delta} W(s) + 3\,H_2O(g)$

(d) $4\,NH_2OH(l) + O_2(g) \rightarrow 6\,H_2O(l) + 2\,N_2(g)$

(e) $Al_4C_3(s) + 12\,H_2O(l) \rightarrow 4\,Al(OH)_3(s) + 3\,CH_4(g)$

Hydrogen, the Noble Gases, and the Halogens (Sections 22.2, 22.3, and 22.4)

22.19 *Analyze/Plan.* Use information on the isotopes of hydrogen in Section 22.2 to list their symbols, names, and relative abundances. *Solve.*

(a) ${}^{1}_{1}H$-protium; ${}^{2}_{1}H$-deuterium; ${}^{3}_{1}H$-tritium

(b) The order of abundance is proteum > deuterium > tritium.

(c) ${}^{3}_{1}H$-tritium is radioactive.

(d) ${}^{3}_{1}H \rightarrow {}^{3}_{2}He + {}^{0}_{-1}e$

22.21 *Analyze/Plan.* Consider the electron configuration of hydrogen and the Group 1A elements. *Solve.*

Like other elements in group 1A, hydrogen has only one valence electron and its most common oxidation number is +1.

22.23 *Analyze/Plan.* Use information on the descriptive chemistry of hydrogen given in Section 22.2 to complete and balance the equations. *Solve.*

(a) $NaH(s) + H_2O(l) \rightarrow NaOH(aq) + H_2(g)$

(b) $Fe(s) + H_2SO_4(aq) \rightarrow Fe^{2+}(aq) + H_2(g) + SO_4^{2-}(aq)$

(c) $H_2(g) + Br_2(g) \rightarrow 2\,HBr(g)$

(d) $2\,Na(l) + H_2(g) \rightarrow 2\,NaH(s)$

(e) $PbO(s) + H_2(g) \xrightarrow{\Delta} Pb(s) + H_2O(g)$

22.25 *Analyze/Plan*. If the element bound to H is a nonmetal, the hydride is molecular. If H is bound to a metal with integer stoichiometry, the hydride is ionic; with noninteger stoichiometry, the hydride is metallic. *Solve.*

(a) ionic (metal hydride)

(b) molecular (nonmetal hydride)

(c) metallic (nonstoichiometric transition metal hydride)

22.27 Vehicle fuels produce energy via combustion reactions. The reaction $H_2(g) + 1/2\ O_2(g) \rightarrow H_2O(g)$ is very exothermic, producing 242 kJ per mole of H_2 burned. The only product of combustion is H_2O, a nonpollutant (but like CO_2, a greenhouse gas).

22.29 *Analyze/Plan*. Consider the periodic properties of Xe and Ar. *Solve.*

Xenon is larger, and can more readily accommodate an expanded octet. More important is the lower ionization energy of xenon; because the valence electrons are a greater average distance from the nucleus, they are more readily promoted to a state in which the Xe atom can form bonds with fluorine.

22.31 *Analyze/Plan*. Follow the rules for assigning oxidation numbers in Section 4.4 and the logic in Sample Exercise 4.8. *Solve.*

(a) $Ca(O\mathbf{Br})_2$, Br, +1 (b) $H\mathbf{Br}O_3$, Br, +5 (c) $\mathbf{Xe}O_3$, Xe, +6

(d) $\mathbf{Cl}O_4^-$, Cl, +7 (e) $H\mathbf{I}O_2$, I, +3 (f) $\mathbf{IF}_5$; I, +5; F, –1

22.33 *Analyze/Plan*. Review the nomenclature rules and ion names in Section 2.8, as well as the rules for assigning oxidation numbers in Section 4.4. *Solve.*

(a) iron(III) chlorate, Cl, +5 (b) chlorous acid, Cl, +3

(c) xenon hexafluoride, F, –1 (d) bromine pentafluoride; Br, +5; F, –1

(e) xenon oxide tetrafluoride, F, –1 (f) iodic acid, I, +5

22.35 *Analyze/Plan*. Consider intermolecular forces and periodic properties, including oxidizing power, of the listed substances. *Solve.*

(a) Van der Waals intermolecular attractive forces increase with increasing numbers of electrons in the atoms.

(b) F_2 reacts with water: $F_2(g) + H_2O(l) \rightarrow 2\ HF(aq) + 1/2\ O_2(g)$. That is, fluorine is too strong an oxidizing agent to exist in water.

(c) HF has extensive hydrogen bonding.

(d) Oxidizing power is related to electronegativity. Electronegativity decreases in the order given.

Oxygen and the Other Group 6A Elements (Sections 22.5 and 22.6)

22.37 *Analyze/Plan*. Use information on the descriptive chemistry of oxygen given in Section 22.5 to complete and balance the equations. *Solve.*

(a) $2\ HgO(s) \xrightarrow{\Delta} 2\ Hg(l) + O_2(g)$

(b) $2\ Cu(NO_3)_2(s) \xrightarrow{\Delta} 2\ CuO(s) + 4\ NO_2(g) + O_2(g)$

(c) $PbS(s) + 4\,O_3(g) \rightarrow PbSO_4(s) + 4\,O_2(g)$

(d) $2\,ZnS(s) + 3\,O_2(g) \xrightarrow{\Delta} 2\,ZnO(s) + 2\,SO_2(g)$

(e) $2\,K_2O_2(s) + 2\,CO_2(g) \rightarrow 2\,K_2CO_3(s) + O_2(g)$

(f) $3\,O_2(g) \xrightarrow{h\nu} 2\,O_3(g)$

22.39 *Analyze/Plan.* Oxides of metals are bases, oxides of nonmetals are acids, oxides that act as both acids and bases are amphoteric, and oxides that act as neither acids nor bases are neutral. *Solve.*

(a) acidic (oxide of a nonmetal)

(b) acidic (oxide of a nonmetal)

(c) amphoteric

(d) basic (oxide of a metal)

22.41 *Analyze/Plan.* Follow the rules for assigning oxidation numbers in Section 4.4 and the logic in Sample Exercise 4.8. *Solve.*

(a) $H_2\mathbf{Se}O_3$, +4 (b) $KH\mathbf{S}O_3$, +4 (c) $H_2\mathbf{Te}$, –2 (d) $C\mathbf{S}_2$, –2

(e) $Ca\mathbf{S}O_4$, +6 (f) $Cd\mathbf{S}$, –2 (g) $Zn\mathbf{Te}$, –2

Oxygen (a group 6A element) is in the –2 oxidation state in compounds (a), (b), and (e).

22.43 *Analyze/Plan.* The half-reaction for oxidation in all these cases is:

$H_2S(aq) \rightarrow S(s) + 2\,H^+ + 2\,e^-$ [The product could be written as $S_8(s)$, but this is not necessary. In fact it is not necessarily the case that S_8 would be formed, rather than some other allotropic form of the element.] Combine this half-reaction with the given reductions to write complete equations. The reduction in (c) happens only in acid solution. The reactants in (d) are acids, so the medium is acidic. *Solve.*

(a) $2\,Fe^{3+}(aq) + H_2S(aq) \rightarrow 2\,Fe^{2+}(aq) + S(s) + 2\,H^+(aq)$

(b) $Br_2(l) + H_2S(aq) \rightarrow 2\,Br^-(aq) + S(s) + 2\,H^+(aq)$

(c) $2\,MnO_4^-(aq) + 6\,H^+(aq) + 5\,H_2S(aq) \rightarrow 2\,Mn^{2+}(aq) + 5\,S(s) + 8\,H_2O(l)$

(d) $2\,NO_3^-(aq) + H_2S(aq) + 2\,H^+(aq) \rightarrow 2\,NO_2(aq) + S(s) + 2\,H_2O(l)$

22.45 *Analyze/Plan.* For each substance, count valence electrons, draw the correct Lewis structure, and apply the rules of VSEPR to decide electron domain geometry and geometric structure. *Solve.*

(a) $[:\ddot{O}-\ddot{Se}-\ddot{O}:]^{2-}$ with $:\ddot{O}:$ below Se — trigonal pyramidal

(b) Cl–S–S–Cl — Bent (free rotation around S–S bond)

(c) O, Cl, O–H bonded to S — tetrahedral

22.47 *Analyze/Plan.* Use information on the descriptive chemistry of sulfur given in Section 22.6 to complete and balance the equations. *Solve.*

(a) $SO_2(s) + H_2O(l) \rightarrow H_2SO_3(aq) \rightleftharpoons H^+(aq) + HSO_3^-(aq)$

(b) $ZnS(s) + 2\ HCl(aq) \rightarrow ZnCl_2(aq) + H_2S(g)$

(c) $8\ SO_3^{2-}(aq) + S_8(s) \rightarrow 8\ S_2O_3^{2-}(aq)$

(d) $SO_3(aq) + H_2SO_4(l) \rightarrow H_2S_2O_7(l)$

Nitrogen and the Other Group 5A Elements (Sections 22.7 and 22.8)

22.49 *Analyze/Plan.* Follow the rules for assigning oxidation numbers in Section 4.4 and the logic in Sample Exercise 4.8. *Solve.*

(a) $NaNO_2$, +3 (b) NH_3, –3 (c) N_2O, +1 (d) NaCN, –3

(e) HNO_3, +5 (f) NO_2, +4 (g) N_2, 0 (h) BN, –3

22.51 *Analyze/Plan.* For each substance, count valence electrons, draw the correct Lewis structure, and apply the rules of VSEPR to decide electron domain geometry and geometric structure. *Solve.*

(a) :Ö═N̈—Ö—H ⟷ :Ö—N̈═Ö—H

The molecule is bent around the central oxygen and nitrogen atoms; the four atoms need not lie in a plane. The right-most form does not minimize formal charges and is less important in the actual bonding model. The oxidation state of N is +3.

(b) [:N̈═N═N̈:]⁻ ⟷ [:N≡N—N̈:]⁻ ⟷ [:N̈—N≡N:]⁻

The molecule is linear. The oxidation state of N is –1/3.

(c) $[H_2N{-}NH_3]^+$ (Lewis structure: H—N—N: with two H on each N; left N bears H above and below, right N bears H above and below, bracketed with + charge)

The geometry is tetrahedral around the left nitrogen, trigonal pyramidal around the right. The oxidation state of N is –2.

(d) $[NO_3]^-$ (Lewis structure: :Ö: above N, :Ö—N═Ö, bracketed with – charge)

(three equivalent resonance forms) The ion is trigonal planar. The oxidation state of N is +5.

22.53 *Analyze/Plan.* Use information on the descriptive chemistry of nitrogen given in Section 22.7 to complete and balance the equations. *Solve.*

(a) $Mg_3N_2(s) + 6\ H_2O(l) \rightarrow 2\ NH_3(g) + 3\ Mg(OH)_2(s)$

Because $H_2O(l)$ is a reactant, the state of NH_3 in the products could be expressed as $NH_3(aq)$.

(b) $2\ NO(g) + O_2(g) \rightarrow 2\ NO_2(g)$, redox reaction

(c) $N_2O_5(g) + H_2O(l) \rightarrow 2\ H^+(aq) + 2\ NO_3^-(aq)$

(d) $NH_3(aq) + H^+(aq) \rightarrow NH_4^+(aq)$

(e) $N_2H_4(l) + O_2(g) \rightarrow N_2(g) + 2\ H_2O(g)$, redox reaction

22.55 *Analyze/Plan*. Follow the method for writing balanced half-reactions given in Section 20.2 and Sample Exercises 20.2. *Solve*.

(a) $HNO_2(aq) + H_2O(l) \rightarrow NO_3^-(aq) + 3\ H^+(aq) + 2\ e^-$, $E^\circ_{red} = 0.96\ V$

(b) $N_2(g) + H_2O(l) \rightarrow N_2O(g) + 2\ H^+(aq) + 2\ e^-$, $E^\circ_{red} = 1.77\ V$

22.57 *Analyze/Plan*. Follow the rules for assigning oxidation numbers in Section 4.4 and the logic in Sample Exercise 4.8. *Solve*.

(a) $H_3\mathbf{P}O_3$, +3 (b) $H_4\mathbf{P}_2O_7$, +5 (c) $\mathbf{Sb}Cl_3$, +3

(d) $Mg_3\mathbf{As}_2$, −3 (e) $\mathbf{P}_2O_5$, +5 (f) $Na_3\mathbf{P}O_4$, +5

22.59 *Analyze/Plan*. Consider the structures of the compounds of interest when explaining the observations. *Solve*.

(a) Phosphorus is a larger atom and can more easily accommodate five surrounding atoms and an expanded octet of electrons than nitrogen can. Also, P has energetically "available" 3d orbitals that participate in the bonding, but nitrogen does not.

(b) Only one of the three hydrogens in H_3PO_2 is bonded to oxygen. The other two are bonded directly to phosphorus and are not easily ionized because the P–H bond is not very polar.

(c) PH_3 is a weaker base than H_2O (PH_4^+ is a stronger acid than H_3O^+). Any attempt to add H^+ to PH_3 in the presence of H_2O merely causes protonation of H_2O.

(d) Refer to the structures of white and red phosphorus in Figure 22.26. White phosphorus consists of P_4 molecules, with P–P–P bond angles of 60°. Each P atom has four VSEPR pairs of electrons, so the predicted electron pair geometry is tetrahedral and the preferred bond angle is 109°. Because of the severely strained bond angles in P_4 molecules, white phosphorus is highly reactive. Red phosphorus is a chain of groups of four P atoms. It has fewer severely strained P–P–P bond angles and is less reactive than white phosphorus.

22.61 *Analyze/Plan*. Use information on the descriptive chemistry of phosphorus given in Section 22.8 to complete and balance the equations. *Solve*.

(a) $2\ Ca_3(PO_4)_2(s) + 6\ SiO_2(s) + 10\ C(s) \xrightarrow{\Delta} P_4(g) + 6\ CaSiO_3(l) + 10\ CO(g)$

(b) $PBr_3(l) + 3\ H_2O(l) \rightarrow H_3PO_3\ (aq) + 3\ HBr(aq)$

(c) $4\ PBr_3\ (g) + 6\ H_2(g) \rightarrow P_4\ (g) + 12\ HBr(g)$

Carbon, the Other Group 4A Elements, and Boron (Sections 22.9, 22.10, and 22.11)

22.63 *Analyze/Plan.* Review the nomenclature rules and ion names in Section 2.8. *Solve.*

(a) HCN (b) $Ni(CO)_4$ (c) $Ba(HCO_3)_2$ (d) CaC_2 (e) K_2CO_3

22.65 *Analyze/Plan.* Use information on the descriptive chemistry of carbon given in Section 22.9 to complete and balance the equations. *Solve.*

(a) $ZnCO_3(s) \xrightarrow{\Delta} ZnO(s) + CO_2(g)$

(b) $BaC_2(s) + 2\,H_2O(l) \rightarrow Ba^{2+}(aq) + 2\,OH^-(aq) + C_2H_2(g)$

(c) $2\,C_2H_2(g) + 5\,O_2(g) \rightarrow 4\,CO_2(g) + 2\,H_2O(g)$

(d) $CS_2(g) + 3\,O_2(g) \rightarrow CO_2(g) + 2\,SO_2(g)$

(e) $Ca(CN)_2(s) + 2\,HBr(aq) \rightarrow CaBr_2(aq) + 2\,HCN(aq)$

22.67 *Analyze/Plan.* Use information on the descriptive chemistry of carbon given in Section 22.9 to complete and balance the equations. *Solve.*

(a) $2\,CH_4(g) + 2\,NH_3(g) + 3\,O_2(g) \xrightarrow[\text{cat}]{800°C} 2\,HCN(g) + 6\,H_2O(g)$

(b) $NaHCO_3(s) + H^+(aq) \rightarrow CO_2(g) + H_2O(l) + Na^+(aq)$

(c) $2\,BaCO_3(s) + O_2(g) + 2\,SO_2(g) \rightarrow 2\,BaSO_4(s) + 2\,CO_2(g)$

22.69 *Analyze/Plan.* Follow the rules for assigning oxidation numbers in Section 4.4 and the logic in Sample 4.8. *Solve.*

(a) $H_3\mathbf{B}O_3$, +3 (b) $\mathbf{Si}Br_4$, +4 (c) $\mathbf{Pb}Cl_2$, +2

(d) $Na_2\mathbf{B}_4O_7 \cdot 10H_2O$, +3 (e) $\mathbf{B}_2O_3$, +3 (f) $\mathbf{Ge}O_2$, +4

22.71 *Analyze/Plan.* Consider periodic trends within a family, particularly metallic character, as well as descriptive chemistry in Sections 22.9 and 22.10. *Solve.*

(a) Tin; see Table 22.8. The filling of the 4f subshell at the beginning of the sixth row of the periodic table increases Z and Z_{eff} for later elements. This causes the ionization energy of Pb to be greater than that of Sn.

(b) Carbon, silicon, and germanium; these are the nonmetal and metalloids in group 4A. They form compounds ranging from XH_4 (–4) to XO_2 (+4). The metals tin and lead are not found in negative oxidation states.

(c) Silicon; silicates are the main component of sand.

22.73 *Analyze/Plan.* Consider the structural chemistry of silicates discussed in Section 22.10 and shown in Figure 22.32. *Solve.*

(a) Tetrahedral

(b) Metasilicic acid will probably adopt the single-strand silicate chain structure shown in Figure 22.32(b). The empirical formula shows 3 O and 2 H atoms per Si atom. The chain has the same Si to O ratio as metasilicic acid. Furthermore, in the chain structure, there are two terminal (not bridging) O atoms on each Si. These can accommodate the 2 H atoms associated with each Si atom of the acid. The sheet structure does not fulfill these requirements.

22.75 *Analyze/Plan*. In silicate anions, the oxidation number of silicon is +4 and that of oxygen is –2 (Section 22.10). Determine the charge on the silicate anion, then balance the charges of the cations and anions in the minerals. *Solve.*

(a) x = 2. The charge on $Si_2O_7^{6-}$ anion is 6–, that on Zn^{2+} cation is 2+. Two Ca^{2+} cations are required to balance charge.

(b) x = 2. The charge on $Si_2O_5^{2-}$ anion is 2–, that on Al^{3+} cation is 3+. Two OH^- anions are required to balance charge.

22.77 (a) Diborane (Figure 22.34 and below) has bridging H atoms linking the two B atoms. The structure of ethane shown below has the C atoms bound directly, with no bridging atoms.

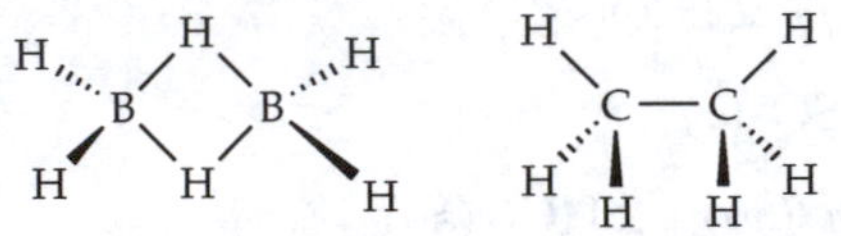

(b) B_2H_6 is an electron deficient molecule. It has 12 valence electrons, whereas C_2H_6 has 14 valence electrons. The 6 valence electron pairs in B_2H_6 are all involved in B–H σ bonding, so the only way to satisfy the octet rule at B is to have the bridging H atoms shown in Figure 22.34.

(c) A hydride ion, H^-, has two electrons, whereas an H atom has one. The term *hydridic* indicates that the H atoms in B_2H_6 have more than the usual amount of electron density for a covalently bound H atom.

Additional Exercises

22.79 (a) False. $H_2(g)$ and $D_2(g)$ are composed of different isotopes of hydrogen. Allotropes are composed of atoms of a single element bound into different structures.

(b) True. An interhalogen is a compound formed from atoms of two or more halogens.

(c) False. MgO(s) is a basic anhydride; it is the oxide of a metal.

(d) True. $SO_2(g)$ is the oxide of a nonmetal.

(e) True. A condensation reaction is the combination of two molecules to form a large molecule and a small one such as H_2O or HCl.

(f) True. The nucleus of tritium contains one proton and two neutrons.

(g) False. Disproportionation is an oxidation–reduction process where the same element is both oxidized and reduced. In this reaction, sulfur is oxidized and oxygen is reduced.

22.82 (a) $H_2SO_4 - H_2O \rightarrow SO_3$

(b) $2\,HClO_3 - H_2O \rightarrow Cl_2O_5$

(c) $2\,HNO_2 - H_2O \rightarrow N_2O_3$

(d) $H_2CO_3 - H_2O \rightarrow CO_2$

(e) $2\,H_3PO_4 - 3H_2O \rightarrow P_2O_5$

22.85 (a) PO_4^{3-}, +5; NO_3^-, +5

(b) The Lewis structure for NO_4^{3-} would be:

$$\left[\begin{array}{ccc} & :\ddot{O}: & \\ & | & \\ :\ddot{O}- & N & -\ddot{O}: \\ & | & \\ & :\ddot{O}: & \end{array}\right]^{3-}$$

The formal charge on N is +1 and on each O atom is –1. The four electronegative oxygen atoms withdraw electron density, leaving the nitrogen deficient. Because N can form a maximum of four bonds, it cannot form a π bond with one or more of the O atoms to regain electron density, as the P atom in PO_4^{3-} does. Also, the short N–O distance would lead to a tight tetrahedron of O atoms subject to steric repulsion.

Integrative Exercises

22.91 (a) $$100.0 \times 10^3 \text{ g FeTi} \times \frac{1 \text{ mol FeTi}}{103.7 \text{ g FeTi}} \times \frac{1 \text{ mol H}_2}{1 \text{ mol FeTi}} \times \frac{2.016 \text{ g H}_2}{1 \text{ mol H}_2} = 1944.1 = 1.94 \times 10^3 \text{ g H}_2$$

(b) $$V = \frac{1944.1 \text{ g H}_2}{2.016 \text{ g/mol H}_2} \times \frac{0.08206 \text{ L-atm}}{\text{mol-K}} \times \frac{273 \text{ K}}{1 \text{ atm}} = 21{,}603 = 2.16 \times 10^4 \text{ L H}_2$$

(c) $2\,H_2(g) + O_2(g) \rightarrow 2\,H_2O(l)$

$\Delta H^\circ = 2\Delta H^\circ_f\, H_2O(l) - 2\Delta H^\circ_f\, H_2(g) - \Delta H^\circ_f\, O_2(g)$

$\Delta H^\circ = 2(-285.83) - 2(0) - (0) = -571.66 \text{ kJ}$

$$1944.1 \text{ g H}_2 \times \frac{1 \text{ mol H}_2}{2.016 \text{ g H}_2} \times \frac{-571.66 \text{ kJ}}{2 \text{ mol H}_2} = -275{,}636 = -2.76 \times 10^5 \text{ kJ}$$

The minus sign indicates that energy is produced.

22.93 (a) $H_2(g) + 1/2\,O_2(g) \rightarrow H_2O(l);\ \Delta H = -285.83 \text{ kJ/mol } H_2$

$CH_4(g) + 2\,O_2(g) \rightarrow CO_2(g) + 2\,H_2O(l)$

$\Delta H = 2(-285.83) - 393.5 - (-74.8) = -890.4 \text{ kJ/ mol } CH_4$

(b) for H_2: $$\frac{-285.83 \text{ kJ}}{1 \text{ mol H}_2} \times \frac{1 \text{ mol H}_2}{2.0159 \text{ g H}_2} = -141.79 \text{ kJ/g H}_2$$

for CH_4: $$\frac{-890.4 \text{ kJ}}{1 \text{ mol CH}_4} \times \frac{1 \text{ mol CH}_4}{16.043 \text{ g CH}_4} = -55.50 \text{ kJ/g CH}_4$$

(c) Find the number of moles of gas that occupy 1 m^3 at STP:

$$n = \frac{1 \text{ atm} \times 1 \text{ m}^3}{273 \text{ K}} \times \frac{1 \text{ mol-K}}{0.08206 \text{ L-atm}} \times \left[\frac{100 \text{ cm}}{1 \text{ m}}\right]^3 \times \frac{1 \text{ L}}{10^3 \text{ cm}^3} = 44.64 \text{ mol}$$

for H_2: $$\frac{-285.83 \text{ kJ}}{1 \text{ mol H}_2} \times \frac{44.64 \text{ mol H}_2}{1 \text{ m}^3 \text{ H}_2} = -1.276 \times 10^4 \text{ kJ/m}^3 \text{ H}_2$$

for CH_4: $$\frac{-890.4 \text{ kJ}}{1 \text{ mol CH}_4} \times \frac{44.64 \text{ mol CH}_4}{1 \text{ m}^3 \text{ CH}_4} = -3.975 \times 10^4 \text{ kJ/m}^3 \text{ CH}_4$$

22.95 (a) First, calculate the molar solubility of Cl_2 in water.

$$n = \frac{1\ \text{atm}\ (0.310\ \text{L})}{\frac{0.08206\ \text{L-atm}}{1\ \text{mol-K}} \times 273\ \text{K}} = 0.01384 = 0.0138\ \text{mol Cl}_2$$

$$M = \frac{0.01384\ \text{mol}}{0.100\ \text{L}} = 0.1384 = 0.138\ M$$

$[Cl^-] = [HOCl] = [H^+]$ Let this quantity = x. Then, $\frac{x^3}{(0.1384 - x)} = 4.7 \times 10^{-4}$

Assuming that x is small compared with 0.1384:

$x^3 = (0.1384)(4.7 \times 10^{-4}) = 6.504 \times 10^{-5}$; $x = 0.0402 = 0.040\ M$

We can correct the denominator using this value, to get a better estimate of x:

$$\frac{x^3}{0.1384 - 0.0402} = 4.7 \times 10^{-4};\ x = 0.0359 = 0.036\ M$$

One more round of approximation gives x = 0.0364 = 0.036 *M*. This is the equilibrium concentration of HClO.

(b) From the equilibrium reaction in part (a), $[H^+] = 0.036\ M$. $pH = -\log[H^+] = 1.4$

The HOCl produced by this equilibrium will ionize slightly to produce additional $H^+(aq)$. However, the K_a value for HOCl is small, 3.0×10^{-8}, and the acid ionization will be suppressed by the presence of $H^+(aq)$ from the solubility equilibrium. $[H^+]$ from ionization of HOCl will be small compared to 0.036 *M* and will not significantly impact the pH.

22.98 (a) $SO_2(g) + 2\ H_2S(aq) \rightarrow 3\ S(s) + 2\ H_2O(g)$ or, if we assume S_8 is the product,

$8\ SO_2(g) + 16\ H_2S(aq) \rightarrow 3\ S_8(s) + 16\ H_2O(g)$.

(b) Assume that all S in the coal becomes SO_2 upon combustion, so that

1 mol S (coal) = 1 mol SO_2; 1 ton = 2000 lb; 760 torr = 1.00 atm

$$4000\ \text{lb coal} \times \frac{0.035\ \text{lb S}}{1\ \text{lb coal}} \times \frac{453.6\ \text{g S}}{1\ \text{lb S}} \times \frac{1\ \text{mol S (coal)}}{32.07\ \text{g S}} \times \frac{1\ \text{mol SO}_2}{1\ \text{mol S (coal)}} \times \frac{2\ \text{mol H}_2\text{S}}{1\ \text{mol SO}_2}$$

$$= 3960 = 4.0 \times 10^3\ \text{mol H}_2\text{S}$$

$$V = \frac{3960\ \text{mol} \times (0.08206\ \text{L-atm/mol-K}) \times 300\ \text{K}}{1.00\ \text{atm}} = 97{,}496 = 9.7 \times 10^4\ \text{L}$$

(c) $$3960\ \text{mol H}_2\text{S} \times \frac{3\ \text{mol S}}{2\ \text{mol H}_2\text{S}} \times \frac{32.07\ \text{g S}}{1\ \text{mol S}} = 1.9 \times 10^5\ \text{g S}$$

This is about 210 lb S per ton of coal combusted. (However, two-thirds of this comes from the H_2S, which presumably at some point was also obtained from coal.)

22.100 The reactions can be written as follows:

$$H_2(g) + X(\text{std state}) \rightarrow H_2X(g) \qquad \Delta H_f^\circ$$
$$2\,H(g) \rightarrow H_2(g) \qquad \Delta H_f^\circ(H\text{–}H)$$
$$X(g) \rightarrow X(\text{std state}) \qquad \Delta H_3$$
$$\text{Add: } 2\,H(g) + X(g) \rightarrow H_2X(g) \qquad \Delta H = \Delta H_f^\circ + \Delta H_f^\circ(H\text{–}H) + \Delta H_3$$

These are all the necessary ΔH values. Thus,

Compound	**ΔH**	**D H–X**
H_2O	ΔH = –241.8 kJ – 436 kJ – 248 kJ = –926 kJ	463 kJ
H_2S	ΔH = –20.17 kJ – 436 kJ – 277 kJ = –733 kJ	367 kJ
H_2Se	ΔH = +29.7 kJ – 436 kJ – 227 kJ = –633 kJ	317 kJ
H_2Te	ΔH = +99.6 kJ – 436 kJ – 197 kJ = –533 kJ	267 kJ

The average H–X bond energy in each case is just half of ΔH. The H–X bond energy decreases steadily in the series. The origin of this effect is probably the increasing size of the orbital from X with which the hydrogen 1s orbital must overlap.

22.103 First write the balanced equation to give the number of moles of gaseous products per mole of hydrazine.

(A) $(CH_3)_2NNH_2 + 2\,N_2O_4 \rightarrow 3\,N_2(g) + 4\,H_2O(g) + 2\,CO_2(g)$

(B) $(CH_3)HNNH_2 + 5/4\,N_2O_4 \rightarrow 9/4\,N_2(g) + 3\,H_2O(g) + CO_2(g)$

In case (A) there are nine moles gas per one mole $(CH_3)_2NNH_2$ plus two moles N_2O_4. The total mass of reactants is 60 + 2(92) = 244 g. Thus, there are

$$\frac{9\text{ mol gas}}{244\text{ g reactants}} = \frac{0.0369\text{ mol gas}}{1\text{ g reactants}}$$

In case (B) there are 6.25 moles of gaseous product per one mole $(CH_3)HNNH_2$ plus 1.25 moles N_2O_4. The total mass of this amount of reactants is 46.0 + 1.25(92.0) = 161 g.

$$\frac{6.25\text{ mol gas}}{161\text{ g reactants}} = \frac{0.0388\text{ mol gas}}{1\text{ g reactants}}$$

Thus the methylhydrazine (B) has marginally greater thrust.

22.104 (a) $3\,B_2H_6(g) + 6\,NH_3(g) \rightarrow 2\,(BH)_3(NH)_3(l) + 12\,H_2(g)$

$3\,LiBH_4(s) + 3\,NH_4Cl(s) \rightarrow 2\,(BH)_3(NH)_3(l) + 9\,H_2(g) + 3\,LiCl(s)$

(b) 30 valence e^-, 15 e^- pairs

The structure with nonbonded pairs minimizes formal charge, but these electrons are almost certainly delocalized about the six-membered ring, mimicking the bonding in benzene.

(c) $$n = \frac{PV}{RT} = \frac{1.00\text{ atm} \times 2.00\text{ L}}{273\text{ K}} \times \frac{\text{mol-K}}{0.08206\text{ L-atm}} = 0.08929 = 8.93 \times 10^{-2}\text{ mol NH}_3$$

$$0.08929\text{ mol NH}_3 \times \frac{2\text{ mol (BH)}_3\text{(NH)}_3}{6\text{ mol NH}_3} \times \frac{80.50\text{ g (BH)}_3\text{(NH)}_3}{1\text{ mol (BH)}_3\text{(NH)}_3}$$

$$= 2.3956 = 2.40\text{ g (BH)}_3\text{(NH)}_3$$

23 Transition Metals and Coordination Chemistry

Visualizing Concepts

23.2 *Analyze.* Given the formula of a coordination compound, draw the structure, determine the coordination number, coordination geometry, oxidation state of the metal, and number of unpaired electrons.

Plan. From the formula, determine the identity of the ligands and the number of coordination sites they occupy. From the total coordination number, decide on a likely geometry. Use ligand and overall complex charges to calculate the oxidation number of the metal. Refer to the d-orbital energy level diagram that corresponds to the structure of the compound and the field strength of the ligands to determine the number of unpaired electrons.

Solve. The ligands are $2Cl^-$, one coordination site each, and en, ethylenediamine, two coordination sites, for a coordination number of 4. This coordination number has two possible geometries, tetrahedral and square planar. Pt is one of the metals known to adopt square planar geometry when CN = 4.

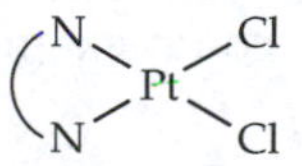

N Cl
Pt
N Cl

(a) Coordination number is 4

(b) Coordination geometry is square planar

(c) The oxidation state of Pt is +2. $Pt(en)Cl_2$ is a neutral compound, the en ligand is neutral, and the $2Cl^-$ ligands are each –1, so the oxidation state of Pt must be +2, Pt(II).

(d) There are no unpaired electrons. Square planar d^8 complexes are usually low spin, especially with heavier metals like Pt.

23.6 *Analyze.* Given four structures, decide which are chiral.

Plan. Chiral molecules have nonsuperimposable mirror images. Draw the mirror image of each molecule and visualize whether it can be rotated into the original molecule. If so, the complex is not chiral. If the original orientation cannot be regenerated by rotation, the complex is chiral. *Solve.*

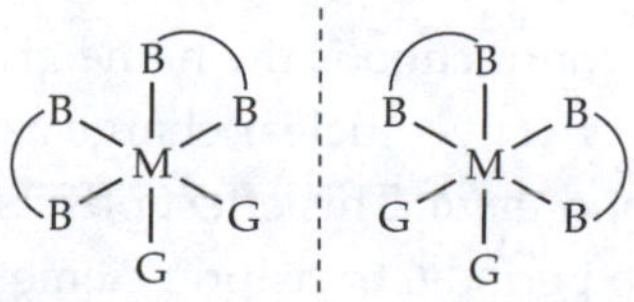

B B
B B B B
M M
B G G B
G G
(1) (1) mirror

The two orientations are not superimposable and molecule (1) is chiral.

The two orientations are superimposable. Rotate the right-most structure 90° counterclockwise about the B-M-B axis to align the G's; the bidentate ligands then also overlap. Molecule (2) is not chiral.

The two orientations are not superimposable and molecule (3) is chiral.

The two orientations are not superimposable and molecule (4) is chiral.

23.8 *Analyze.* Fit the crystal field splitting diagram to the complex description in each part.

Plan. Determine the number of d-electrons in each transition metal. On the splitting diagrams match the d-orbital splitting patterns to complex geometry and electron pairing to the definition of high-spin and low-spin.

Solve. Octahedral complexes have the 3 lower, 2 higher splitting pattern, whereas tetrahedral complexes have the opposite 2 lower, 3 higher pattern. Low spin complexes favor electron pairing because of large d-orbital splitting. High-spin complexes have maximum occupancy because of small orbital splitting.

(a) Fe^{3+}, 5 d-electrons; weak field: spins unpaired; octahedral: 3 lower, 2 higher d-splitting ∴ diagram (4)

(b) Fe^{3+}, 5 d-electrons; strong field: spins paired; octahedral: 3 lower, 2 higher d-splitting ∴ diagram (1)

(c) Fe^{3+}, 5 d-electrons; tetrahedral: 2 lower, 3 higher d-splitting ∴ diagram (3)

(d) Ni^{2+}, 8 d-electrons; tetrahedral: 2 lower, 3 higher d-splitting ∴ diagram (2)

Check. Diagram (2) was the remaining choice for (d) and it fits the description.

The Transition Metals (Section 23.1)

23.11 Trend (c). The lanthanide contraction is the name given to the decrease in atomic size because of the build-up in effective nuclear charge as we move through the lanthanides (elements 57–71) and beyond them. This effect offsets the expected increase in atomic size going from period 5 to period 6 transition elements.

23.13 (a) Ti^{2+}, $[Ar]3d^2$ (b) Ti^{4+}, [Ar] (c) Ni^{2+}, $[Ar]3d^8$ (d) Zn^{2+}, $[Ar]3d^{10}$

23.15 (a) Ti^{3+}, $[Ar]3d^1$ (b) Ru^{2+}, $[Kr]4d^6$ (c) Au^{3+}, $[Xe]4f^{14}5d^8$ (d) Mn^{4+}, $[Ar]3d^3$

23.17 *Analyze/Plan*. Consider the definitions of paramagnetic and diamagnetic. *Solve*.

The unpaired electrons in a paramagnetic substance cause it to be weakly attracted into a magnetic field. (A diamagnetic material, where all electrons are paired, is very weakly repelled by a magnetic field.)

23.19 *Analyze/Plan*. Consider the orientation of spins in various types of magnetic materials as shown in Figure 23.6.

The diagram shows a material with misaligned spins that become aligned in the direction of an applied magnetic field. This is a paramagnetic material.

Transition-Metal Complexes (Section 23.2)

23.21 (a) Primary valence. In Werner's theory, primary valence is the charge of the metal cation at the center of the complex. "Oxidation state" is a broader term than "ionic charge," but Werner's complexes contain metal ions where cation charge and oxidation state are equal.

(b) Coordination number. In Werner's theory, secondary valence is the number of atoms bound or coordinated to the central metal ion.

(c) NH_3 can serve as a ligand because it has an unshared electron pair, whereas BH_3 does not. Ligands act as a Lewis base in metal-ligand interactions. As such, they must possess at least one unshared electron pair. BH_3, with fewer than 8 electrons about B, has no unshared electron pair and cannot act as a ligand. In fact, BH_3 acts as a Lewis acid, an electron pair acceptor, because it is electron-deficient.

23.23 *Analyze/Plan*. Follow the logic in Sample Exercises 23.1 and 23.2. *Solve*.

(a) This compound is electrically neutral, and the NH_3 ligands carry no charge, so the charge on Ni must balance the –2 charge of the 2 Br^- ions. The charge and oxidation state of Ni is +2.

(b) Because there are 6 NH_3 molecules in the complex, the likely coordination number is 6. In some cases Br^- acts as a ligand, so the coordination number could be other than 6.

(c) Assuming that the 6 NH_3 molecules are the ligands, 2 Br^- ions are not coordinated to the Ni^{2+}, so 2 mol AgBr(s) will precipitate.

23.25 *Analyze/Plan*. Count the number of donor atoms in each complex, taking the identity of polydentate ligands into account. Follow the logic in Sample Exercise 23.2 to obtain oxidation numbers of the metals.

(a) Coordination number = 4, oxidation number = +2

(b) 5, +4

(c) 6, +3

(d) 5, +2

(e) 6, +3

(f) 4, +2

23 Transition Metals and Coordination Chemistry

Solutions to Red Exercises

Common Ligands in Coordination Chemistry (Section 23.3)

23.27 (a) $CH_3CH_2NH_2$, 20 e^-, 10 e^- pr

```
    H   H   H
    |   |   |
H — C — C — N:
    |   |   |
    H   H   H
```

monodentate ligand, only N atom has a nonbonded pair of electrons

(b) $P(CH_3)_3$, 26 e^-, 13 e^- pr

```
    H       H
    |       |
H — C — P̈ — C — H
    |   |   |
    H   |   H
    H — C — H
        |
        H
```

monodentate ligand, only P atom has a nonbonded pair of electrons

(c) CO_3^{2-}, 24 e^-, 12 e^- pr

```
[   :Ö:   ]2-     [   :Ö:   ]2-     [   :O:   ]2-
[    |    ]       [    |    ]       [    ‖    ]
[    C    ]  <->  [    C    ]  <->  [    C    ]
[  /   ⫽  ]       [  ⫽   \  ]       [  /   \  ]
[:Ö:    O:]       [O:    :Ö:]       [:Ö:   :Ö:]
```

either monodentate or bidentate

[All three O atoms are possible bonding sites, but it is not geometrically possible for all three O atoms to be bound to the same metal ion.]

(d) C_2H_6, 14 e^-, 7 e^- pr

```
    H   H
    |   |
H — C — C — H
    |   |
    H   H
```

unlikely to act as a ligand, no nonbonded pairs of electrons

23.29 *Analyze/Plan.* Given the formula of a coordination compound, determine the number of coordination sites occupied by the polydentate ligand. The coordination number of the complexes is probably 4 or 6. Note the number of monodentate ligands and determine the number of coordination sites occupied by the polydentate ligands. *Solve.*

(a) *ortho*-Phenanthroline, *o*-phen, is bidentate. The complex is 6-coordinate, there are 4 monodentate NH_3 ligands, so *o*-phen occupies 2 sites.

(b) Oxalate, $C_2O_4^{2-}$, is bidentate. The complex is 6-coordinate, there are 4 monodentate H_2O ligands, so oxalate occupies 2 coordination sites.

(c) Ethylenediaminetetraacetate, EDTA, is hexadentate. The complex is probably 6-coordinate octahedral with EDTA occupying all six coordination sites.

(d) Ethylenediamine, en, is bidentate. The complex is 4-coordinate and each en ligand occupies 2 coordination sites.

23.31 *Analyze/Plan.* Anions and polar molecules (with nonbonded electron pairs) are most likely to act as ligands in a metal complex (Solution 23.22). *Solve.*

(a) CH_3CN, polar molecule with nonbonded electron pair

(b) H^-, anion

(c) CO, polar molecule with a nonbonded electron pair

23.33 False. The ligand shown in the figure does not typically act as a bidentate ligand for a single metal center. The entire molecule is planar; there is no "bend" in the central 6-membered ring that includes the two N atoms. The benzene rings on either side of the two N atoms inhibit their approach in the correct orientation for chelation. It might act as a bridging ligand between two metal centers, but again, the benzene rings would create a significant amount of steric hindrance.

Nomenclature and Isomerism in Coordination Chemistry (Section 23.4)

23.35 *Analyze/Plan.* Given the name of a coordination compound, write the chemical formula. Refer to Tables 23.4 and 23.5 to find ligand formulas. Place the metal complex (metal ion + ligands) inside square brackets and the counterion (if there is one) outside the brackets. *Solve.*

(a) $[Cr(NH_3)_6](NO_3)_3$ (b) $[Co(NH_3)_4CO_3]_2SO_4$ (c) $[Pt(en)_2Cl_2]Br_2$

(d) $K[V(H_2O)_2Br_4]$ (e) $[Zn(en)_2][HgI_4]$

23.37 *Analyze/Plan.* Follow the logic in Sample Exercise 23.3, paying attention to naming rules in Section 23.4. *Solve.*

(a) tetraamminedichlororhodium(III) chloride

(b) potassium hexachlorotitanate(IV)

(c) tetrachlorooxomolybdenum(VI)

(d) tetraaqua(oxalato)platinum(IV) bromide

23.39 *Analyze/Plan.* Consider the coordination number and geometry of each of the complexes, along with the definitions of the various types of isomerism. Use this information to decide which of the complexes could exhibit isomerism of the specified type. *Solve.*

Complex 1 has a coordination number of 6 and octahedral geometry about the metal. There are 4 monodentate ligands of one kind and two of another.

Complex 2 has a coordination number of 4 and square planar geometry. There are two monodentate ligands of one kind and two of another.

Complex 3 has a coordination number of 6 and octahedral geometry about the metal. There are 2 bidentate ligands and two monodentate ligands.

(a) Complexes 1, 2, and 3 can have geometric isomers. These are different arrangements of the same set of ligands. All three complexes have cis-trans isomers, where a pair of ligands is either opposite or adjacent to each other.

(b) Only complex 2 can have linkage isomers. Nitrite ion, NO_2^-, can coordinate through either N or O. It is the only ligand in the three complexes that has this ability.

(c) Only the cis geometric isomer of complex 3 can have optical isomers. These are isomers, with the same arrangement of bonds, that are mirror images of each other and cannot be superimposed.

(d) Only complex 1 can have coordination sphere isomers. This is where an anion can either be a ligand or a counterion. Complex 1 is the only example with a counterion that can also be a ligand.

23.41 Yes. A tetrahedral complex of the form MA_2B_2 would have neither structural isomers nor stereoisomers. For a tetrahedral complex, no differences in connectivity are possible for a single central atom, so the terms cis and trans do not apply. No optical isomers with tetrahedral geometry are possible because M is not bound to four different groups. The complex must be square planar with cis and trans geometric isomers.

23.43 *Analyze/Plan*. Follow the logic in Sample Exercise 23.4 and 23.5. *Solve.*

(a) No geometric isomers

(b) Two geometric isomers, cis and trans

cis trans

(c) Three geometric isomers: cis and trans; the cis isomer has enantiomers

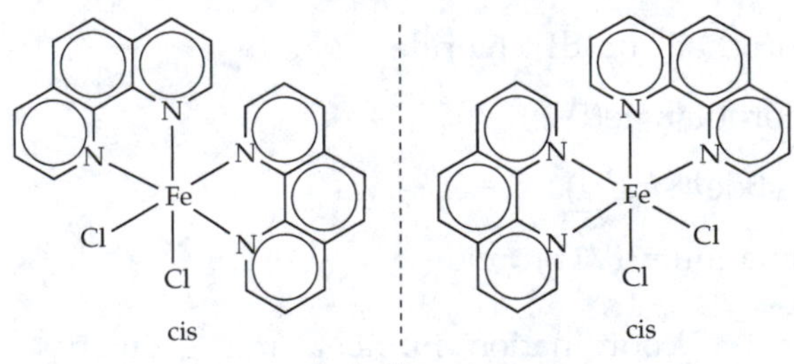

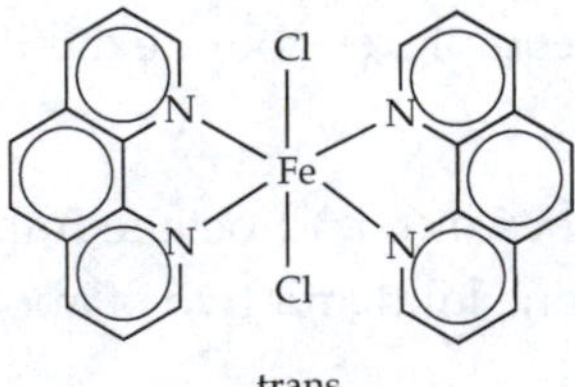

(The three isomeric complex ions in part (c) each have a 1+ charge.)

23.45 *Analyze/Plan*. Consider the geometry and arrangement of ligands in each complex. Decide whether the structure has a nonsuperimposable mirror image. *Solve.*

(a) Not chiral. The tetrahedral Zn^{2+} is not bonded to four different ligands.

(b) Not chiral. One mirror image can be superimposed on the other. [See similar trans structure in Solution 23.43 (c)]

(c) Chiral, has an optical isomer. [See similar cis structures in Solution 23.43 (c)]

23 Transition Metals and Coordination Chemistry

Solutions to Red Exercises

Color and Magnetism in Coordination Chemistry; Crystal-Field Theory (Sections 23.5 and 23.6)

23.47 (a) Visible light with a wavelength of 610 nm is orange. If the complex absorbs orange light, it will appear blue.

(b) E(J/photon) = $h\nu = hc/\lambda$.

$$E = \frac{6.626 \times 10^{-34}\ \text{J-s}}{610\ \text{nm}} \times \frac{2.998 \times 10^{8}\ \text{m}}{\text{s}} \times \frac{1\text{nm}}{1 \times 10^{-9}\ \text{m}} = 3.257 \times 10^{-19} = 3.26 \times 10^{-19}\ \text{J}$$

(c) Change J/photon to kJ/mol.

$$\frac{3.259 \times 10^{-19}\ \text{J}}{\text{photon}} \times \frac{1\ \text{kJ}}{1000\ \text{J}} \times \frac{6.022 \times 10^{23}\ \text{photons}}{\text{mol}} = 196\ \text{kJ/mol}$$

23.49 *Analyze/Plan.* Given the formula of a coordination complex, determine the oxidation state and electron configuration of the central metal ion. If necessary, use a d-orbital energy level diagram appropriate for the geometry of the complex to decide if the metal ion has unpaired electrons. If so, it is paramagnetic.

(a) Zn^{2+}, $[Ar]3d^{10}$. There are no unpaired electrons, so the complex is diamagnetic.

(b) Pd^{2+}, $[Kr]4d^{8}$. The complex is probably square planar. Square planar complexes with 8 d-electrons are usually diamagnetic, especially with a heavy metal center like Pd.

(c) V^{3+}, $[Ar]3d^{2}$. There are 2 d-electrons, so the complex is paramagnetic. The complex is octahedral, but the two electrons would be unpaired in any of the d-orbital energy level diagrams.

(d) Ni^{2+}, $[Ar]3d^{8}$. The complex is paramagnetic. The geometry is octahedral and there are two unpaired electrons.

23.51 An electron in a d orbital with lobes that point directly at the ligands will have higher energy than an electron in a d orbital with lobes that do not point directly at the ligands.

23.53 (a)

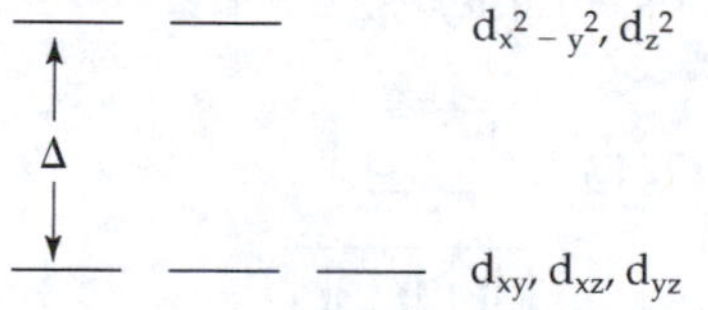

$d_{x^2-y^2}$, d_{z^2} (upper two levels)

Δ

d_{xy}, d_{xz}, d_{yz} (lower three levels)

(b) The magnitude of Δ and the energy of the d-d transition for a d^1 complex are equal.

(c)
$$\frac{6.626 \times 10^{-34}\ \text{J-s}}{545\ \text{nm}} \times \frac{2.998 \times 10^{8}\ \text{m}}{\text{s}} \times \frac{1\ \text{nm}}{1 \times 10^{-9}\ \text{m}} \times \frac{1\ \text{kJ}}{1000\ \text{J}} \times \frac{6.022 \times 10^{23}\ \text{photons}}{\text{mol}}$$

$$= 220\ \text{kJ/mol}$$

23.55 *Analyze/Plan.* Determine the oxidation state of the copper ions in each mineral from their molecular formulas. Write the appropriate electron configuration(s). Consider the relationship between d-orbital electron configuration, the color of a complex, the wavelength of absorbed light, and the magnitude of the crystal field splitting Δ. *Solve.*

(a) Both minerals contain Cu^{2+} ions. The electron configuration of Cu^{2+} is $[Ar]3d^9$.

(b) Azurite will probably have the larger Δ. Malachite appears green and absorbs red. Azurite appears blue and absorbs orange. Orange light has a wavelength range of 580 to 650 nm, whereas red light has wavelengths between 650 and 750 nm. The shorter wavelengths of the orange light absorbed by azurite correspond to higher-energy electron transitions and larger Δ values.

23.57 *Analyze/Plan*. Determine the charge on the metal ion, subtract it from the row number (3-12) of the transition metal, and the remainder is the number of d-electrons. *Solve.*

(a) Ti^{3+}, d^1 (b) Co^{3+}, d^6 (c) Ru^{3+}, d^5

(d) Mo^{5+}, d^1 (e) Re^{3+}, d^4

23.59 Yes. A weak-field ligand leads to a small Δ value and a small d-orbital splitting energy. If the splitting energy of a complex is smaller than the energy required to pair electrons in an orbital, the complex is high-spin.

23.61 *Analyze/Plan*. Follow the logic in Sample Exercise 23.7. *Solve.*

(a) Mn: $[Ar]4s^2 3d^5$; Mn^{2+}: $[Ar]3d^5$

(b) Ru: $[Kr]5s^1 4d^7$; Ru^{2+}: $[Kr]4d^6$

(c) Rh: $[Kr]5s^1 4d^8$; Rh^{2+}: $[Kr]4d^7$

(a) upper: [ ][ ]; lower: [↑↓][↑↓][↑] — 1 unpaired electron

(b) upper: [ ][ ]; lower: [↑↓][↑↓][↑↓] — 0 unpaired electrons

(c) upper: [↑][ ]; lower: [↑↓][↑↓][↑↓] — 1 unpaired electron

23.63 *Analyze/Plan*. All complexes in this exercise are six-coordinate octahedral. Use the definitions of high-spin and low-spin along with the orbital diagram from Sample Exercise 23.7 to place electrons for the various complexes. *Solve.*

(a) upper: [↑][ ]; lower: [↑][↑][↑] — d^4, high spin

(b) upper: [↑][↑]; lower: [↑][↑][↑] — d^5, high spin

(c) upper: [ ][ ]; lower: [↑↓][↑↓][↑↓] — d^6, low spin

(d) upper: [ ][ ]; lower: [↑↓][↑↓][↑] — d^5, low spin

(e) upper: [ ][ ]; lower: [↑][↑][↑] — d^3

(f) upper: [↑][↑]; lower: [↑↓][↑↓][↑↓] — d^8

23 Transition Metals and Coordination Chemistry

Solutions to Red Exercises

23.65 *Analyze/Plan*. Follow the ideas but reverse the logic in Sample Exercise 23.7. *Solve.*

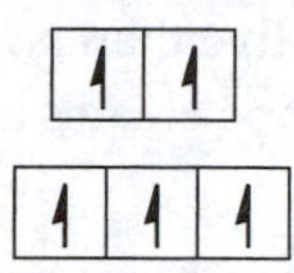

high spin

Additional Exercises

23.67 The paper clip must contain a significant amount of Ni, a ferromagnetic metal. At ambient temperature, the paper clip is below its Curie temperature, behaves ferromagnetically, and is strongly attracted to the permanent magnet. The lighter heats the left paperclip above its Curie temperature (354 °C), and it switches from from ferromagnetic to paramagnetic behavior. That is, below its Curie temperature, the spins of the unpaired electrons in Ni are perfectly aligned and the clip is strongly attracted to the permanent magnet. Above the Curie temperature, the unpaired spins become randomly aligned, and the paper clip loses most of its attraction for the permanent magnet.

23.69 $[Pt(NH_3)_6]Cl_4$; $[Pt(NH_3)_4Cl_2]Cl_2$; $[Pt(NH_3)_3Cl_3]Cl$; $[Pt(NH_3)_2Cl_4]$; $K[Pt(NH_3)Cl_5]$

23.71 (a) $[Co(NH_3)_4(H_2O)_2]^{2+}$ — octahedral

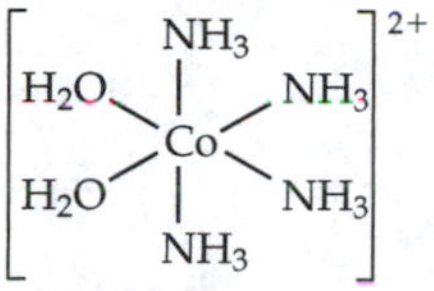

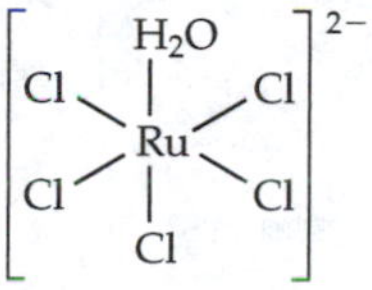

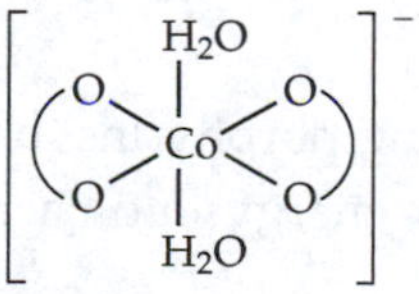

(b) $[Ru(H_2O)Cl_5]^{2-}$ — octahedral

(c) $[Co(H_2O)_2(C_2O_4)_2]^{-}$ — octahedral

(d) $[Ru(en)_2Cl_2]$ — octahedral

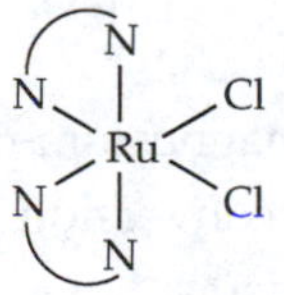

(a) *cis*-tetraamminediaquacobalt(II) nitrate

(b) sodium aquapentachlororuthenate(III)

(c) ammonium *trans*-diaquabisoxalatocobaltate(III)

(d) *cis*-dichlorobisethylenediammineruthenium(II)

23.73 (a) Valence electrons: 2P + 6C + 16H = 10 + 24 + 16 = 50 e^-, 25 e^- pr

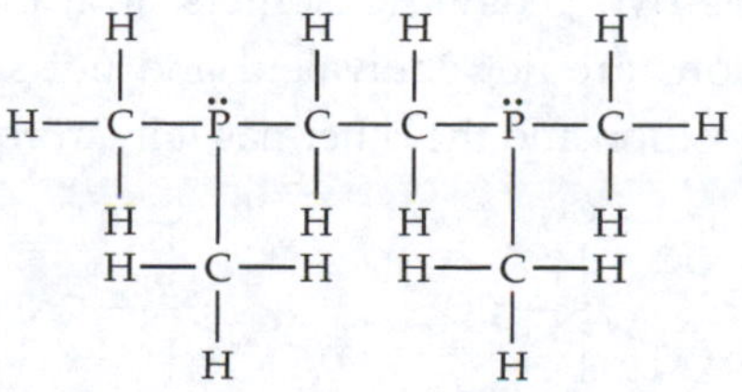

Both dmpe and en are bidentate ligands. The dmpe ligand binds through P, whereas en binds through N. Phosphorus is less electronegative than N, so dmpe is a stronger electron pair donor and Lewis base than en. Dmpe creates a stronger ligand field and is higher on the spectrochemical series.

Structurally, P has a larger covalent radius than N, so M–P bonds are longer than M–N bonds. This is convenient because the two $-CH_3$ groups on each P atom in dmpe create more steric hindrance (bumping with adjacent atoms) than the H atoms on N in en.

(b) CO and dmpe are neutral, 2 CN^- = 2–, 2 Na^+ = 2+. The ion charges balance, so the oxidation state of Mo is zero.

(c) The symbol P⌒P represents the bidentate dmpe ligand.

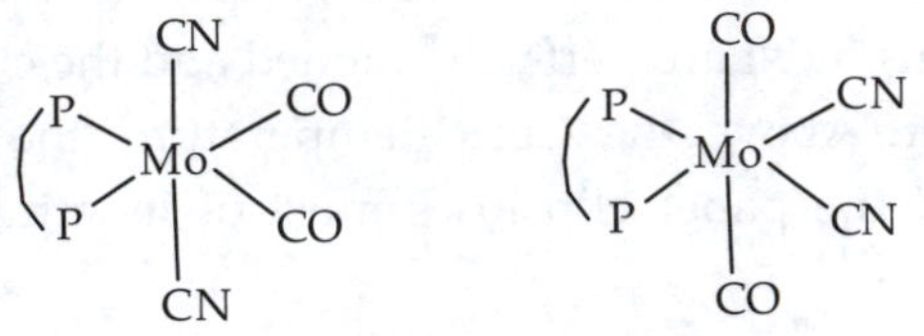

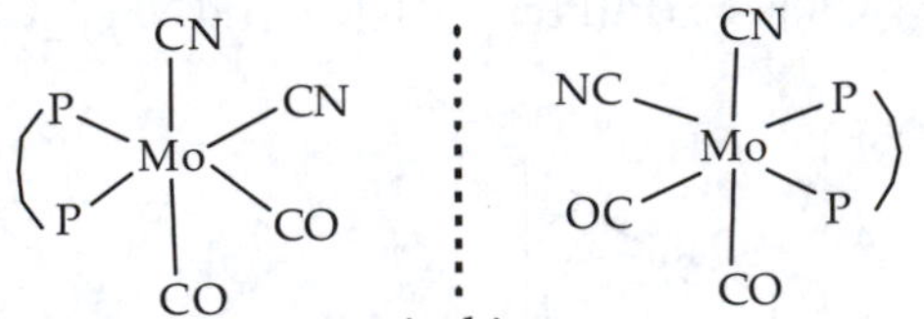

optical isomers

23.76 (a) Iron. Hemoglobin is the iron-containing protein that transports O_2 in human blood.

(b) Magnesium. Chlorophylls are magnesium-containing porphyrins in plants. They are the key components in the conversion of solar energy into chemical energy that can be used by living organisms.

(c) Iron. Siderophores are iron-binding compounds or ligands produced by a microorganism. They compete on a molecular level for iron in the medium outside the organism and carry needed iron into the cells of the organism.

(d) Copper. Hemocyanine is a copper-containing protein responsible for oxygen transport in the blue blood of certain marine animals.

23.78 (a) pentacarbonyliron(0)

(b) Because CO is a neutral molecule, the oxidation state of iron must be zero.

(c) $[Fe(CO)_4CN]^-$ has two geometric isomers. In a trigonal bipyramid, the axial and equatorial positions are not equivalent and not superimposable. One isomer has CN in an axial position and the other has it in an equatorial position.

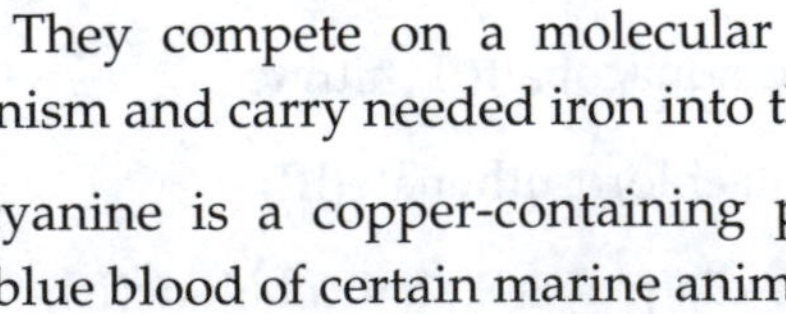

23 Transition Metals and Coordination Chemistry

Solutions to Red Exercises

23.80 (a)

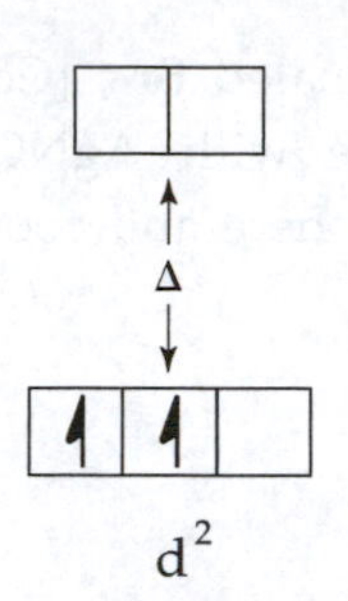

(b) These complexes are colored because the crystal-field splitting energy, Δ, is in the visible portion of the electromagnetic spectrum. Visible light with $\lambda = hc/\Delta$ is absorbed, promoting one of the d-electrons into a higher energy d-orbital. The remaining wavelengths of visible light are reflected or transmitted; the combination of these wavelengths is the color we see.

(c) $[V(H_2O)_6]^{3+}$ will absorb light with higher energy. H_2O is in the middle of the spectrochemical series, and causes a larger Δ than F^-, a weak-field ligand. Because Δ and λ are inversely related, larger Δ corresponds to higher energy and shorter λ.

23.82 According to the spectrochemical series, the order of increasing Δ for the ligands is $Cl^- < H_2O < NH_3$. (The tetrahedral Cl^- complex will have an even smaller Δ than an octahedral one.) The smaller the value of Δ, the longer the wavelength of visible light absorbed. The color of light absorbed is the complement of the observed color. A blue complex absorbs orange light (580 to 650 nm), a pink complex absorbs green light (490 to 560 nm), and a yellow complex absorbs violet light (400 to 430 nm). Because $[CoCl_4]^{2-}$ absorbs the longest wavelength, it appears blue. $[Co(H_2O)_6]^{2+}$ absorbs green and appears pink, and $[Co(NH_3)_6]^{3+}$ absorbs violet and appears yellow.

23.87 (a) $[Fe(CO)_2(CN)_4]^{2-}$ (structure: Fe with two trans CO ligands, top and bottom, and four CN ligands: NC, CN, NC, CN)

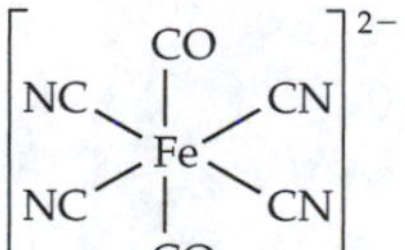

(b) sodium dicarbonyltetracyanoferrate(II)

(c) +2, 6 d-electrons

(d) We expect the complex to be low spin. Cyanide (and carbonyl) are high on the spectrochemical series, which means the complex will have a large Δ splitting characteristic of low spin complexes.

Integrative Exercises

23.93 (a) Both compounds have the same general formulation, so Co is in the same (+3) oxidation state in both complexes.

(b) Cobalt(III) complexes are generally inert; that is, they do not rapidly exchange ligands inside the coordination sphere. Therefore, the ions that form precipitates in these two cases are probably outside the coordination sphere. The dark violet compound A forms a precipitate with $BaCl_2(aq)$ but not $AgNO_3(aq)$, so it has

SO_4^{2-} outside the coordination sphere and coordinated Br^-, $[Co(NH_3)_5Br]SO_4$. The red-violet compound B forms a precipitate with $AgNO_3(aq)$ but not $BaCl_2(aq)$ so it has Br^- outside the coordination sphere and coordinated SO_4^{2-}, $[Co(NH_3)_5SO_4]Br$.

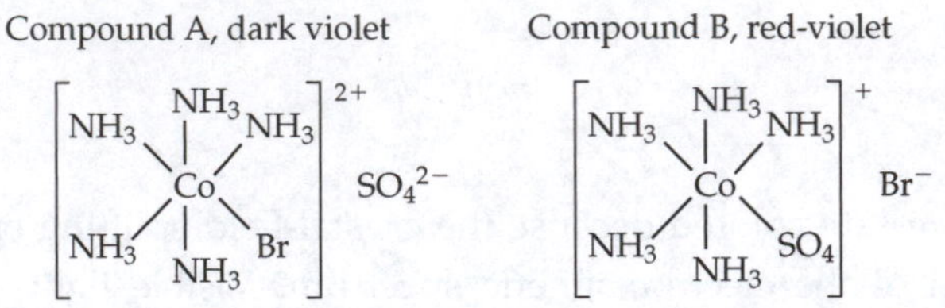

Compound A, dark violet Compound B, red-violet

(c) Compounds A and B have the same formula but different properties (color, chemical reactivity), so they are isomers. They vary by which ion is inside the coordination sphere, so they are *coordination sphere isomers*.

(d) Compound A is an ionic sulfate and compound B is an ionic bromide, so both are strong electrolytes. According to the solubility guidelines in Table 4.1, both should be water-soluble.

23.96 First determine the empirical formula, assuming that the remaining mass of complex is Pd.

$$37.6 \text{ g Br} \times \frac{1 \text{ mol Br}}{79.904 \text{ g Br}} = 0.4706 \text{ mol Br}; 0.4706/0.2361 = 2$$

$$28.3 \text{ g C} \times \frac{1 \text{ mol C}}{12.01 \text{ g C}} = 2.356 \text{ mol C}; 2.356/0.2361 = 10$$

$$6.60 \text{ g N} \times \frac{1 \text{ mol N}}{14.01 \text{ g N}} = 0.4711 \text{ mol N}; 0.4711/0.2361 = 2$$

$$2.37 \text{ g H} \times \frac{1 \text{ mol H}}{1.008 \text{ g H}} = 2.351 \text{ mol H}; 2.351/0.2361 = 10$$

$$25.13 \text{ g Pd} \times \frac{1 \text{ mol Pd}}{106.42 \text{ g Pd}} = 0.2361 \text{ mol Pd}; 0.2361/0.2361 = 1$$

The chemical formula is $[Pd(NC_5H_5)_2Br_2]$. This should be a neutral square-planar complex of Pd(II), a nonelectrolyte. Because the dipole moment is zero, we can infer that it must be the trans isomer.

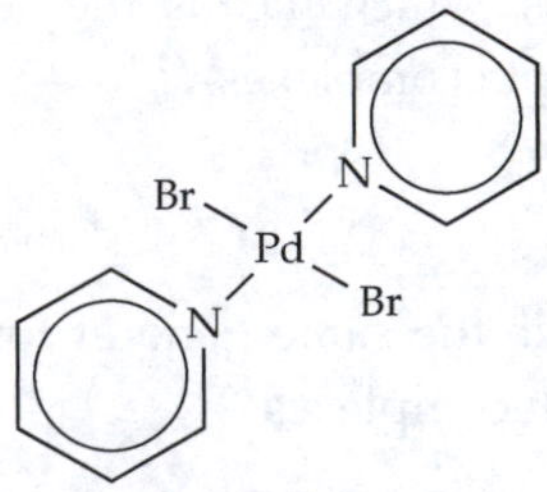

23.98 Calculate the concentration of Mg^{2+} alone, and then the concentration of Ca^{2+} by difference. $M \times L = mol$

$$\frac{0.0104 \text{ mol EDTA}}{1 \text{ L}} \times 0.0187 \text{ L} \times \frac{1 \text{ mol } Mg^{2+}}{1 \text{ mol EDTA}} \times \frac{24.31 \text{ g } Mg^{2+}}{1 \text{ mol } Mg^{2+}} \times \frac{1000 \text{ mg}}{\text{g}}$$

$$\times \frac{1}{0.100 \text{ L } H_2O} = 47.28 = 47.3 \text{ mg } Mg^{2+}/\text{L}$$

$0.0104\ M \text{ EDTA} \times 0.0315 \text{ L} = \text{mol } (Ca^{2+} + Mg^{2+})$

$0.0104\ M \text{ EDTA} \times 0.0187 \text{ L} = \text{mol } Mg^{2+}$

$0.0104\ M \text{ EDTA} \times 0.0128 \text{ L} = \text{mol } Ca^{2+}$

$$0.0104\ M \text{ EDTA} \times 0.0128 \text{ L} \times \frac{1 \text{ mol } Ca^{2+}}{1 \text{ mol EDTA}} \times \frac{40.08 \text{ g } Ca^{2+}}{1 \text{ mol } Ca^{2+}} \times \frac{1000 \text{ mg}}{\text{g}} \times \frac{1}{0.100 \text{ L } H_2O}$$

$$= 53.35 = 53.4 \text{ mg } Ca^{2+}/\text{L}$$

23.100 $$\frac{182 \times 10^3 \text{ J}}{1 \text{ mol}} \times \frac{1 \text{ mol}}{6.022 \times 10^{23} \text{ molecules}} = 3.022 \times 10^{-19} = 3.02 \times 10^{-19} \text{ J/photon}$$

$\Delta E = h\nu = 3.02 \times 10^{-19} \text{ J}; \nu = \Delta E/h$

$\nu = 3.022 \times 10^{-19} \text{ J}/6.626 \times 10^{-34} \text{ J-s} = 4.561 \times 10^{14} = 4.56 \times 10^{14} \text{ s}^{-1}$

$$\lambda = \frac{2.998 \times 10^8 \text{ m/s}}{4.561 \times 10^{14} \text{ s}^{-1}} = 6.57 \times 10^{-7} \text{ m} = 657 \text{ nm}$$

We expect that this complex will absorb in the visible, at around 660 nm. It will thus exhibit a blue-green color (Figure 23.25).

24 The Chemistry of Life: Organic and Biological Chemistry

Visualizing Concepts

24.1 *Analyze/Plan.* Follow the logic in Sample Exercise 24.1 to name each compound. Decide which structures are the same compound. *Solve.*

(a) 2,2,4-trimethylpentane (b) 3-ethyl-2-methylpentane

(c) 2,3,4-trimethylpentane (d) 2,3,4-trimethylpentane

Structures (c) and (d) are the same molecule.

Introduction to Organic Compounds; Hydrocarbons (Sections 24.1 and 24.2)

24.7 (a) False. Butane is an alkane; it contains carbon atoms that are sp^3 hybridized.

(b) False. Cyclohexane is a saturated hydrocarbon, whereas benzene is aromatic.

(c) True.

(d) False. Olefin is another name for alk**ene**.

24.9 *Analyze/Plan.* Given a condensed structural formula, determine the bond angles and hybridization about each carbon atom in the molecule. Visualize the number of electron domains about each carbon. State the bond angle and hybridization based on electron domain geometry. *Solve.*

```
     H            H   O
     |            |   ||
  H—C—C≡C—C—C—OH
   5 |  4   3  2 |   1
     H            H
```

C1 has trigonal planar electron domain geometry, 120° bond angles, and sp^2 hybridization. C3 and C4 have linear electron domain geometry, 180° bond angles, and sp hybridization. C2 and C5 both have tetrahedral electron domain geometry, 109° bond angles, and sp^3 hybridization.

24.11 *Analyze/Plan.* For each molecule, count the number of carbon atoms in the root name and in the substituent group(s). *Solve.*

(a) One. The root "meth" indicates one carbon atom. Methane is the simplest alkane.

(b) Ten. The root "dec" indicates 10 carbon atoms.

(c) Seven. The root "hex" indicates 6 carbon atoms, plus 1 in the "methyl" substituent.

(d) Five. Although this is a common name (not an IUPAC name), the root "pent" still indicates 5 total carbon atoms.

(e) Two. Again a common name, acetylene is the simplest two-carbon alkyne. The IUPAC name is ethyne.

24 Organic and Biological Chemistry

Solutions to Red Exercises

24.13 (a) True.

(b) True.

(c) False. Alkenes contain carbon-carbon double bonds.

(d) False. Alkynes contain carbon-carbon triple bonds.

(e) True.

(f) False. Cyclohexane is a saturated hydrocarbon; benzene is aromatic.

(g) True.

24.15 *Analyze/Plan.* Follow the rules for naming alkanes given in Section 24.2 and illustrated in Sample Exercise 24.1. *Solve.*

(a) 2-methylhexane

(b) 4-ethyl-2,4-dimethyldecane

(c)

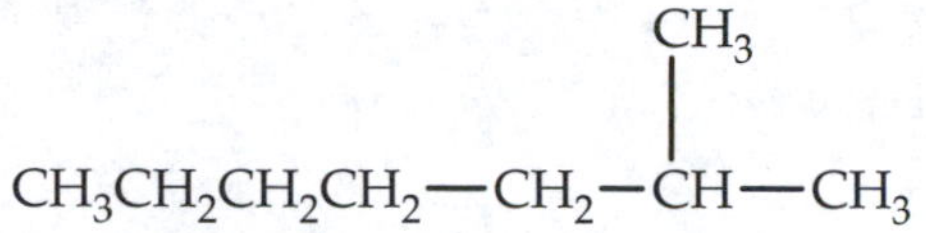

```
                                  CH3
                                  |
CH3CH2CH2CH2—CH2—CH—CH3
```

(d)

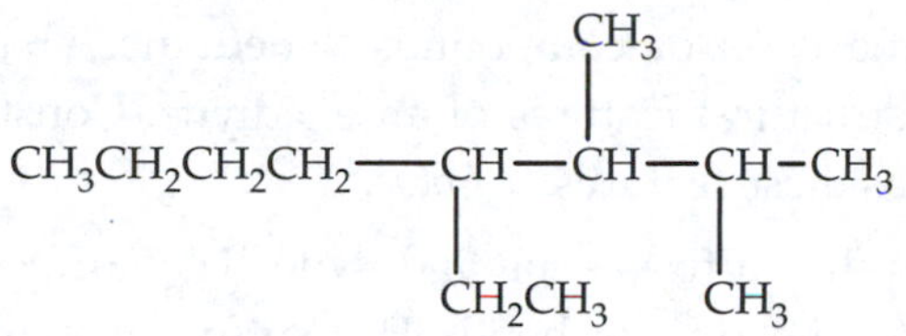

```
                              CH3
                              |
CH3CH2CH2CH2—CH—CH—CH—CH3
                     |             |
                   CH2CH3     CH3
```

(e)

```
          CH2        CH3
        /      \    /
     CH2        CH
      |          |
     CH2        CH
        \      /    \
          CH2        CH3
```

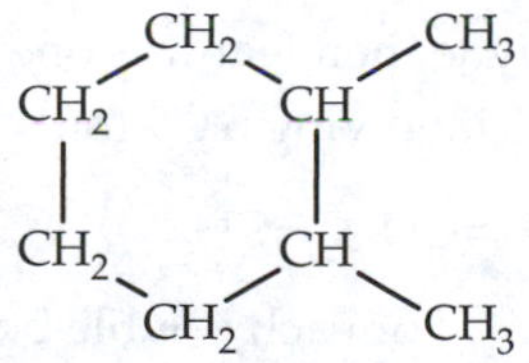

24.17 *Analyze/Plan.* Follow the rules for naming alkanes given in Section 24.2 and illustrated in Sample Exercise 24.1. *Solve.*

(a) 2,3-dimethylheptane

(b)

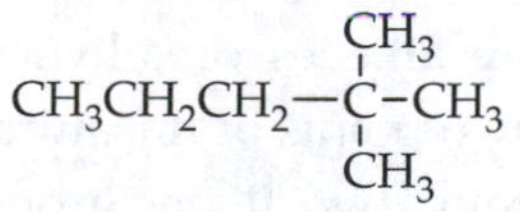

```
                 CH3
                 |
CH3CH2CH2—C—CH3
                 |
                 CH3
```

(c)

```
H3C      CH3
    \   /
      C
    /   \
H2C      CH2
 |        |
H2C      CH2
    \   /
     CH
     |
     CH2CH3
```

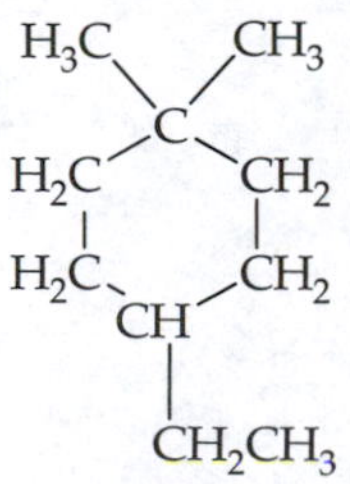

(d) 2,2,5-trimethylhexane

(e) 3-ethylheptane

24.19 Assuming that each component retains its effective octane number in the mixture (and this isn't always the case), we obtain: octane number = 0.35(0) + 0.65(100) = 65.

24 Organic and Biological Chemistry — Solutions to Red Exercises

Alkenes, Alkynes, and Aromatic Hydrocarbons (Section 24.3)

24.21 (a) C_4H_6 is an unsaturated hydrocarbon. The maximum number of hydrogen atoms for 4 C atoms in a saturated alkane is $[(2 \times 4) + 2] = 10$. C_4H_6 does not contain the maximum possible hydrogen atoms and is unsaturated.

(b) Yes, all alkynes are unsaturated. The presence of a triple bond means that the alkyne carbon atoms are not bound to the maximum possible number of hydrogen atoms.

24.23 *Analyze/Plan*. Consider the definition of the stated classification and apply it to a compound containing five C atoms. *Solve*.

(a) $CH_3CH_2CH_2CH_2CH_3$, C_5H_{12}

(b) 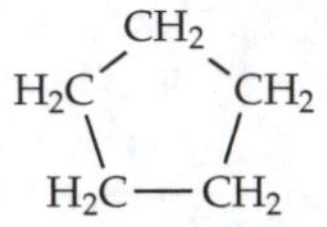 cyclopentane ring: CH_2 / H_2C, CH_2 / $H_2C—CH_2$, C_5H_{10}

(c) $CH_2=CHCH_2CH_2CH_3$, C_5H_{10}

(d) $HC\equiv CCH_2CH_2CH_3$, C_5H_8

24.25 *Analyze/Plan*. We are given the class of compounds "enediyne." Based on organic nomenclature, determine the structural features of an enediyne. Construct a molecule with 6 C atoms in a row that has these features. *Solve*.

The term "enediyne" contains the suffixes –ene and –yne. The suffix –ene is used to name alkenes, molecules with one double bond. The suffix –yne is used to name alkynes, molecules with one triple bond. An enediyne then features one double and two triple bonds. Possible arrangements of these bonds involving 6 C atoms in a row are:

$CH_2=CH–C\equiv C–C\equiv CH$ $\quad\quad$ $CH\equiv C–CH=CH–C\equiv CH$

Check. The formula of a saturated alkane is C_nH_{2n+2}. For each double bond subtract 2 H atoms, for each triple bond subtract 4 H atoms. A saturated 6 C alkane has 14 H atoms. For the enediyne, subtract (2 + 4 + 4 =) 10 H atoms. The molecular formula is C_6H_4. That is the molecular formula of each structure above.

24.27 *Analyze/Plan*. Follow the logic in Sample Exercise 24.3.
Solve. There are many correct structures that are alkenes or alkynes and have the molecular formula C_6H_{10}. The molecule will have two points of unsaturation. Molecules can include various arrangements of one alkyne group, two alkene groups, or one cyclic mono-alkene. A few of the possibilities are shown below.

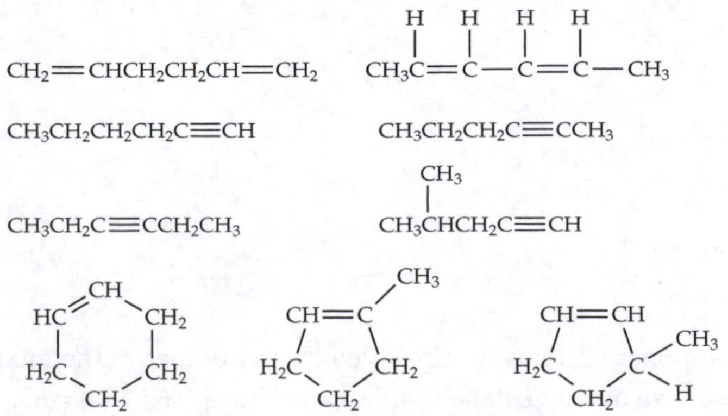

$CH_2=CHCH_2CH_2CH=CH_2$ $\quad\quad$ $CH_3C(H)=C(H)—C(H)=C(H)—CH_3$

$CH_3CH_2CH_2CH_2C\equiv CH$ $\quad\quad$ $CH_3CH_2CH_2C\equiv CCH_3$

$CH_3CH_2C\equiv CCH_2CH_3$ $\quad\quad$ $CH_3CH(CH_3)CH_2C\equiv CH$

Cyclohexene: CH=CH, HC / CH_2, H_2C / CH_2, CH_2 $\quad$ 1-methylcyclopentene: CH=C(CH_3), H_2C, CH_2, CH_2 $\quad$ 3-methylcyclopentene: CH=CH, H_2C, C(CH_3)(H), CH_2

24.29 *Analyze/Plan*. Follow the logic in Sample Exercises 24.1 and 24.4. *Solve*.

(a) 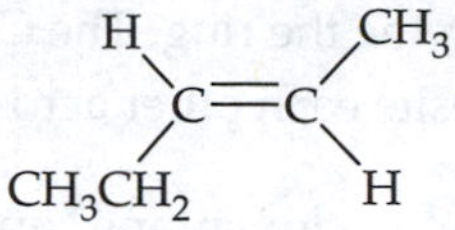 H and CH_3CH_2 on one C, CH_3 and H on the other: $(H)(CH_3CH_2)C{=}C(CH_3)(H)$

(b) 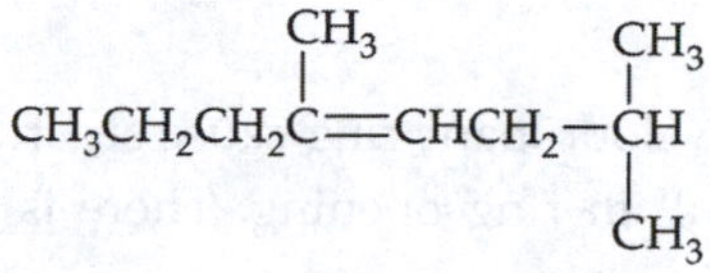 $CH_3CH_2CH_2C(CH_3){=}CHCH_2{-}CH(CH_3)_2$

(c) *cis*-6-methyl-3-octene

(d) *para*-dibromobenzene (or 1,4-dibromobenzene)

(e) 4,4-dimethyl-1-hexyne

24.31 (a) True

(b) True. (Alkenes *can* have cis and trans isomers, but whether they *do* have them depends on the groups bonded to the sp^2 C atoms of the alkene.)

(c) False. The geometry of the alkyne functional group is linear.

24.33 *Analyze/Plan*. In order for geometrical isomerism to be possible, the molecule must be an alkene with two different groups bound to each of the alkene C atoms. *Solve*.

(a) 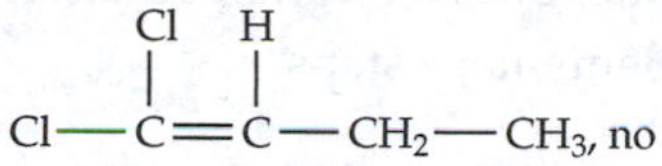 $Cl{-}C(Cl){=}C(H){-}CH_2{-}CH_3$, no

(b) 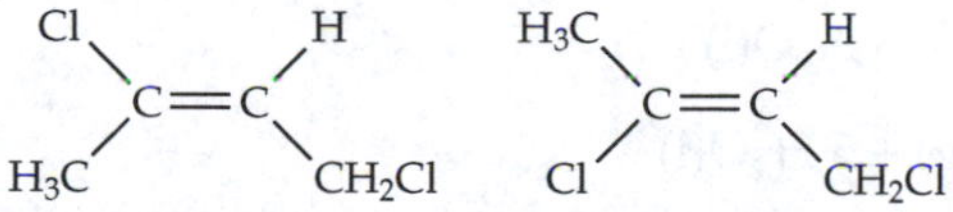 $(Cl)(H_3C)C{=}C(H)(CH_2Cl)$ and $(H_3C)(Cl)C{=}C(H)(CH_2Cl)$

(c) no, not an alkene

(d) no, not an alkene

24.35 (a) True

(b) *Plan*. Draw the condensed structural formula of 2-pentene. Consider the part of the molecule that is likely to react with Br_2. *Solve*.

$$CH_3CH_2CH{=}CH{-}CH_3 + Br_2 \rightarrow CH_3CH_2CH(Br)CH(Br)CH_3$$

2-pentene → 2,3-dibromopentane

This is an addition reaction. The π bond is broken and a Br atom adds to each of the C atoms involved in the π bond.

(c) *Plan*. Draw the structure of benzene. The term *para* means the Cl atoms will be opposite each other across the benzene ring in the product. *Solve*.

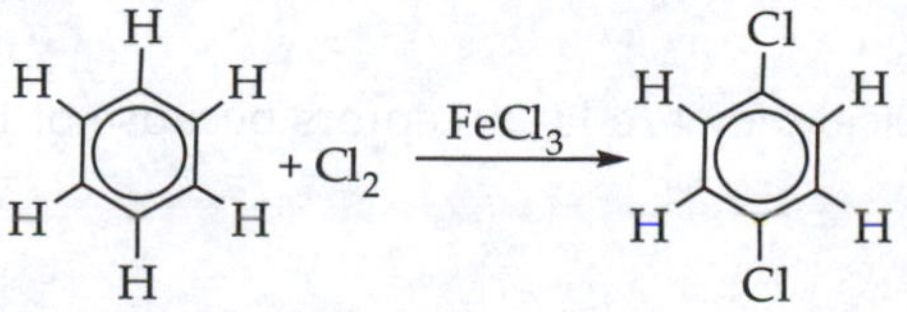

Benzene (C_6H_6, six H) $+ Cl_2 \xrightarrow{FeCl_3}$ 1,4-dichlorobenzene (four H, two Cl para)

$$C_6H_6 + Cl_2 \xrightarrow{FeCl_3} C_6H_4Cl_2$$

This is a substitution reaction. None of the double bonds in the benzene ring are broken. Each Cl atom has replaced an H atom on the the ring. The Cl atoms have substituted for two of the H atoms that are opposite each other across the ring.

24.37 (a) *Plan.* Consider the structures of cyclopropane, cyclopentane, and cyclohexane. *Solve.*

The small 60° C–C–C angles in the cyclopropane ring cause strain that provides a driving force for reactions that result in ring opening. There is no comparable strain in the five- or six-membered rings.

(b) *Plan.* First form an alkyl halide: $C_2H_4(g) + HBr(g) \rightarrow CH_3CH_2Br(l)$; then carry out a Friedel-Crafts reaction. *Solve.*

$$C_6H_6 + CH_3CH_2Br \xrightarrow{AlCl_3} C_6H_5CH_2CH_3 + HBr$$

24.39 Yes, this information suggests (but does not prove) that the reactions proceed in the same manner. That the rate laws are both first order in both reactants and second order overall indicates that the activated complex in the rate-determining step in each mechanism is bimolecular and contains one molecule of each reactant. This is usually an indication that the mechanisms are the same, but it does not rule out the possibility of different fast steps, or a different order of elementary steps.

24.41 *Analyze/Plan.* Both combustion reactions produce CO_2 and H_2O:

$$C_3H_6(g) + 9/2\, O_2(g) \rightarrow 3\, CO_2(g) + 3\, H_2O(l)$$

$$C_5H_{10}(g) + 15/2\, O_2(g) \rightarrow 5\, CO_2(g) + 5\, H_2O(l)$$

Thus, we can calculate the ΔH_{comb} / CH_2 group for each compound. *Solve.*

$$\frac{\Delta H_{comb}}{CH_2\ \text{group}} = \frac{2089\ \text{kJ/mol}\ C_3H_6}{3\ CH_2\ \text{groups}} = \frac{696.3\ \text{kJ}}{\text{mol}\ CH_2};\ \frac{3317\ \text{kJ/mol}\ C_5H_{10}}{5\ CH_2\ \text{groups}} = 663.4\ \text{kJ/mol}\ CH_2$$

$\Delta H_{comb}/CH_2$ group for cyclopropane is greater because C_3H_6 contains a strained ring. When combustion occurs, the strain is relieved and the stored energy is released during the reaction.

Functional Groups and Chirality (Sections 24.4 and 24.5)

24.43 *Analyze/Plan.* Match the structural features of various functional groups shown in Table 24.6 to the molecular structures in this exercise. *Solve.*

(a) (iii)

(b) (i)

(c) (ii) Amines are organic bases; they are H^+ acceptors because of the the lone pair of electrons on the N atom.

(d) (iv)

(e) (v)

24 Organic and Biological Chemistry

Solutions to Red Exercises

24.45 *Analyze/Plan*. Given the name of a compound, write its molecular formula. Identify the structure of the functional group present in the isomer. Finally, draw the structural formula of a molecule that contains the new functional group and has the same molecular formula as the parent compound. *Solve.*

(a) The formula of acetone is C_3H_6O. An aldehyde contains the group —C(=O)H

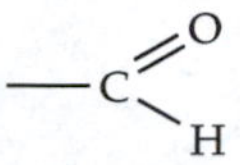

An aldehyde that is an isomer of acetone is propionaldehyde (or propanal):

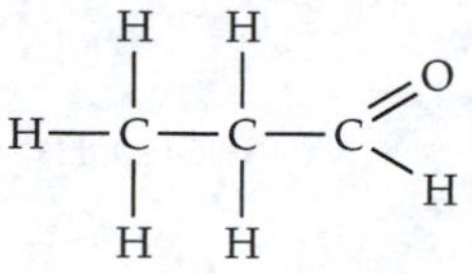

```
    H  H
    |  |      O
H—C—C—C⟨
    |  |      H
    H  H
```

(b) The formula of 1-propanol is C_3H_8O. An ether contains the group –O–. An ether that is an isomer of 1-propanol is ethylmethyl ether:

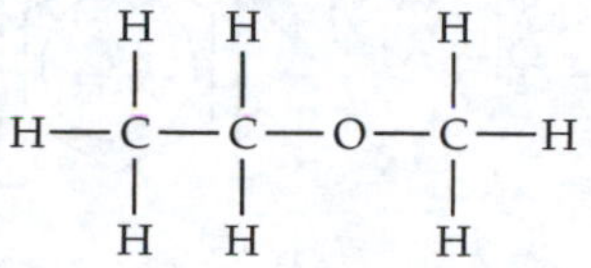

```
    H  H       H
    |  |       |
H—C—C—O—C—H
    |  |       |
    H  H       H
```

24.47 *Analyze/Plan.* From the hydrocarbon name, deduce the number of C atoms in the acid; one carbon atom is in the carboxyl group. *Solve.*

(a) meth = 1 C atom

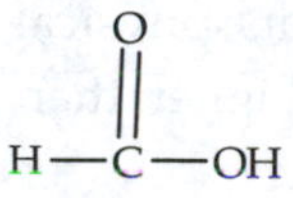

```
    O
    ||
H—C—OH
```

(b) pent = 5 C atoms

```
                O
                ||
CH3CH2CH2CH2C—OH
```

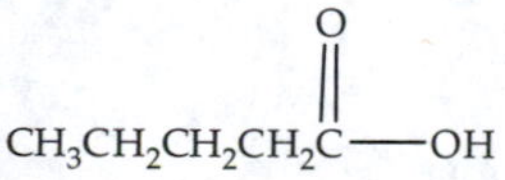

or

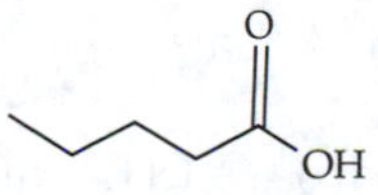

(c) dec = 10 C atoms in backbone

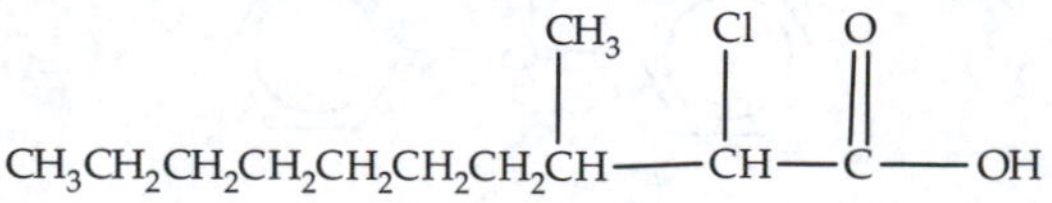

```
                      CH3    Cl   O
                       |      |   ||
CH3CH2CH2CH2CH2CH2CH2CH—CH—C—OH
```

or

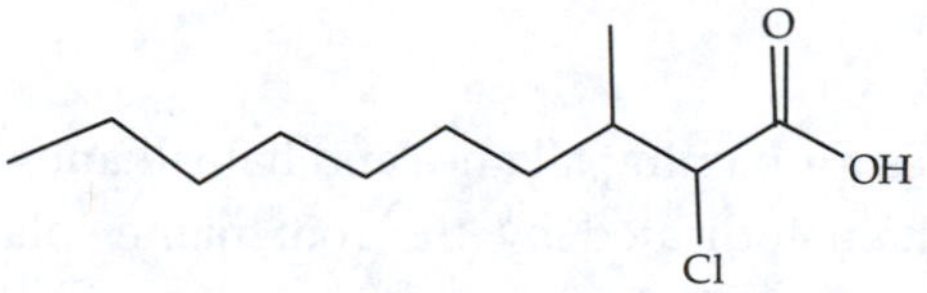

24.49 *Analyze/Plan.* In a condensation reaction between an alcohol and a carboxylic acid, the alcohol loses its –OH hydrogen atom and the acid loses its –OH group. The alkyl group from the acid is attached to the carbonyl group and the alkyl group from alcohol is attached to the ether oxygen of the ester. The name of the ester is the alkyl group from the alcohol plus the alkyl group from the acid plus the suffix *-oate.* *Solve.*

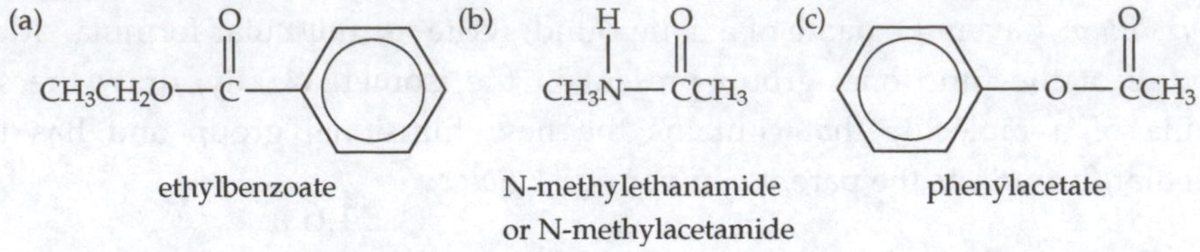

(a) $CH_3CH_2O{-}C(=O){-}C_6H_5$ ethylbenzoate

(b) $CH_3N(H){-}C(=O)CH_3$ N-methylethanamide or N-methylacetamide

(c) $C_6H_5{-}O{-}C(=O)CH_3$ phenylacetate

24.51 *Analyze/Plan.* Follow the logic in Sample Exercise 24.6. *Solve.*

(a)

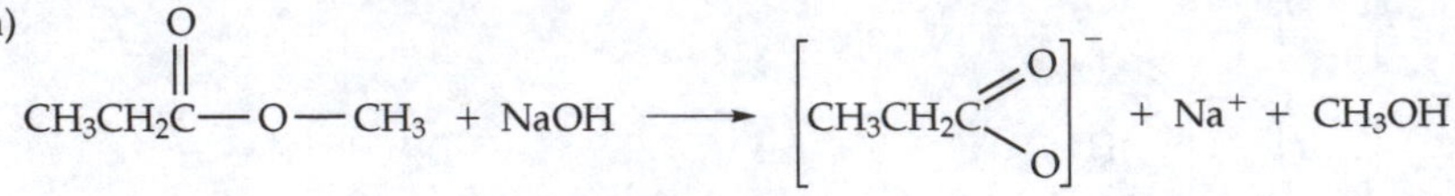

$$CH_3CH_2C(=O){-}O{-}CH_3 + NaOH \longrightarrow [CH_3CH_2CO_2]^- + Na^+ + CH_3OH$$

(b)

$$CH_3C(=O){-}O{-}C_6H_5 + 2NaOH \longrightarrow [CH_3CO_2]^- + 2Na^+ + [C_6H_5O]^-$$

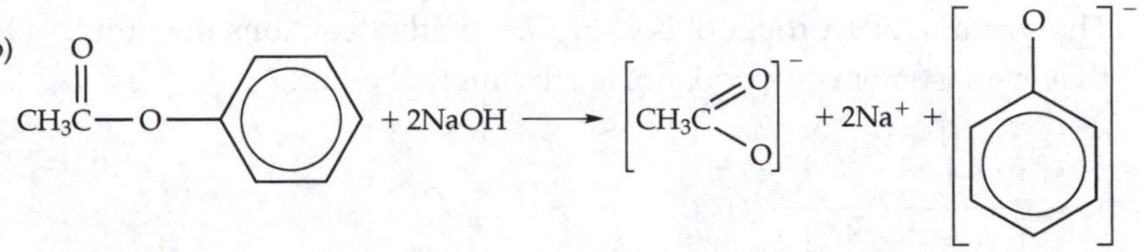

24.53 High melting and boiling points are indicators of strong intermolecular forces in the bulk substance. The strongest intermolecular force among neutral covalent molecules is hydrogen bonding. The carboxyl group of acetic acid has —OH, which acts as a donor, and —C=O, which acts as an acceptor in hydrogen bonding. We expect acetic acid to be a strongly hydrogen-bonded substance, as shown by its physical properties. The melting and boiling points of acetic acid are somewhat higher than those of water, another substance that experiences strong hydrogen bonding.

24.55 *Analyze/Plan.* Follow the logic in Sample Exercise 24.2, incorporating functional group information from Table 24.6. *Solve.*

(a) $CH_3CH_2CH_2CH(OH)CH_3$　　(b) $CH_3CH(OH)CH_2OH$

(c) 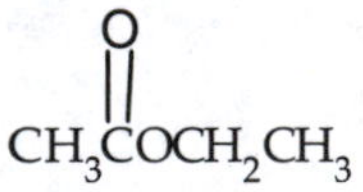 $CH_3C(=O)OCH_2CH_3$　　(d) 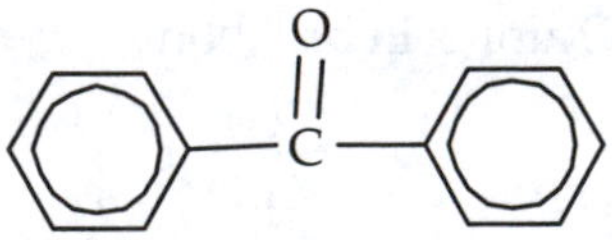 $C_6H_5{-}C(=O){-}C_6H_5$

(e) $CH_3OCH_2CH_3$

24.57 *Analyze/Plan.* Review the rules for naming alkanes and haloalkanes; draw the structure. That is, draw the carbon chain indicated by the root name, place substituents, fill remaining positions with H atoms. Each C atom attached to four different groups is chiral. *Solve.*

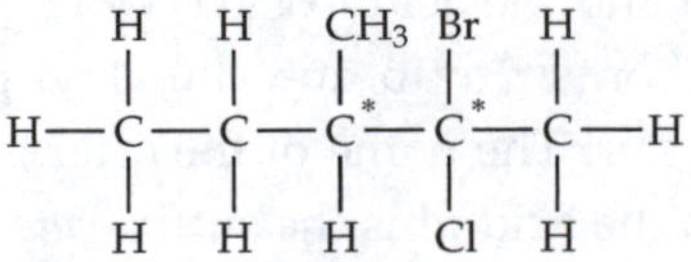

$$H{-}CH_2{-}CH_2{-}C^{*}H(CH_3){-}C^{*}H(Br)(Cl)\ldots$$

$$\text{H}_3\text{C}{-}\text{CH}_2{-}\overset{*}{\text{C}}\text{H}(\text{CH}_3){-}\overset{*}{\text{C}}(\text{Br})(\text{Cl}){-}\text{CH}_3$$

The correct choice is (c); there are two chiral carbon atoms in the molecule. C2 is obviously attached to four different groups. C3 is chiral because the substituents on C2 render the C1-C2 group different than the C4-C5 group.

Introduction to Biochemistry; Proteins (Sections 24.6 and 24.7)

24.59 (a)

$$H_2N-\underset{\large H}{\overset{\large R}{\underset{|}{\overset{|}{C}}}}-COOH$$

(b) In forming a protein, amino acids undergo a condensation reaction between the amino group and carboxylic acid:

$$-\overset{O}{\overset{\|}{C}}-OH + H-\underset{\large H}{\underset{|}{N}}-CH_2\overset{O}{\overset{\|}{C}}-OH \longrightarrow -\overset{O}{\overset{\|}{C}}-\underset{\large H}{\underset{|}{N}}-CH_2-\overset{O}{\overset{\|}{C}}-OH + H_2O$$

(c) The bond that links amino acids in proteins is called the *peptide* bond.

$$-\overset{O}{\overset{\|}{C}}-\underset{\large H}{\underset{|}{N}}-$$

24.61 *Analyze/Plan.* Two dipeptides are possible. Either peptide can have the terminal carboxyl group or the terminal amino group. *Solve.*

histadylaspartic acid, His-Asp or HD

$$(\text{imidazolium ring: } HN^{+},\ NH)-CH_2-\overset{\overset{+}{N}H_3}{\overset{|}{CH}}-\underset{O}{\underset{\|}{C}}-\overset{H}{N}-\underset{\underset{\large C(=O)O^-}{|}}{\underset{CH_2}{\underset{|}{CH}}}-C(=O)O^-$$

aspartylhistidine, Asp-His or DH

$$^-O(O=)C-CH_2-\overset{\overset{+}{N}H_3}{\overset{|}{CH}}-\underset{O}{\underset{\|}{C}}-\overset{H}{N}-\underset{\underset{\large (\text{imidazolium ring: } NH,\ \overset{+}{N}H)}{|}}{\underset{CH_2}{\underset{|}{CH}}}-C(=O)O^-$$

24 Organic and Biological Chemistry

Solutions to Red Exercises

24.63 *Analyze/Plan*. Follow the logic in Sample Exercise 24.7. *Solve*.

(a)

```
                O         O        O
                ||        ||       ||
   H3N+CH2CNHCH2CNHCHCO −
                          |
                         CH2
                          |
                       (imidazole ring: NH, N=)
```

$H_3\overset{+}{N}CH_2\overset{O}{\overset{\|}{C}}NHCH_2\overset{O}{\overset{\|}{C}}NH\underset{CH_2}{CH}\overset{O}{\overset{\|}{C}}O^-$

NH

N

Gly-Gly-His

(b) Three tripeptides are possible: Gly-Gly-His, GGH; Gly-His-Gly, GHG; His-Gly-Gly, HGG

24.65 (a) True.

(b) False. Alpha helix and beta sheet are examples of secondary structure. They are different configurations of the protein chain.

(c) False.

Carbohydrates and Lipids (Sections 24.8 and 24.9)

24.67 (a) True. A disaccharide is composed of two sugar units. Sugars are carbohydrates.

(b) False. Sucrose is a disaccharide.

(c) True. Well, most carbohydrates have this general formula.

24.69 (a) The empirical formula of cellulose is $C_6H_{10}O_5$.

(b) The six-membered ring form of glucose forms the monomer unit that is the basis of the polymer cellulose. In cellulose, glucose monomer units are joined by β linkages.

(c) Ether linkages connect the glucose monomer units in cellulose.

24.71 (a) Yes, D-mannose is a sugar; it is a polyhydroxy aldehyde.

(b) In the linear form of mannose, there are four chiral carbon atoms, C2, C3, C4, and C5. The two terminal carbon atoms, C1 and C6, are not chiral.

(c) Both the α (left) and β (right) forms are possible.

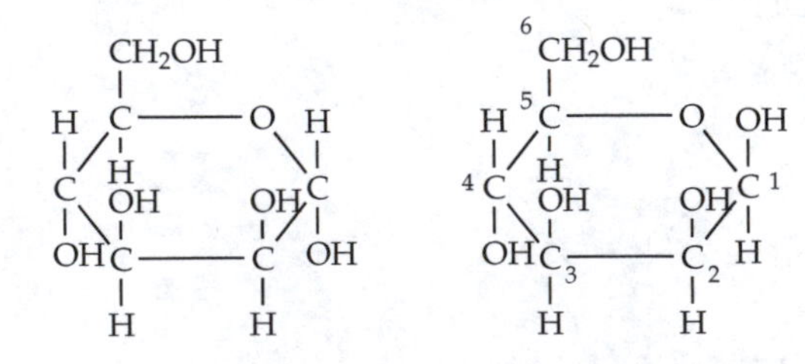

24.73 (a) False. Fat molecules contain ester functional groups.

(b) True. See Figure 24.24.

(c) True. See Figure 24.24

Nucleic Acids (Section 24.10)

24.75 Dispersion forces increase as molecular size (and molar mass) increases. The larger purines (2 rings vs. 1 ring for pyrimidines) have larger dispersion forces.

24.77 *Analyze/Plan.* Consider the structures of the organic bases in Section 24.10. The first base in the sequence is attached to the sugar with the free phosphate group in the 5′ position. The last base is attached to the sugar with a free –OH group in the 3′ position. *Solve.*

The DNA sequence is 5′–TACG–3′.

24.79 *Analyze/Plan.* Recall that there is complimentary base pairing in nucleic acids because of hydrogen bond geometry. The DNA pairs are A–T and G–C. (The RNA pairs are A–U and G–C.)

Solve. From the single strand sequence, formulate the complimentary strand. Note that 3′ of the complimentary strand aligns with 5′ of the parent strand.

5′–GCATTGGC–3′

3′–CGTAACCG–5′

Additional Exercises

24.81

$H_2C{=}CH{-}C({=}O){-}H$

(three-membered ring: $HC{=}C(H){-}C(H)(OH)$, i.e. H–C=C–C(H)–OH ring)

$H{-}C{\equiv}C{-}CH_2OH$

$O{=}C{=}CH{-}CH_3$

$HC{\equiv}COCH_3$

(three-membered ring: $C({=}O)$ bonded to $H_2C{-}CH_2$)

Structures with the –OH group attached to an alkene carbon atom are not included. These molecules are called "vinyl alcohols" and are not the major form at equilibrium.

24.90 (a) The molecule has one ketone and two alcohol functional groups. There are no chiral centers, no carbon atoms attached to four different groups.

(b) The molecule has one ketone and three alcohol functional groups. There is one chiral center. The carbon bearing the secondary –OH has four different groups attached, and is thus chiral.

(c) The molecule has a carboxylic acid and an amine functional group. It is an amino acid, shown in its neutral (not zwitterion) form. There are two chiral centers. The carbon bearing the $-NH_2$ group and the carbon bearing the $-CH_3$ group are both chiral.

24 Organic and Biological Chemistry

Solutions to Red Exercises

Integrative Exercises

24.95

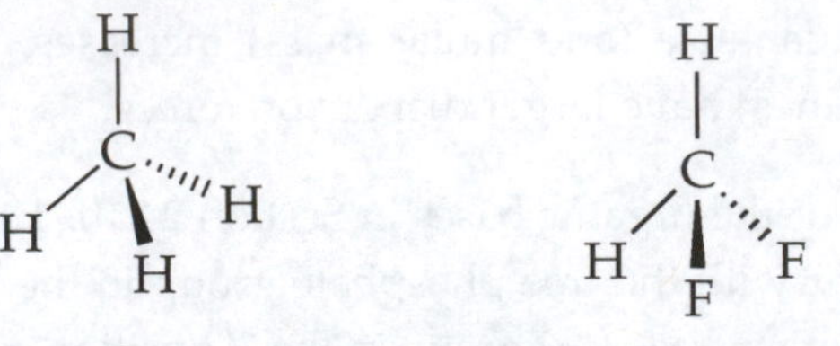

methane difluoromethane

CH_3CH_2OH — ethanol

CH_3—O—CH_3 — dimethyl ether

(a) Methane boils at –128 °C. Methane is nonpolar and has the smallest molar mass. It experiences only weak dispersion forces and has the lowest boiling point.

(b) Difluoromethane boils at –52 °C. It is somewhat polar and experiences weak dipole-dipole and dispersion forces. Although difluoromethane has a larger molar mass than dimethyl ether, it has approximately spherical shape, which reduces the strength of its dispersion forces.

(c) Dimethyl ether boils at –25 °C. It is somewhat polar and experiences weak dipole-dipole and dispersion forces. Its shape is bent, which enables stronger dispersion forces than the distorted spherical shape of difluoromethane. [This explanation of the relative boiling points of dimethyl ether and difluoromethane is a rationalization based on the actual boiling points of the compounds. It is difficult to distinguish the two based on simple principles of relative strengths of intermolecular forces.]

(d) Ethanol boils at 78 °C. It is an alcohol that participates in hydrogen bonding. This is the strongest kind of intermolecular force and explains the highest boiling point.

24.99 The reaction is: $2\ NH_2CH_2COOH(aq) \rightarrow NH_2CH_2CONHCH_2COOH(aq) + H_2O(l)$

$\Delta G° = (-488) + (-237.13) - 2(-369) = 12.87 = 13\ kJ$

NOTES